ADVANCES IN STATISTICS

ADVANCES IN STATISTICS

Proceedings of the Conference in Honor of
Professor Zhidong Bai on His 65th Birthday

National University of Singapore 20 July 2008

editors

Zehua Chen
Jin-Ting Zhang
National University of Singapore, Singapore

Feifang Hu
University of Virginia, USA

World Scientific

NEW JERSEY • LONDON • SINGAPORE • BEIJING • SHANGHAI • HONG KONG • TAIPEI • CHENNAI

Published by

World Scientific Publishing Co. Pte. Ltd.

5 Toh Tuck Link, Singapore 596224

USA office: 27 Warren Street, Suite 401-402, Hackensack, NJ 07601

UK office: 57 Shelton Street, Covent Garden, London WC2H 9HE

Library of Congress Cataloging-in-Publication Data

Advances in statistics : proceedings of the conference in honor of Professor Zhidong Bai on his 65th birthday, National University of Singapore, 20 July 2008 / edited by Zehua Chen, Jin-Ting Zhang & Feifang Hu.

p. cm.

Includes bibliographical references and index.

ISBN-13: 978-981-279-308-9 (hardcover : alk. paper)

ISBN-10: 981-279-308-9 (hardcover : alk. paper)

1. Bai, Zhidong. 2. Mathematicians--Singapore--Biography--Congresses. 3. Statisticians--Singapore--Biography--Congresses. 4. Mathematical statistics--Congresses. I. Bai, Zhidong. II. Chen, Zehua. III. Zhang, Jin-Ting, 1964– IV. Hu, Feifang, 1964–

QA29.B32A39 2008

519.5--dc22

2007048506

British Library Cataloguing-in-Publication Data

A catalogue record for this book is available from the British Library.

Printed in Singapore by B & JO Enterprise

PREFACE

In August, 2006, while Professor Xuming He, University of Illinois, and Professor Feifang Hu, University of Verginia, were visiting the Institute of Mathematical Sciences, NUS, we had a dinner together. Besides Xuming, Feifang and myself, in presence at the dinner were Professor Louis Chen, Director of the Institute of Mathematical Sciences, Professors Anthony Kuk and Kwok Pui Choi, the head and deputy head of the Department of Statistics & Applied Probability, NUS. The idea of a mini-conference in honour of Professor Zhidong Bai on his 65th birthday was conceived during the dinner. Louis suggested for me to take the lead to organise this conference. I felt obliged. Zhidong and I have been long time friends and colleagues. I first met Zhidong in 1986 when Zhidong together with Professors Xiru Chen and Lincheng Zhao visited the University of Wisconsin-Madison while I was still a PhD student there. After Zhidong joined the NUS, we became colleagues, co-authors and close friends. It is indeed my honour to play a leading role in the organizing of this event. A organizing committee was formed afterwards. It consisted of Feifang Hu, Jin-Ting Zhang, Ningzhong Shi and myself. Jin-Ting is a professor of the National University of Singapore and Ningzhong is a professor of the Northeast China Normal University. It was decided to have a proceedings of the mini-conference published. Xuming later suggested to publish a volume of Zhidong's selected papers. This led to the current book.

The book consists of two parts. The first part is about Zhidong's life and his contributions to Statistics and Probability. This part contains an interview with Zhidong conducted by Dr. Atanu Biswas, Indian Statistical Institute, and seven short articles on Zhidong's contributions. These articles are written by Zhidong's long term collaborators and coauthors who together give us a whole picture of Zhidong's extraordinary career. The second part is a collection of Zhidong's selected seminal papers.

Zhidong has a legendary life. He was adopted into a poor peasant family at birth. He spent his childhood during the Chinese resistance war against Japan. He had an incomplete elementary education under extremely backward conditions. Yet he managed to enter one of the most prestigious universities in China, the University of Science and Technology of China (USTC). After graduation from USTC, he worked as a truck driver's team leader and was completely detached from the academics for ten years during the Cultural Revolution of China. However, he successfully got admitted into the graduate program of USTC in 1978 when China restarted its tertiary education after ten years interruption, and became one of the first 18 PhDs in China's history four year later. In 1984, he went to the United States. He soon had his academic power felt. He became elected as a fellow of the Third World Academy of Sciences in 1989, elected as a Fellow of the Institute of Mathematical Statistics in 1990. Zhidong has worked as researcher and professor at University of Pittsburgh, Temple University, Sun-Yat-Sen University at Taiwan, National University of Singapore and Northeast China Normal University. He has published three monographs and over 160 research papers. He has produced numerous graduate students. Zhidong's life and his career would inspire many young researchers and statisticians.

Zhidong's research interests are broad. He has made great contributions in various areas such as random matrix theory, Edgeworth expansion, M-estimation, model selection, adaptive design in clinical trials, applied probability in algorithms, small area estimation, time series, and so on. The selected papers are samples among many of Zhidong's important papers in these areas. These selected papers not only present Zhidong's research achievements but also an image of a great researcher. Zhidong is not a trendy statistician. He enjoys tackling hard problems. As long as those problems are of scientific interest, he does not care too much about whether papers can be produced from them for the purposes of "survival" such as tenure, promotion, etc. Such character is well demonstrated in his dealing with the circular law in the theory of large dimensional random matrix. It was an extremely difficult problem. He delved into it for thirteen years until his relentless effort eventually bore fruit. Zhidong has left his marks in Statistics, indelible ones.

This book provides an easy access to Zhidong's important works. It serves as a useful reference for the researchers who are working in the relavent areas.

Finally, I would like to thank the following persons for their contribution to the book: Biswas, A., Silverstein, J., Babu, G. J., Kundu, D., Zhao, L., Hu, F., Wu, Y. and Yao, J.

The permissions for the reprinting of the selected papers are granted by Institute of Mathematical Statistics, Statistica Sinica and Indian Statistical Institute. Their permissions are acknowledged with great appreciation.

The editing of this book is a joint effort by Feifang Hu, Jin-Ting Zhang and myself.

Zehua Chen
(Chairman, Organizing Committee
for the Conference on Advances in Statistics
in Honor of Professor Zhidong Bai on His 65th Birthday)

Singapore
30 September 2007

CONTENTS

Part B Selected Papers of Professor Bai 43

PART A

Professor Bai's Life and His Contributions

A CONVERSATION WITH PROFESSOR ZHIDONG BAI

Atanu Biswas

Applied Statistics Unit
Indian Statistical Institute
203 B.T. Raod, Kolkata 700 108, India
atanu@isical.ac.in

Zhidong Bai is a Professor in Department of Statistics and Applied Probability, National University of Singapore. He also holds an appointment in the School of Mathematics and Statistics, Northeast Normal University, China. He has a very illustrious career which in many aspects resembles a story book. Currently he is in the editorial boards of Sankhya, Journal of Multivariate Analysis, Statistica Sinica and the Journal of Statistical Planning and Inference. Atanu Biswas is an Associate Professor in the Applied Statistics Unit, Indian Statistical Institute, Kolkata. Dr. Biswas visited Professor Bai during January-February 2006 for some collaborative research when Professor Bai was in National University of Singapore. During that visit, Dr. Biswas had the opportunity to talk with Professor Bai in a casual mood, which reveals a really interesting career of a strong mathematical statistician. Dr. Jian Tao of the Northeast Normal University, China, was present during the full conversation.

Biswas: A very different question to start with. Professor Bai, most of the Chinese names have meaning. What does the name Zhidong mean?

Bai: This is an interesting question. Zhidong means Towards east.

Biswas: That is interesting. Could you tell me something about your childhood? How did you grow up?

Bai: I was born in 1943, in the cold Hebei province of Northern China. Hebei means North of the Yellow river. My hometown was in Laoting county.

Biswas: That was the war time, not a very cool and normal surrounding, I suppose.

Bai: Right. The Chinese resistance war against Japan war and the second world war were going on.

Biswas: So how was your time? Any memory of those war days? You were really young at that time.

Bai: Sure, I was too young. But, I still remember, it was a run-away time. People hid quite often in not-easy-to-find places in the countryside out of the fear of the Japanese.

Biswas: Could you tell me something about your family, your parents?

Bai: I was adopted by a poor peasant family. I have no information about my biological parents. My father was working secretly for the Eighth Army Group led by the Chinese Communist Party at that time. I still remember he ran away from home frequently to escape from the Japanese. At those days, we knew nothing about the Communist Party, we simply called Ba Lu (meaning Eighth Army Group) for any people associated with the Eighth Army Group.

Biswas: Could you now talk about your school days?

Bai: I went to elementary school in 1950. The school I attended was originally established by the Ba Lu people during the war time under very poor condition. It was originally a temple with a big hall. The classes for all grades were conducted in the same hall at the same time. You could hear all the other classes. The teachers were not formally educated. They were demobilized soldiers from the Cimmunist let army. They acquired their knowledge in the army. There were no teaching facilities except the big hall. No tables, no chairs, no papers, no text books, nothing at all. The pupils had to carry their own stools from home every day to sit on. They had also to carry a small stone slate with their homework done on it because of lack of papers. The stone slate had to be cared so that what was written on it did not get erased. I came out of this school in 1957.

Biswas: That is very interesting. Any memory about your childhood plays?

Bai: I liked to play table tennis (Ping Pong), which was very popular in China even at that time. Since there was no Ping Pong table, we used a prostrate stone-made monument on the ground, which was about 2 meters long, as our Ping Pong table. But we really had fun.

Biswas: What was your next school?

Bai: I was admitted to a Junior High School in 1957. It was 5 kilometers from my home. The school was established in 1956, I was in the second batch of the students. I graduated from the school in 1960.

Biswas: What did you learn there?

Bai: Eucleadian Geometry and logical thinking, among others.

Biswas: Any special teacher you remember?

Bai: Yes, there was one excellent teacher. That is Teacher Zhang Jinglin.

Biswas: What about your Senior High School?

Bai: My senior high school was in the capital of Laoting county, 8 kilometers from my home. I got admitted into the school in 1960. I stayed in the school dormitory. This was the first time I left my family and lived alone. I still remember vividly the joy of "go-home-week", which was only once a month. I studied there for 3 years. I got very good trainings in mathematical and logical thinking, writing and so on. I learned the first time the notion of mathematical limit in that school, which amazed me and had an effect on my later research interests. I also learnt some elementary courses in Physics, Chemistry, Nature and Culture, and so on. One of the teachers whose name I still remember, Teacher Sun Jing Hua, left me with a deep impression. Sun Jing Hua did not have much formal education. He was a student in a High School run by the Ba Lu people during the resistance war against Japan. After two years study there, he together with all teachers and students of that school joined the Eighth Army Group collectively due to the war situation. He remained in the army until 1949. Then he quitted from the Army and became a teacher of my senior high school. He educated himself by self studying while teaching, and soon became a well-established teacher. My impression of Teacher Sun Jing Hua had a certain influence on my later life.

Biswas: Then, the University. How was that?

Bai: I joined the University of Science and Technology of China (USTC) in 1963. At that time the USTC was located in Beijing.

Biswas: You studied mathematics, right?

Bai: Yes, the first two years were interlinked with the mathematics department. From the third year onwards I studied statistics and operation research. I was in the statistics group. I had a broad training in mathematics and statistics in those five years. I studied Mathematical Analysis, Advanced Algebra, ODE, PDE, Probability, Real and Complex Analysis, Measure theory, Functional analysis, Matrix Theory, Ordinary Physics, Applied Statistics, Theoretical Statistics, Advanced Statistics, which covered Lehmanns book.

Biswas: You were the best student in the class, I suppose.

Bai: I am one of the three best students in a class of 37.

Biswas: Then you graduated in 1968.

Bai: Yes, graduated, but without a degree. There was no degree system in existence at that time in China.

Biswas: What next?

Bai: After graduation I went to Xinjiang Autonomous Region, west of China, and started my job as a truck driver's team leader. My job was to supervise the truck drivers.

Biswas: Could you continue study or research during that time?

Bai: No way. It was during the Cultural Revolution. I remained in this job for 10 years, from 1968 to 1978.

Biswas: You were married in this period. Right?

Bai: I married in 1972, and my two sons were born in 1973 and 1975.

Biswas: How did you shift to academics?

Bai: In 1978, China restarted tertiary education after ten years interuption. I seized the opportunity to take an examination and get admitted into the graduate program of the USTC as a graduate student. I completed my graduate thesis in 1980. But there was still no degree system in existance until then. No degree was conferred to me at that time. However, the China government began seriously to consider the establishment of the degree system. It was approved by the State Coucil (the China cabnet) in 1982 that the degree system be adopted by the China Academy of Sciences as a trial. I was then conferred the Ph.D degree. I was among the very first batch of Ph.Ds in China, which consists of only 18 people. There were 3 among the 18 are in Statistics. I was one of them.

Biswas: I am a bit puzzled. How was that possible? You were out of touch of academics for 10 years. Then you had to recapture everything when you came back. How could you finish your thesis within 2 years then?

Bai: To recapture I had to read something, but that was easy. I found everything I learned 10 years ago was getting fresh after a quick glance at it. And writing the thesis was not at all difficult as I just compiled 15 papers of mine to form the thesis.

Biswas: When did you write these 15 papers?

Bai: Within these 2 years. Of course these were in Chinese, and not all of them were published at that time, half published and half pending publication.

Biswas: This is beyond my comment. Could you tell me something about your thesis, and about your supervisor?

Bai: The title of my thesis is: "Independence of random variables and applications". I had two advisors: Yin Yong Quan and Chen Xiru. None of them had Ph.D degree, because of the reason mentioned earlier.

Biswas: What next? Did you start your academic career then?

Bai: I started teaching in USTC in 1981 as a Lecturer for three years, then I moved to the United States in August 1984.

Biswas: That must be a new beginning.

Bai: True.

Biswas: Tell me the story.

Bai: My advisor Yin Yong Quan had been in good terms with P.R. Krishnaiah. Krishnaiah came to know about me from Yong Quan, and invited me to visit him at the University of Pittsburgh. I went there as a research associate.

Biswas: Did you face any serious problem with English at that stage? I understand that you did not have much training in English in China.

Bai: I did have some problem with my English, and the problem continued for many years. At the beginning, I could not understand Krishnaiah when we talked face to face, but quite stangely I could understand him over phone. I attributed this to the fact that my English training is obtained mainly through listening to the radio.

Biswas: What about your research there?

Bai: I collaborated with the group of Krishnaiah on signal processing, image processing and model selection. In collaboration with a guy named Reddy from the medical school, I worked on some cardiological problem to construct the shape of the heart, to be pricise, the left ventricle, using two orthogonal pictures. It was altogether a completely new experience to me. I had quite some number of papers with Krishnaiah, a large number of unpublished technical reports also. Unfortunately Krishnaiah passed away in 1987 and C.R. Rao took over his Center of Multivariate Analysis. Then I started collaborating with C.R. Rao. I worked in collaboration with C. R. Rao until 1990. It was a different and fruitful experience. Rao's working style was different. Quite often we tackled the same problem from different angles and arrived at the same results.

Biswas: How many papers have you coauthored with C.R. Rao?

Bai: Roughly around 10.

Biswas: How do you compare your research experience in China with that in the US?

Bai: In China we did statistical research just by doing mathematics, paper to paper. But, in the US I observed that most of statistical research is motivated by real problems. It was interesting.

Biswas: What next?

Bai: I joined Temple University in 1990 as an Associate Professor and stayed there until 1994. My family moved to the US in that period. There was a teachers strike in Temple during my first year there, and the University lost about one-third of the students. As a consequence, some new recruits had to go. I was one of them. I moved to Taiwan in 1994.

Biswas: Thats interesting. How was your experience in Taiwan being a mainland Chinese?

Bai: People there were friendly. I was in Kao Hsiung, southern Taiwan, during 1994-1997, as a professor.

Biswas: When did you move to Singapore?

Bai: In 1997. I could not work there for too long since I was holding a Chinese passport. So I had to leave Taiwan. Singapore was a good choice. I joined the National University of Singapore as a Professor in 1997 and remained there since.

Biswas: Now let's talk something about your reseach area.

Bai: Spectral analysis of large dimensional random matrices is my biggest area of research. I have about 30 papers published in this area, some in IPES journal. For one of these papers I worked for 13 years from 1984 to 1997, which was eventually published in Annals of Probability. It was the hardest problem I have ever worked on. The problem is: Consider an n by $n$random matrix of i.i.d entries $X = (X_{ij})$, where $EX_{ij} = 0, EX_{ij}^2 \leq 1$. If $\lambda_1, \ldots, \lambda_n$ are the eigenvalues of $X/\sqrt{n}$, the famous conjecture is that the empirical spectral distribution constructed from $\lambda_1, \ldots, \lambda_n$ tends to the uniform distribution over the unit disk in the complex plane, i.e., $\frac{1}{\pi}I\{x^2 + y^2 \leq 1\}$. We derived the limiting spectral density, which is a circular law.

I've written about 10 papers on Edgeworth expansion. Some of them were jointly with Jogesh Babu.

I did some works on model selection, as well, mostly jointly with Krishnaiah. Mostly AIC based, the penalty is C_n multiplying the number of parameters, where C_n satisfies $C_n/\log\log n \to \infty$ and $C_n/n \to 0$. Then, with probability 1, the true model is eventually selected. The paper was published in JMA which is the most cited among my papers.

Recently I have been doing some works on adaptive allocation, some works with Hu, Rosenberger, now with you.

There are about 10 papers on Applied Probability in Algorithms. I did some interesting works on record problem, on maximum points, with H.K. Hwang of Academia Sinica in Taiwan.

There are a few works on small area estimation and time series as well.

Biswas: Who is your favourite coauthor, except me?

Bai: Silverstein. Besides, I enjoyed working with C.R. Rao in Statistics, and with Jogesh Babu in Mathematics.

Biswas: What is your view on theoretical research?

Bai: I believe that research should be practical problem oriented. To me theoretical research is an art, an entertainment. But, practical research is for the benefit of the people. This is some sort of push and energy to do something. But, there should be some freedom to do something from your mind.

Biswas: I know that you are a strong follower of Chinese culture.

Bai: Certainly, the Chinese cultute, the festivals, the Chinese medicines.

Biswas: What is your general view on research?

Bai: Research in Universities are of two types: "interesting research" and "survival research". Interesting research are those which you do from your own interest. Survival research are those you do for your mere survival, to get promotion, to get your contract renewed and so on. This is the major portion of now-a-days research.

Biswas: How many "survival papers" you have writen?

Bai: Roughly around two thirds of my about 160 published papers.

Biswas: What is your view on the future direction of research in statistics?

Bai: I think new theories are to be developed for high dimensional data analysis. Random matrix theory is one of them. Classical large sample theory assumes the dimension is fixed while the sample size goes to infinity. This assumption is not realistic nowadays. You may see that for human DNA sequence, the dimension may be as high as several millions. If you want a sample with size as large as its dimension, you need to collect data from half of the world population. It is impossible. Then how can you assume p is fixed and n tends to infinity? Now-a-days, big impact in statistics comes from modern computer technique. It helps us to store data, to analyze data. But the classical statistical theory still works for large data set, especially with large dimension? Now, consider a simple problem: Suppose $X_{ij} \sim N(0,1)$. Denote the $p \times p$ sample covariance matrix by S_n. If we consider p fixed and n large, then $\sqrt{\frac{n}{2p}} \log |S_n| \to N(0,1)$ in distribution. But, if $p = [cn]$, we have $\sqrt{\frac{n}{2p}} \log |S_n| \to -\infty$. More preciously, one may show that $(1/p) \log |S_n| \to d(c) < 0$ and $\sqrt{\frac{n}{2p}}(\log |S_n| - pd(c))$ tends to normal. Now, suppose $n = 10^3, p = 10$. Now, it is the problem of interpretation. One can as well put the relationship in many other forms, $p = n^{1/3}$, or $p = cn^{1/2}$. Then which assumption and which limiting result you need to use? Assume p is fixed (as suggested by all current statistics textbooks)? Or assume $p/n \to 0.01$? Simulation results show that the empirical density of $\sqrt{\frac{n}{2p}} \log |S_n|$ is very skewed to the left. Therefore, I would strongly suggest to use the CLT for linear spectral statistics of large sample covariance matrix. Again, in one of my recent work, I noticed that the rounding in computation by a computer results in inconsistent estimation. For example, suppose the data comes from $N(\mu, 1)$. We just use the t-test to test the true hypothesis. When the data were rounded, surprisingly, when n is large, the t-test eventually reject the true hypothesis! In the statistical problems in old days, the sample size was not large and hence the rounding errors were not a problem. But today, it is no longer the case! It has been a very serious problem now!

Biswas: What is your idea about Bayesian philosophy?

Bai: To be honest, I could not understand that philosophy. I am not a Bayesian. I like Lehmanns idea of average risk.

Biswas: Is Lehmann your favorite statistician?

Bai: You are right.

Biswas: Tell me something about your family, your wife.

Bai: My wife worked in China for some years, before she went to the US to join me. She managed my kids quite efficiently, she is a good manager of my family. My two sons are now well-settled in the US, one is a Professor in Electrical Engineering, and the other works in FDA.

Biswas: Where is Professor Bai after 10 years from now?

Bai: No idea.

Biswas: Thank you, Professor Bai.

Bai: My pleasure.

PROFESSOR Z.D. BAI: MY FRIEND, PHILOSOPHER AND GUIDE

D. KUNDU

Indian Institute of Technology Kanpur
Department of Mathematics & Statistics
Pin 208016, INDIA
E-mail: kundu@iitk.ac.in

I am very happy to know that a group of friends and colleagues of Professor Z.D. Bai is planning to organize a conference and want to publish a volume of his selected papers to celebrate his 65th birthday. A really genius and a rare quality scientist of Professor Bai's stature definitely deserves this honor from his peer. I am really thankful to Professor Zehua Chen for his invitation to write a few words regarding the contribution of Professor Z.D. Bai in the area of *Statistical Signal Processing* on this occasion. It is indeed a great pleasure for me to write about one of my best friend, philosopher and guide in true spirit.

I first met Professor Bai in 1984, at the University of Pittsburgh, when I had joined in the Department of Mathematics and Statistics as a Ph.D. student. Professor Bai was also working in the same Department at that time with Professor C.R. Rao. If I remember it correctly, at the beginning although we were in the same Department, we hardly used to interact. I was busy with my course work and he was busy with his own research. Moreover, I believe since both of us were new to US, we were more comfortable mingling with our fellow country men only.

But things changed completely in 1986, when I completed my comprehensive examination and started working under Professor Rao on my research problem. Professor Rao had given me a problem on *Statistical Signal Processing* and told me to discuss it with Professor Bai, since Professor Bai was working with Professor Krishnaiah and Professor Rao in this area during that time. Although, they had just started working in this field, but they had already developed some important results. I was really lucky to have access to those unpublished work. Several Electrical Engineers were working in the area of *Signal Processing* for quite sometime but among the Statisticians it was a completely new field. Two of their classic papers Zhao *et al.*,[8,9] in the area of *Statistical Signal Processing* had already appeared by that time in the *Journal of Multivariate Analysis*. In these two papers they had studied the estimation procedures of the different parameters of the difficult Directions of Arrivals (DOA) model, which is very useful in the area of *Array Processing*. This particular model has several applications in Radar, Sonar and Satellite communications. The above two papers were a really important mixture between multivarite analysis and model selection, which led to very useful applications in the area of *Statistical Signal Processing.* Many authors had discussed about this model prior to them, but in my opinion these two papers for the first time made the proper theoretical developments. These two papers had generated at least four Ph.D. thesis in this area. Later, Professor Bai jointly with Professor Rao in Bai and Rao[4] had developed an efficient spectral analytic method for the estimation of the different parameters of the DOAs model.

Another important area where Professor Bai had laid his hands on was the estimation of the parameters on the sum of superimposed exponential signals and estimating the number of sinusoidal components, in presence of noise. This problem was an important and old problem but it did not have satisfactory solutions for quite sometime. In the mid-eighties, this problem had attracted a lot of attention among the Signal Processors, because the sum of superimposed exponential model forms a building block for different signal processing models. Several Linear Prediction methods were used by different authors to solve this problem. Unfortunately all the methods were lacking the consistency properties, which were overlooked by most of the authors. Professor Bai along with his colleagues had developed in Bai *et al.*[2] a completely new estimation procedure known as EquiVariance Linear Prediction (EVLP) method, which was very simple to implement and enjoyed strong consistency properties as well. Interestingly, this is the first time they showed how to estimate the number of components and the other unknown parameters simultaneously. Since the EVLP is very easy to use, it has been used quite effectively for on-line implementation purposes. Later, they had further improved their methods in Bai *et al.*[5,6] which are now well recognized in the *Statistical Signal Processing* community. In the mean time they had also made an important contribution in deriving the properties of the maximum likelihood estimators of the unknown parameters in the sum of sinusoidal model in Bai *et al.*,[1] which was theoretically a very challenging problem. He along with his colleagues have further generalized these results to the multivariate case also in Kundu *et al.*[7] and Bai *et al.*,[3] which has several applications in colored texture imaging. Some of these results have been further generalized by others even for the colored noise.

Fortunately, I am associated with him for more than twenty years. I really feel that he has made some fundamental contributions in this area of *Statistical Signal Processing*, which may not be that well known to the Statisticians. Professor Bai is a rare mixture of very strong theoretical knowledge with applied mind and finally of course a wonderful human being. Last time I met him was almost 6 years back in a conference in US, but we are in constant touch through e-mail and whenever I need to discuss any problem I just write to him, and I know I will get a reply immediately. It is a real pleasure to have a friend and teacher like Professor Bai and I wish him a very long, happy, prosperous and fruitful life.

References

1. Bai, Z. D., Chen, X. R., Krishnaiah, P. R., Wu, Y. H., Zhao, L. C. (1992), "Strong consistency of maximum likelihood parameter estimation of superimposed exponential signals in noise", *Theory Probab. Appl.* 36, no. 2, 349–355.
2. Bai, Z. D., Krishnaiah, P. R. and Zhao, L. C. (1987), " On estimation of the number of signals and frequencies of multiple sinusoids", *IEEE Conference Proceedings*, CH2396-0/87/, 1308 - 1311.
3. Bai, Z.D., Kundu, D. and Mitra, A. (1999), "A Note on the consistency of the Multidimensional Exponential Signals", *Sankhya, Ser A.* , Vol. 61, 270-275.
4. Bai, Z.D. and Rao, C. R. (1990), "Spectral analytic methods for the estimation of the number of signals and direction of arrival", *Spectral Analysis in One or Two Dimensions*, 493 - 507, Oxford & IBH Publishing Co., New Delhi.
5. Bai, Z. D., Rao, C. R., Wu, Y. Zen, M.; Zhao, L. C. (1999), " The simultaneous estimation of the number of signals and frequencies of multiple sinusoids when some observations are missing. I. Asymptotics", *Proc. Natl. Acad. Sci. USA*, vol. 96, no. 20, 11106–11110.
6. Bai, Z. D., Rao, C. R., Chow, M. and Kundu, D. (2003), "An efficient algorithm for estimating the parameters of superimposed exponential signals", *Journal of Statistical Planning and Inference*, vol. 110, 23 - 34.
7. Kundu, D. , Bai, Z.D. and Mitra. A. (1996), "A Theorem in Probability and its Applications in Multidimensional Signal Processing", *IEEE Trans. on Signal Processing* , Vol. 44, 3167 - 3169.
8. Zhao, L.C., Krishnaiah, P.R. and Bai, Z.D. (1986a), "On detection of the number of signals in presence of white noise", *Journal of Multivariate Analysis*, vol. 20, no. 1, 1-25.
9. Zhao, L.C., Krishnaiah, P.R. and Bai, Z.D. (1986b), "On detection of the number of signals when the noise covariance matrix is arbitrary", *Journal of Multivariate Analysis*, vol. 20, no. 1, 26-50.

COLLABORATION WITH A DEAR FRIEND AND COLLEAGUE

Jack W. Silverstein
Department of Mathematics, North Carolina State University,
Raleigh, North Carolina 27695-8205, USA
**E-mail: jack@math.ncsu.edu*
www.math.ncsu.edu/~jack

It was in 1984 that my friend Yong-Quan Yin, who was working at University of Pittsburgh with P.R. Krishnaiah, told me his student, Zhidong Bai, in China was coming to Pittsburgh to work with them. The three produced some great results on eigenvalues of large dimensional random matrices. On several occasions I was asked to referee their papers. Via email Yong-Quan, Zhidong, and I produced a paper in the late 80's proving the finiteness of the fourth moment of the random variables making up the classic sample covariance matrix is necessary to ensure the almost sure convergence of the largest eigenvalue. I would consider this to be the beginning of our long-term collaboration. But it would be several years before we had our next result.

It was only in March of 1992 that Zhidong and I finally met. I invited him to give a talk in our probability seminar. During dinner I told him about the simulations I ran showing eigenvalues of general classes of large sample covariance matrices behaving in a much more orderly way than what the known results at the time would indicate. From these results, concerning the convergence of the empirical distribution of the eigenvalues as the dimension increases, one can only conclude that the proportion of eigenvalues appearing in intervals outside the support of the limiting distribution would go to zero. Simulations reveal that no eigenvalues appear at all in these intervals. Moreover, the number of eigenvalues on either side of an interval outside the support matches exactly with those of the corresponding eigenvalues of the population matrix. The mathematical proof of this would be very important to applications. We shook hands, pledging the formation of a partnership to prove this phenomenon of exact separation.

It took a while, but we did it in two papers, last one appearing in 1999. But this is only two of several things we have worked on throughout the years. It takes lots of email exchanges, and countless hours of working together, one on one. I visited Zhidong many times wherever he was, Taiwan, Singapore, China. He comes and stays with me whenever he can. Our collaborative efforts have so far produced six papers and a book.

And it goes on. Together we are a formidable team. We tackle tough problems.

This past summer a recent Ph.D. from Russia related to me the comment her advisor, a well-known probabilist, gave her upon her asking him whether a certain open question in random matrices will ever be solved. He said "if it ever is solved it will be done by Bai and Silverstein."

It is a shear delight working with Zhidong. He's extremely bright, can see things very clearly. I truly admire his insights. Solid first class mathematician.

I consider Zhidong to be my closest friend. We have helped each other out during some rough periods in our lives. I'm expecting our friendship and academic partnership will go on for a long time. Lots of open questions out there on random matrices.

My Collaborated Works with Bai are given in the references.

References

1. *Spectral Analysis of Large Dimensional Random Matrices*, (Science Press, Beijing, 2006).
2. (with Y.Q. Yin) "A note on the largest eigenvalue of a large dimensional sample covariance matrix" *Journal of Multivariate Analysis* **26**(2) (1988), pp. 166-168.
3. "On the empirical distribution of eigenvalues of a class of large dimensional random matrices" *Journal of Multivariate Analysis* **54**(2) (1995), pp. 175-192.
4. "No eigenvalues outside the suppport of the limiting spectral distribution of large dimensional random matrices" *Annals of Probability* **26**(1) (1998), pp. 316-345.
5. "Exact separation of eigenvalues of large dimensional sample covariance matrices" *Annals of Probability* **27**(3) (1999), pp.1536-1555.
6. "CLT of linear spectral statistics of large dimensional sample covariance matrices" *Annals of Probability* **32**(1A) (2004), pp. 553-605.
7. "On the signal-to-interference-ratio of CDMA systems in wireless communications" *Annals of Applied Probability* **17**(1) (2007), pp. 81-101.

EDGEWORTH EXPANSIONS: A BRIEF REVIEW OF ZHIDONG BAI'S CONTRIBUTIONS

G. J. Babu
Department of Statistics, The Pennsylvania State University, University Park, PA 16803, USA
Email: babu@stat.psu.edu

Professor Bai's contributions to Edgeworth Expansions are reviewed. Author's collaborations with Professor Bai on the topic are also discussed.

Keywords: Edgeworth expansions; Lattice distributions; Local expansions; Partial Cramér's condition; Bayesian bootstrap.

I have the pleasure of collaborating with Professor Bai Zhidong on many papers including three[2–4] on Edgeworth expansions.

The earliest work of Bai on Edgeworth expansions that I came across is the English translation[10] of his joint work with Lin Cheng, which was first published in Chinese. They investigate expansions for the distribution of sums of independent but not necessarily identically distributed random variables. The expansions are obtained in terms of truncated moments and characteristic functions. From this, they derive an ideal result for non-uniform estimates of the residual term in the expansion. In addition they also derive the non-uniform rate of the asymptotic normality of the distribution of the sum of independent but identically distributed random variables, extending some of the earlier work by A. Bikyalis[13] and L. V. Osipov.[17] Few years later Bai[7] obtains Edgeworth expansions for convolutions by providing bounds for the approximation of $\psi * F_n$ by $\psi * U_{kn}$, where F_n denotes the distribution function of the sum of n independent random variables, ψ is a function of bounded variation and U_{kn} denotes the "formal" Edgeworth expansion of F_n up to the kth order.

Many important statistics can be written as functions of sample means of random vectors. Bhattacharya and Ghosh[11] made fundamental contributions to the theory of Edgeworth expansions for functions of sample means of random vectors. Their results are derived under Cramér's condition on the joint distribution of all the components of the vector variable. However, in many practical situations, such as ratio statistics[6] and survival analysis, only one or a few of the components satisfy Cramér's condition while the rest do not. Bai along with Rao[8] established Edgeworth expansions for functions of sample means when only the partial Cramér's condition is satisfied. Bai & Rao[9] derived Edgeworth expansions on ratio's of sample means, where one of the variables is counting (lattice) variable. Such ratios arise in survival analysis in measuring and comparing the risks of exposure of individuals to hazardous environments. Bai in collaboration with Babu[3] has developed Edgeworth expansions under a partial Cramér's condition, extending the results of Bai & Rao.[8,9] But the results of[6,8] require moments higher than the ones appearing in the expansions. However, in,[3] the conditions on the moments are relaxed to the minimum needed to define the expansions. The results generalize Hall's[16] work on expansions for student's t-statistic under minimal moment conditions, and partially some of the derivations of.[12,14,15]

In the simple errors-in-variables models, a pair (X_i, Y_i) of attributes are measured on the i-th individual with error (δ_i, ϵ_i), where $E(\delta_i) = E(\epsilon_i) = 0$, and $X_i - \delta_i$ and $Y_i - \epsilon_i$ are related by a linear equation. That is, $X_i = v_{in} + \delta_i$ and $Y_i = \omega + \beta v_{in} + \epsilon_i$, where v_{in} are unknown nuisance parameters. Various estimators of the slope parameter β are derived by Bai & Babu[2] under additional assumptions. Even though the residuals in these errors-in-variables models are assumed to be independent and identically distributed random variables, the statistics of interest turn out to be functions of means of independent, but not identically distributed, random vectors. They also demonstrate that the bootstrap approximations of the sampling distributions of these estimators correct for the skewness. The bootstrap distributions are shown to approximate the sampling distributions of the studentized estimators better than the classical normal approximation.

Babu & Bai[4] take the results of Babu & Singh,[5] on Edgeworth expansions for statistics based on samples from finite populations, to a new direction by developing mixtures of global and local Edgeworth expansion for functions of random vectors. Edgeworth expansions are obtained for

$$P\Big\{\sum_{j=1}^{N} a_{j,N}(Z_j - E(Z_j)) \in H, \sum_{j=1}^{N} Z_j = n\Big\}$$

as a combination of global and local expansions, where $\{Z_j\}$ is an i.i.d. sequence of random variables with a lattice distribution and $\{a_{j,N}\}$, is an array of constants. From this, expansions for conditional probabilities

$$P\Big\{\sum_{j=1}^{N} a_{j,N}(Z_j - E(Z_j)) \in H \mid \sum_{j=1}^{N} Z_j = n\Big\}$$

are derived using local expansions for $P\{\sum_{j=1}^{N} Z_j = n\}$. In the case of absolutely continuous Z_1, the expansions are derived for $(\sum_{j=1}^{N} a_{j,N} Z_j)/(\sum_{j=1}^{N} Z_j)$. These results are then applied to obtain Edgeworth expansions for bootstrap distributions, for Bayesian bootstrap distributions, and for the distribution of statistics based on samples from finite populations. The Bayesian bootstrap is shown to be second-order correct for smooth positive '*priors*', whenever the third cumulant of the '*prior*' is equal to the third power of its standard deviation. As a consequence, it is easy to conclude that among the standard gamma '*priors*', the only one that leads to second order correctness is the one with mean 4. Similar results are established for the weighted bootstrap when the weights are constructed from random variable with a lattice distribution.

References

1. Babu, G. J., and Singh, K. On Edgeworth expansions in the mixture cases. *Ann. Statist.*, **17** (1989), no. 1, pp. 443–447.
2. Babu, G. J., and Bai, Z. D. Edgeworth expansions for errors-in-variables models. *J. Multivariate Anal.*, **42** (1992), no. 2, pp. 226–244.
3. Babu, G. J., and Bai, Z. D. Edgeworth expansions of a function of sample means under minimal moment conditions and partial Cram?'s condition. *Sankhyā Ser. A*, **55** (1993), no. 2, pp. 244–258.
4. Babu, G. J., and Bai, Z. D. Mixtures of global and local Edgeworth expansions and their applications. *J. Multivariate Anal.*, **59** (1996), no. 2, pp. 282–307.
5. Babu, G. J., and Singh, K. Edgeworth expansions for sampling without replacement from finite populations. *J. Multivariate Anal.*, **17** (1985), no. 3, pp. 261–278.
6. Babu, G. J., and Singh, K. On Edgeworth expansions in the mixture cases. *Annals of Statistics*, **17** (1989), pp. 443–447.
7. Bai, Z. D. Edgeworth expansion for convolutions. *J. Combin. Inform. System Sci.*, **16** (1991), no. 2-3, pp. 190–206.
8. Bai, Z. D., and Rao, C. R. Edgeworth expansion of a function of sample means. *Ann. Statist.*, **19** (1991), no. 3, pp. 1295–1315.
9. Bai, Z. D., and Rao, C. R. A note on Edgeworth expansion for ratio of sample means. *Sankhyā Ser. A*, **54** (1992), no. 3, 3pp. 09–322.
10. Bai, Z. D., and Zhao, L. C. Edgeworth expansions of distribution functions of independent random variables. *Sci. Sinica Ser. A*, **29** (1986), no. 1, pp. 1–22.
11. Bhattacharya, R. N., and Ghosh, J. K. On the validity of the formal Edgeworth expansions. *Ann. Statist.*, **6** (1978), pp. 435–451.
12. Bhattacharya, R. N., and Ghosh, J. K. On moment conditions for valid formal Edgeworth expansions. *J. Multivariate Anal.*, **27** (1988), no. 1, pp. 68–79.
13. Bikjalis, A. Estimates of the remainder term in the central limit theorem. (Russian) *Litovsk. Mat. Sb.*, **6** (1966), pp. 323–346.
14. Chibisov, D. M. Asymptotic expansion for the distribution of statistics admitting a stochastic expansion - I. *Theory Probab. Appl.*, **25** (1980), no. 4, pp. 732–744.
15. Chibisov, D. M. Asymptotic expansion for the distribution of statistics admitting a stochastic expansion - II. *Theory Probab. Appl.*, **26** (1981), no. 1, pp. 1–12.
16. Hall, P. Edgeworth Expansions for student's t statistic under minimal moment conditions. *Ann. Statist.*, **15** (1987), pp. 920-931.
17. Osipov, L. V. Asymptotic expansions in the central limit theorem. (Russian) *Vestnik Leningrad. Univ.*, **22** (1967), no. 19, pp. 45–62.

BAI'S CONTRIBUTION TO M-ESTIMATION AND RELEVANT TESTS IN LINEAR MODELS

Lincheng Zhao

Department of Statistics and Finance
University of Science and Technology of China
Hefei, China
E-mail: lczhao@ustc.edu.cn

In this paper, we briefly survey some contributions of Zhidong Bai to asymptotic theory on M-estimation in a linear model as well as on the relevant test criteria in ANOVA.

As a general approach on statistical data analysis, asymptotic theory of M-estimation in regression models has received extensive attention. In recent years, Bai and some of his coworkers worked on this field and obtained some important results. In this paper we briefly introduce some of them and the related work in the literature. As a special case, the minimum L_1-norm (ML_1N) estimation, also known as the least absolute deviations (LAD) estimation, plays an important role and is of special interest. Considering this point, we will pay much attention to them as well.

Consider the linear model

$$Y_i = x_i'\beta + e_i, \quad i = 1, \cdots, n, \tag{1}$$

where x_i is a known p-vector, β is the unknown p-vector of regression coefficients and e_i is an error variable. We shall assume $e_1, \cdots, e_n$ are i.i.d. variables with a common distribution function F throughout this paper unless there is some other statement. An M-estimate $\hat{\beta}_n$ of β is defined by minimizing

$$\sum \rho(Y_i - x_i'\beta) \tag{2}$$

for a suitable function ρ, or by solving an estimating equation of the type

$$\sum \psi(Y_i - x_i'\beta)x_i \tag{3}$$

for a suitable function ψ. Hereafter for simplicity we always write $\sum$ for $\sum_{i=1}^n$. The well-known least-squares (LS) estimate and the LAD estimate of β can be obtained by taking $\rho(u) = u^2$ and $\rho(u) = |u|$, respectively. Especially, the LAD estimate of β is defined as any value $\tilde{\beta}$ which satisfies

$$\sum |Y_i - x_i'\tilde{\beta}| = \min_{\beta} \sum |Y_i - x_i'\beta|. \tag{4}$$

There is considerable literature on the asymptotic theory of M-estimation, starting with the seminal work of Huber (1973) (see Huber (1981) for details and relevant references to earlier work). The references is also made to Yohai and Maronna (1979), Bai et al. (1992), Rao and Toutenburg (1995), Chen and Zhao (1996), Jurečková and Sen (1996), and Zhao (2000).

Throughout this paper, we assume that ρ is a nonmonotonic convex function, and ψ is a non-trivial nondecreasing function, and that p is fixed as $n \to \infty$.

Write

$$S_n = \sum x_i x_i', \qquad d_n^2 = \max_{1 \le i \le n} x_i' S_n^{-1} x_i. \tag{5}$$

Here, we assume that $S_{n_0} > 0$ for some integer n_0 and that $n \ge n_0$.

Many authors have attempted to give a proof that the LAD estimate $\tilde{\beta}_n$ of β, when suitably standardized, tends to normal N(0; I_p) in distribution as $n \to \infty$, where I_p is the identity matrix of order p. First such an attempt was made by Bassett and Koenker (1978). References are also made to Amemiya (1982), Bloomfield and Steiger (1983), Dupačová (1987) and others. For detailed comments on these references, see for instance Chen et al. (1990), Rao and Toutenburg (1995), and Chen and Zhao (1996).

An important contribution of Bai and his coworkers in this respect is that, they gave a rigorous proof of the asymptotic normality of $\tilde{\beta}_n$ under very general conditions in Chen, Bai et al. (1990) (also refer to Bai et al. (1987), the pre-printed version). In the special case that $e_1, e_2, \cdots$ are i.i.d., their result can be formulated as follows.

Theorem 1. *Suppose that in model (1), med*$(e_1) = 0$ *and the following conditions are satisfied:*

(i) $F'(u)$ *exists for* u *in some vicinity of 0, and is continuous at 0 and* $f(0) = F'(0) > 0$.

(ii) $d_n \to 0$ *as* $n \to \infty$.

Then as $n \to \infty$,

$$2f(0) S_n^{1/2} (\tilde{\beta}_n - \beta) \to N(0, I_p) \;\; \textit{in distribution}.$$

This result was also obtained in the later work of Pollard (1991).

For establishing the asymptotic normality of M-estimation of regression coefficients, Bai et al. (1992) considered the following multivariate linear model:

$$Y_i = X_i'\beta + \mathcal{E}_i, \quad i = 1, 2, \cdots, \tag{6}$$

where $\mathcal{E}_i$ are iid p-vectors with a common distribution F, X_i are $m \times p$ given matrices. In model (6), we are interested in the M-estimate $\hat{\beta}_n$ of β defined by minimizing

$$\sum \rho(Y_i - X_i'\beta) \tag{7}$$

for a given convex function ρ of p variables in this subsection.

Let $\psi(u)$ be a choice of a subgradient of ρ at $u = (u_1, \cdots, u_p)'$. (A p-vector $\psi(u)$ is said to be a subgradient of ρ at u, if $\rho(z) \geq \rho(u) + (z-u)'\psi(u)$ for any $z \in R^p$.) Note that if ρ is differentiable at u according to the usual definition, ρ has a unique subgradient at u and vice versa. In this case,

$$\psi(u) = \nabla\rho(u) := (\partial\rho/\partial u_1, \cdots, \partial\rho/\partial u_p)'.$$

Denote by $\mathcal{D}$ the set of points where ρ is not differentiable. This is, in fact, the set of points where ψ is discontinuous, which is the same for all choice of ψ. It is well known that $\mathcal{D}$ is topologically an F_σ set of Lebesgue measure zero (refer to Rockafellar, 1970, Section 25, p. 218).

Bai et al. (1992) made the following assumptions:

(M1) $F(\mathcal{D}) = 0$.

(M2) $E\psi(\mathcal{E}_1 + u) = \Lambda u + o(||u||)$ as $||u|| \to 0$, where $\Lambda > 0$ is a $p \times p$ constant matrix.

(M3) $E||\psi(\mathcal{E}_1 + u) - \psi(\mathcal{E}_1)||^2 < \infty$ for all sufficiently small $||u||$, and is continuous at $u = 0$.

(M4) $E\psi(\mathcal{E}_1)\psi'(\mathcal{E}_1) := \Gamma > 0$.

(M5) $S_n = \sum X_i X_i' > 0$ for n large, and

$$d_n^2 := \max_{1\leq i\leq n} tr(X_i' S_n^{-1} X_i) \to 0 \quad \text{as } n \to \infty. \tag{8}$$

They established the following:

Theorem 2. *Under assumptions (M1)-(M5),*

$$T_n^{-1/2} K_n(\hat\beta_n - \beta) \to N(0, I_p) \ \textit{in distribution, as } n \to \infty, \tag{9}$$

where

$$T_n = \sum X_i \Gamma X_i', \quad K_n = \sum X_i \Lambda X_i'. \tag{10}$$

In many situations we are interested in testing the linear hypothesis

$$\mathrm{H}_0 : H'(\beta - b) = 0 \quad \textit{against} \ \mathrm{H}_1 : H'(\beta - b) \neq 0 \tag{11}$$

where H' and b are known $q \times p$ matrix of rank q and p-vector, respectively ($0 < q < p$). Put

$$\inf_{\beta\in \mathrm{H}_0} \sum \rho(Y_i - x_i'\beta) := \sum \rho(Y_i - x_i'\beta_n^*) := M_n^*, \tag{12}$$

$$\inf_{\beta\in R^p} \sum \rho(Y_i - x_i'\beta) := \sum \rho(Y_i - x_i'\hat\beta_n) := \hat M_n. \tag{13}$$

To test hypothesis (11), Schrader and Hettmansperger (1980) studied the asymptotic distribution of $M_n^* - \hat M_n$ under H_0 and the three conditions of Huber (1973), where ρ is assumed to be a nonmonotonic convex function and to possess bounded derivatives of sufficiently high order. Based on this, they proposed some related test statistics. McKean and Schrader (1987), Koenker (1987) and Bai et al. (1990) studied the case where $\rho(u) = |u|$.

There are also other test criteria, for example, Wald's test criterion W_n, Rao's score-type criterion R_n (refer to Rao, 1948, Sen (1982), Singer and Sen (1985), Zhao and Chen (1991), Chen and Zhao (1996)).

$$\begin{aligned} W_n &= \hat{\beta}(n)' H_n H_n' \hat{\beta}(n) = (\hat{\beta}_n - b)' H (H' S_n^{-1} H)^{-1} H' (\hat{\beta}_n - b), \\ R_n &= \xi_n(\beta_n^*)' \xi_n(\beta_n^*), \\ R_n^* &= \xi_n(\beta_n^*)' H_n H_n' \xi_n(\beta_n^*). \end{aligned} \tag{14}$$

where β_n^* and $\hat{\beta}_n$ are defined in (12) and (13), and

$$\begin{aligned} &\hat{\beta}(n) = S_n^{1/2}(\hat{\beta}_n - b), \qquad H_n = S_n^{-1/2} H (H' S_n^{-1} H)^{-1/2} : p \times q, \\ &\xi_n(\beta_n^*) = S_n^{-1/2} \sum x_i \psi(Y_i - x_i' \beta_n^*) \end{aligned} \tag{15}$$

and $H_n' H_n = I_q$.

Bai et al. (1992) established the limiting distributions of $M_n^* - \hat{M}_n$ and W_n under H_0 in the more general multivariate linear model (6).

Now we consider a special case of model (6), the standard multivariate linear model

$$Y_i = B' x_i + \mathcal{E}_i, \quad i = 1, \cdots, n, \tag{16}$$

where $\mathcal{E}_i$ are i.i.d. p-vectors with a common distribution F, x_i are given m-vectors, B is an unknown $m \times p$ matrix of regression coefficients. It is interesting to test the hypothesis

$$\mathrm{H}_0 : H' B = C_0, \tag{17}$$

where H and C_0 are known $m \times q$ matrix of rank q and $q \times p$ matrix.

By using M-method, Bai et al. (1993) developed MANOVA-type analysis leading to some test criteria based on the roots of a determinantal equation for testing H_0.

Denote by $\hat{B}_n$ and B_n^* any values of B which minimize

$$\sum \rho(Y_i - B' x_i), \tag{18}$$

respectively, without any restraint and subject to restraint (17), where ρ is a given convex function of p variables.

As before, let $\psi(u)$ be a choice of a subgradient of ρ at $u \in R^p$, and assume that (M1)-(M4) and the following (M5$'$) are satisfied.

(M5$'$) $S_n = \sum x_i x_i' > 0$ for large n and

$$d_n^2 := \max_{1 \le i \le n} x_i' S_n^{-1} x_i \to 0 \ \ as \ \ n \to \infty.$$

For testing H_0, Bai et al. (1993) proposed two alternative test criteria. One is based on the roots of the determinantal equation

$$|W_n - \theta \hat{\Lambda}_n^{-1} \hat{\Gamma}_n \hat{\Lambda}_n^{-1}| = 0, \tag{19}$$

where

$$W_n = (H'\hat{B}_n - C_0)'(H'S_n^{-1}H)^{-1}(H'\hat{B}_n - C_0) \tag{20}$$

is the Wald-type statistics, and $(\hat{\Lambda}_n, \hat{\Gamma}_n)$ is a consistent estimate of (Λ, Γ), the matrix parameters are defined in (M2) and (M4), respectively. Another test is based on the roots of the determinantal equation

$$|R_n - \theta\hat{\Gamma}_n| = 0, \tag{21}$$

where

$$R_n = \xi(B_n^*)'S_n^{-1}\xi(B_n^*) \quad with \quad \xi(B) = \sum x_i\psi'(Y_i - B'x_i) \tag{22}$$

is the Rao's score-type statistic, and $\hat{\Gamma}_n$ is a consistent estimate of Γ. The asymptotic distribution of the roots of (19) or (21) is the same as that in the normal theory, and hence the test proposed by Fisher and Hsu (see for instance Rao, 1973, pp. 556-560) can be used.

Consider a sequence of local alternatives to the null hypothesis $H'B = C_0$, say

$$H'B - C_0 = H'\Delta_n, \tag{23}$$

where Δ_n is a known $m \times p$ matrix such that

$$||S_n^{1/2}\Delta_n|| = O(1). \tag{24}$$

Write

$$\begin{aligned}
x_{in} &= S_n^{-1/2}x_i, \qquad H_n = S_n^{-1/2}H(H'S_n^{-1}H)^{-1/2},\\
U_n' &= \Lambda^{-1}\sum \psi(\mathcal{E}_i)x_{in}'H_n := (u_{1n}, \cdots, u_{qn}) : p \times q,\\
V_n' &= \sum \psi(\mathcal{E}_i)x_{in}'H_n := (v_{1n}, \cdots, v_{qn}) : p \times q,\\
\Theta_n &= H_n'S_n^{1/2}\Delta_n = (H'S_n^{-1}H)^{-1/2}H'\Delta_n,
\end{aligned}$$

where $\sum x_{in}x_{in}' = I_m$ and $H_n'H_n = I_q$.

It is easily seen that $u_{1n}, \cdots, u_{qn}$ are asymptotically independent with common limiting distribution $\mathrm{N}_p(0, \Lambda^{-1}\Gamma\Lambda^{-1})$, so that the limiting distribution of $U_n'U_n$ is central Wishart on q degrees of freedom, $\mathrm{W}_p(q, \Lambda^{-1}\Gamma\Lambda^{-1})$. Similarly $v_{1n}, \cdots, v_{qn}$ are asymptotically independent with common limiting distribution $\mathrm{N}_p(0, \Gamma)$, so that the limiting distribution of $V_n'V_n$ is central Wishart on q degrees of freedom, $\mathrm{W}_p(q, \Gamma)$.

Bai et al. (1993) established the following theorems concerning the asymptotic distributions of W_n and R_n under the null hypothesis and also under the sequence of alternative hypotheses (23).

Theorem 3. *Assume that under model (16), (M1)-(M4), (M5$'$), (23) and (24) are satisfied. Then*

$$W_n = (U_n + \Theta_n)'(U_n + \Theta_n) + o_p(1) \quad as \ n \to \infty.$$

Especially, if H_0 *holds or* $||S_n^{1/2}\Delta_n|| \to 0$ *as* $n \to \infty$, *the asymptotic distribution of* W_n *is the central Wishart,* $W_p(q, \Lambda^{-1}\Gamma\Lambda^{-1})$. *If* Θ_n *has a limit* $\Theta \neq 0$ *as* $n \to \infty$, *the asymptotic distribution of* W_n *is the noncentral Wishart,* $W_p(q, \Lambda^{-1}\Gamma\Lambda^{-1}, \Theta'\Theta)$.

Theorem 4. *Under the conditions of Theorem 3,*

$$R_n = (V_n + \Theta_n\Lambda)'(V_n + \Theta_n\Lambda) + o_p(1) \quad as \ n \to \infty.$$

Especially, if H_0 *holds or* $||S_n^{1/2}\Delta_n|| \to 0$ *as* $n \to \infty$, *the asymptotic distribution of* R_n *is the central Wishart,* $W_p(q, \Gamma)$. *If* Θ_n *has a limit* $\Theta \neq 0$ *as* $n \to \infty$, *the asymptotic distribution of* R_n *is the noncentral Wishart,* $W_p(q, \Gamma, \Lambda\Theta'\Theta\Lambda)$.

Note that the local power for the sequence of alternatives considered depends on the magnitude of the roots of the equation

$$|\Theta'\Theta - \alpha\Lambda^{-1}\Gamma\Lambda^{-1}| = 0$$

for the test based on W_n and

$$|\Lambda\Theta'\Theta\Lambda - \alpha\Gamma| = 0$$

for the test based on R_n. Since the roots of these two equations are the same, the two alternative tests are equally efficient asymptotically.

In practical applications, we need to estimate the nuisance matrix parameters Γ and Λ. A natural estimate of Γ is

$$\hat{\Gamma} = n^{-1}\sum \psi(Y_i - \hat{B}_n'x_i)\psi'(Y_i - \hat{B}_n'x_i).$$

To estimate Λ , we take a $p \times p$ nonsingular matrix Z consisting of $\zeta_1, \cdots, \zeta_p$ as its columns, take $h = h_n > 0$ such that

$$h_n/d_n \to \infty, \quad h_n \to 0, \quad and \quad \liminf_{n\to\infty} nh_n^2 > 0, \tag{25}$$

define

$$\eta_{kn} = (2nh)^{-1}\sum\{\psi(Y_i - \hat{B}_n'x_i + h\zeta_k) - \psi(Y_i - \hat{B}_n'x_i - h\zeta_k)\}, \quad k = 1, \cdots, p,$$
$$A_n = (\eta_{1n}, \cdots, \eta_{pn})Z^{-1}$$

and use

$$\hat{A}_n = (A_n + A_n')/2$$

as an estimate of Λ. Bai et al. (1993) established the following:

Theorem 5. *Assume that (M1)-(M4), (M5′) hold in model (16), and* B *is the true parameter. Then*

$$\hat{\Gamma} \to \Gamma \quad in\ pr., \quad as\ n \to \infty.$$

Furthermore, if (25) also holds, then

$$\hat{\Lambda}_n \to \Lambda \quad in\ pr., \quad as\ n \to \infty.$$

References

1. Amemiya, T.(1982). Two stage least absolute deviations estimators. *Econometrika*, **50**, 689-711.
2. Bai, Z.D., Chen, X.R., Wu, Y.H., Zhao, L.C.(1987). Asymptotic normality of minimum L_1-norm estimates in linear models. Technical Report, 87-35, Center for Multivariate Analysis, University of Pittsburgh.
3. Bai, Z.D., Rao, C.R., Wu, Y.(1992). M-estimation of multivariate linear regression parameters under a convex discrepancy function. *Statist. Sinica*, **2**, 237-254.
4. Bai, Z.D., Rao, C.R., Yin, Y.Q.(1990). Least absolute deviations analysis of variance. *Sankhyā*, Ser. A, **52**, 166-177.
5. Bai, Z.D., Rao, C.R., Zhao, L.C.(1993). MANOVA type test under a convex discrepancy function for the standard multivariate linear model. *J. Statist. Plann. Inference*, **36**, 77-90.
6. Bassett, G., Koenker, R.(1978). Asymptotic theory of least absolute error regression. *J. Amer. Statist. Assoc.* **73**, 618-622.
7. Bloomfield, P., Steiger, W.L.(1983). *Least Absolute Deviations.* Birkhauser, Boston.
8. Chen, X.R., Bai, Z.D., Zhao, L.C., Wu, Y.H.(1990). Asymptotic normality of minimum L_1-norm estimates in linear models. *Sci. China*, Ser. A, Chinese Edition: **20**, 162-177; English Edition: **33**, 1311-1328.
9. Chen, X.R., Zhao, L.C.(1996). *M-Methods in Linear Model.* Shanghai Scientific & Technical Publishers, Shanghai, (in Chinese).
10. Dupačová, J.(1987). Asymptotic properties of restricted L_1-estimates of regression. In: Dodge, Y. (Ed.), *Statistical Data Analysis Based on the L_1-Norm and Related Methods.* North-Holland, Amsterdam, pp. 263-274.
11. Huber, P.J.(1973). Robust regression. *Ann. Statist.* **1**, 799-821.
12. Huber, P.J.(1981). *Robust Statistics.* Wiley, New York.
13. Jurečková, J., Sen, P.K.(1996). *Robust Statistical Procedures: Asymptotics and Interrelations.* Wiley, New York.
14. Koenker, R.W.(1987). A comparison of asymptotic testing methods for L_1-regression. In: Dodge, Y. (Ed.), *Statistical Data Analysis Based on the L_1-Norm and Related Methods.* North-Holland, Amsterdam, pp. 287-295.
15. McKean, J.W., Schrader, R.M.(1987). Least absolute errors analysis of variance. In: Dodge, Y. (Ed.), *Statistical Data Analysis Based on the L_1-Norm and Related Methods.* North-Holland, Amsterdam, pp. 297-305.
16. Pollard, D.(1991). Asymptotics for least absolute deviation regression estimators. *Econometric Theory*, **7**, 186-199.
17. Rao, C.R.(1948). Tests of significance in multivariate analysis. *Biometrika*, **35**, 58-79.
18. Rao, C.R.(1973). *Linear Statistical Inference and Its Applications*, 2nd Edition. Wiley, New York.
19. Rao, C.R., Toutenburg, H.(1995). *Linear Models, Least Squares and Alternatives.* Springer, New York.
20. Rockafellar, R.T.(1970). *Convex Analysis.* Princeton University Press, Princeton, NJ.
21. Schrader, R.M., Hettmansperger, T.P.(1980). Robust analysis of variance based upon a likelihood ratio criterion. *Biometrika*, **67**, 93-101.
22. Sen, P.K.(1982). On M tests in linear models. *Biometrika*, **69**, 245-248.
23. Singer, J.M., Sen, P.K.(1985). M-methods in multivariate linear models. *J. Multivariate Anal.* **17**, 168-184.
24. Yohai, V.J., Maronna, R.A.(1979). Asymptotic behavior of M-estimators for the linear model. *Ann. Statist.* **7**, 258-268.
25. Zhao, L. C. (2000). Some contributions to M-estimation in linear models. *J. Statist.*

Planning and Inference, Vol. 88, 189-203.

26. Zhao, L.C., Chen, X.R.(1991). Asymptotic behavior of M-test Statistics in linear models. *J. Combin. Inform. System Sci.* **16**, 234-248.

PROFESSOR BAI'S MAIN CONTRIBUTIONS ON RANDOMIZED URN MODELS

FEIFANG HU*
Department of Statistics, University of Virginia,
Charlottesville, Virginia, 22904, USA
** E-mail: fh6e@virginia.edu*
www.stat.virginia.edu/hu.html

In the area of response-adaptive design based on randomized urn models, Professor Bai's research has been focused on providing a mathematically rigorous of generalized Friedman's urn model. In a series of papers, matrix recursions and martingale theory were introduced to study randomized urn models. Based on these techniques, some fundamental questions were answered.

Keywords: Generalized Friedman's urn model; Martingale; Matrix.

1. Theoretical Properties of Urn Models

Urn models have a long history in probability literature and induce many useful and interesting stochastic processes. A large class of response-adaptive randomization procedures are based on randomized urn models (Hu and Rosenberger, 2006). In randomized urn models, a ball is randomly drawn from the urn and then balls are replaced according to some probability system. It is important to obtain the theoretical properties of the urn composition and the proportion of balls drawn (this is the allocation proportion of different treatments of a response-adaptive randomization procedure, when it is used in clinical trials.).

For generalized Friedman's urn model, Athreya and Karlin (1967, 1968) investigated the asymptotic limit and asymptotic distribution of the urn composition. Of more interest is the limiting distribution of the allocation proportions. Athreya and Karlin (1967) stated, "... It is suggestive to conjecture that the allocation proportions properly normalized is asymptotically normal. This problem is open.". In their papers, they use the technique of embedding the urn process in a continuous-time branching process. After that, the technique of embedding became the standard tool to deal with randomized urn models.

Bai and Hu (1999) first introduced the technique of matrix recursions to study the asymptotic properties of randomized urn models. By combining this technique with martingale theory, the asymptotic distribution of the urn composition were obtained under very generalized conditions. Further, Bai and Hu (2005) obtained the asymptotic distribution of the allocation proportions by using the technique of matrix recursions and martingale theory. This completely proved the conjecture of Athreya and Karlin (1967). The technique of matrix recursions and martingale theory, which is used in Bai, Hu and Shen (2002), Bai, Hu and Rosenberger (2002), Hu and Zhang (2004a, 2004b) and Zhang, Hu and Cheung (2006), is now a standard tool in studying the asymptotic properties of response adaptive design.

There are three important advantages of this technique:
(i) It is easy to understand and can be implemented to different urn models.
(ii) It can be applied to the case with heterogeneous responses as well as other no standard cases. For heterogeneous responses, the theoretical proof can be found in Bai and Hu (1999, 2005) by using this technique. It is interest to notice that the Athreya and Karlin's embedding technique does not work under heterogeneity of responses.
(iii) Based on this technique, Bai and Hu (2005) also obtained the asymptotic variance-covariance matrix of the allocation proportions explicitly. This asymptotic variance-covariance matrix plays an important role in comparing different response adaptive designs (Hu and Rosenberger, 2003).

2. Accommodating Heterogeneity

In clinical trials, the patients are selected sequentially. Heterogeneity can occur because of a covariate or a time trend. Examples of heterogeneity have been described in Altman and Royston (1988) as well as Hu and Rosenberger (2000). For example, they described one clinical trial where characteristics of patients changed over the course of recruitment, and therefore the probability of response to treatments differed among patients recruited at different times. To accommodate heterogeneity of responses in response-adaptive randomization procedures based on randomized urn models, one has to solve following problems. What are the properties of randomized urn models under heterogeneity of responses? How is statistical inference performed after using randomized urn models with heterogeneous responses?

For generalized Friedman's urn, Bai and Hu (1999, 2005) studied the asymptotic properties under following statistical model. Assume the responses of i-th patient,

$$X_i \sim f(\cdot, \boldsymbol{\theta}(i)), i = 1, ..., n.$$

The parameters $\boldsymbol{\theta}(i)$ are different for different patient. To discuss the properties of randomized urn model, one needs the following assumption:

$$\sum_{i=1}^{\infty} \frac{\|\boldsymbol{\theta}(i) - \boldsymbol{\theta}\|_2}{\sqrt{i}} < \infty,$$

for some fixed parameter $\boldsymbol{\theta}$. Here $\|.\|_2$ is the L_2-norm of a vector. After Bai and Hu (1999), this heterogeneity model is used in other response-adaptive designs (Zhang, Chan, Cheung and Hu, 2007). Based on the results of Bai and Hu (1999), Hu and Rosenberger (2000) can then apply weighted likelihood techniques for statistical inference.

3. Delayed Responses

In practice, the responses are usually not available immediately. We call this delayed responses. There is no logistical difficulty in incorporating delayed responses into urn model. One can just update the urn when responses become available (Hu and Rosenberger, 2006). Bai, Hu and Rosenberger (2002) were the first to explore the effects of delayed responses theoretically. They proposed a very general framework for delayed response under the generalized Friedman's urn. The delay mechanism can depend on the patient's entry time, treatment assignment, and response. This framework is commonly used response adaptive randomization procedures (Hu and Rosenberger, 2006).

Bai, Hu and Zhang (2002) obtained the Gaussian approximation theorems of generalized Friedman's urn model with two treatments (two types of balls). This introduced Gaussian approximations and the law of the iterated logarithm to study the properties of urn models. In another paper, Bai, Hu and Shen (2002) proposed an urn model for comparing K-treatment.

References

1. F. Hu and W. F. Rosenberger. *The Theory of Response-Adaptive Randomization in Clinical Trials.* John Wiley and Sons. Wiley Series in Probability and Statistics (2006).
2. K. B. Athreya and S. Karlin. Limit theorems for the split times of branching processes. *Journal of Mathematics and Mechanics* 17, 257-277 (1967).
3. K. B. Athreya and S. Karlin.. Embedding of urn schemes into continuous time Markov branching processes and related limit theorems. *Annals of Mathematical Statistics* 39, 1801-1817 (1968).
4. Z. D. Bai and F. Hu. Asymptotic Theorems for urn models with nonhomogeneous generating matrix. *Stochastic Processes and Their Applications*, 80, 87-101 (1999).
5. Z. D. Bai and F. Hu. Asymptotics in randomized urn models. *Annals of Applied Probability.* 15, 914-940 (2005).
6. Z. D. Bai, F. Hu and L. Shen. An Adaptive Design for Multi-Arm Clinical Trials. *Journal of Multivariate Analysis,* 81, 1-18 (2002).
7. Z. D. Bai, F. Hu and W. F. Rosenberger. Asymptotic properties of adaptive designs for clinical trials with delayed response. *Annals of Statistics*, 30, 122-139 (2002).
8. F. Hu and L. X. Zhang. Asymptotic properties of doubly adaptive biased coin designs for multi-treatment clinical trials. *Annals of Statistics* . 32, 268-301 (2004a).
9. F. Hu and L. X. Zhang. Asymptotic normality of adaptive designs with delayed response. *Bernoulli* . 10, 447-463 (2004b).
10. L. X. Zhang, F. Hu and S. H. Cheung. Asymptotic theorems of sequential estimation-adjusted urn models. *Annals of Applied Probability.* 16, 340-369 (2006).
11. F. Hu and W. F. Rosenberger. Optimality, variability, power: Evaluating response-adaptive randomization procedures for treatment comparisons. *Journal of the American Statistical Association*, 98, 671-678 (2003).
12. D. G. Altman and J. P. Royston. The hidden effect of time. *Statistics in Medicine.* 7, 629-637 (1988).
13. F. Hu and W. F. Rosenberger. Analysis of time trends in adaptive designs with application to a neurophysiology experiment. *Statistics in Medicine.* 19, 2067-2075 (2000).
14. L. X. Zhang, W. S. Chan, S. H. Cheung and F. Hu. Generalized drop-the-loser rule with delayed response. *Statistica Sinica.* 17, 387-409 (2007).

15. Z. D. Bai, F. Hu and L. X. Zhang. The Gaussian approximation theorems for urn models and their applications. *Annals of Applied Probability,* 12, 1149-1173 (2002).

PROFESSOR BAI'S CONTRIBUTIONS TO M-ESTIMATION

Y. WU

Department of Mathematics and Statistics, York University
Toronto, Ontario M3J 1P3, Canada
** E-mail: wuyh@mathstat.yorku.ca*

M-estimation, as a generalization of maximum likelihood estimation, plays an important and complementary role in robust statistics. Dr. Bai has made prestigious contributions to this area. In this article, as a coauthor and good friend, I will describe some of his major contributions.

1. Introduction

It is often that the underlying distribution of data is not exactly as assumed and/or the data are subject to error from various sources with complex or un-expressible likelihood functions. In such a situation, robust statistics should be more preferable. M-estimation, which is a wonderful generalization of maximum likelihood estimation, plays an important and complementary role in robust statistics. It has received considerable attention in the literature and gained significant progress in the past four decades. Among his many prestigious contributions to Probability and Statistics, Dr. Bai's contributions to the area of M-estimation is extremely outstanding and significant. I will briefly describe some of his main contributions to M-estimation in the following five categories:

- The case that the discrepancy function is a difference of two convex functions (Section 2);
- Minimum L_1-norm estimation (Section 3);
- Recursive algorithm (Section 4);
- Solvability of an equation arising in the study of M-estimates (Section 5);
- General M-estimation (Section 6).

2. The case that the discrepancy function is a difference of two convex functions

Consider a general multivariate regression model

$$\boldsymbol{y}_i = X_i\boldsymbol{\beta} + \boldsymbol{\varepsilon}_i, \quad i = 1, \ldots, n, \tag{1}$$

where $\boldsymbol{y}_i$ is an m-vector of observations, X_i is an $m \times p$ given matrix, $\boldsymbol{\beta}$ is a p-vector of unknown parameters, and $\boldsymbol{\varepsilon}_i$ is an m-vector of unobservable random error suitably centered and having an m-variate distribution. When $m = 1$, the model (1) reduces to the usual univariate regression model

$$y_i = \boldsymbol{x}_i'\boldsymbol{\beta} + \varepsilon_i, \quad i = 1, \ldots, n, \tag{2}$$

where $\boldsymbol{x}_i$ is a p-vector and $\boldsymbol{x}_i'$ denotes the transpose of $\boldsymbol{x}_i$.

For the model (2), (Huber, 1964, 1973) introduced and named as M-estimate of $\boldsymbol{\beta}$ which is defined as a value of $\boldsymbol{\beta}$ minimizing

$$\sum_{i=1}^{n} \rho(y_i - \boldsymbol{x}_i'\boldsymbol{\beta}) \tag{3}$$

for a suitably chosen function ρ, or the value of $\boldsymbol{\beta}$ satisfying the estimating equation

$$\sum_{i=1}^{n} \psi(y_i - \boldsymbol{x}_i'\boldsymbol{\beta})\boldsymbol{x}_i = \mathbf{0} \tag{4}$$

for a suitably chosen ψ-function. A natural method of obtaining the estimating equation (4) is equating zero to the derivative of (3) with respect of $\boldsymbol{\beta}$ when ρ is continuously differentiable. However, in general one can use any suitably chosen function ψ and set up the equation (4). In papers (Bai, Rao and Wu, 1991, 1992), a general theory is developed for M-estimation using a convex discrepancy function for ρ in (3), which covers most useful cases considered in the literature such as least squares (LS), least absolute deviations, least distances (LD), mixed LS and LD and L_q-norm.

Some advantages with a convex discrepancy function are the existence of a unique global minimizer $\hat{\boldsymbol{\beta}}$ of (3) and the simplicity of conditions for establishing asymptotic results for inference on $\boldsymbol{\beta}$. However some of the well known discrepancy functions suggested for minimizing the effects of outliers are not convex, but needed very restrictive conditions to guarantee the asymptotic results.

In the paper,[4] Bai, jointly with Rao and Wu, have developed a general theory of M-function under what appear to be a necessary set of assumptions to develop a satisfactory asymptotic theory, which includes the case that the discrepancy function is a difference of two convex functions. An appropriate criterion was also developed for tests of hypotheses concerning regression parameters. Note that almost every practically useable dispersion function can be expressed as a difference of two convex functions.

3. Minimum L_1-norm estimation

Consider the model (2), the minimum L_1-norm estimation is defined as any $\hat{\boldsymbol{\beta}}$ such that

$$\sum_{i=1}^{n} |y_i - \boldsymbol{x}_i'\hat{\boldsymbol{\beta}}| = \min \sum_{i=1}^{n} |y_i - \boldsymbol{x}_i'\boldsymbol{\beta}|$$

when minimized with respect to $\boldsymbol{\beta}$, which is a special case of M-estimation with $\rho(\cdot) = |\cdot|$. By using the gradually densified partition method, Bai, jointly with Chen,

Zhao and Wu, proved asymptotical normality of the minimum L_1-norm estimation under conditions which are currently known as the weakest in the paper.[8] They also showed that all earlier proofs of this result by Basset and Koenker, Bloomfield and Steiger, Amemiya, Jurečkova and Dupačov contain flaws or undue restrictions on independent variables. The same method was also adopted in (Bai and He, 1999).

4. Recursive algorithm

A serious disadvantage of M-estimation, comparing to the least squares method, is due to the difficulty in the computation of the M-estimates because almost no M-estimators have close forms. Sometimes the Newton approach can be applied to the computation of M-estimates. Even when the Newton approach is applicable, it is usually selective for the initial values. Moreover, in many real applications to prediction, control and target tracking, it is often necessary to re-estimate the parameters as new portion of data are observed. It will be onerous and gaucherie to recalculate the estimates from enlarged data set. And meanwhile. we need a large space to store the historical data. Thus, there is a great need of developing a recursive algorithm which updates M-estimates easily and uses only the previous estimates and newly observed data.

Let $\rho(u)$ be a nonnegative function on $[0, \infty)$, and $\rho(u) = 0$ if and only if $u = 0$. Assume that in the model (1), $(X_i, \boldsymbol{\varepsilon}_i)$, $i = 1, 2, \cdots$, are independently and identically distributed and X_i is independent of $\boldsymbol{\varepsilon}_i$. The M-estimates of regression coefficients $\boldsymbol{\beta}$ and scatter parameters V for the model (1) are defined as the solutions of the following minimization problem:

$$\sum_{i=1}^{n}[\rho(\|\boldsymbol{y}_i - X_i\hat{\boldsymbol{\beta}}_n\|_{\hat{V}_n}) + \log(\det(\hat{V}_n))] = \min,$$

where $\det(V)$ denotes the determinant of a positive definite matrix V, $\|\boldsymbol{y}\|_V^2 = \boldsymbol{y}'V^{-1}\boldsymbol{y}$.

Suppose that ρ is continuously differentiable. Then $(\hat{\boldsymbol{\beta}}_n, \hat{V}_n)$ is the solution to the following equations

$$\begin{cases} \sum\limits_{i=1}^{n} X_i'V^{-1}(\boldsymbol{y}_i - X_i\boldsymbol{\beta})u_1(\|\boldsymbol{y}_i - X_i\boldsymbol{\beta}\|_V) = 0, \\ \sum\limits_{i=1}^{n} [(\boldsymbol{y}_i - X_i\boldsymbol{\beta})(\boldsymbol{y}_i - X_i\boldsymbol{\beta})'u_2(\|\boldsymbol{y}_i - X_i\boldsymbol{\beta}\|_V^2) - V] = 0, \end{cases} \tag{5}$$

where $u_1(t) = t^{-1}d\rho(t)/dt$, and $u_2(t) = u_1(\sqrt{t})/2$.

When $u_1(t)$ and $u_2(t)$ are determined by the same ρ, it is difficult to keep the robustness of the M-estimates for both regression coefficients and scatter parameters simultaneously. In light of (Maronna, 1976), (5) may be extended to allow that u_1 and u_2 are chosen independently.

Motivated by (Englund, 1993), Bai (jointly with Wu) proposed the following recursive algorithm for the multivariate linear regression model (1) in the paper:[5]

$$\begin{cases} \hat{\boldsymbol{\beta}}_{n+1} = \hat{\boldsymbol{\beta}}_n + S_{n+1}^{-1} a_n \boldsymbol{h}_1(\hat{\boldsymbol{\beta}}_n, \hat{V}_n, X_{n+1}, \boldsymbol{y}_{n+1}), \\ \hat{V}_{n+1} = \hat{V}_n + (n+1)^{-1} H_2(\hat{\boldsymbol{\beta}}_n, \hat{V}_n, X_{n+1}, \boldsymbol{y}_{n+1}), \end{cases} \tag{6}$$

where $S_n = \sum_{i=1}^n X_i' X_i$,

$$\boldsymbol{h}_1(\boldsymbol{\beta}, V, X, \boldsymbol{\varepsilon}) = X' \widetilde{V}^{-1} (\boldsymbol{\varepsilon} - X\boldsymbol{\beta}) u_1(\|\boldsymbol{\varepsilon} - X\boldsymbol{\beta}\|_V),$$
$$H_2(\boldsymbol{\beta}, V, X, \boldsymbol{\varepsilon}) = (\boldsymbol{\varepsilon} - X\boldsymbol{\beta})(\boldsymbol{\varepsilon} - X\boldsymbol{\beta})' u_2(\|\boldsymbol{\varepsilon} - X\boldsymbol{\beta}\|_V^2) - V,$$

$\boldsymbol{\beta}_0$ and $V_0 > 0$ are arbitrary, $\{a_n\}$ satisfies certain conditions, and $\widetilde{V}$ is a Lipschitz continuous $m \times m$ matrix function of V defined as follows:

> Let λ_i and $\boldsymbol{\alpha}_i$ be the i-th eigenvalue and corresponding eigenvector of V respectively. Then
>
> $$\widetilde{V} = \sum_{i=1}^m \tilde{\lambda}_i \boldsymbol{\alpha}_i \boldsymbol{\alpha}_i',$$
>
> where $\tilde{\lambda}_i = (\delta_1 \vee \lambda_i) \wedge \delta_2$ and δ_1, δ_2 ($0 < \delta_1 < \delta_2 < \infty$) are two constants selected properly.

It is noted that when $X_i = I$, $i = 1, 2, \ldots$, this recursive algorithm reduces to the one given by (Englund, 1993) for the multivariate location models.

Bai (jointly with Wu) showed that $\hat{\boldsymbol{\beta}}_n$ and $\hat{V}_n$ are strongly consistent under mild conditions in the paper.[5] They successfully removed one critical but not verifiable condition imposed by (Englund, 1993). Motivated by (Bai and Wu, 1993), a series of papers on recursive M-estimation have been published, including the paper,[13] in which several new recursive algorithms were developed to compute M-estimates of regression coefficients and scatter parameters in multivariate linear models, the strong consistency and asymptotic normality of the recursive M-estimators were demonstrated and a promising recursive algorithm was provided.

5. Solvability of an equation arising in the study of M-estimates

Much work on the properties of M-estimators relies on the assumption that finding a minimum can, at least asymptotically, be replaced by finding a root of a estimating equation via the derivative of the discrepancy function, even if there are isolated points where the derivative does not exist. In the paper,[7] Bai, jointly with Wu, Chen and Miao, showed that for some discrepancy functions, the probability that no such root exists will tend to a positive value as the sample size tends to infinity under some mild assumptions on the distribution of the sample. This probability is even very close to 1 in some situations. This discovery is very important since it points out a common misuse in the study of M-estimation.

6. General M-estimation

By noticing that parameter estimation in all linear and nonlinear regression models, AR time series, EIVR models can be unified, Bai (jointly with Wu) proposed a general form of M-estimation given below and studied its asymptotic behavior in the paper,[6] which includes all those estimation as its special cases.

Let $\{\boldsymbol{y}_1, \ldots, \boldsymbol{y}_n, \ldots\}$ be a sequence of random vectors and for each $\boldsymbol{\beta} \in \Omega$, $\{\rho_1(\boldsymbol{y}_1, \boldsymbol{\beta}), \ldots, \rho_n(\boldsymbol{y}_n, \boldsymbol{\beta}), \ldots\}$ be a sequence of dispersion functions which are differentiable about $\boldsymbol{\beta}$ for almost all $\boldsymbol{y}$'s, where Ω is an open convex subset of R^p known as the parameter space. Then the general M-estimate $\hat{\boldsymbol{\beta}}$ is defined as any value of $\boldsymbol{\beta}$ minimizing

$$\sum_{i=1}^{n} \rho_i(\boldsymbol{y}_i, \boldsymbol{\beta}).$$

Let $\boldsymbol{\psi}_i(\boldsymbol{y}_i, \boldsymbol{\beta})$ denote the derivative of $\rho_i(\boldsymbol{y}_i, \boldsymbol{\beta})$ about $\boldsymbol{\beta}$ if the derivative exists and $\mathbf{0}$ otherwise. Then, $\hat{\boldsymbol{\beta}}$ satisfies

$$\sum_{i=1}^{n} \boldsymbol{\psi}_i(\boldsymbol{y}_i, \hat{\boldsymbol{\beta}}) = \mathbf{0}, \tag{7}$$

if the left side of (7) is continuous at $\hat{\boldsymbol{\beta}}$, or

$$\left| \sum_{i=1}^{n} \boldsymbol{\psi}_i(\boldsymbol{y}_i, \hat{\boldsymbol{\beta}}) \right| = \min_{\boldsymbol{\beta}} \left| \sum_{i=1}^{n} \boldsymbol{\psi}_i(\boldsymbol{y}_i, \boldsymbol{\beta}) \right|,$$

otherwise.

Assume that there exists a vector $\boldsymbol{\beta}_0 \in \Omega$ and for each i there are an nonnegative definite matrix G_i and a function $\boldsymbol{\eta}_i(\boldsymbol{\beta})$ such that

$$E\boldsymbol{\psi}_i(\boldsymbol{y}_i, \boldsymbol{\beta}) = G_i(\boldsymbol{\beta} - \boldsymbol{\beta}_0) + \boldsymbol{\eta}_i(\boldsymbol{\beta}) \tag{8}$$

with $\boldsymbol{\eta}_i(\boldsymbol{\beta}) = \boldsymbol{o}(Q_i(\boldsymbol{\beta} - \boldsymbol{\beta}_0))$ as $\boldsymbol{\beta} \to \boldsymbol{\beta}_0$, where Q_i is some nonnegative definite matrix. The dispersion functions $\{\rho_i\}$ considered may be convex or differences of convex functions, the later covers almost all useable choices of the dispersion functions. Under very general assumptions, the limiting properties of the general M-estimates are investigated and a general theory is established. This contribution is substantial since one does not need to spend time to find out the asymptotic properties of an estimation if it is a special case of this general estimation.

The following are some examples where the derivatives of non-convex dispersion functions are expressed as a difference of derivatives of two convex functions:

(1) $\psi(x) = 2x/(1+x^2)$. Set $\psi_1(x) = 2x/(1+x^2)$ for $|x| \le 1$ and $=\text{sign}(x)$ for $|x| > 1$, whereas $\psi_2(x) = 0$ for $|x| \le 1$ and $=\text{sign}(x) - 2x/(1+x^2)$ for $|x| > 1$. Both $\psi_1(\cdot)$ and $\psi_2(\cdot)$ are derivative functions of two convex functions and one can verify that $\psi(x) = \psi_1(x) - \psi_2(x)$.

(2) Hampel's ψ, i.e., for constants, $0 < a < b < c$, $\psi(x) = x$ for $|x| \le a$, $=$ asign(x) for $a < |x| \le b$, $=$ asign$(x)(c - |x|)/(c-b)$ for $b < |x| \le c$ and $= 0$ otherwise. Let $\psi_1(x) = x$ for $|x| \le a$ and $=$ asign(x) for $|x| > a$ whereas $\psi_2(x) = 0$ for $|x| \le b$, $=$ asign$(x)(|x| - b)/(c-b)$ for $b < |x| \le c$ and $=$ asign(x), otherwise. Both $\psi_1(\cdot)$ and $\psi_2(\cdot)$ are derivative functions of two convex functions and it can be seen that $\psi(x) = \psi_1(x) - \psi_2(x)$.

Set $G(n) = \sum_{i=1}^n G_i$ and $Q = E(\sum_{i=1}^n \boldsymbol{\psi}_i(\boldsymbol{y}_i, \boldsymbol{\beta}_0))(\sum_{i=1}^n \boldsymbol{\psi}_i(\boldsymbol{y}_i, \boldsymbol{\beta}_0))'$. Define $\boldsymbol{\Delta}_i(\boldsymbol{y}_i, \boldsymbol{\beta}) = \boldsymbol{\psi}_i(\boldsymbol{y}_i, \boldsymbol{\beta}) - \boldsymbol{\psi}_i(\boldsymbol{y}_i, \boldsymbol{\beta}_0)$ and $\boldsymbol{\Delta} = \sum_{i=1}^n \boldsymbol{\Delta}_i$. By (Bai and Wu, 1997), the general M-estimation has the following the asymptotic properties under some mild conditions:

(1) There exists a local minimizer $\hat{\boldsymbol{\beta}}$ such that

$$\hat{\boldsymbol{\beta}} \to \boldsymbol{\beta}_0 \quad \text{in probability.}$$

(2) For any $\mu > 0$,

$$\sup_{|Q^{1/2}(\boldsymbol{\beta}-\boldsymbol{\beta}_0)|<\mu} |Q^{-1/2}(\boldsymbol{\Delta} - G(\boldsymbol{\beta} - \boldsymbol{\beta}_0))| \to 0 \quad \text{in probability.}$$

(3) $Q^{-1/2}G(\hat{\boldsymbol{\beta}} - \boldsymbol{\beta}_0) \to N(\mathbf{0}, I)$.

Several applications of the general M-estimation are given in (Bai and Wu, 1997). Here is another example: The paper[14] has proposed an M-estimation of the parameters in an undamped exponential signal model. However its asymptotic behavior is hardly to show. By (Bai and Wu, 1997), the M-estimation is successfully proved to be consistent under mild conditions.

References

1. Z. D. Bai, Z.D. and X. He, *Ann. Statist.* **27**, 1616 (1999).
2. Z. D. Bai, Z.D., C. R. Rao and Y. Wu, Y, in *Probability, Statistics and Design of Experiments, R.C. Bose Symposium Volumn* (Wiley Eastern, 1991).
3. Z. D. Bai, Z.D., C. R. Rao and Y. Wu, Y, *Statistica Sinica*, **2**, 237 (1992).
4. Z. D. Bai, Z.D., C. R. Rao and Y. Wu, Y, in *Robust inference, Handbook of Statist. Vol. 15, 1-19* (North-Holland, Amsterdam, 1997).
5. Z. D. Bai and Y. Wu, *Sankhyā, Ser. B*, **55**, 199 (1993).
6. Z. D. Bai and Y. Wu, *J. Multivariate Anal.* **63**, 119 (1997).
7. Z. D. Bai, Y. Wu, X. R. Chen and B. Q. Miao, *Comm. Statist. Theory Methods*, **19**, 363 (1990).
8. X. R. Chen, Z. D. Bai, L. C. Zhao and Y. Wu, *Sci. China, Ser. A*, **33**, 1311 (1990).
9. J. -E. Englund *J. Multivariate Anal.* **45**, 257 (1993).
10. P. J. Huber *Ann. Math. Statist.* **35**, 73 (1964).
11. P. J. Huber *Ann. Statist.* **1**, 799 (1973).
12. R. A. Maronna, R. A. *Ann. Statist.*, **4**, 51 (1976).
13. B. Q. Miao and Y. Wu, *J. of Multivariate Anal.*, **59**, 60 (1996).
14. Y. Wu and K. Tam, *IEEE Trans. Signal Processing*, **49**, 373 (2001).

ON PROFESSOR BAI'S MAIN CONTRIBUTIONS TO THE SPECTRAL THEORY ON RANDOM MATRICES

Jian-Feng YAO

IRMAR, Universit de Rennes 1,
Campus de Beaulieu, F-35042 Rennes, France
** E-mail: jian-feng.yao@.univ-rennes1.fr*

The aim of the spectral theory of large dimensional random matrices (RMT) is to investigate the limiting behaviour of the eigenvalues $(\lambda_{n,j})$ of a random matrix (A_n) when its sizes tend to infinity. Of particular interest are the empirical spectral distribution (ESD) $F_n := n^{-1} \sum_j \delta_{\lambda_{n,j}}$, the extreme eigenvalues $\lambda_{\max}(A_n) = \max_j \lambda_{n,j}$ and $\lambda_{\min}(A_n) = \min_j \lambda_{n,j}$, or the spacings $\{\lambda_{n,j} - \lambda_{n,j-1}\}$. The main underlying mathematical problems for a given class of random matrices (A_n) are the following: a) find a limiting spectral distribution (LSD) G to which converges the sequence of ESD (F_n) ; b) find the limits of the extreme eigenvalues $\lambda_{\max}(A_n)$ and $\lambda_{\min}(A_n)$; c) quantify the rates of the convergences a) and b). d) find second order limit therems such as central limit theorems for the convergences a) and b).

Professor Bai, one of the world leading experts of the field, has brought several major contributions to the theory.

a). Limiting spectral distributions

This problem covers the beginning age of the RMT when E. Wigner discovers the famous semi-circular law in 1950's with his pioneering work on energy level distributions in nuclear physics. The class of ramdom matrices he considered is now called Wigner matrices which are Hermitian or real symmetrical. Later, Marčenko and Pasttur (1967) establishes the existence of a LSD for several other classes of RM's including sample covariance matrices.

These problems also monopolized Bai's time for mathematics when, in the middle of 1980's, he started his research on the RMT in collaboration with his teacher Yong-Quan Yin, and sometimes P.R. Krishnaiah in Pittsburgh. Let $X_n = \{X_{ij}\}$ be an $p \times n$ matrix of i.i.d. standardized complex-valued random variables so that the sample covariance matrix is defined as $S_n = n^{-1} X_n X_n^*$. In a series of papers, see Yin, Bai and Krishnaiah (1983), Bai, Yin and Krishnaiah (1986, 1987), Bai and Yin (1986), the existence of a LSD is proved for products $A_n = S_n T_n$ of S_n with an independent and positive definite Hermitian matrix T_n. This class of RM's includes the widely-used F-matrix in multivariate data analysis.

Important contributions from Bai on this topic result from a series of collaborations with his freind J.W. Silverstein on the class of generalised sample covariance matrices. In Silverstein and Bai (1995), the LSD is found for a sequence of affinely perturbed sample covariance matrices of the form $A_n = B_n + n^{-1}Y_nT_nY_n^*$, where $Y_n = X_n^T$, (T_n) is a sequence of diagonal matrices and B_n a sequence of Hermitian matrices both with a converging ESD. Although this result already appears in Marčenko and Pastur (1967), Bai and Silverstein provided a new method of proof which will be also benefical for their next findings.

One breaking through result is about the spectrum separation. Let $B_p = n^{-1}T_p^{1/2}X_nX_n^*T_p^{1/2}$ where X_n is as before but with finite fourth moment, and $T_p^{1/2}$ is a Hermitian square root of the nonnegative definite Hermitian matrix T_p. In Bai and Silverstein (1998), it was shown that if $p/n \to y > 0$ and (T_p) has a proper LSD, then with probability 1 no eigenvalues lie in any interval which is outside the support of the LSD of (B_p) (known to exist) for all large p. Moreover, for these p the interval corresponds to one that separates the eigenvalues of T_p. Furthermore in Bai and Silverstein (1999), the exact separation of eigenvalues is proved, that is, with probability 1, the number of eigenvalues of B_p and T_p lying on one side of their respective intervals are identical for all large p.

Another fundamental contribution from Bai is on the circular law which states that, the ESD of $A_n = n^{-1/2}(\xi_{ij})_{1\le i,j\le n}$ where $\{\xi_{ij}\}$ is a doubly infinite array of i.i.d. complex random variables of mean 0 and variance 1, converges to the uniform distribution on the unit disk of the complex plan. There is some controverse on the history of this problem, since in Girko (1984), a proof of the circular law was proposed. However, according to many experts of the field, Girko's proof was flawed (see e.g. Geman, 1980). Indeed, Bai (1997) first gave a rigorous proof of the circular law assuming $E[|\xi_{11}|^6] < \infty$ and that both $\Im(\xi_{11})$ and $\Re(\xi_{11})$ have a bounded density function. Recently, in the book Bai and Silverstein (2006), the moment condition is weakened to $E[|\xi_{11}|^{2+\eta}] < \infty$ for some $\eta > 0$. Whether or not one can take $\eta = 0$ still remains an open problem.

b). Limits of extreme eigenvalues

During his Pittsburgh period, Bai investigated the problem of limits of extreme eigenvalues of the sample covariance S_n. Very few was known on the problem at that time: indeed Geman (1980) proved that $\lambda_{\max}(S_n)$ converges almost surely to $\sigma^2(1+\sqrt{y})^2$ where $\sigma^2 = E[|X_{11}|^2]$ is the common variance of the observed variables and $y > 0$ is the limit of column-row ratio p/n, under a restriction on the growth of the moments : for each k, $E[|X_{11}|^k] \le k^{\alpha k}$ for some constant α. A major achievement on the problem was made by Yin, Bai and Krishnaiah (1988) where the above convergence of $\lambda_{\max}(S_n)$ was established without Geman's restrictive assumption and assuming $E[|X_{11}|^4] < \infty$ only. This result is indeed optimal: actually in Bai, Silverstein and Yin (1988), the authors proved that almost surely, $\limsup_n \lambda_{\max}(S_n) = \infty$ if $E[|X_{11}|^4] = \infty$. On the other hand, for the smallest eigenvalue $\lambda_{\min}(S_n)$, its convergence to the left edge $\sigma^2(1-\sqrt{y})^2$ (assuming $y<1$) is also established in Bai and Yin (1993).

Such achievements from Bai and Yin are made possible by their introduction of sophisticated arguments from graph theory on one hand, and of general truncation technique on the variables under suitable assumptions on the other hand. As one proof more of the power of their advances, Bai and Yin (1988) establishes necessary and sufficient conditions for the almost sure converhence of $\lambda_{\max}(n^{-1/2}W_n)$ for a $n \times n$ real (symmetrical) Wigner matrix $W_n = (w_{ij})$.

c). Convergences rates of an ESD F_n to its LSD

In 1993, two papers of Bai in the Annals of Probability again attest to his rich creativity. At that time, for the sequence of real Wigner matrix $(n^{-1/2}W_n)$ as well as for sample covariances matrices (S_n), the problem of convergence rates of F_n to their LSD G, namely the Wigner's semi-circular law and the Marčenko-Pastur law respectively, was entirely virgin. In Bai (1993a, b), Bai creates a methodology for the estimation of $\|E(F_n - G)\|_\infty$ by establishing two fundamental inequalities: the first one gives, for two probability distribution functions H_1 and H_2, an upper bound for $\|H_1 - H_2\|_\infty$ in terme of the integrals of their Stieltjes transforms, and the second one compare $\|H_1 - H_2\|_\infty$ to the Lévy distance $L(H_1, H_2)$. In a sense, the creation of this methodology is more important than the convergence rates established in these papers, which equal $O(n^{-1/4})$ and $O(n^{-5/48})$, respectively. Based on this methodology, the above rates in expectation are since then successively improved, together with rate estimation for other convergence type like a.s. convergence or convergence in probability, in Bai, Miao and Tsay (1997, 1999, 2002) for Wigner matrices and in Bai, Miao and Yao (2003) for sample covariance matrices.

d). CLT's for smooth integrals of F_n

One of the current problems in RMT deals with second order limit theorems. As an example, the celebrated Tracy-Widom laws determine the limiting distribution of $c_n(\lambda_{\max}(A_n) - b)$ for a suitable scaling sequence (c_n) and a point limit value b when the ensemble (A_n) is the sample covariance matrices (S_n) or the Wigner matrices $((n^{-1/2}W_n))$.

On the other hand, it is worth considering the limiting behavior of the stochastic process $G_n(x) = n(F_n(x) - G(x))$, $x \in \mathbb{R}$. Unfortunately, this process does not have a weak limit. In Bai and Silverstein (2004) for generalised sample covariance matrices (B_p), and in Bai and Yao (2005) for Wigner matrices $((n^{-1/2}W_n))$, a methodology is developed to get CLT's for s-dimensional vectors of integrals $\{\sqrt{n}[F_n(f_k) - G(f_k)],\ 1 \le k \le s\}$ for any given set of analytic functions (f_k). The CLT's provided in Bai and Silverstein (2004) have attracted a considerable attention from researchers in application fields, since such CLTs form a basis for statistical inference when many applications in high-dimensional data analysis rely on the principal component analysis or spectral statistics of large sample covariance matrices, see e.g. Tulino and Verdú (2004).

To summerize, Professor Bai has achieved impressive contributions to the RMT. The above shortly described results have solved old problems and open a lot of new directions for future research in the field. Perhaps more importantly, Professor Bai has introduced several new mathematical techniques such as a refined analysis of Stieltjes transforms, rank inequalities or general truncation techniques, which now form an important piece of the modern toolbox for the spectral analysis of random matrices.

Acknowledgments

Advice from Professors Jack Silverstein and Peter Forrester on this review is greatly appeciated.

References

1. Bai, Z. D. and Silverstein, J. W. *Random Matrix Theory*, Science Press, Beijing, (2006).
2. Bai, Z. D. and Yao, J. On the convergence of the spectral empirical process of Wigner matrices. *Bernoulli*, 11, 1059-1092, (2005).
3. Cui, W., Zhao, L. and Bai, Z. D. On asymptotic joint distributions of eigenvalues of random matrices which arise from components of covariance model. *J. Syst. Sci. Complex.* 18, 126-135, (2005).
4. Bai, Z. D. and Silverstein, J. W. CLT for linear spectral statistics of large-dimensional sample covariance matrices. *Ann. Probab.* 32, 553-605, (2004).
5. Bai, Z. D., Miao, B. and Yao, J. F. Convergence rates of spectral distributions of large sample covariance matrices, *SIAM J. Matrix Anal. Appl.*, 25,105–127, (2003).
6. Bai, Z. D., Miao, B. and Tsay, J. Convergence rates of the spectral distributions of large Wigner matrices, *Int. Math. J.*, 1, 65-90, (2002).
7. Bai, Z. D. and Silverstein, J. W., Exact separation of eigenvalues of large-dimensional sample covariance matrices, *Ann. Probab.*, 27, 1536-1555, (1999).
8. Bai, Z. D. Methodologies in spectral analysis of large-dimensional random matrices, a review, *Statist. Sinica*, 9, 611-677, (1999).
9. Bai, Z. D., Miao, B. and Tsay, J. Remarks on the convergence rate of the spectral distributions of Wigner matrices,*J. Theoret. Probab.*, 12,301–311, (1999).
10. Bai, Z. D. and Hu, F. Asymptotic theorems for urn models with nonhomogeneous generating matrices,*Stochastic Process. Appl.*,80, 87–101, 1999,
11. Bai, Z. D. and Silverstein, J. W. No eigenvalues outside the support of the limiting spectral distribution of large-dimensional sample covariance matrices, *Ann. Probab.*, 26, 316–345, (1998).
12. Bai, Z. D. and Miao, B. and Tsay, J. A note on the convergence rate of the spectral distributions of large random matrices, *Statist. Probab. Lett.*, 34,95–101, (1997).
13. Bai, Z. D. Circular law, *Ann. Probab.*, 25,494–529, (1997).
14. Silverstein, J. W. and Bai, Z. D. On the empirical distribution of eigenvalues of a class of large-dimensional random matrices, *J. Multivariate Anal.*, 54,175–192, (1995).
15. Bai, Z. D. and Yin, Y. Q., Limit of the smallest eigenvalue of a large-dimensional sample covariance matrix, *Ann. Probab.*, 21,1275–1294,(1993).

16. Bai, Z. D., Convergence rate of expected spectral distributions of large random matrices. I. Wigner matrices, *Ann. Probab.*, 21,625–648, (1993a).
17. Bai, Z. D., Convergence rate of expected spectral distributions of large random matrices. II. Sample covariance matrices, *Ann. Probab.*,21,649–672, (1993b).
18. Bai, Z. D., Silverstein, J. W. and Yin, Y. Q., A note on the largest eigenvalue of a large-dimensional sample covariance matrix, *J. Multivariate Anal.*, 26,166–168,(1988).
19. Bai, Z. D., A note on asymptotic joint distribution of the eigenvalues of a noncentral multivariate F matrix, *J. Math. Res. Exposition*, 8,291–300,(1988).
20. Bai, Z. D. and Yin, Y. Q., Necessary and sufficient conditions for almost sure convergence of the largest eigenvalue of a Wigner matrix, *Ann. Probab.*,16,1729–1741,(1988).
21. Bai, Z. D. and Yin, Y. Q., Convergence to the semicircle law, *Ann. Probab.*, 16,863–875, (1988).
22. Yin, Y. Q., Bai, Z. D. and Krishnaiah, P. R., On the limit of the largest eigenvalue of the large-dimensional sample covariance matrix, *Probab. Theory Related Fields*, 78,509–521,(1988).
23. Bai, Z. D., Yin, Y. Q. and Krishnaiah, P. R., On limiting empirical distribution function of the eigenvalues of a multivariate F matrix, *Teor. Veroyatnost. i Primenen.*, 32,537–548,(1987).
24. Bai, Z. D., Krishnaiah, P. R. and Liang, W. Q., On asymptotic joint distribution of the eigenvalues of the noncentral MANOVA matrix for nonnormal populations, *Sankhyā Ser. B*, 48,153–162,(1986).
25. Bai, Z. D. and Yin, Y. Q., Limiting behavior of the norm of products of random matrices and two problems of Geman-Hwang, *Probab. Theory Related Fields*, 73,555–569, (1986).
26. Zhao, L. C., Krishnaiah, P. R. and Bai, Z. D., On detection of the number of signals when the noise covariance matrix is arbitrary, *J. Multivariate Anal.*, 20,26–49,(1986).
27. Bai, Z. D., Yin, Y. Q. and Krishnaiah, P. R., On limiting spectral distribution of product of two random matrices when the underlying distribution is isotropic, *J. Multivariate Anal.*, 19,189–200,(1986).
28. Yin, Y. Q. and Bai, Z. D., Spectra for large-dimensional random matrices, in *Random matrices and their applications (Brunswick, Maine, 1984)*, Contemp. Math., 50, 161–167, (1986).
29. Bai, Z. D., A note on the limiting distribution of the eigenvalues of a class of random matrices, *J. Math. Res. Exposition*, 5,113–118,(1985).
30. Yin, Y. Q., Bai, Z. D. and Krishnaiah, P. R., Limiting behavior of the eigenvalues of a multivariate F matrix, *J. Multivariate Anal.*, 13,508–516, (1983).
31. Marčenko, V. A. and Pastur, L. A., *Distribution of eigenvalues in certain sets of random matrices,Mat. Sb. (N.S.)*, 72 (114),507–536, (1967).
32. Geman, S., A limit theorem for the norm of random matrices, *Ann. Probab.*, 8,252–261,(1980).
33. Edelman, A., The probability that a random real Gaussian matrix has k real eigenvalues, related distributions, and the circular law, *J. Multivariate Anal.*, 60,203–232,(1997).
34. Girko, V. L., The circular law, *Teor. Veroyatnost. i Primenen.*, 29, 669–679, (1984).
35. Tulino, Antonia M. and Verdú, Sergio, *Random Matrix Theory and Wireless Communications*, Now Publishers Inc., (2004).

PART B

Selected Papers of Professor Bai

Sankhyā : The Indian Journal of Statistics
1993, Volume 55, Series A, Pt. 2, pp. 244–258.

EDGEWORTH EXPANSIONS OF A FUNCTION OF SAMPLE MEANS UNDER MINIMAL MOMENT CONDITIONS AND PARTIAL CRAMER'S CONDITION

By GUTTI JOGESH BABU*
The Pennsylvania State University
and
Z. D. BAI**
Temple University

SUMMARY. A wide class of statistics can be expressed as smooth functions of sample means of random vectors. Edgeworth expansions of such statistics are generally obtained under Cramer's condition. In many practical situations, like in the case of ratio statistics, only one of the components of the random vector satisfies the Cramer's condition, while the rest do not. Edgeworth expansions are established under partial Cramer's condition. Further the conditions on the moments are relaxed to the minimum needed to define the expansions.

1. Introduction

Many important statistics car be written as functions of means of random vectors $\boldsymbol{Z}_i$. Bhattacharya and Ghosh (1978) made fundamental contributions to Edgeworth expansions of such statistics. In the case of student's t-statistic and many others, the highest order of moments involved in the actual expansion is much smaller than the order of moments assumed finite in Bhattacharya and Ghosh (1978). Chibisov (1980, 1981) obtained Edgeworth expansions for polynomial type statistics under weaker moment assumptions. Hall (1987) obtained expansions for student's t-statistic under best possible moment conditions. Hall uses spcial methods to treat t-statistic. Bhattacharya and Ghosh (1988) made an attempt to generalize Hall's work to a wide class of statistics. Their method still needs existence of finite moments of order higher than those involved in the expansion. All these results assume Cramer's condition on the distribution of $\boldsymbol{Z}_1$. However, in many practical

Paper received. December 1991 ; revised January 1993.

AMS (1980) *subject classifications.* 60F05, 62F20.

Keywords and phrases. Asymptotic expansions, Cramer's condition, Edgeworth expansion, Student's t-statistic.

*Research supported in part by NSA Grant MDA-904-90-H-1001 and by NSF Grants DMS-9007717 and DMS-9208066.

**Research supported by U.S. Army Research Office under Grant DAAL03-89-K-0139.

situations, like ratio statistic and in survival analysis, only few of the components of Z_1 satisfy Cramer's condition and the others do not. Babu and singh (1989) considered the case of bivariate random variables with one component continuous and the other lattice. For applications to survival analysis see Babu (1991a, b) and Bai and Rao (1991) generalized these results. But still needing the existence of moments with order higher than the ones required in the expansions. In the present paper, we combine the benefits of improvements in both the directions mentioned above and obtain expansions under minimal moment conditions and minimal smoothness conditions. The results generalize the work of Hall (1987), partially the results of Bhattacharya and Ghosh (1988) and some of the work of Chibisov (1980 and 1981). Incidentally, we note here that the proof of Lemma 5.1 on page 3 of Chibisov (1981) which is essential in the proof of the main results of Chibisov (1980) seems to be incorrect. The lemma is stated on page 742 of Chibisov (1980). The inequality

$$|P_1(y)| \leqslant C_{l,1}(e^y+e^{-y})$$

is not correct as $P_1(y) = -y\sigma_n^{-2} = -yn^a$, $a > r-2 \geqslant 1$ and $|y| \leqslant n^{1+\delta/2}$.

2. Preliminaries and the statements of the main results

Let $\{Z_j\}$ be a sequence of i.i.d (independent identically distributed) k-variate random vectors, with common mean vector μ and dispersion Σ. Let H be a real valued measurable function on R^k which is differentiable in a neighborhood of μ. Let $l = (l_1 \dots, l_k) = \text{grad}\,[H(\mu)]$ denote the vector of first order partial derivatives of H at μ. Suppose $l \neq 0$,

$$\sigma^2 = l \Sigma l' \qquad \dots (2.1)$$

and F_n denotes the distribution of

$$W_n = \sqrt{n}\,\sigma^{-1}(H(\bar{Z}_n) - H(\mu)), \qquad \dots (2.2)$$

where

$$\bar{Z}_n = \frac{1}{n}\sum_{j=1}^{n} Z_j = (\bar{Z}_1, \dots, \bar{Z}_k), \qquad \dots (2.3)$$

and

$$Z_j = Z_{j1}, \dots, Z_{jk}).$$

Let

$$\psi_{nm}(x) = \Phi(x) + \sum_{j=1}^{m-2} n^{-j/2} Q_j(x)\varphi(x), \qquad \dots (2.4)$$

denote the $m-1$ term formal Edgeworth expansion of F_n, where Φ and φ denote the distribution and density of the standard normal variable, Q_j is a polynomial of degree not exceeding $3j-1$, whose coefficients are determined by the moments of Z_1 of order up to j and by the partial derivatives of H at μ of order up to $j+1$. Some of these moments may not appear in the expression if one or more of the partial derivatives of H vanish.

We shall establish the validity of the formal Edgeworth expansion for F_n under some of the following assumptions.

Assumption 1. H is $\eta(m)$-times continuously differentiable in a neighborhood of μ, where $m \geq 3$ and $\eta(m) \geq m-1$ are integers. The grad $[H(\mu)] = \boldsymbol{l}$ satisfies $l_1 \neq 0$ and $l_{p+1} = \ldots = l_k = 0$, for some positive integer $p < k$. Further, $E|Z_{1j}^m| < \infty$, for $j = 1, \ldots, p$,

Assumption 2. $\limsup_{|t|\to\infty} E|E(\exp(itZ_{11})|Z_{12}, \ldots, Z_{1k})| < 1$. ... (2.5)

Assumption 3. $E|Z_{1j}|^{\delta_j} < \infty$ for $j = p+1, \ldots, k$, where $\delta_j = m\theta_j/(2\theta_j - 1)$, $\theta_j \geq 2$. Further, for $\alpha = (\alpha_1, \ldots, \alpha_k)$.

$$D^\alpha(H(x) - H(\mu))|_{x=\mu} = 0, \qquad \ldots \quad (2.6)$$

wherever $\alpha_1 = \ldots = \alpha_p = 0$, and $\alpha_{p+1} < \theta_{p+1}, \ldots, \alpha_k < \theta_k$.

Assumption 4. $E|Z_{1j}|^{m/2} < \infty$ for $j = p+1, \ldots, k$ and (2.6) holds, whenever $\alpha_1 = \ldots = \alpha_p = 0, \alpha_{p+1} \leq m-1, \ldots, \alpha_k \leq m-1$.

Note that Assumption 2 may not hold even under Cramér's condition. In that sense the results do not completely cover the case considered by Bhattacharya and Ghosh (1978), just as the main result of Bai and Rao (1991) does not completely extend the results known under the usual Cramer's condition.

In this paper we establish the following two main theorems. Without loss of generality, we can assume that $\mu = \mathbf{0}$ and $H(\mu) = 0$.

Theorem 1. *Suppose Assumptions* 1, 2, *and* 3 *hold with* $\eta(m) = m-1$. *Then we have*

$$\|F_n - \psi_{mn}\| = \sup_x |F_n(x) - \psi_{mn}(x)| = o(n^{-(m-2)/2}). \qquad \ldots \quad (2.7)$$

Theorem 2. *Suppose Assumptions* 1, 2 *and* 4 *hold* $\eta(m)$ *equal to the smallest odd integer not less than* $m-1$, *and the remainder term* R *in the Taylor series expansion of* H,

$$H(x) = \sum_{|\alpha| \leq \eta(m)} \eta_\alpha x^\alpha + R(x)$$

satisfies

$$|R(x)| \leq \rho(\|x\|)\left(\sum_{j=1}^{p} |x_j|^{\eta(m)} + \sum_{j=1}^{p} |x_j| \sum_{j=p+1}^{k} |x_j|^{\eta(m)-1}\right), \qquad \ldots \quad (2.8)$$

where $\rho(t) \to 0$ *as* $t \to 0$. *Then* (2.7) *holds.*

Remark 2.1. Let $(Z_{11}^*, \ldots, Z_{1p}^*) = (Z_{11}, \ldots, Z_{1p})A$, where A is a non-singular $p \times p$ matrix. From the proofs one can see that the main results of this paper still hold if the condition (2.5) is replaced by a weaker assumption that

$$\limsup_{|t| \to \infty} E \left| E(\exp(it(Z_{11}^* + \ldots + Z_{s1}^*)) \,\Big|\, \sum_{i=1}^{s} (Z_{i2}^*, \ldots, Z_{ik}^*) \right| < 1, \qquad \ldots \quad (2.9)$$

for some $s \geqslant 1$.

Remark 2.2. It is well known that the Student's statistic is determined by the sample means of X_i and X_i^2. By Lemma 2.2 of Bhattacharya and Ghosh (1978), if the distribution of X has an absolutely continuous part, then so does the distribution of $(X_1+X_2, X_1^2+X_2^2)$. Hence the condition (2.9) holds for $Z_{i1} = X_i$ and $Z_{i2} = X_i^2$ with $s = 2$ and A is the identity matrix.

Remark 2.3. Theorem 1 is established when $\delta_j = m$ for all j in Bai and Rao (1991). Bhattacharya and Ghosh (1988) considered the case when (2.6) holds whenever $\alpha_1 = \ldots = \alpha_p = 0, \alpha_{p+1} < 2, \ldots, \alpha_k < 2$ under the existence of the $(m-1)$th moments for $Z_{1p+1}, \ldots, Z_{1k}$. The student's t-statistic can be written as a smooth function H, of the sample mean of (X_i, X_i^2), for which (2.8) and the condition on the derivatives given in Assumption 4 hold. Hence if the distribution of X has an absolutely continuous part, then by remark 2.2, the conclusion of Theorem 2 holds for t-statistic. Consequently, Theorem 2 implies the result of Hall (1987). Without condition (2.8). one requires moments slightly higher than $m/2$ for $Z_{1j}, p+1 \leqslant j \leqslant k$. The moment restrictions of Assumption 3 are the best possible as demonstrated in Chibisov (1980) for polynomial statistics.

3. Proofs of the Main Theorems

Throughout this paper, we use $\{\epsilon_n\}$ as a generic sequence satisfying

$$\epsilon_n = o(n^{-(m-2)/2}),$$

and the letter c with or without arguments is used as a generic positive constant.

To avoid complex notations, we provide the proof for the case $p = 1$. Without loss of generality, we assume that $l_1 = 1$ and $\mathrm{var}(Z_{11}) = 1$.

The lemmas required in the proof are given in the Appendix.

Proof of Theorem 1. By implicit function theorem, there exists a constant η^*, an open interval θ containing the origin, and a function g such that $x = H(u, v)$ holds if and only if $u = g(x, v)$, whenever $\max(|u|, |v|) \leqslant \eta^*$

and $x \in \theta$. By Lemma 2, the function g is $m-1$ times cortinuously differentialbe in the region $\{(x, v) : x \in \theta,\ |v| < \eta^*\}$ and $D^\alpha g(0) = 0$ whenever $\alpha_1 = 0$ and $\alpha_2 < \theta_2$. Furthermore

$$H(u, v) \leqslant x \text{ if and only if } u \leqslant g(x, v), \quad \ldots \quad (3.1)$$

in the region $\{(u, v, x) : \max(|u|, |v|) < \eta^*, x \in \theta\}$. Choose positive $\eta < \eta^*$ satisfying $(-\eta, \eta) \subset \theta$. Define

$$\theta = \theta_2,\ \delta = \delta_2 \text{ and } b = m/2\delta = 1-1/(2\theta),$$

$$U_i = Z_{i1} I(|Z_{i1}| < \sqrt{n}),\ \bar{U} = \frac{1}{n} \sum_{i=1}^{n} U_i,$$

$$W_i = Z_{i2} I(|Z_{i2}| < n^b),\ \bar{W} = \frac{1}{n} \sum_{i=1}^{n} W_i,$$

$$V_i = E(U_i | W_i),\ \bar{V} = \frac{1}{n} \sum_{i=1}^{n} V_i$$

$$D_n = \{W_1, \ldots, W_n\},$$

$$n_0 = \sup\{n : (\log n)^3 > \eta \sqrt{n}\},$$

$$G_n = \{x : |x| \leqslant (\log n)^2\}, \quad \ldots \quad (3.2)$$

$$A_0 = \{U_i = Z_{i1},\ W_i = Z_{i2}, \text{ for } i = 1, 2, \ldots, n\},$$

$$A_1 = \{10 |\bar{U}| \leqslant \eta\},$$

and

$$A_2 = \{10 |\bar{W}| \leqslant \eta,\ 10 |\bar{V}| \leqslant \eta\}. \quad \ldots \quad (3.3)$$

Since $E(U_i) = E(V_i) = \epsilon_n$, for any even integer $\nu \geqslant m$, we have by Lemma 4 that

$$E(\bar{W}^\nu) + E(\bar{V}^\nu + (E(\bar{U}^\nu) \leqslant c\, n^{-\nu/2} + \epsilon_n n^{\nu(b-1)} \leqslant \epsilon_n.$$

So by Markov inequality,

$$\sum_{j=0}^{2} P(A_j^c) \leqslant \epsilon_n. \quad \ldots \quad (3.4)$$

The proof is divided into two main parts. The first part of the proof consists of approximating the conditional distribution of $\sqrt{n}\, \bar{U}$ given D_n, by its formal Edgeworth expansion in L_1-norm. The second part, which makes essential use of smoothness of g, consists of showing that the expected value of the conditional Edgeworth expansion is approximated by an expression involving only $\{E(E_{11}^i E_{12}^j) : i + mj/\delta \leqslant m\}$ and the derivatives $D^\alpha H$ with $|\alpha| \leqslant m-1$.

From now on we assume that $n \geqslant n_0$. By (3.1) we have on $\bigcap_{i=0}^{2} A_i$,

$$H(\bar{Z}_1, \bar{Z}_2) \leqslant x/\sqrt{n} \text{ if and only if } \bar{U} \leqslant g(x/\sqrt{n}, \bar{W}), \quad \ldots \quad (3.5)$$

for all $x \in G_n$. By (3.4) and (3.5), we have

$$\sup\{|F_n(x)-F_{n1}(x)| : x \in G_n\} \leqslant \epsilon_n, \quad \ldots \quad (3.6)$$

where

$$F_{n1}(x) = P((\bar{U} \leqslant g(x/\sqrt{n}, \bar{W})) \cap A_2). \quad \ldots \quad (3.7)$$

Boundedness of F_n and F_{n1} and the inequality (3.4) are used to omit the effect of the events A_0 and A_1 in arriving at the inequality (3.6).

To get the conditional Edgeworth expansion let

$$B_n^2 = \sum_{i=2}^{n} \text{var}(U_i | W_i),$$

$$\beta = E(\text{var}(U_1 | W_1)),$$

$$\tilde{F}_n(y) = P(n(\bar{U}-\bar{V}) \leqslant yB_n | D_n), \quad \ldots \quad (3.8)$$

$$A_3 = \{ B_n^2 - n\beta | < n\beta/4\}, \quad \ldots \quad (3.9)$$

$$y'(x) = n(g(x/\sqrt{n}, \bar{W}) - \bar{V})/B_n, \quad \ldots \quad (3.10)$$

and Δ be such that

$$\sum_{i=2}^{n} E\{(V_i - U_i)^2 I(|U_i - V_i| \geqslant (12\Delta)^{-1} | D_n\} \leqslant \frac{1}{8} B_n^2. \quad \ldots \quad (3.11)$$

Note that Δ is a random variab e and that for $x \in G_n$,

$$F_{n1}(x) = E\tilde{F}_n[y'(x) I(A_2)\}].$$

By Lemma 4, as in the proof of the inequality (3.4), we have

$$P(A_3^c) \leqslant \epsilon_n. \quad \ldots \quad (3.12)$$

It follows from the arguments of Hall (1987, Page 924) that there exists a constant $\Delta_1 > 0$ such that

$$P(A_4^c) \leqslant \epsilon_n, \quad \ldots \quad (3.13)$$

where

$$A_4 = \{\Delta \geqslant \Delta_1\}. \quad \ldots \quad (3.14)$$

Let v_{nj} and $\gamma_{j\nu}$ respectively denote the conditional characteristic function and ν-th cumulant of U_j given W_j. Define

$$\lambda_{\nu n} = n^{(\nu-2)/2} B_n^{-\nu} \sum_{j=1}^{n} \gamma_{j\nu},$$

$$h_{\nu s} = \Sigma' \prod_{i=1}^{\nu} (k_i\,!)^{-1} \left(\frac{\lambda_{i+2,n}}{(i+2)\,!}\right)^{k_i},$$

$$Q_{\nu n}(y) = \sum_{s=1}^{\nu} (-1)^{\nu} \Phi^{(\nu+2s)}(y) h_{\nu s},$$

and

$$U_{mn}(y) = \phi(y) + \sum_{\nu=1}^{m-2} n^{-\nu/2} Q_{\nu n}(y), \qquad \dots \quad (3.15)$$

where Σ' denotes the sum over all nonnegative integers

$$\sum_{i=1}^{\nu} k_i = s \text{ and } \sum_{i=1}^{\nu} ik_i = \nu.$$

Using the arguments of Hall (1982, page 42), we get that for $|t| \leqslant \Delta B_n$,

$$\prod_{i=1}^{n} |v_{ni}(t/B_n)| \leqslant \exp\{-5t^2/24\}. \qquad \dots \quad (3.16)$$

Since $|U_i - V_i| \leqslant 2\sqrt{n}$, following the lines of the proof given in Bai and Zhao (1986) and using (3.16), we get that

$$\|\tilde{F}_n - U_{mn}\| I\Big(\bigcap_{j=2}^{4} A_j\Big) \leqslant c(m)\Big((B_n\Delta_2)^{-1} + \int_{\Delta_1 \leqslant |t| < \Delta_2} |t|^{-1} \prod_{j=1}^{n} |v_{nj}(t)| dt\Big), \qquad \dots \quad (3.17)$$

where

$$\Delta_2^{-1} = B_n^{-m} \sum_{j=1}^{n} E(|U_j - V_j|^{m+1} | D_n).$$

By Hölder inequality, we have on A_3, that

$$\Delta_2 \leqslant cn^{(m-2)/2} \qquad \dots \quad (3.18)$$

and clearly

$$\begin{aligned} E[(B_n\Delta_2)^{-1} I(A_3)] &\leqslant cn^{-(m-2)/2} E(|U_1 - V_1|^{m+1}) \\ &\leqslant cn^{-(m-1)/2} E(|Z_{11}|^{m-1} I(|Z_{11}| \leqslant \sqrt{n})) \\ &\leqslant \epsilon_n. \end{aligned} \qquad \dots \quad (3.19)$$

In case $m = 3$, then using Berry-Esseen Lemma, Δ_2 may be chosen as $A_n\sqrt{n}$ for some $A_n \to \infty$. By Assumption 2, there exists a constant $\rho < 1$ such that for all $|t| > \Delta_1$,

$$\begin{aligned} E(|v_{n1}(t)|) &\leqslant E|E(\exp(itU_1)|Z_{12})| + P(A_0^c) \\ &\leqslant E|E(\exp(itZ_{11})|Z_{12})| + 2P(A_0^c) \\ &\leqslant \rho, \end{aligned} \qquad \dots \quad (3.20)$$

for all large n. From (3.17)—(3.20) we get

$$E\Big\{\sup_y \tilde{F}_n(y) - U_{mn}(y)| I\Big(\bigcap_{i=2}^{4} A_i\Big)\Big\} \leqslant \epsilon_n. \qquad \dots \quad (3.21)$$

Since $\tilde{F}_n$ and U_{mn} are bounded, by (3.4), (3.6), (3.12), (3.13), and (3.21) we have

$$\sup\{|F_n(x) - F_{n2}(x)| : x \in G_n\} \leqslant \epsilon_n, \qquad \dots \quad (3.22)$$

where for $x \in G_n$,

$$F_{n2}(x) = E[U_{mn}(y'(x)) I(A_2 \cap A_3)]. \qquad \dots \quad (3.23)$$

We now arrive at the second part of the proof and concentrate on F_{n2}. Define

$$y_1(x) = \sqrt{n}(x(x-\sqrt{n}\,\bar{V})/B_n,$$

$$y(x) = (x-\sqrt{n}\,V)\beta^{-1/2},$$

$$B(n) = (B_n^2-n\beta)/(n\beta)$$

$$R_1(x) = y'(x)-y_1(x)-\sigma(x)$$

and

$$A_5 = \{20|\sigma(x)|+20|R_1(x)| \leqslant \eta, \text{ for all } x \in G_n\} \quad \dots \quad (3.24)$$

where

$$\sigma(x) = n\left(\sum_{i=1}^{m-1}\sum_{j=0\vee(2-i)}^{m-1-i} d_{ij}\,\bar{W}^j\,(x/\sqrt{n})^i+\sum_{j=\theta}^{m-1} d_{0j}\,\bar{W}^j\right)/B_n \quad \dots \quad (3.25)$$

and

$$d_{ij} = \frac{1}{i\,!\,j\,!}\,\frac{\partial^{i+j}}{\partial u^i\,\partial v^j}\,g(u,v)\Big|_{(u,v)=0}$$

Since g has continuous derivatives of all orders $|\alpha| \leqslant m-1$, we have by roting $\theta \leqslant m-1$ and using Lemma 4(*iii*), that

$$E|\bar{W}|^{m-1} \leqslant cn^{-(m-1)/2}$$

and on A_3 for some $\delta(n)\to 0$,

$$|R_1(x)| \leqslant \delta(n)\sqrt{n}(|x/\sqrt{n}|^{m-1}+|\bar{W}|^{m-1}), \quad \dots \quad (3.26)$$

and A_2

$$|\sigma(x)| \leqslant c[|x\bar{W}|+\sqrt{n}|\bar{W}|^\theta+n^{-1/2}\log^2 n], \quad \dots \quad (3.27)$$

hold uniformly for $x \in G_n$. By applying Lemma 4, we have for even integers, $j > (m-2)/(\theta-1)$ and $\nu \geqslant m$,

$$E|\sqrt{n}\,\bar{W}^\theta|^j \leqslant c(\epsilon_n+n^{-j(\theta-1)/2}) \leqslant \epsilon_n, \quad \dots \quad (3.28)$$

and

$$E|\bar{W}|^\nu \leqslant cn^{-(m-1)/2}. \quad \dots \quad (3.29)$$

By Markov inequality it follows that

$$P(A_5^c) \leqslant \epsilon_n. \quad \dots \quad (3.30)$$

Since Φ has all the derivatives and $|n^{-\nu/2}\,h_{\nu s}| \leqslant c$ for all $s \leqslant \nu \leqslant m-2$, we have by Taylor series expansion

$$U_{mn}(y'(x)) = \sum_{j=0}^{m-2}\frac{1}{j\,!}\,(\sigma(x))+R_1(x))^j\,U_{mn}^{(j)}(y_1(x))+R_2(x),$$

$$= \sum_{j=0}^{m-2}\frac{1}{j\,!}(\sigma(x))^j\,U_{mn}^{(j)}(y_1(x))+R_3(x)+R_2(x),$$

where, on $A_3 \cap A_5$

$$|R_2(x)| \leqslant c|R_1(x)+\sigma(x)|^{m-1}.$$

and

$$|R_3(x)| \leqslant c(|R_1(x)U^{(1)}_{mn}(y_1(x))|+R_1(x)^2)$$
$$\leqslant c\delta(n)\sqrt{n}(|\overline{V}|^{m-1}+|\overline{W}|^{m-1}+n^{-(m-1)/2})+R_1(x)^2. \quad \ldots \quad (3.31)$$

Hence by (3.26)—(3.31), we have that for $j = 2, 3$

$$E \sup\{|R_j(x)| : x \in G_n\} \leqslant \epsilon_n.$$

Observe that

$$y_1(x) = y(x)(1+a_n),$$

where

$$a_n = (1+B(n))^{-1/2}-1 = d_{nm}+R_0,$$

and

$$d_{nm} = \sum_{j=1}^{m-2} c_j B(n)^j. \quad \ldots \quad (3.32)$$

The terms R_0 and a_n satisfy,

$$|R_0| \leqslant c|B(n)|^{m-1} \quad \ldots \quad (3.33)$$

and

$$P(10|a_n| > \eta) \leqslant \epsilon_n. \quad \ldots \quad (3.34)$$

Here Lemma 4 and Markov inequality are used in establishing (3.34). Another application of Taylor series expansion and (3.33) gives

$$U_{mn}(y'(x)) = \sum_{j=0}^{m-2}\sum_{i=0}^{m-2} \frac{1}{i!j!} d^i_{nm}(\sigma(x))^j b_n(x, i, j)+R_4(x)$$

$$= V_{mn}(y(x))+R_4(x), \text{ (say)}, \quad \ldots \quad (3.35)$$

where

$$b_n(x, i, j) = \frac{\partial^{i+j}}{\partial w^i \partial v^j} U_{mn}(w(1+v))\bigg|_{w=y(x),\, v=0}$$

and on $A_3 \cap A_5$,

$$|R_4(x)| \leqslant c|B(n)|^{m-1}.$$

By Lemma 4(iii),

$$E\left(|R_5(x)|I\left(\bigcap_{i=2}^{5} A_i\right)\right) \leqslant \epsilon_n. \quad \ldots \quad (3.36)$$

Since $y^j \Phi^{(i)}(y)$ can be written as a linear combination of $\{\Phi^{(\vartheta)}(y), 1 \leqslant \vartheta \leqslant i+j\}$, by using (3.25) and (3.32), $V_{mn}(y(x))$ can be written as a linear combination of terms of the form

$$n^{-w/2}\Phi^{(\vartheta)}(y(x))(\sqrt{n}\,\overline{V})^i(\sqrt{n}\overline{W})^j(\sqrt{n}\,B(n))^r h_{vs}$$

with $w \geqslant v$. Now using the arguments following Remark 3.1 of Bai and Rao (1991), and Lemma 4, we conclude that uniformly for $x \in G_n$,

$$E\left(U_{mn}(y'(x))I\left(\bigcap_{i=2}^{5} A_i\right)\right) = \psi_{mn}(x)+\epsilon_n, \quad \ldots \quad (3.37)$$

where

$$\psi_{mn}(x) = \Phi(x) + \sum_{\nu=1}^{m-2} n^{-\nu/2} \varphi(x) Q_\nu(x),$$

and the coefficients of Q_ν involve only the terms $\{E(Z_{11}^i Z_{12}^j) : i+(jm/\delta) \leqslant v+2\}$. clearly

$$\sup_{x \geqslant (\log n)^2} |\psi_{mn}(-x)| + |1-\psi_{mn}(x)| \leqslant \epsilon_n \qquad \ldots \quad (3.38)$$

and hence

$$\sup_{x \geqslant (\log n)^2} (F_n(-x)+1-F_n(x)) \leqslant F_n(-(\log n)^2)+1-F_n((\log n)^2)$$
$$\leqslant 2 \sup_{x \epsilon G_n} |F_n(x) - \psi_{mn}(x)| + \epsilon_n.$$

The Theorem now follows from (3.30)—(3.38) and (3.22).

Remark 3.1. In the general case, one has to consider,

$$y_1(x) = \sqrt{n}[l_1^{-1}\left(x - \sum_{j=2}^{p} l_j \bar{Z}_j\right) - E(\bar{Z}_1 | \bar{Z}_2, \ldots, \bar{Z}_k)]/B_n.$$

Proof of Theorem 2 is similar. The only differences are that the expressions (3.25), (3.27) and (3.31) should be replaced, using Lemma 3, by

$$\sigma(x) = n\left(\sum_{i=1}^{\eta(m)} \sum_{j=0 \vee (2-i)}^{\eta(m)-i} d_{ij}(x/\sqrt{n})^i (\bar{W})^j\right) \Big/ B_n,$$

$$|R_1(x)| \leqslant \delta(n)\sqrt{n}(|x/\sqrt{n}|^{\eta(m)} + |x/\sqrt{n}|\,|\bar{W}|^{\eta(m)-1}),$$

and

$$|R_3(x)| \leqslant c\delta(n)\sqrt{n}(|\bar{V}|^{m-1} + |\bar{V}|\,|\bar{W}|^{(\eta(m)-1} + n^{-(m-1)/2}) + R_1(x)^2 \quad \ldots \quad (3.39)$$

for some $\delta(n) \to 0$. To estimate $\sup_{x \epsilon G_n} E|R_3(x)|$, we need to use Lemma 4 to the term $|\bar{V}|\,|\bar{W}|^{\eta(m)-1}$ in (3.39). This is the only place where the assumption that $\eta(m)$ is an odd integer not smaller than $m-1$, is required.

Theorem 3. *Suppose* $m = 3$. *Then Theorems* 1 *and* 2 *hold under the assumption that for some* $\nu \geqslant 1$, *for all* $|t| \neq 0$,

$$E\left|E\left(exp\left(it \sum_{j=1}^{\nu} Z_{j1}\right)\middle| \sum_{j=1}^{\nu} Z_{j2}, \ldots, \sum_{j=1}^{\nu} Z_{jk}\right)\right| < 1,$$

instead of Assumption 2.

Proof. Proof is similar to that of Theorem 1. Using Lemma 3 of Bai and Zhao (1986) and the argumeents of Hall (1982, p. 42), one can choose $\Delta_2 = \sqrt{n}a_n$, for $a_n \to \infty$.

Appendix

For $x = (x_1, \ldots, x_k)$ and a vector of nonnegative integers $\alpha = (\alpha_1, \ldots, \alpha_k)$, let $|\alpha| = \alpha_1 + \ldots + \alpha_k$ and $D^\alpha g(x) = \dfrac{\partial^{|\alpha|} g(x)}{\partial x_1^{\alpha_1} \ldots \partial x_k^{\alpha_k}}$. For $r \geqslant 1$ and a function h^*, let

$$\partial^r(h^*, x, g) = \frac{\partial^r}{\partial g^r} h^*(g, x_1, \ldots, x_k)\,|_{g=g(x)}.$$

and

$$h^*(x, \alpha) = \left.\frac{\partial^{|\alpha|} h^*(g, x)}{\partial x_1^{\alpha_1} \ldots \partial x_k^{\alpha_k}}\right|_{g=g(x)}$$

Lemma 1. *Let h^* and g be two functions defined on R^{k+1} and R^k, respectively. Let $H(x) = h^*(g(x), x)$. Suppose h and g have derivatives of all orders $|\alpha| \leqslant m$. Then*

$$D^\alpha H = \partial^1(h_*, x, g) D^\alpha g + P_\alpha + h^*(x, \alpha) \qquad \ldots \quad (A.1)$$

where P_α is a polynomial in $\{D^\beta g : |\beta| < |\alpha|\}$, $P_\alpha(0) = 0$, whose coefficients may depends on the partial derivatives of h^*.

Proof. Clearly,

$$\frac{\partial H}{\partial x_i} = \partial^1(h^*, x, g)\, \frac{\partial g}{\partial x_i} + \frac{\partial h^*}{\partial x_i},$$

which shows (A.1) is true for $|\alpha| = 1$. Now, suppose that (A.1) is true for some $|\alpha| < m$. We shall prove that (A.1) is true for $\gamma = \alpha + e_i$, where e_i is the vector whose ith component is one and the rest are all zero. By assumed equality (A.1), we have

$$\begin{aligned} D^\gamma H &= \partial^1(h^*, x, g)\, D^\gamma g + \partial^2(h^*, x, g) D^{e_i} g D^\alpha g + D^{e_i} P_\alpha + h^*(x, \gamma) \\ &= \partial^1(h^*, x, g) D^\gamma g + P_\gamma + h^*(x, \gamma). \end{aligned}$$

It is easy to verify that P_γ is a polynomial in $\{D^\beta g : |\beta| < |\gamma|\}$. And the lemma is proved by induction.

Now, let h be a function on R^k with $h(0) = 0$ and has m continuous derivatives and let $g = g(x)$ be the solution to the equation

$$h(g, x_2, \ldots, x_k) = x_1$$

in some neighborhood of the origin. Then by implicit function theorem, g has derivatives of all orders $|\alpha| \leqslant m$.

Lemma 2. *Let h be a function having derivatives of all order $|\alpha| \leqslant m$, $\frac{\partial h}{\partial x_1} \neq 0$ and $h(\boldsymbol{0}) = 0$. Suppose $D^\alpha h(\boldsymbol{0}) = 0$, whenever $\alpha_1 = \ldots = \alpha_p = 0$, $\alpha_{p+1} < \theta_{p+1}, \ldots,$* and *$\alpha_k < \theta_k$, $\alpha = (\alpha_1, \ldots, \alpha_k)$, $1 \leqslant p < k$. Then $D^\alpha g(\boldsymbol{0}) = 0$ whenever $\alpha_1 = \ldots = \alpha_p = 0, \alpha_{p+1} < \theta_{p+1}, \ldots, \alpha_k < 0_k$.*

Proof. The result follows by induction from Lemma 1.

Lemma 3. *Suppose that, in addition to the assumptions of Lemma 2, for any $|\alpha| = m$,*

$$|D^\alpha h| \leqslant \epsilon(\|\boldsymbol{x}\|) \left[\sum_{i=1}^{p} |x_i|^m + \sum_{i=1}^{p} |x_i| \sum_{j=p+1}^{k} |x_j|^{m-1} \right],$$

where $\epsilon(t) \to 0$, as $t \to 0$. Then for any $|\alpha| = m$,

$$|D^\alpha g| \leqslant \eta(\|\boldsymbol{x}\|) \left[\sum_{i=1}^{p} |x|^m + \sum_{i=1}^{p} |x_i| \sum_{j=p+1}^{k} |x_j|^{m-1} \right], \quad \ldots \quad (A.2)$$

where $\eta(t) \to 0$, as $t \to 0$.

Proof. Let $h(\boldsymbol{x}) = P(\boldsymbol{x}) + R_h(\boldsymbol{x})$ and $g(\boldsymbol{x}) = Q(z) + R_g(\boldsymbol{x})$, where P and Q are the polynomial of Taylor expansions of h and g respectively, and R_h and R_g are their remainder terms. By definition of g we know that

$$P(\boldsymbol{x})|_{x_1 = g(\boldsymbol{x})} - x_1$$

contains only terms of degrees higher than m and each term contains at least one factor among $x_1, \ldots, x_p$. Hence we have

$$|P(\boldsymbol{x})|_{x_1 = Q(\boldsymbol{x})} - x_1| \leqslant \eta_1(\|\boldsymbol{x}\|) \left[\sum_{i=1}^{p} |x_i|^m + \sum_{i=1}^{p} |x_i| \sum_{j=p+1}^{k} |x_j|^{m-1} \right], \quad \ldots \quad (A.3)$$

where $\eta_1(t) \to 0$ as $t \to 0$. In

$$P(\boldsymbol{x})|_{x_1 = g(\boldsymbol{x})} - P(\boldsymbol{x})|_{x_1 = Q(\boldsymbol{x})} - \frac{\partial h}{\partial g} R_g(\boldsymbol{x}),$$

each term contains a factor $R_g(\boldsymbol{x})$ and at least one other factor of $x_1, \ldots, x_k$ or $R_g(\boldsymbol{x})$. Thus we have

$$|P(\boldsymbol{x})|_{x_1 = g(\boldsymbol{x})} - P(\boldsymbol{x})|_{x_1 = Q(\boldsymbol{x})} - \frac{\partial h}{\partial g} R_g(\boldsymbol{x})| \leqslant \eta_2(\|\boldsymbol{x}\|)\, |R_g(\boldsymbol{x})| \quad \ldots \quad (A.4)$$

where $\eta_2(\|t\|) \to 0$ as $t \to 0$. Similarly we have

$$|R_h(\boldsymbol{x})|_{x_1 = Q(\boldsymbol{x})}| \leqslant \eta_3(\|\boldsymbol{x}\|) \left[\sum_{i=1}^{p} |x_i|^m + \sum_{i=1}^{p} |x_i| \sum_{j=p+1}^{k} |x_j|^{m-1} \right], \quad \ldots \quad (A.5)$$

and

$$|R_h(\boldsymbol{x})|_{x_1 = g(\boldsymbol{x})} - R_h(\boldsymbol{x})|_{x_1 = Q(\boldsymbol{x})}| \leqslant \eta_4(\|\boldsymbol{x}\|)\, |R_g(\boldsymbol{x})|. \quad \ldots \quad (A.6)$$

where $\eta_3(\|t\|)\to 0$, $\eta_4(\|t\|)\to 0$ as $t\to 0$. Then (A.2) follows from (A.3)–(A.6) by noting that $\frac{\partial h}{\partial g}\neq 0$ and $h(\boldsymbol{x})|_{x_1=g(\boldsymbol{x})}=x_1$.

Lemma 4. *Let $\{(X'_{j1},\dots,X'_{jr}), j=1,2,\dots\}$ be a sequence of iid random vectors with mean zero and let $m\geqslant 3$. For $i=1,2,\dots,r$, suppose $\beta_i>0$, $E|X'_{1i}|^{\delta_i}<\infty$, for some $1\leqslant\delta_i\leqslant m$ and $b_i=m/(2\delta_i)$. Let $X_{ji}=X'_{ji}I(|X'_{ij}|\leqslant\beta_i n^{b_i})$.*

(i) *Then for any nonnegative integers α_i,*

$$\left|E\prod_{i=1}^{r}\left(\sum_{j=1}^{n}X_{ji}\right)^{\alpha_i}\right|\leqslant c(n^{\alpha/2}+o(n^{b-(m-2)/2})), \qquad \dots \text{(A.7)}$$

where $\alpha=\sum_{i=1}^{r}\alpha_i$, $b=\sum_{i=1}^{r}\alpha_i b_i$ and c is a constant depending only on β_i,δ_i, m and $E|X_{1i}|^{\delta_i}$, $i=1,2,\dots,r$.

(ii) *If $\sum_{(b_i>1/2)}\alpha_i$ is even and all b_i which are greater than $1/2$ are equal to each other, or all α_i whose corresponding b_i is greater than $1/2$ are even, then the absolute sign can be taken inside the expectation.*

(iii) *If $r=1$, (A.7) holds for any positive integer α_1, when the absolute value sign is taken inside the expectation.*

Proof. We provide the proof for the case $r=2$. Proof for the general case is similar. For ease of notation, let $\alpha=\alpha_1\geqslant 1$, $\gamma=\alpha_2\geqslant 1$, $\beta_i=1$, $X=X_{11}$ and $Y=X_{12}$. Let c denote a generic constant depending only on α,γ,m and $E|X|^{\alpha_1}$ and $E|Y|^{\alpha_2}$. We have

$$\left|E\left(\sum_{j=1}^{n}X_{j1}\right)^{\alpha}\left(\sum_{j=1}^{n}X_{j2}\right)^{\gamma}\right|\leqslant c\sum_1\sum{}^{*} n^{T+S+U}\left|\prod_{t=1}^{T}E(X^{i_t})\prod_{s=1}^{S}E(Y^{j_s})\prod_{u=1}^{U}E(X^{h_u}Y^{g_u})\right| \qquad \dots \text{(A.8)}$$

where Σ_1 denotes the sum over positive integers $T\leqslant\alpha$, $S\leqslant\gamma$ and $U\leqslant\min(\alpha,\gamma)$, and Σ^* denotes the sum over all positive integers i_t, j_s, h_u, g_u satisfying

$$\sum_{t=1}^{T}i_t+\sum_{u=1}^{U}h_u=\alpha \text{ and } \sum_{s=1}^{S}j_s+\sum_{u=1}^{U}g_u=\gamma.$$

Note that for any integer $a\geqslant 1$, we have

$$|E(X^a)|\begin{cases}=o(n^{ab_1-(m/2)}) & \text{if } a=1 \text{ or } a>\delta_1,\\ \leqslant cn^{ab_1-1} & \text{otherwise,}\end{cases} \qquad \dots \text{(A.9)}$$

$$|E(Y^a)|\begin{cases}=o(n^{ab_2-(m/2)}) & \text{if } a=1 \text{ or } a>\delta_2,\\ \leqslant cn^{ab_2-1} & \text{otherwise.}\end{cases} \qquad \dots \text{(A.10)}$$

For any integers $a, b \geqslant 1$, we can choose $p, q > 1$ such that 1) $ap > \delta_1$ and $bq > \delta_2$ if $ab_1+bb_2 > m/2$ or equivalently $a/\delta_1+b/\delta_2 > 1$, and 2) $ap \leqslant \delta_1$ and $bq \leqslant \delta_2$ otherwise. An application of Hölder's inequality gives

$$E|X^aY^b| \leqslant (E|X|^{ap})^{1/p}(E|Y|^{bq})^{1/q}$$

$$\begin{cases} = o(n^{ab_1+bb_2-(m/2)}), & \text{if } ab_1+bb_2 > m/2, \\ \leqslant c \leqslant n^{ab_1+bb_2-1}, & \text{otherwise.} \end{cases} \quad \ldots \quad (A.11)$$

We split the left hand side of (A.8) into two parts, the first part consists of sum of those terms for which $i_t = 1$ or $> \delta_1$ for some $t \leqslant T$, or $j_s = 1$ or $> \delta_2$ for some $s \leqslant S$, or $b_1h_u+b_2g_u > m/2$, and the second part consists of the following upper bound for the first part

$$o(n^{\alpha b_1+\gamma b_2(m-2)2}).$$

For the second part we have $2T+U \leqslant \alpha$ and $2S+U \leqslant \gamma$, which imply $T+S+U \leqslant (\alpha+\gamma)/2$. Hence the second part is bounded by $cn^{(\alpha+\gamma)/2}$. The same proof gives the result when $\alpha = 0$ or $\gamma = 0$. This proves (A.7).

Suppose now that $b_1 = 1/2$, α is odd and γ is even. Then by Cauchy inequality, we have

$$E\left|\sum_{j=1}^{n} X_{ji}\right|^{\alpha}\left(\sum_{j=1}^{n} X_{j2}\right)^{\gamma} \leqslant \left(E\left(\sum_{j=1}^{n} X_{j1}\right)^{\alpha-1}\left(\sum_{j=1}^{n} X_{j2}\right)^{\gamma} E\left(\sum_{j=1}^{n} X_{j1}\right)^{\alpha+1}\left(\sum_{j=1}^{n} X_{j2}\right)^{\gamma}\right)^{1/2}$$

$$\leqslant c(n^{(\alpha+\gamma-1/2}+o(n^{(\alpha-1)/2+\gamma b_2-(m-2)/2}))^{1/2}$$

$$\times (n^{(\alpha+\gamma+1)/2}+o(n^{(\alpha+1)/2+\gamma b_2-(m-2)/2}))^{1/2}$$

$$\leqslant c(n^{(\alpha+\gamma)/2}+o(n^{\alpha/2+\gamma b_2-(m-2)/2})).$$

Part (iii) follows from the proof of the Proposition of Babu (1980). This completes the proof.

Remark A.1. The proof of the lemma also establishes that

$$E\prod_{i=1}^{r}\left(\frac{1}{n}\sum_{i=1}^{n} X_{ij}\right)^{\alpha_i} = P_n(\alpha_1, \alpha_2, \ldots, \alpha_r)+o(n^{-(m-2/2)}), \quad \ldots \quad (A.12)$$

where $P_n(\boldsymbol{\alpha})$ is a polynomial of $\left\{E\prod_{i=1}^{r} X_{i1}^{t_i} : t_1 \geqslant 0, \sum_{b=1}^{r} t_i\delta_i^{-1} \leqslant 1.\right\}$

We believe that (A.7) holds in general when the absoluto value sign is taken inside the expectation. If this is indeed the case, then Theorem 2 holds with $\eta(m) = m-1$.

References

Babu, G. J. (1980). An inequality for the moments of sums of truncated φ-mixing random variables. *Sankhyā* A, **42**, 1-8.

——— (1991a). Asymptotic theory for estimators under random censorship. *Probability Theory and Related Fields*, **90**, 275-290.

——— (1991b). Edgeworth expansions for statistics which are functions of lattice and non-lattice variables. *Statistics and Probability Letters*, **12**, 1-7.

Bai, Z. D. and Rao, C. R. (1991). Edgeworth expansins of a function of sample means. *Ann. Statist.*, **19**, 1295-1315.

Bai, Z. D. and Zhao, L. C. (1986). Edgeworth expansions of distribution functions of independent variables. *Scientia Sinica* A, **29**, 1-22.

Bhattachrya, R. N. and Ghosh, J. K. (1978). On the validity of the formal Edgeworth expansion. *Ann. Statist.*, **6**, 435-451.

——— (1988). On moment conditions for valid formal Edgeworth expansion. *J. Multivariate Anlal.* **27**, 68-79.

Chibisov, D. M. (1980). An asymptotic expansion for the distribution of a statistic admitting a stochastic expansion. I. *Theory of Probab. and its Appl.* **25**, 732-744.

——— (1981). An Asymptotic expansion for the distribution of a statistic admitting a stochastic expansion. II. *Theory of Probab. and its Appl.* **26**, 1-12.

Hall, P. (1982). *Rates of the Convergence in the Central Limit Theorem*, Pitman, London.

——— (1987). Edgeworth expansion for Student's *t*-statistic under minimal moment conditions. *Ann. Probab*, **15**, 920-931.

Petrov, V. V. (1975). *Sums of Independent Random Variables*, Springer-Verlag, Translated from Russian.

Department of Statistics
The Pennsylvania State University
University Park, PA 16802
U.S.A.

Department of Statistics
Temple University
Philadelphia, PA 19122
U.S.A.

The Annals of Probability
1993, Vol. 21, No. 2, 625–648

CONVERGENCE RATE OF EXPECTED SPECTRAL DISTRIBUTIONS OF LARGE RANDOM MATRICES. PART I. WIGNER MATRICES

BY Z. D. BAI

Temple University

In this paper, we shall develop certain inequalities to bound the difference between distributions in terms of their Stieltjes transforms. Using these inequalities, convergence rates of expected spectral distributions of large dimensional Wigner and sample covariance matrices are established. The paper is organized into two parts. This is the first part, which is devoted to establishing the basic inequalities and a convergence rate for Wigner matrices.

1. Introduction. Let W_n be an $n \times n$ symmetric matrix. Denote its eigenvalues by $\lambda_1 \leq \cdots \leq \lambda_n$. Then its spectral distribution is defined by

$$F_n(x) = \frac{1}{n}\#\{i: \lambda_i \leq x\},$$

where $\#\{Q\}$ denotes the number of entries in the set Q. The interest in the spectral analysis of high dimensional random matrices is to investigate limiting theorems for spectral distributions of high-dimensional random matrices with nonrandom limiting spectral distributions.

Research on the limiting spectral analysis of high-dimensional random matrices dates back to Wigner's (1955, 1958) semicircular law for a Gaussian (or Wigner) matrix; he proved that the expected spectral distribution of a high-dimensional Wigner matrix tends to the so-called semicircular law. This work was generalized by Arnold (1967) and Grenander (1963) in various aspects. Bai and Yin (1988a) proved that the spectral distribution of a sample covariance matrix (suitably normalized) tends to the semicircular law when the dimension is relatively smaller than the sample size. Following the work by Pastur (1972, 1973), the asymptotic theory of spectral analysis of high-dimensional sample covariance matrices was developed by many researchers including Bai, Yin and Krishnaiah (1986), Grenander and Silverstein (1977), Jonsson (1982), Wachter (1978), Yin (1986) and Yin and Krishnaiah (1983). Also, Bai, Yin and Krishnaiah (1986, 1987), Silverstein (1985a), Wachter (1980), Yin (1986) and Yin and Krishnaiah (1983) investigated the limiting spectral distribution of the multivariate F matrix, or more generally, of products of random

Received December 1990; revised January 1992.
AMS 1991 *subject classifications*. 60F15, 62F15.
Key words and phrases. Berry–Esseen inequality, convergence rate, large dimensional random matrix, Marchenko–Pastur distribution, sample covariance matrix, semicircular law, spectral analysis, Stieltjes transform, Wigner matrix.

matrices. In recent years, Voiculescu (1990, 1991) investigated the convergence to the semicircular law associated with free random variables.

In applications of the asymptotic theorems of spectral analysis of high-dimensional random matrices, two important problems arose after the limiting spectral distribution was found. The first is the bound on extreme eigenvalues; the second is the convergence rate of the spectral distribution, with respect to sample size. For the first problem, the literature is extensive. The first success was due to Geman (1980), who proved that the largest eigenvalue of a sample covariance matrix converges almost surely to a limit under a condition of existence of all moments of the underlying distribution. Yin, Bai and Krishnaiah (1988) proved the same result under existence of the fourth moment, and Bai, Silverstein and Yin (1988) proved that the existence of the fourth moment is also necessary for the existence of the limit. Bai and Yin (1988b) found necessary and sufficient conditions for almost sure convergence of the largest eigenvalue of a Wigner matrix. Bai and Yin (1990), Silverstein (1985b) and Yin, Bai and Krishnaiah (1983) considered the almost sure limit of the smallest eigenvalue of a covariance matrix. Some related works can be found in Geman (1986) and Bai and Yin (1986).

The second problem, the convergence rate of the spectral distributions of high-dimensional random matrices, is of practical interest, but has been open for decades. The principal approach to establishing limiting theorems for spectral analysis of high-dimensional random matrices is to show that each moment (with fixed order) of the spectral distribution tends to a nonrandom limit; this proves the existence of the limiting spectral distribution by the Carleman criterion. This method successfully established the limiting theorems for spectral distributions of high-dimensional Wigner matrices, sample covariance matrices and multivariate F matrices. However, this method cannot give a convergence rate.

This paper develops a new methodology to establish convergence rates of spectral distributions of high-dimensional random matrices. The paper is written in two parts: In Part I, we shall mainly consider the convergence rate of empirical spectral distributions of Wigner matrices. The convergence rate for sample covariance matrices will be discussed in Part II. The organization of Part I is as follows: In Section 2, basic concepts of Stieltjes transforms are introduced. Three inequalities to bound the difference between distribution functions in terms of their Stieltjes transforms are established. This paper involves a lot of computation of matrix algebra and complex-valued functions. For completeness, some necessary results in these areas are included in Section 3. Some lemmas are also presented in this section. Theorem 2.1 is used in Section 4 to establish a convergence rate for the expected spectral distribution of high-dimensional Wigner matrices.

The rate for Wigner matrices established in this part of the paper is $O(n^{-1/4})$. From the proof of the main theorem, one may find that the rate may be further improved to $O(n^{-1/3+\eta})$ by expanding more terms and assuming the existence of higher moments of the underlying distributions. However, it is not known whether we can get improvements beyond the order of $O(n^{-1/3})$, say $O(n^{-1/2})$ or $O(n^{-1})$, as conjectured in Section 4.

2. Inequalities of distance between distributions in terms of their Stieltjes transforms. Suppose that F is a function of bounded variation. Then its Stieltjes transform is defined by

$$s(z) = \int_{-\infty}^{\infty} \frac{1}{x-z}\, dF(x), \tag{2.1}$$

where $z = u + iv$ is a complex variable. It is well known [see Girko (1989)] that the following inversion formula holds: For any continuity points $x_1 \le x_2$ of F,

$$F(x_2) - F(x_1) = \lim_{v \downarrow 0} \frac{1}{\pi} \int_{x_1}^{x_2} \operatorname{Im}(s(u+iv))\, du, \tag{2.2}$$

where $\operatorname{Im}(\cdot)$ denotes the imaginary part of a complex number. From this, it is easy to show that if $\operatorname{Im}(s(z))$ is continuous at $z = x + i0$, then F is differentiable at x and its derivative is given by

$$F'(x) = \frac{1}{\pi} \operatorname{Im}(s(x+i0)). \tag{2.3}$$

This formula gives an easy way to extract the density function from its corresponding Stieltjes transform.

Also, one can easily verify the continuity theorem for Stieltjes transforms; that is, $F_n \to_w F$ if and only if $s_n(z) \to s(z)$ for all $z = u + iv$ with $v > 0$, where s_n and s are the Stieltjes transforms of the distributions F_n and F, respectively. Due to this fact, it is natural to ask whether we can establish a Berry–Esseen type inequality to evaluate the closeness between distributions in terms of their Stieltjes transforms. The first attempt was made by Girko (1989) who established an inequality by integrating both sides of Berry–Esseen's basic inequality. Unfortunately, the justification of the exchange of integration signs in his proof is not obvious. More importantly, Girko's inequality seems too complicated to apply. We establish the following basic inequality.

THEOREM 2.1. *Let F be a distribution function and let G be a function of bounded variation satisfying $\int |F(x) - G(x)|\, dx < \infty$. Denote their Stieltjes transforms by $f(z)$ and $g(z)$, respectively. Then we have*

$$\begin{aligned} \|F - G\| &:= \sup_x |F(x) - G(x)| \\ &\le \frac{1}{\pi(2\gamma - 1)} \left[\int_{-\infty}^{\infty} |f(z) - g(z)|\, du \right. \\ &\qquad \left. + \frac{1}{v} \sup_x \int_{|y| \le 2va} |G(x+y) - G(x)|\, dy \right], \end{aligned} \tag{2.4}$$

where $z = u + iv$, $v > 0$, *and* a *and* γ *are constants related to each other by*

$$\gamma = \frac{1}{\pi}\int_{|u|<a}\frac{1}{u^2+1}\,du > \frac{1}{2}. \tag{2.5}$$

PROOF. Write $\Delta = \sup_x|F(x) - G(x)|$. Without loss of generality, we can assume that $\Delta > 0$. Then, there is a sequence $\{x_n\}$ such that $F(x_n) - G(x_n) \to \Delta$ or $-\Delta$.

We shall first consider the case that $F(x_n) - G(x_n) \to \Delta$. For each x, we have

$$\begin{aligned}
\frac{1}{\pi}\int_{-\infty}^{\infty}|f(z) - g(z)|\,du &\geq \frac{1}{\pi}\int_{-\infty}^{x}\operatorname{Im}(f(z) - g(z))\,du \\
&= \frac{1}{\pi}\int_{-\infty}^{x}\left[\int_{-\infty}^{\infty}\frac{v\,d(F(y) - G(y))}{(y-u)^2 + v^2}\right]du \\
&= \frac{1}{\pi}\int_{-\infty}^{x}\left[\int_{-\infty}^{\infty}\frac{2v(y-u)(F(y) - G(y))\,dy}{((y-u)^2 + v^2)^2}\right]du \qquad (2.6)\\
&= \frac{1}{\pi}\int_{-\infty}^{\infty}(F(y) - G(y))\left[\int_{-\infty}^{x}\frac{2v(y-u)\,du}{((y-u)^2 + v^2)^2}\right]dy \\
&= \frac{1}{\pi}\int_{-\infty}^{\infty}\frac{(F(x - vy) - G(x - vy))\,dy}{y^2+1}.
\end{aligned}$$

Here, the second equality follows from integration by parts while the third follows from Fubini's theorem due to the integrability of $|F(y) - G(y)|$. Since F is nondecreasing, we have

$$\begin{aligned}
&\frac{1}{\pi}\int_{|y|<a}\frac{(F(x - vy) - G(x - vy))\,dy}{y^2+1} \\
&\qquad \geq \gamma(F(x - va) - G(x - va)) \\
&\qquad\quad - \frac{1}{\pi}\int_{|y|<a}|G(x - vy) - G(x - va)|\,dy \qquad (2.7)\\
&\qquad \geq \gamma(F(x - va) - G(x - va)) \\
&\qquad\quad - \frac{1}{\pi v}\sup_x\int_{|y|<2va}|G(x+y) - G(x)|\,dy.
\end{aligned}$$

Take $x = x_n + va$. Then, (2.6) and (2.7) imply that

$$
\begin{aligned}
&\frac{1}{\pi}\int_{-\infty}^{\infty}|f(z) - g(z)|\,du \\
&\qquad \geq \gamma(F(x_n) - G(x_n)) \\
&\qquad\quad - \frac{1}{\pi v}\sup_x \int_{|y|<2va} |G(x+y) - G(x)|\,dy - (1-\gamma)\Delta \\
&\qquad \to (2\gamma - 1)\Delta - \frac{1}{\pi v}\sup_x \int_{|y|<2va} |G(x+y) - G(x)|\,dy,
\end{aligned}
\tag{2.8}
$$

which implies (2.4).

Now we consider the case that $F(x_n) - G(x_n) \to -\Delta$. Similarly, we have, for each x, that

$$
\begin{aligned}
&\frac{1}{\pi}\int_{-\infty}^{\infty}|f(z) - g(z)|\,du \\
&\qquad \geq \frac{1}{\pi}\int_{-\infty}^{\infty}\frac{(G(x - vy) - F(x - vy))\,dy}{y^2 + 1} \\
&\qquad \geq \gamma(G(x + va) - F(x + va)) \\
&\qquad\quad - \frac{1}{\pi v}\sup_x \int_{|y|<2va} |G(x+y) - G(x)|\,dy - (1-\gamma)\Delta.
\end{aligned}
\tag{2.9}
$$

By taking $x = x_n - va$, we have

$$
\begin{aligned}
&\frac{1}{\pi}\int_{-\infty}^{\infty}|f(z) - g(z)|\,du \\
&\qquad \geq \gamma(G(x_n) - F(x_n)) \\
&\qquad\quad - \frac{1}{\pi v}\sup_x \int_{|y|<2va} |G(x+y) - G(x)|\,dy - (1-\gamma)\Delta \\
&\qquad \to (2\gamma - 1)\Delta - \frac{1}{\pi v}\sup_x \int_{|y|<2va} |G(x+y) - G(x)|\,dy,
\end{aligned}
\tag{2.10}
$$

which implies (2.4) for the latter case. This completes the proof of Theorem 2.1. □

REMARK 2.1. In the proof of Theorem 2.1, one may find that the following version is stronger than Theorem 2.1:

$$
\begin{aligned}
\|F - G\| \leq \frac{1}{\pi(2\gamma - 1)}\bigg[&\int_{-\infty}^{\infty}|\mathrm{Im}(f(z) - g(z))|\,du \\
&+ \frac{1}{v}\sup_x \int_{|y|\leq 2va} |G(x+y) - G(x)|\,dy\bigg].
\end{aligned}
\tag{2.11}
$$

However, in application of the inequalities, we did not find any significant superiority of (2.11) over (2.4).

Sometimes the functions F and G may have light tails or both may even have bounded support. In such cases, we may establish a bound for $\|F - G\|$ by means of the integral of the absolute difference of their Stieltjes transforms on only a finite interval. We have the following theorem.

THEOREM 2.2. *Under the assumptions of Theorem* 2.1, *we have*

$$\|F - G\| \le \frac{1}{\pi(1-\kappa)(2\gamma-1)}\left[\int_{-A}^{A}|f(z) - g(z)|\,du + 2\pi v^{-1}\int_{|x|>B}|F(x) - G(x)|\,dx + v^{-1}\sup_x \int_{|y|\le 2va}|G(x+y) - G(x)|\,dy\right], \tag{2.12}$$

where A and B are positive constants such that $A > B$ *and*

$$\kappa = \frac{4B}{\pi(A-B)(2\gamma-1)} < 1. \tag{2.13}$$

The following corollary is immediate.

COROLLARY 2.3. *In addition to the assumptions of Theorem* 2.1, *assume further that, for some constant* $B > 0$, $F([-B, B]) = 1$ *and* $|G|((-\infty, -B)) = |G|((B, \infty)) = 0$, *where* $|G|((a, b))$ *denotes the total variation of the signed measure G on the interval* (a, b). *Then, we have*

$$\|F - G\| \le \frac{1}{\pi(1-\kappa)(2\gamma-1)}\left[\int_{-A}^{A}|f(z) - g(z)|\,du + v^{-1}\sup_x \int_{|y|\le 2va}|G(x+y) - G(x)|\,dy\right], \tag{2.14}$$

where A, B and κ *are defined in* (2.13).

REMARK 2.2. The benefit of using Theorem 2.2 and Corollary 2.3 is that we need only estimate the difference of Stieltjes transforms of the two distributions of interest on a fixed interval. When Theorem 2.2 is applied to establish the convergence rate of the spectral distribution of a sample covariance matrix in Section 4, it is crucial to the proof of Theorem 4.1 that A is independent of the sample size n. It should also be noted that the integral limit A in Girko's (1989) inequality should tend to infinity with a rate of A^{-1} faster than the convergence rate to be established. Therefore, our Theorem 2.2 and Corollary 2.3 are much easier to use than Girko's inequality.

PROOF OF THEOREM 2.2. Using the notation given in the proof of Theorem 2.1, we have

$$
\begin{aligned}
&\int_A^\infty |f(z) - g(z)|\, du \\
&= \int_A^\infty \left| \int_{-\infty}^\infty \frac{(F(x) - G(x))\, dx}{(x-z)^2} \right| du \\
&\le \int_A^\infty \left| \int_{-B}^B \frac{(F(x) - G(x))\, dx}{(x-z)^2} \right| du \\
&\quad + \int_{-\infty}^\infty \left| \int_{|x|>B} \frac{(F(x) - G(x))\, dx}{(x-z)^2} \right| du \qquad (2.15)\\
&\le 2B\Delta \int_A^\infty \frac{du}{(u-B)^2 + v^2} + \pi v^{-1} \int_{|x|>B} |F(x) - G(x)|\, dx \\
&\le \frac{2B\Delta}{A-B} + \frac{\pi}{v} \int_{|x|>B} |F(x) - G(x)|\, dx.
\end{aligned}
$$

By symmetry, we get the same bound for $\int_{-\infty}^{-A} |f(z) - g(z)|\, du$. Substituting the above inequality into (2.4), we obtain (2.12) and the proof is complete. □

3. Preliminaries.

3.1. *The notation* $\sqrt{z}$. We need first to clarify the notation $\sqrt{z}$, $z = u + iv$, $(v \neq 0, i = \sqrt{-1})$. Throughout this paper, $\sqrt{z}$ denotes the square root of z with a positive imaginary part. In fact, we have the following expressions:

$$\sqrt{z} = \operatorname{sign}(v) \frac{|z| + z}{\sqrt{2(|z| + u)}}, \tag{3.1}$$

or

$$\operatorname{Re}(\sqrt{z}) = \frac{1}{\sqrt{2}} \operatorname{sign}(v) \sqrt{\sqrt{u^2 + v^2} + u} = \frac{v}{\sqrt{2(\sqrt{u^2 + v^2} - u)}}$$

and

$$\operatorname{Im}(\sqrt{z}) = \frac{1}{\sqrt{2}} \sqrt{\sqrt{u^2 + v^2} - u} = \frac{|v|}{\sqrt{2(\sqrt{u^2 + v^2} + u)}},$$

where $\operatorname{Re}(\cdot)$ and $\operatorname{Im}(\cdot)$ denote the real and imaginary parts of a complex number indicated in the parentheses. If z is a real number, define $\sqrt{z} = \lim_{v \downarrow 0} \sqrt{z + iv}$. Then the definition agrees with the arithmetic square root of positive numbers. However, under this definition, the multiplication rule for square roots fails; that is, in general, $\sqrt{z_1 z_2} \neq \sqrt{z_1}\sqrt{z_2}$. The introduction of the definition (3.1) is merely for convenience and definiteness.

3.2. *Stieltjes transform of the semicircular law.* By definition,

$$\begin{aligned} s(z) &= \frac{1}{2\pi}\int_{-2}^{2}\frac{\sqrt{4-x^2}}{x-z}\,dx \\ &= \frac{1}{\pi}\int_{0}^{\pi}\frac{\sin^2\theta}{\cos\theta-(1/2)z}\,d\theta \quad \text{by } x = 2\cos\theta \\ &= \frac{1}{2\pi}\int_{0}^{2\pi}\frac{\sin^2\theta}{\cos\theta-(1/2)z}\,d\theta \\ &= -\frac{1}{4\pi i}\oint_{|\zeta|=1}\frac{(\zeta^2-1)^2}{\zeta^2(\zeta^2-\zeta z+1)}\,d\zeta \quad \text{by } \zeta = \exp\{i\theta\}. \end{aligned}$$

Now, we apply the residue theorem to evaluate the integral. First, we note that the function $(\zeta^2-1)^2/\zeta^2(\zeta^2-\zeta z+1)$ has three singular points, 0 and $\zeta_{1,2} = (1/2)(z \pm \sqrt{z^2-4})$, with residues z and $\pm\sqrt{z^2-4}$. Here $\zeta_{1,2}$ are in fact the roots of the quadratic equation $\zeta^2 - z\zeta + 1 = 0$. Thus, $\zeta_1\zeta_2 = 1$. Applying the formula (3.1) to the square root of $z^2 - 4 = (u^2 - v^2 - 4) + 2uvi$, one finds that the real parts of z and $\sqrt{z^2-4}$ have the same sign while their imaginary parts are positive. Hence, both the real and imaginary parts of ζ_1 have larger absolute values than those of ζ_2. Therefore, ζ_1 is outside the unit circle while ζ_2 is inside. Hence, we obtain

$$s(z) = -\tfrac{1}{2}\left(z - \sqrt{z^2-4}\right). \tag{3.2}$$

Noting that $s(z) = -\zeta_2$, we have

$$|s(z)| < 1. \tag{3.3}$$

3.3. *Integrals of the square of the absolute value of Stieltjes transforms.*

LEMMA 3.1. *Suppose that $\phi(x)$ is a bounded probability density supported on a finite interval (A, B). Then,*

$$\int_{-\infty}^{\infty}|s(z)|^2\,du < 2\pi^2 M_\phi,$$

where $s(z)$ is the Stieltjes transform of ϕ and M_ϕ the upper bound of ϕ.

PROOF. We have

$$\begin{aligned} I &:= \int_{-\infty}^{\infty}|s(z)|^2\,du \\ &= \int_{-\infty}^{\infty}\int_A^B\int_A^B\frac{\phi(x)\phi(y)\,dx\,dy}{(x-z)(y-\bar z)}\,du \\ &= \int_A^B\int_A^B\phi(x)\phi(y)\,dx\,dy\int_{-\infty}^{\infty}\frac{1}{(x-z)(y-\bar z)}\,du \quad \text{(Fubini's theorem)} \\ &= \int_A^B\int_A^B\frac{2\pi i}{y-x+2vi}\phi(x)\phi(y)\,dx\,dy, \quad \text{(residue theorem)}. \end{aligned}$$

Note that

$$\int_A^B \int_A^B \mathrm{Re}\left(\frac{1}{y - x + 2vi} \right) \phi(x)\phi(y)\, dx\, dy$$
$$= \int_A^B \int_A^B \left(\frac{y - x}{(y - x)^2 + 4v^2} \right) \phi(x)\phi(y)\, dx\, dy = 0, \quad \text{by symmetry.}$$

We finally obtain that

$$I = -2\pi \int_A^B \int_A^B \mathrm{Im}\left(\frac{1}{y - x + 2vi} \right) \phi(x)\phi(y)\, dx\, dy$$
$$= 4\pi v \int_A^B \int_A^B \left(\frac{1}{(y - x)^2 + 4v^2} \right) \phi(x)\phi(y)\, dx\, dy$$
$$\leq 4\pi v M_\phi \int_{-\infty}^{\infty} \int_A^B \phi(y) \left(\frac{1}{w^2 + 4v^2} \right) dw\, dy \quad \text{by setting } w = x - y$$
$$= 2\pi^2 M_\phi.$$

The proof is complete. □

REMARK 3.1. The assumption that ϕ has finite support has been used in the verification of the conditions of Fubini's theorem.

Applying this lemma to the semicircular law, we get the following corollary.

COROLLARY 3.2. We have

$$\int |s(z)|^2\, du \leq 2\pi. \tag{3.4}$$

3.4. *Some algebraic formulae used in this paper.* In this paper, certain algebraic formulae are used. Some of them are well known and will be listed only. For the others, brief proofs will be given. Most of the known results can be found in Xu (1982).

3.4.1. *Inverse matrix formula.* Let A be an $n \times n$ nonsingular matrix. Then

$$A^{-1} = \frac{1}{\det(A)} A^*,$$

where A^* is the adjoint matrix of A, that is, the transposed matrix of cofactors of order $n - 1$ of A and $\det(A)$ denotes the determinant of the matrix A. By this formula, we have

$$\mathrm{tr}(A^{-1}) = \sum_{k=1}^{n} \det(A_k)/\det(A), \tag{3.5}$$

where A_k is the kth major submatrix of order $n-1$ of the matrix A, that is, the matrix obtained from A by deleting the kth row and column.

3.4.2. If A is nonsingular, then

$$\det\begin{bmatrix} A & B \\ C & D \end{bmatrix} = \det(A)\det(D - CA^{-1}B), \tag{3.6}$$

which follows immediately from the fact that

$$\begin{bmatrix} I & 0 \\ -CA^{-1} & I \end{bmatrix}\begin{bmatrix} A & B \\ C & D \end{bmatrix} = \begin{bmatrix} A & B \\ 0 & D - CA^{-1}B \end{bmatrix}.$$

3.4.3. If both A and A_k are nonsingular and if we write $A^{-1} = [a^{kl}]$, then

$$a^{kk} = \frac{1}{a_{kk} - \alpha_k' A_k^{-1}\beta_k}, \tag{3.7}$$

where a_{kk} is the kth diagonal entry of A, A_k the major submatrix of order $n-1$ as defined in Section 3.4.1, α_k the vector obtained from the kth row of A by deleting the kth entry and β_k the vector from the kth column by deleting the kth entry. Then, (3.7) follows from (3.5) and (3.6).

If A is an $n \times n$ symmetric nonsingular matrix and all its major submatrices of order $(n-1)$ are nonsingular, then from (3.5) and (3.7), it follows immediately that

$$\operatorname{tr}(A^{-1}) = \sum_{k=1}^{n} \frac{1}{a_{kk} - \alpha_k' A_k^{-1}\alpha_k}. \tag{3.8}$$

3.4.4. Use the notation of Section 3.4.3. If A and A_k are nonsingular symmetric matrices, then

$$\operatorname{tr}(A^{-1}) - \operatorname{tr}(A_k^{-1}) = \frac{1 + \alpha_k' A_k^{-2}\alpha_k}{a_{kk} - \alpha_k' A_k^{-1}\alpha_k}. \tag{3.9}$$

This is a direct consequence of the following well-known formula for a nonsingular symmetric matrix:

$$\Sigma^{-1} = \begin{bmatrix} \Sigma_{11}^{-1} + \Sigma_{11}^{-1}\Sigma_{12}(\Sigma_{22} - \Sigma_{21}\Sigma_{11}^{-1}\Sigma_{12})^{-1}\Sigma_{21}\Sigma_{11}^{-1} & -\Sigma_{11}^{-1}\Sigma_{12}(\Sigma_{22} - \Sigma_{21}\Sigma_{11}^{-1}\Sigma_{12})^{-1} \\ -(\Sigma_{22} - \Sigma_{21}\Sigma_{11}^{-1}\Sigma_{12})^{-1}\Sigma_{21}\Sigma_{11}^{-1} & (\Sigma_{22} - \Sigma_{21}\Sigma_{11}^{-1}\Sigma_{12})^{-1} \end{bmatrix},$$

where $\Sigma = \begin{bmatrix} \Sigma_{11} & \Sigma_{12} \\ \Sigma_{21} & \Sigma_{22} \end{bmatrix}$ is a partition of the symmetric matrix Σ.

3.4.5. If real symmetric matrices A and B are commutative and such that $A^2 + B^2$ is nonsingular, then the complex matrix $A + iB$ is nonsingular and

$$(A + iB)^{-1} = (A - iB)(A^2 + B^2)^{-1}. \tag{3.10}$$

This can be directly verified.

3.4.6. Let $z = u + iv$, $v > 0$, and let A be an $n \times n$ real symmetric matrix. Then

$$|\operatorname{tr}(A - zI_n)^{-1} - \operatorname{tr}(A_k - zI_{n-1})^{-1}| \leq v^{-1}. \tag{3.11}$$

PROOF. By (3.9), we have

$$\operatorname{tr}(A - zI_n)^{-1} - \operatorname{tr}(A_k - zI_{n-1})^{-1} = \frac{1 + \alpha_k'(A_k - zI_{n-1})^{-2}\alpha_k}{a_{kk} - z - \alpha_k'(A - zI_{n-1})^{-1}\alpha_k}.$$

If we denote $A_k = E' \operatorname{diag}[\lambda_1 \cdots \lambda_{n-1}]E$ and $\alpha_k' E' = (y_1, \ldots, y_{n-1})$, where E is an $(n-1) \times (n-1)$ (real) orthogonal matrix, then we have

$$\begin{aligned} |1 + \alpha_k'(A_k - zI_{n-1})^{-2}\alpha_k| &= \left|1 + \sum_{l=1}^{n-1} y_l^2(\lambda_l - z)^{-2}\right| \\ &\leq 1 + \sum_{l=1}^{n-1} y_l^2\left((\lambda_l - u)^2 + v^2\right)^{-1} \\ &= 1 + \alpha_k'\left((A_k - uI_{n-1})^2 + v^2 I_{n-1}\right)^{-1}\alpha_k. \end{aligned}$$

On the other hand, by (3.10) we have

$$\begin{aligned} &\operatorname{Im}\left(a_{kk} - z - \alpha_k'(A - zI_{n-1})^{-1}\alpha_k\right) \\ &\quad = v\left(1 + \alpha_k'\left((A_k - uI_{n-1})^2 + v^2 I_{n-1}\right)^{-1}\alpha_k\right). \end{aligned} \tag{3.12}$$

From these estimates, (3.11) follows. □

3.5. *A lemma on empirical spectral distributions.*

LEMMA 3.3. *Let W_n be an $n \times n$ symmetric matrix and W_{n-1} be an $(n-1) \times (n-1)$ major submatrix of W_n. Denote the spectral distributions of W_n and W_{n-1} by F_n and F_{n-1}, respectively. Then, we have*

$$\|nF_n - (n-1)F_{n-1}\| \leq 1.$$

PROOF. Denote the eigenvalues of the matrices W_n and W_{n-1} by $\lambda_1 \leq \cdots \leq \lambda_n$ and $\mu_1 \leq \cdots \leq \mu_{n-1}$, respectively. Then, the lemma follows from the following well-known fact:

$$\lambda_1 \leq \mu_1 \leq \lambda_2 \leq \cdots \leq \mu_{n-1} \leq \lambda_n.$$ □

4. Convergence rates of expected spectral distributions of Wigner matrices. In this section, we shall apply the inequality of Theorem 2.1 to establish a convergence rate of the spectral distributions of high dimensional Wigner matrices. A Wigner matrix $W_n = (x_{ij}(n))$, $i, j = 1, \ldots, n$, is defined to be a symmetric matrix with independent entries on and above the diagonal. Throughout this section, we shall drop the index n from the entries of W_n and assume that the following conditions hold:

$$
(4.1)\qquad
\begin{aligned}
&\text{(i)} && Ex_{ij} = 0, && \text{for all } 1 \le i \le j \le n;\\
&\text{(ii)} && Ex_{ij}^2 = 1, && \text{for all } 1 \le i < j \le n;\\
& && Ex_{ii}^2 = \sigma^2, && \text{for all } 1 \le i \le n;\\
&\text{(iii)} && \sup_n \sup_{1\le i\le j\le n} Ex_{ij}^4 \le M < \infty.
\end{aligned}
$$

Denote by F_n the empirical spectral distribution of $(1/\sqrt{n})W_n$. Under the conditions given in (4.1), it is well known that $F_n \to_w F$ in probability, where F is the limiting spectral distribution of F_n, known as Wigner's semicircular law, that is,

$$
(4.2)\qquad F(x) = \frac{1}{2\pi}\int_{-\infty}^{x} \sqrt{4-y^2}\, I_{[-2,2]}(y)\, dy.
$$

If W_n is the $n \times n$ submatrix of the upper-left corner of an infinite dimensional random matrix $[x_{ij}, i, j = 1, 2, \dots]$, then the convergence is almost sure (a.s.) [see Girko (1975) or Pastur (1972)].

In this section, we shall establish the following theorem.

THEOREM 4.1. *Under assumptions* (4.1), *we have*

$$
(4.3)\qquad \|EF_n - F\| = O(n^{-1/4}).
$$

REMARK 4.1. In Section 3 of Girko (1989), an estimate of the difference between the expected Stieltjes transform of the spectral distribution F_n of Wigner matrices and that of the limiting spectral distribution F is established. In his proof, some arguments are not easily verifiable. If the proof is correct, then his result implies

$$
\|EF_n - F\| = O(n^{-\gamma/14}), \quad \text{for some } 0 < \gamma < 1,
$$

by applying Theorem 2.1. The result of Theorem 4.1 is stronger than that implied by Girko's Theorem 3.1.

REMARK 4.2. It may be of greater interest to establish a convergence rate of $\|F_n - F\|$. This is under further investigation. In the proof of Theorem 4.1, one may find that the terms in the expansion of the Stieltjes transform of EF_n have a step-decreasing rate of n^{-1} if the estimation of the remainder term is not taken into account. Thus, we may conjecture that the ideal convergence rate of $\|EF_n - F\|$ is $O(n^{-1})$. Based on experience [say, for functions of sample means, the rate of expected bias is of $O(1/n)$, but $\sqrt{n}(f(\bar{X}_n) - f(\mu)) \to N(0, \sigma^2)$], one may conjecture that of $\|F_n - F\|$ is $O_p(n^{-1/2})$. But I was told through private communication that J. W. Silverstein conjectured that the rate for both cases is $O(n^{-1})$.

The proof of Theorem 4.1 is somewhat tedious. We first prove a preliminary result and then refine it.

PROPOSITION 4.2. *Under the assumptions of Theorem* 4.1, *we have*

$$\|EF_n - F\| = O(n^{-1/6}). \tag{4.4}$$

PROOF. It is shown in (3.2) that the Stieltjes transform of F is given by

$$s(z) = -\tfrac{1}{2}\{z - \sqrt{z^2 - 4}\}. \tag{4.5}$$

Let u and $v > 0$ be real numbers and let $z = u + iv$. Set

$$s_n(z) = \int_{-\infty}^{\infty} \frac{1}{x - z}\, dE\,F_n(x) = \frac{1}{n} E\,\mathrm{tr}\left(\frac{1}{\sqrt{n}} W_n - zI_n\right)^{-1}. \tag{4.6}$$

Then, by the inverse matrix formula [see (3.8)], we have

$$\begin{aligned} s_n(z) &= \frac{1}{n} \sum_{k=1}^{n} E \frac{1}{(1/\sqrt{n})x_{kk} - z - (1/n)\alpha'(k)((1/\sqrt{n})W_n(k) - zI_{n-1})^{-1}\alpha(k)} \\ &= \frac{1}{n} \sum_{k=1}^{n} E \frac{1}{\varepsilon_k - z - s_n(z)} = -\frac{1}{z + s_n(z)} + \delta, \end{aligned} \tag{4.7}$$

where $\alpha'(k) = (x_{1k}, \dots, x_{k-1,k}, x_{k+1,k}, \dots, x_{nk})$, $W_n(k)$ is the matrix obtained from W_n by deleting the kth row and kth column,

$$\varepsilon_k = \frac{1}{\sqrt{n}} x_{kk} - \frac{1}{n}\alpha'(k)\left(\frac{1}{\sqrt{n}} W_n(k) - zI_{n-1}\right)^{-1} \alpha(k) + s_n(z) \tag{4.8}$$

and

$$\delta = \delta_n = -\frac{1}{n} \sum_{k=1}^{n} E \frac{\varepsilon_k}{(z + s_n(z))(z + s_n(z) - \varepsilon_k)}. \tag{4.9}$$

Solving (4.7), we obtain

$$s_{(1),(2)}(z) = -\tfrac{1}{2}\left(z - \delta \pm \sqrt{(z + \delta)^2 - 4}\right). \tag{4.10}$$

We claim that

$$s_n(z) = s_{(2)}(z) = -\tfrac{1}{2}\left(z - \delta - \sqrt{(z + \delta)^2 - 4}\right). \tag{4.11}$$

Note that

$$\mathrm{Im}(z + s_n(z)) = v\left(1 + E\int \frac{1}{(x - u)^2 + v^2}\, dF_n(x)\right) \ge v, \tag{4.12}$$

which immediately yields

$$|z + s_n(z)|^{-1} \le v^{-1}. \tag{4.13}$$

By definition, it is obvious that

$$|s_n(z)| \le v^{-1}. \tag{4.14}$$

Hence by (4.7),

$$|\delta| \le 2/v. \tag{4.15}$$

We conclude that (4.11) is true for all $v > \sqrt{2}$ because, by (4.15), $\operatorname{Im}(s_{(1)}(z)) \le -(1/2)(v - |\delta|) \le -(1/2)(v - 2/v) < 0$, which contradicts the fact that $\operatorname{Im}(s_n(z)) > 0$. By definition, $s_n(z)$ is a continuous function of z on the upper half-plane $\{z = u + iv\colon v > 0\}$. By (4.13), $(z + s_n(z))^{-1}$ is also continuous. Hence, δ, and consequently, $s_{(1)}(z)$ and $s_{(2)}(z)$ are continuous on the upper half-plane. Therefore, to prove $s_n(z) \ne s_{(1)}(z)$, or equivalently the assertion (4.11), it is sufficient to show that the two continuous functions $s_{(1)}(z)$ and $s_{(2)}(z)$ cannot be equal at any point on the upper half-plane. If $s_{(1)}(z) = s_{(2)}(z)$ for some $z = u + iv$ with $v > 0$, then the square root in (4.11) should be zero, that is, $\delta = \pm 2 - z$. This implies that $s_n(z) = \pm 1 - z$, which contradicts the fact that $\operatorname{Im}(s_n(z)) > 0$. This completes the proof of our assertion (4.11).

Comparing (4.5) and (4.11), we shall prove Proposition 4.2 by the following steps: Prove $|\delta|$ is "small" for both its absolute value and for the integral of its absolute value with respect to u. Then, find a bound of $s_n(z) - s(z)$ in terms of δ. First, let us begin to estimate $|\delta|$.

Applying (3.12), we have

$$|z + s_n(z) - \varepsilon_k|^{-1} \le v^{-1}. \tag{4.16}$$

By (4.9), we have

$$|\delta| \le \frac{1}{n} \sum_{k=1}^{n} \left[\frac{|E\varepsilon_k|}{|z + s_n(z)|^2} + E\left| \frac{\varepsilon_k^2}{(z + s_n(z))^2 (z + s_n(z) - \varepsilon_k)} \right| \right]$$

$$\le (z + s_n(z))^{-2} \left[\frac{1}{n} \sum_{k=1}^{n} \left(|E\varepsilon_k| + v^{-1} E|\varepsilon_k^2| \right) \right] \tag{4.17}$$

$$\le \frac{1}{nv^2} \left(\sum_{k=1}^{n} |E\varepsilon_k| + v^{-1} \sum_{k=1}^{n} E|\varepsilon_k^2| \right). \tag{4.18}$$

Recalling the definition of ε_k in (4.8) and applying (3.11), we obtain

$$\begin{aligned} |E\varepsilon_k| &= \frac{1}{n} \left| E\left[\operatorname{tr}\left(\frac{1}{\sqrt{n}} W_n - zI_n \right)^{-1} - \operatorname{tr}\left(\frac{1}{\sqrt{n}} W_n(k) - zI_{n-1} \right)^{-1} \right] \right| \\ &\le \frac{1}{nv}. \end{aligned} \tag{4.19}$$

Now, we begin to estimate $E|\varepsilon_k|^2$. By (4.8), we have

$$
\begin{aligned}
E|\varepsilon_k|^2 &= E|\varepsilon_k - E\varepsilon_k|^2 + |E(\varepsilon_k)|^2 \\
&= \frac{\sigma^2}{n} + \frac{1}{n}E\left|\alpha_k'\left(\frac{1}{\sqrt{n}}W_n(k) - zI_{n-1}\right)^{-1}\alpha_k\right. \\
&\qquad\left. -\operatorname{tr}\left(\frac{1}{\sqrt{n}}W_n(k) - zI_{n-1}\right)^{-1}\right|^2 \\
&\quad + E\left|\frac{1}{n}\operatorname{tr}\left(\frac{1}{\sqrt{n}}W_n(k) - zI_{n-1}\right)^{-1}\right. \\
&\qquad\left. -\frac{1}{n}E\operatorname{tr}\left(\frac{1}{\sqrt{n}}W_n(k) - zI_{n-1}\right)^{-1}\right|^2 \\
&\quad + |E\varepsilon_k|^2.
\end{aligned}
\tag{4.20}
$$

Let

$$
(r_{ij}(k)) = \left(\frac{1}{\sqrt{n}}W_n(k) - zI_{n-1}\right)^{-1}.
$$

Then, we have

$$
\begin{aligned}
&E\left|\frac{1}{n}\alpha_k'\left(\frac{1}{\sqrt{n}}W_n(k) - zI_{n-1}\right)^{-1}\alpha_k - \frac{1}{n}\operatorname{tr}\left(\frac{1}{\sqrt{n}}W_n(k) - zI_{n-1}\right)^{-1}\right|^2 \\
&= E\left|\frac{1}{n}\sum_{jl} r_{jl}(k)\left(x_{jk}x_{lk} - E(x_{jk}x_{lk})\right)\right|^2 \\
&\le \frac{2M}{n^2}\sum_{i,j} E|r_{ij}^2(k)| \\
&= \frac{2M}{n^2}E\operatorname{tr}\left(\frac{1}{\sqrt{n}}W_n(k) - zI_{n-1}\right)^{-1}\left(\frac{1}{\sqrt{n}}W_n(k) - \bar{z}I_{n-1}\right)^{-1} \\
&= \frac{2M}{n^2}E\operatorname{tr}\left(\left(\frac{1}{\sqrt{n}}W_n(k) - uI_{n-1}\right)^2 + v^2I_{n-1}\right)^{-1} \\
&\le \frac{2M}{nv^2}.
\end{aligned}
\tag{4.21}
$$

Let $\gamma_k(k) = 0$ and for $d \neq k$, let

$$\gamma_d(k) = E_{d-1}\operatorname{tr}\left(\frac{1}{\sqrt{n}}W_n(k) - zI_{n-1}\right)^{-1} - E_d\operatorname{tr}\left(\frac{1}{\sqrt{n}}W_n(k) - zI_{n-1}\right)^{-1}$$
$$= E_{d-1}\sigma_d(k) - E_d\sigma_d(k),$$

where

$$(4.22)\quad \sigma_d(k) = \left[\operatorname{tr}\left(\frac{1}{\sqrt{n}}W_n(k) - zI_{n-1}\right)^{-1} - \left(\frac{1}{\sqrt{n}}W(d,k) - zI_{n-2}\right)^{-1}\right],$$

E_d denotes the conditional expectation given $\{x_{ij},\ d+1 \le i \le j \le n\}$, $\alpha(d,k)$ is the vector obtained from the dth column of W_n by deleting the dth and kth entries and $W_n(d,k)$ the matrix obtained from W_n by deleting the dth and kth columns and rows. By (3.11), we have

$$(4.23)\qquad |\sigma_d(k)| \le v^{-1},$$

which implies that

$$(4.24)\quad \begin{aligned} &E\left|\frac{1}{n}\operatorname{tr}\left(\frac{1}{\sqrt{n}}W_n(k) - zI_{n-1}\right)^{-1} - \frac{1}{n}E\operatorname{tr}\left(\frac{1}{\sqrt{n}}W_n(k) - zI_{n-1}\right)^{-1}\right|^2 \\ &\le n^{-2}\sum_{d=1}^{n} E|\gamma_d^2(k)| \le n^{-1}v^{-2}. \end{aligned}$$

By (4.8), (4.19)–(4.21) and (4.24), we obtain, for all large n,

$$(4.25)\qquad E|\varepsilon_k|^2 \le \frac{\sigma^2}{n} + \frac{2M}{nv^2} + \frac{1}{nv^2} + \frac{1}{n^2v^2} < \frac{2M+5}{nv^2},$$

where M is the upper bound of the fourth moments of the entries of W_n in (4.1).

Take $v = ((2M+6)/n)^{1/6}$ and assume $n > 2M+6$. From (4.18), (4.19) and (4.24)–(4.25), we conclude that

$$(4.26)\qquad |\delta| < \frac{2M+6}{nv^5} = v.$$

By (4.7), (4.17), (4.19) and (4.25), for large n so that $v < 1/3$, we have

$$\begin{aligned} \int_{-\infty}^{\infty}|\delta|\,du &\le \frac{2M+6}{nv^3}\left[\int_{-\infty}^{\infty}|z + s_n(z)|^{-2}\,du\right] \\ &\le \frac{4M+12}{nv^3}\left[\int_{-\infty}^{\infty}|s_n(z)|^2\,du + \int_{-\infty}^{\infty}|\delta|^2\,du\right] \\ &\le \frac{4M+12}{nv^3}\left[\int_{-\infty}^{\infty}|s_n(z)|^2\,du + v\int_{-\infty}^{\infty}|\delta|\,du\right]. \end{aligned}$$

By the simple fact that $y \le ax + by$ implies $ax + by \le (a/(1-b))x$ for positive a, $b < 1$, x and y, we get, for large n so that $nv^2 > (2M+7)(4M+12)$,

$$\begin{aligned} \int_{-\infty}^{\infty} |\delta|\, du &\le \frac{4M+12}{nv^3(1-(4M+12)/(nv^2))} \int_{-\infty}^{\infty} |s_n(z)|^2\, du \\ &\le \frac{4M+14}{nv^3} E \int_{-\infty}^{\infty}\int_{-\infty}^{\infty} \frac{1}{(x-u)^2+v^2}\, du\, dF_n(x) \qquad (4.27) \\ &\le \frac{4M+14}{nv^4} \le 2v^2. \end{aligned}$$

Now, we proceed to estimate $|s_n(z) - s(z)|$. By (4.5) and (4.11),

$$|s_n(z) - s(z)| \le \frac{1}{2}|\delta| \left[1 + \frac{|2z+\delta|}{\left|\sqrt{z^2-4} + \sqrt{(z+\delta)^2 - 4}\right|}\right]. \qquad (4.28)$$

Since the signs of the real parts of $\sqrt{z^2-4}$ and $\sqrt{(z+\delta)^2-4}$ [see (3.1)] are $\mathrm{sign}(uv) = \mathrm{sign}(u)$ and $\mathrm{sign}((u + \mathrm{Re}(\delta))(v + \mathrm{Im}(\delta))) = \mathrm{sign}(u + \mathrm{Re}(\delta))$, we conclude that for $|u| > v$ the real and imaginary parts of $\sqrt{z^2-4}$ and $\sqrt{(z+\delta)^2-4}$ have the same signs. Hence, by (4.28) we have

$$|s_n(z) - s(z)| \le \frac{1}{2}|\delta| \left(1 + \frac{2|u| + 3v}{\sqrt{|u^2 - v^2 - 4|}}\right). \qquad (4.29)$$

For $|u| > 4$ and n such that $v < 1/3$, we have

$$\frac{2|u| + 3v}{\sqrt{|u^2 - v^2 - 4|}} \le \frac{8 + 3v}{\sqrt{12 - v^2}} < 3,$$

which, together with (4.29), implies that

$$|s_n(z) - s(z)| \le 2|\delta|. \qquad (4.30)$$

For $|u| \le v$, by (4.26) we have $|\sqrt{(z+\delta)^2 - 4} - 2i| \le (9/2)v^2$. Similarly we have $|\sqrt{z^2-4} - 2i| \le 2v^2$. Therefore, we have for all n such that $v < 1/3$,

$$|s_n(z) - s(z)| \le \frac{1}{2}|\delta| \left(1 + \frac{5v}{4 - 7v^2}\right) \le 2|\delta|. \qquad (4.31)$$

Summing up (4.29)–(4.31), we get that for n so large that $v < 1/3$,

$$|s_n(z) - s(z)| \leq \begin{cases} \frac{1}{2}|\delta|\left[1 + \frac{2|u| + 3v}{\sqrt{|u^2 - v^2 - 4|}}\right], & \text{if } v < |u| \leq 4, \\ 2|\delta|, & \text{otherwise.} \end{cases} \tag{4.32}$$

Finally, by (4.28), (4.27) and (4.32),

$$\begin{aligned} &\int_{-\infty}^{\infty} |s_n(z) - s(z)|\, du \\ &= \left\{\int_{[|u| \leq v]} + \int_{[v < |u| \leq 4]} + \int_{[|u| > 4]}\right\} |s_n(z) - s(z)|\, du \\ &\leq 2\int_{-\infty}^{\infty} |\delta|\, du + 4\int_{-4}^{4} |\delta| \frac{1}{\sqrt{|u^2 - v^2 - 4|}}\, du \\ &\leq 4v^2 + \eta v, \end{aligned} \tag{4.33}$$

where

$$\eta = 4 \sup_{v<1} \int_{-4}^{4} \frac{du}{\sqrt{|u^2 - v^2 - 4|}}.$$

Note that the density function of the semicircular law is bounded by $1/\pi$. An application of Theorem 2.1 completes the proof of Proposition 4.2. □

Now, we are in position to prove Theorem 4.1. The basic approach to prove Theorem 4.1 is similar to that in the proof of Proposition 4.2. The only work needed to do is to refine the estimates of $E|\varepsilon_k^2|$ and the integral of $|\delta|$ by using the preliminary result of Proposition 4.2.

PROOF OF THEOREM 4.1. Denote by Δ_1 the initial estimate of the convergence rate of $\|EF_n - F\|$. By Proposition 4.2, we may choose $\Delta_1 = C_0 n^{-1/6}$ for some positive constant $C_0 \geq 1$.

Choose $v = Dn^{-1/4}$, where D is a positive constant to be specified later. Suppose that n is so large that $v < \Delta_1$.

For later use, let us derive an estimate of $|z + s_n(z)|^{-2}$.

For any two Stieltjes transforms $s_1(z)$ and $s_2(z)$ with their corresponding distributions F_1 and F_2, integration by parts yields

$$\begin{aligned} |s_1(z) - s_2(z)| &= \left|\int_{-\infty}^{\infty} \frac{d(F_1(x) - F_2(x))}{x - z}\right| \\ &= \left|\int_{-\infty}^{\infty} \frac{(F_1(x) - F_2(x))\, dx}{(x - z)^2}\right| \leq \frac{\pi\|F_1 - F_2\|}{v}. \end{aligned} \tag{4.34}$$

Then, by (4.7), (3.3) and (4.34), it follows that

$$\begin{aligned}|z + s_n(z)|^{-2} &\le 3|s(z)|^2 + 3|\delta|^2 + 3|s_n(z) - s(z)|^2 \\ &\le 3|\delta|^2 + 4\pi^2 v^{-2}\Delta_1^2.\end{aligned} \tag{4.35}$$

Now, we begin to refine the estimate of $(1/n)\sum_{ij} E|r_{ij}^2(k)|$. By (4.21) and Lemma 3.3, we obtain that, for large n,

$$\begin{aligned}\frac{1}{n}\sum_{i,j} E|r_{ij}^2(k)| &= \frac{n-1}{n}\int_{-\infty}^{\infty}\frac{1}{(x-u)^2+v^2}\,dE\,F_{n-1}^{(k)}(x) \\ &\le \left[\int_{-\infty}^{\infty}\frac{1}{(x-u)^2+v^2}\,dF(x)\right. \\ &\quad + \left|\int_{-\infty}^{\infty}\frac{2(x-u)(EF_n(x)-F(x))}{\left((x-u)^2+v^2\right)^2}\,dx\right| \\ &\quad \left. + \left|\int_{-\infty}^{\infty}\frac{2(x-u)E(((n-1)/n)F_{n-1}^{(k)}(x)-F_n(x))}{\left((x-u)^2+v^2\right)^2}\,dx\right|\right] \\ &\le \frac{1}{\pi}\int_{-\infty}^{\infty}\frac{1}{(x-u)^2+v^2}\,dx + \left(\Delta_1+\frac{1}{n}\right)\int_{-\infty}^{\infty}\frac{2|x-u|\,dx}{\left((x-u)^2+v^2\right)^2} \\ &= v^{-1} + 2\left(\Delta_1 + \frac{1}{n}\right)v^{-2} < \frac{4\Delta_1}{v^2},\end{aligned} \tag{4.36}$$

where $F_{n-1}^{(k)}$ denotes the spectral distribution of the matrix $(1/\sqrt{n})W_n(k)$.

To get a refinement of the estimate of $E|\gamma_d^2(k)|$, we introduce the following notation. Set

$$R_d(k) = \frac{1}{\sqrt{n}}W_n(d,k) - zI_{n-2},$$

$$s_{nd(k)}(z) = \frac{1}{n}E\left(\operatorname{tr}\left(R_d^{-1}(k)\right)\right),$$

$$\varepsilon_d(k) = \frac{1}{\sqrt{n}}x_{dd} - \alpha'(d,k)R_d^{-1}(k)\alpha(d,k) + s_{nd}(z)$$

and

$$R_d^{-2}(k) = \left(\rho_{ij}(d,k)\right),$$

where $W_n(d,k)$ and $\alpha(d,k)$ are defined below (4.22).

By (4.22), we have

$$\sigma_d(k) = -\frac{1 + (1/n)\alpha'(d,k)R_d^2(k)\alpha(d,k)}{z + s_{nd(k)}(z) - \varepsilon_d(k)}.$$

Then, rewrite $\sigma_d(k)$ [see (4.22)] in the following form:

$$\begin{aligned}\sigma_d(k) &= -\frac{1 + n^{-1}\operatorname{tr}(R_d^{-2}(k))}{z + s_{nd(k)}(z)} \\ &\quad - \frac{\varepsilon_d(k)\left(1 + (1/n)\alpha'(d,k)R_d^{-2}(k)\alpha(d,k)\right)}{\left(z + s_{nd(k)}(z)\right)\left(z + s_{nd(k)}(z) - \varepsilon_d(k)\right)} \\ &\quad - \frac{(1/n)\alpha'(d,k)R_d^{-2}(k)\alpha(d,k) - (1/n)\operatorname{tr}(R_d^{-2}(k))}{z + s_{nd(k)}(z)}.\end{aligned} \tag{4.37}$$

Note that

$$E_{d-1}\frac{1 + n^{-1}\operatorname{tr}(R_d^{-2}(k))}{z + s_{nd(k)}(z)} - E_d\frac{1 + n^{-1}\operatorname{tr}(R_d^{-2}(k))}{z + s_{nd(k)}(z)} = 0 \tag{4.38}$$

and by (4.23),

$$\left|\frac{1 + (1/n)\alpha'(d,k)R_d^{-2}(k)\alpha(d,k)}{z + s_{nd(k)}(z) - \varepsilon_d(k)}\right| = |\sigma_d(k)| \le \frac{1}{v}. \tag{4.39}$$

We have

$$E|\gamma_d^2(k)| \le \frac{2E|\varepsilon_d(k)|^2}{v^2|z + s_{nd(k)}(z)|^2} + \frac{4M\sum_{i,j}E|\rho_{ij}(d,k)|^2}{n^2|z + s_{nd(k)}(z)|^2}. \tag{4.40}$$

Similarly to estimating $n^{-1}\Sigma|r_{ij}^2(k)|$ in (4.21), one may obtain

$$\frac{1}{n}\sum_{i,j}|\rho_{ij}^2(d,k)| \le v^{-4}. \tag{4.41}$$

Let $F_{n-2,d,k}$ denote the spectral distribution of the matrix $(1/\sqrt{n})W_n(d,k)$. Then, by Lemma 3.3, we have $\|((n-2)/n)F_{n-2,d,k} - F_n\| \le 2/n$. Therefore, for all n,

$$\begin{aligned}|s_{nd(k)}(z) - s_n(z)| &= \left|\int_{-\infty}^{\infty}\frac{(EF_n(x) - ((n-2)/n)EF_{n-2,d,k}(x))\,dx}{(x-z)^2}\right| \\ &\le \frac{2\pi}{nv}.\end{aligned} \tag{4.42}$$

Thus, by (3.3), (4.7), (4.13), (4.34) and (4.42), we have for all large n,

$$\begin{aligned}|z + s_{nd(k)}(z)|^{-2} &\le 4\left\{\left|\frac{s_n(z) - s_{nd(k)}(z)}{(z + s_n(z))(z + s_{nd(k)}(z))}\right|^2 + \left|\frac{1}{z + s_n(z)} + s_n(z)\right|^2\right.\\ &\qquad \left. + |s_n(z) - s(z)|^2 + |s(z)|^2\right\}\\ &\le 4\left\{\frac{4\pi^2}{n^2v^4}|z + s_{nd(k)}(z)|^{-2} + |\delta|^2 + \frac{\pi^2\Delta_1^2}{v^2} + 1\right\}\\ &\le 5\left(|\delta|^2 + \frac{\pi^2\Delta_1^2}{v^2}\right).\end{aligned} \tag{4.43}$$

Similarly to estimating $E|\varepsilon_k|^2$ in (4.25), one may obtain for large n that

$$E|\varepsilon_d^2(k)| \le \frac{2M + 5}{nv^2}. \tag{4.44}$$

Hence, by (4.40)–(4.44), we have

$$E|\gamma_d^2(k)| \le \frac{40M + 50}{nv^4}\left(|\delta|^2 + \frac{\pi^2\Delta_1^2}{v^2}\right). \tag{4.45}$$

Substituting (4.19), (4.36) and (4.45) into (4.25), we obtain, for all large n,

$$\begin{aligned}E|\varepsilon_k|^2 &\le \frac{\sigma^2}{n} + \frac{8M\Delta_1}{nv^2} + \frac{40M + 50}{n^2v^4}\left(|\delta|^2 + \frac{\pi^2\Delta_1^2}{v^2}\right) + \frac{1}{n^2v^2}\\ &\le \frac{40M + 60}{n^2v^4}|\delta|^2 + \frac{(8M + 3)\Delta_1}{nv^2}.\end{aligned} \tag{4.46}$$

Consequently, by (4.17), (4.19), (4.35) and (4.46),

$$\begin{aligned}|\delta| &\le |z + s_n(z)|^{-2}\left(\frac{1}{n}\sum_{k=1}^{n}\left[|E\varepsilon_k| + v^{-1}E|\varepsilon_k^2|\right]\right)\\ &\le |z + s_n(z)|^{-2}\left[\frac{40M + 50}{n^2v^5}|\delta|^2 + \frac{(8M + 4)\Delta_1}{nv^3}\right]\\ &\le \left[\frac{40M + 50}{n^2v^5}|\delta|^2 + \frac{(8M + 4)\Delta_1}{nv^3}\right]\left(3|\delta|^2 + \frac{4\pi^2\Delta_1^2}{v^2}\right).\end{aligned} \tag{4.47}$$

From (4.15) and (4.47), it follows that

$$|\delta| \le \left[\frac{250(4M+5)}{D^8} + \frac{24(2M+1)\Delta_1}{D^4} + \frac{80(4M+5)\pi^2\Delta_1^2}{n^2v^6}\right]|\delta|$$

$$(4.48) \qquad + \frac{6\pi^2(2M+1)\Delta_1^3}{nv^5}$$

$$\le \frac{300(4M+5)}{D^8}|\delta| + \frac{6\pi^2(2M+1)\Delta_1^3}{nv^5}.$$

If D is chosen so large that $300(4M+5)D^{-8} \le 1/2$, again using the fact used in the proof of (4.27), we obtain

$$(4.49) \qquad |\delta| \le \frac{12\pi^2(2M+1)\Delta_1^3}{nv^5} < v.$$

By (4.5), (4.11) and (4.48) we have, for large n,

$$(4.50) \qquad |s_n(z) - s(z)| = |s(z+\delta) - s(z) + \delta| \le 2 + v \le 3.$$

Thus, by (4.49) and the second inequality of (4.47),

$$(4.51) \qquad |\delta| \le \frac{(8M+5)\Delta_1}{nv^3}|z + s_n(z)|^{-2}.$$

Therefore, by (4.50), we have

$$\int_{-\infty}^{\infty} |\delta|\, du \le \frac{(8M+5)\Delta_1}{nv^3}\int_{-\infty}^{\infty} |z + s_n(z)|^{-2}\, du$$

$$\le 3\frac{(8M+5)\Delta_1}{nv^3}\left[\int_{-\infty}^{\infty} |s_n(z) - s(z)|^2\, du\right.$$

$$(4.52) \qquad \left. + \int_{-\infty}^{\infty} |s(z)|^2\, du + \int_{-\infty}^{\infty} |\delta|^2\, du\right]$$

$$\le 3\frac{(8M+5)\Delta_1}{nv^3}\left[\int_{-\infty}^{\infty} |s_n(z) - s(z)|^2\, du + 2\pi + v\int_{-\infty}^{\infty} |\delta|\, du\right]$$

$$\le 3\frac{(8M+6)\Delta_1}{nv^3}\left[2\pi + 3\int_{-\infty}^{\infty} |s_n(z) - s(z)|\, du\right],$$

where an upper bound for the integral of $|s(z)|^2$ is established in Corollary 3.2 in Section 3.

Recall the proof of the first inequality of (4.33), where the only condition used is that $|\delta| < v$. Therefore, by (4.49) we have

$$\begin{aligned}\int_{-\infty}^{\infty}|s_n(z) - s(z)|\,du &\le 2\int_{-\infty}^{\infty}|\delta|\,du + \eta v\\ &\le \left(\eta + 12\pi\Delta_1(8M+6)D^{-4}\right)v\\ &\quad + 18\frac{(8M+6)\Delta_1}{nv^3}\int_{-\infty}^{\infty}|s_n(z) - s(z)|\,du\\ &\le 2\left(\eta + 12\pi\Delta_1(8M+6)D^{-4}\right)v.\end{aligned} \tag{4.53}$$

Applying Theorem 2.1, the proof is complete. □

Acknowledgments. The author would like to express his thanks to Professor Alan J. Izenman for his help in the preparation of the paper, and also thanks to Professor J. W. Silverstein and an Associate Editor for their helpful comments.

REFERENCES

ARNOLD, L. (1967). On the asymptotic distribution of the eigenvalues of random matrices. *J. Math. Anal. Appl.* **20** 262–268.

BAI, Z. D. (1993). Convergence rate of expected spectral distributions of large random matrices. Part II. Sample covariance matrices. *Ann. Probab.* **21** 649–672.

BAI, Z. D., SILVERSTEIN, J. W. and YIN, Y. Q. (1988). A note on the largest eigenvalue of a large dimensional sample matrix. *J. Multivariate Anal.* **26** 166–168.

BAI, Z. D. and YIN, Y. Q. (1986). Limiting behavior of the norm of products of random matrices and two problems of Geman–Hwang. *Probab. Theory Related Fields* **73** 555–569.

BAI, Z. D. and YIN, Y. Q. (1988a). A convergence to the semicircle law. *Ann. Probab.* **16** 863–875.

BAI, Z. D. and YIN, Y. Q. (1988b). Necessary and sufficient conditions for the almost sure convergence of the largest eigenvalue of Wigner matrices. *Ann. Probab.* **16** 1729–1741.

BAI, Z. D. and YIN, Y. Q. (1990). Limit of the smallest eigenvalue of large dimensional sample covariance matrix. Technical Report 90-05, Center for Multivariate Analysis, Pennsylvania State Univ.

BAI, Z. D., YIN, Y. Q. and KRISHNAIAH, P. R. (1986). On limiting spectral distribution of product of two random matrices when the underlying distribution is isotropic. *J. Multivariate Anal.* **19** 189–200.

BAI, Z. D., YIN, Y. Q. and KRISHNAIAH, P. R. (1987). On limiting spectral distribution function of the eigenvalues of a multivariate F matrix. *Teor. Verojatnost. i Primenen.* (*Theory Probab. Appl.*) **32** 537–548.

GEMAN, S. (1980). A limit theorem for the norm of random matrices. *Ann. Probab.* **8** 252–261.

GEMAN, S. (1986). The spectral radius of large random matrices. *Ann Probab.* **14** 1318–1328.

GIRKO, V. L. (1975). *Random Matrices.* Vishcha Shkola, Kiev (in Russian).

GIRKO, V. L. (1989). Asymptotics of the distribution of the spectrum of random matrices. *Russian Math. Surveys* **44** 3–36.

GRENANDER, U. (1963). *Probabilities on Algebraic Structures.* Almqvist and Wiksell, Stockholm.

GRENANDER, U. and SILVERSTEIN, J. (1977). Spectral analysis of networks with random topologies. *SIAM J. Appl. Math.* **32** 499–519.

JONSSON, D. (1982). Some limit theorems for the eigenvalues of a sample covariance matrix. *J. Multivariate Anal.* **12** 1–38.

PASTUR, L. A. (1972). On the spectrum of random matrices. *Teoret. Mat. Fiz.* **10** 102–112 (*Teoret. Mat. Phys.* **10** 67–74).

PASTUR, L. A. (1973). Spectra of random self-adjoint operators. *Uspehi Mat. Nauk* **28** 4–63 (*Russian Math. Surveys* **28** 1–67).

SILVERSTEIN, J. W. (1985a). The limiting eigenvalue distribution of a multivariate F matrix. *SIAM J. Math. Anal.* **16** 641–646.

SILVERSTEIN, J. W. (1985b). The smallest eigenvalue of a large dimensional Wishart matrix. *Ann. Probab.* **13** 1364–1368.

VOICULESCU, DAN (1990). Non-commutative random variables and spectral problems in free product C^*-algebras. *Rocky Mountain J. Math.* To appear.

VOICULESCU, DAN (1991). Limit laws for random matrices and free products. *Invent. Math.* **104** 201–220.

WACHTER, K. W. (1978). The strong limits of random matrix spectra for sample matrices of independent elements. *Ann Probab.* **6** 1–18.

WACHTER, K. W. (1980). The limiting empirical measure of multiple discriminant ratios. *Ann. Statist.* **8** 937–957.

WIGNER, E. P. (1955). Characteristic vectors bordered matrices with infinite dimensions. *Ann. of Math.* **62** 548–564.

WIGNER, E. P. (1958). On the distributions of the roots of certain symmetric matrices. *Ann. of Math.* **67** 325–327.

XU, Y. C. (1982). *An Introduction to Algebra*. Shanghai Sci. & Tech. Press, Shanghai, China (in Chinese).

YIN, Y. Q. (1986). Limiting spectral distribution for a class of random matrices. *J. Multivariate Anal.* **20** 50–68.

YIN, Y. Q., BAI, Z. D. and KRISHNAIAH, P. R. (1983). Limiting behavior of the eigenvalues of a multivariate F matrix. *J. Multivariate Anal.* **13** 508–516.

YIN, Y. Q., BAI, Z. D. and KRISHNAIAH, P. R. (1988). On the limit of the largest eigenvalue of the large dimensional sample covariance matrix. *Probab. Theory Related Fields* **78** 509–521.

YIN, Y. Q. and KRISHNAIAH, P. R. (1983). A limit theorem for the eigenvalues of product of two random matrices. *J. Multivariate Anal.* **13** 489–507.

DEPARTMENT OF STATISTICS
341 SPEAKMAN HALL
TEMPLE UNIVERSITY
PHILADELPHIA, PENNSYLVANIA 19122

The Annals of Probability
1993, Vol. 21, No. 2, 649–672

CONVERGENCE RATE OF EXPECTED SPECTRAL DISTRIBUTIONS OF LARGE RANDOM MATRICES. PART II. SAMPLE COVARIANCE MATRICES

BY Z. D. BAI

Temple University

In the first part of the paper, we developed certain inequalities to bound the difference between distributions in terms of their Stieltjes transforms and established a convergence rate of expected spectral distributions of large Wigner matrices. The second part is devoted to establishing convergence rates for the sample covariance matrices, for the cases where the ratio of the dimension to the degrees of freedom is bounded away from 1 or close to 1, respectively.

1. Introduction. Basic concepts and literature review in this area have been given in Part I of this paper, and will not be repeated in this part. However, for convenience, a basic inequality needed in the proofs is cited in Section 2. Also, in Section 2, we shall establish some lemmas needed in the proofs of the main theorems. In Section 3, we shall establish the convergence rate for empirical spectral distributions of sample covariance matrices.

Note that the density function of the Marchenko–Pastur law [see (3.2)] is bounded when y, the ratio of the dimension to the degrees of freedom (or sample size), is different from 0 and 1. We may expect to have a similar result as that for Wigner matrices, that is, the order of $O(n^{1/4})$. We prove this result. However, when y is close to one, the density function is no longer bounded. The third term on the right-hand side of (2.12) of Theorem 2.2 in Part I is controlled only by $Av^{1/2}$. This shows that we can only get a rate of the order of $\sqrt{v}$, if we establish similar estimates for the integral of the difference of Stieltjes transforms of the empirical spectral distribution and the limiting spectral distribution. Moreover, its Stieltjes transform and the integral of the absolutely squared Stieltjes transform are not bounded. All these make it more difficult to establish an inequality at an ideal order when y is close to 1. In fact, the rate we actually established in this part is $O(n^{-5/48})$.

2. Preliminaries.

2.1. *A basic inequality from Part I.* We shall use Theorem 2.2 proved in Part I to prove our main results in this part. For reference, this theorem is now stated.

Received December 1990; revised January 1992.

AMS 1991 *subject classifications*. 60F15, 62F15.

Key words and phrases. Berry–Esseen inequality, convergence rate, large dimensional random matrix, Marchenko–Pastur distribution, sample covariance matrix, semicircular law, spectral analysis, Stieltjes transform, Wigner matrix.

THEOREM 2.2 OF PART I. *Let F be a distribution function and let G be a function of bounded variation satisfying $\int|F(x)-G(x)|\,dx<\infty$. Then we have*

$$\|F-G\|\le\frac{1}{\pi(1-\kappa)(2\gamma-1)}\left[\int_{-A}^{A}|f(z)-g(z)|\,du\right.$$

$$+2\pi v^{-1}\int_{|x|>B}|F(x)-G(x)|\,dx \tag{2.1}$$

$$\left.+v^{-1}\sup_x\int_{|y|\neq 2v\tau}|G(x+y)-G(x)|\,dy\right],$$

where γ, κ, τ, A and B are positive constants satisfying $A>B$,

$$\kappa=\frac{4B}{\pi(A-B)(2\gamma-1)}<1 \tag{2.2}$$

and

$$\gamma=\frac{1}{\pi}\int_{|u|<\tau}\frac{1}{u^2+1}\,du>\frac{1}{2}.$$

2.2. *Stieltjes transforms of Marchenko–Pastur distribution.* Assume that $0<y<1$. In a similar manner as we did for the semicircular law in Part I, we obtain

$$s_y(z)=-\frac{1}{4\pi iy}\oint_{|\zeta|=1}\frac{(\zeta^2-1)^2\,d\zeta}{\zeta(\zeta^2+2\alpha'\zeta+1)(\zeta^2+2\alpha\zeta+1)},$$

where $\alpha=(1+y)/2\sqrt{y}$ and $\alpha'=\alpha-z/2\sqrt{y}$. The function

$$(\zeta^2-1)^2/\zeta(\zeta^2+2\alpha'\zeta+1)(\zeta^2+2\alpha\zeta+1)$$

has five singular points:

$$\zeta_1=0,\qquad \zeta_2=-\sqrt{y},\qquad \zeta_3=-1/\sqrt{y}$$

and

$$\zeta_{4,5}=-\left((1+y-z)\pm\sqrt{(1+y-z)^2-4y}\right)/2\sqrt{y}.$$

It is obvious that $|\zeta_3|>1$. By the convention (3.1) for the square root of a complex number given in Section 3.1 of Part I, both the real and imaginary parts of $\sqrt{(1+y-z)^2-4y}$ have the same signs as those of $z-y-1$, respectively. Hence, the absolute value of ζ_5 is greater than that of ζ_4. Since $\zeta_4\zeta_5=1$, we have $|\zeta_5|>1$. By elementary calculus, one finds that the residues of $\zeta_{1,2,4}$ are 1, $-(1-y)/z$ and $-\sqrt{(1+y-z)^2-4y}/z$, respectively. Hence, by the residue theorem, we obtain

$$s_y(z)=-\frac{y+z-1-\sqrt{(1+y-z)^2-4y}}{2yz}. \tag{2.3}$$

Because Marchenko–Pastur distributions are weakly continuous in y, letting $y \uparrow 1$, we obtain

$$s_1(z) = -\frac{z - \sqrt{z^2 - 4z}}{2z}. \tag{2.4}$$

Similarly, one may prove that (2.3) is still true when $y > 1$.

Now, we shall find bounds for $s_y(z)$ for both cases when $0 < \theta \le y \le \Theta < 1$ and when $\Theta \le y \le 1$, respectively. Note that

$$\left(s_y(z) + \frac{1}{z}\right)\left(s_y^*(z) + \frac{1}{z}\right) = \frac{1}{yz^2}$$

and

$$s_y(z) s_y^*(z) = \frac{1}{yz},$$

where

$$s_y^*(z) = -\frac{2}{y + z - 1 - \sqrt{(1 + y - z)^2 - 4y}}.$$

Recalling the convention for the square root of a complex number given in Part I, we find that the real part of $\sqrt{(1 + y - z)^2 - 4y}$ has the same sign as $u - 1 - y$. We conclude that

$$\left|s_y(z) + \frac{1}{z}\right| \le \left|s_y^*(z) + \frac{1}{z}\right|$$

for all z, and

$$|s_y(z)| \le |s_y^*(z)|$$

for all z such that $u \ge 1 + y$ or $u < 1 - y$. Therefore, we have

$$|s_y(z)| < \frac{1 + \sqrt{y}}{|z|\sqrt{y}}, \quad \text{for all } z \tag{2.5}$$

and

$$|s_y(z)| < \frac{1}{\sqrt{y|z|}}, \quad \text{for } z \text{ with } |u - 1| \ge y. \tag{2.6}$$

For the case of $\theta \le y \le \Theta < 1$, if $|u| \le (1 - y)(1 + \sqrt{y})/(1 + 3\sqrt{y})$, then by the fact that

$$s_y(z) = -\frac{2}{y + z - 1 + \sqrt{(1 + y - z)^2 - 4y}}, \tag{2.7}$$

we obtain

$$|s_y(z)| < \frac{2}{1-y-|u|} \tag{2.8}$$
$$\leq \frac{1+3\sqrt{y}}{\sqrt{y}(1-y)}.$$

From (2.5) it is easy to see that (2.8) is still true if $|u| \geq (1-y)(1+\sqrt{y})/(1+3\sqrt{y})$.

For the case of $\Theta \leq y \leq 1$, if $(1-\sqrt{y})^2 \leq u \leq 1+y$, then $(1+y-u)^2 \leq 4y$. Hence, by (2.7) we obtain

$$|s_y(z)| < \frac{2}{\operatorname{Im}\left(\sqrt{(1+y-z)^2 - 4y}\right)}$$
$$= \frac{2\sqrt{2}}{\sqrt{\sqrt{(w^2-v^2-4y)^2 + 4v^2w^2} - w^2 + 4y + v^2}} \tag{2.9}$$
$$\leq \frac{2\sqrt{2}}{\sqrt[4]{(w^2-4y)^2 + 4v^2w^2}}$$
$$\leq \frac{2\sqrt{2}}{\sqrt[4]{8yv^2}} \leq \frac{2}{\sqrt{yv}},$$

where $w = 1+y-u$. If $u < (1-\sqrt{y})^2$ or $u > 1+y$, the estimate (2.6) is true and hence the inequality (2.9) is still true.

2.3. *Integrals of the square of the absolute value of Stieltjes transforms.* Applying Lemma 3.1 of Part I to Marchenko–Pastur (for $0 < y < 1$) laws, we obtain

$$\int |s_y(z)|^2 \, du \leq \frac{2\pi}{\sqrt{y}(1-y)}, \tag{2.10}$$

since the density function has an upper bound $1/(\pi\sqrt{y}(1-y))$.

It should be noted that this bound tends to infinity when y tends to 0 or 1. This is reasonable for the case that $y \to 0$ because the distribution tends to be degenerate. For the case $y \to 1$ or even $y = 1$, the distribution is still continuous, although the density for $y = 1$ is unbounded. One may want to have a finite bound (probably depending on v). We have the following inequality.

LEMMA 2.1. *For any $y \le 1$, we have*

$$\int |s_y(z)|^2\, du \le \frac{2\pi(1+\sqrt{y})}{y\sqrt{v}}. \tag{2.11}$$

PROOF. Using the notation and going through the same lines of the proof of Lemma 3.1 of Part I, we find that

$$\begin{aligned} I &= 4\pi v \int_a^b \int_a^b \left(\frac{1}{(u-x)^2 + 4v^2} \right) \phi(x)\phi(u)\, dx\, du \\ &= 8\pi v \int_a^b \int_x^b \left(\frac{1}{(u-x)^2 + 4v^2} \right) \phi(x)\phi(u)\, dx\, du \quad \text{(by symmetry)} \\ &\le 4vy^{-1}\sqrt{b} \int_0^\infty \int_a^b \phi(x) \left(\frac{1}{\sqrt{w}\,(w^2+4v^2)} \right) dw\, dx \\ &= 2y^{-1}\sqrt{\frac{2b}{v}} \int_0^\infty \left(\frac{1}{\sqrt{w}\,(w^2+1)} \right) dw \\ &= \frac{2\pi(1+\sqrt{y})}{\sqrt{v}\,y}. \end{aligned}$$

Here we used the fact that

$$\int_0^\infty \left(\frac{1}{\sqrt{w}\,(w^2+1)} \right) dw = \frac{\pi}{\sqrt{2}},$$

which may be computed by using the residue theorem and the equality

$$\int_0^\infty \left(\frac{1}{\sqrt{w}\,(w^2+1)} \right) dw = \frac{1}{1-i} \int_{-\infty}^\infty \left(\frac{1}{\sqrt{z}\,(z^2+1)} \right) dz.$$

This completes the proof of Lemma 2.1. □

LEMMA 2.2. *Let G be a function of bounded variation satisfying $\int |G(u)|\, du < \infty$. Let $g(z)$ denote its Stieltjes transform. When $z = u + iv$ with $v > 0$, we have*

$$\int |g(z)|^2\, du \le 2\pi v^{-1} V(G) \|G\|, \tag{2.12}$$

where $V(G)$ denotes the total variation of G and $\|G\| = \sup_x(|G(x)|)$.

PROOF. Going through the same lines of the proof of Lemma 3.1 of Part I, we may obtain

$$\begin{aligned} I &= 4\pi v \int\int \left(\frac{1}{(u-x)^2 + 4v^2} \right) dG(x)\, dG(u) \\ &= 8\pi v \int \left[\int \frac{(u-x)G(x)\,dx}{\left((u-x)^2 + 4v^2\right)^2} \right] dG(u) \quad \text{(integration by parts)} \\ &\le 2\pi v^{-1} V(G) \|G\|. \end{aligned}$$

□

2.4. *Lemmas concerning Lévy distance.*

LEMMA 2.3. *Let $L(F,G)$ be the Lévy distance between the distributions F and G. Then we have*

$$L^2(F,G) \le \int |F(x) - G(x)|\, dx. \tag{2.13}$$

PROOF. Without loss of generality, assume that $L(F,G) > 0$. For any $r \in (0, L(F,G))$, there exists an x such that

$$F(x-r) - r > G(x) \quad [\text{or } F(x+r) + r < G(x)].$$

Then the square between the points $(x - r, F(x - r) - r)$, $(x, F(x - r) - r)$, $(x - r, F(x - r))$ and $(x, F(x - r))$ [or $(x, F(x + r))$, $(x + r, F(x + r))$, $(x, F(x + r) + r)$ and $(x + r, F(x + r) + r)$ for the latter case] is located between F and G (see Figure 1a, b). Then (2.13) follows from the fact that the right-hand side of (2.13) equals the area of the region between F and G. The proof is complete. □

LEMMA 2.4. *If G satisfies $\sup_x |G(x + y) - G(x)| \le D|y|$ for all y, then one may prove that*

$$L(F,G) \le \|F - G\| \le (D+1)L(F,G), \quad \textit{for all } F. \tag{2.14}$$

PROOF. The inequality on the left-hand side is actually true for all distributions F and G. It can follow easily from the argument in the proof of Lemma 2.3. To prove the right-hand-side inequality, let us consider the case where for some x,

$$F(x) > G(x) + \rho,$$

where $\rho \in (0, \|F - G\|)$. Since G satisfies the Lipschitz condition, we have

$$G(x + \rho/(D+1)) + \rho/(D+1) \le G(x) + \rho < F(x),$$

which implies that

$$L(F,G) \ge \frac{\rho}{D+1}.$$

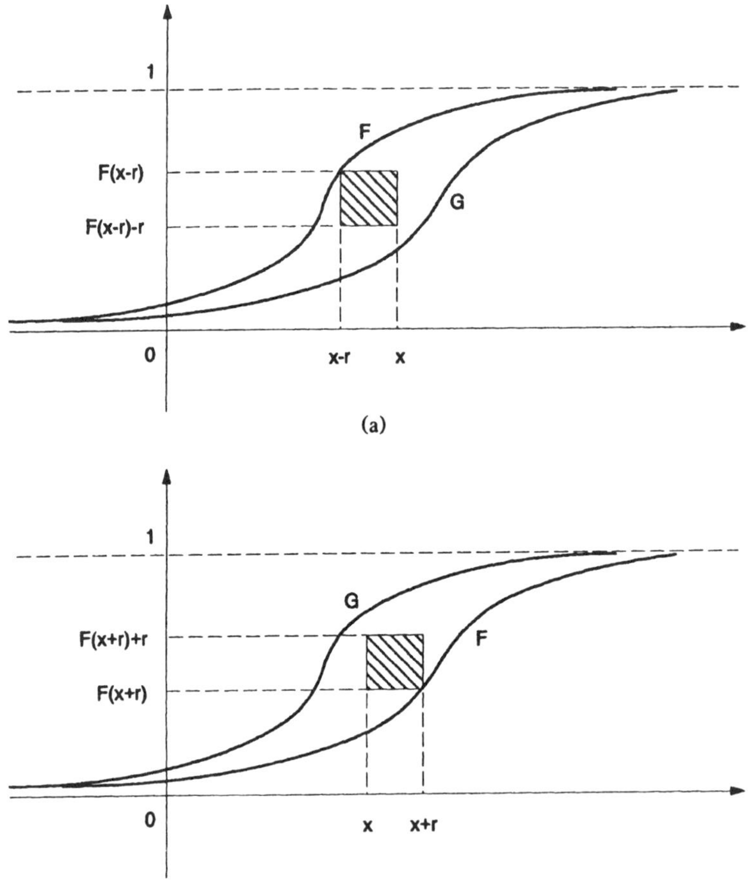

(a)

(b)

FIG. 1.

Then the right-hand-side inequality of (2.14) follows by making $\rho \to \|F - G\|$. The inequality for the other case, that is, $G(x) > F(x) + \rho$, can be similarly proved. □

LEMMA 2.5. *Let F_1, F_2 be distribution functions and let G satisfy $\sup_x |G(x+u) - G(x)| \le D|u|^\beta$, for all u and some $\beta \in (0,1]$. Then*

$$\|F_1 - G\|^{1+1/\beta} \le 2\|F_1 - G\|^{1/\beta}\|F_2 - G\| + 2(2D)^{1/\beta}\int |F_1(x) - F_2(x)|\,dx. \tag{2.15}$$

PROOF. Let $0 < \rho < \|F_1 - G\|$. Then, we may find an x_0 such that $F_1(x_0) - G(x_0) > \rho$ [or $F_1(x_0) - G(x_0) < -\rho$, alternatively]. By the condition on G, for any $x \in [x_0 - (\rho/2D)^{1/\beta}, x_0]$ (or $[x_0, x_0 + (\rho/2D)^{1/\beta}]$ for the other case), we have $|F_1(x) - G(x)| \geq (1/2)\rho$. But for any x in this interval, we have

$$\tfrac{1}{2}\rho \leq |F_1(x) - G(x)| \leq \|F_2 - G\| + |F_1(x) - F_2(x)|.$$

Integrating the above inequality over this interval and then making $\rho \to \|F_1 - G\|$, we obtain (2.15). The proof is complete. □

3. Convergence rates of spectral distributions of sample covariance matrices. Let $W_p = (w_{ij}(n)) = n^{-1}X_pX_p'$: $p \times p$, where $X_p = (x_{ij}(n)$, $i = 1,\dots,p$, $j = 1,\dots,n)$. Throughout this section, we shall drop the index n from the entries of X_p and those of W_p and assume that X_{ij}'s are independent and the following conditions hold:

$$\text{(3.1)} \qquad \begin{aligned} &\text{(i)} \quad Ex_{ij} = 0, \qquad Ex_{ij}^2 = 1, \quad \text{for all } i, j; \\ &\text{(ii)} \quad \sup_n \sup_{i,j} Ex_{ij}^4 I_{[|x_{ij}| \geq M]} \to 0, \quad \text{as } M \to \infty. \end{aligned}$$

The matrix W_p is known as a sample covariance matrix. It should be noted here that the notation W_p is no longer the same as used in the last section. Denote by F_p its empirical spectral distribution. Under the conditions in (3.1), it is well known that $F_p \to_w F_y$ in probability, where $y = \lim_{n\to\infty}(p/n) \in (0, \infty)$ and F_y is the limiting spectral distribution of F_p, known as the Marchenko–Pastur (1967) distribution, which may have a mass of $1 - y^{-1}$ at the origin if $y > 1$ and has a density

$$F_y'(x) = \frac{1}{2xy\pi}\sqrt{4y - (x - y - 1)^2}\, I_{[a,b]}(x), \tag{3.2}$$

with $a = a(y) = (1 - \sqrt{y})^2$, and $b = b(y) = (1 + \sqrt{y})^2$. If X_p is the $p \times n$ submatrix of the upper-left corner of an infinitely dimensional random matrix $[x_{ij},\ i, j = 1, 2, \dots]$, then the above convergence is true a.s. (almost surely) [e.g., Wachter (1978) who actually proved the a.s. version of the convergence under the uniform boundedness of the $(2 + \varepsilon)$th moments of the entries of X_p].

To consider the rate of the convergence, we shall establish the following theorem.

THEOREM 3.1. *Under the assumptions in* (3.1), *we have*

$$\|EF_p - F_{y_p}\| = O(n^{-1/4}), \tag{3.3}$$

for $y_p = p/n \in (\theta, \Theta)$, *where* $0 < \theta < \Theta < 1$ *or* $1 < \theta < \Theta < \infty$.

THEOREM 3.2. *If* $0 < \theta \leq y \leq \Theta < \infty$, *then under the assumptions in* (3.1) *we have*

$$\|EF_p - F_{y_p}\| = O(n^{-5/48}), \tag{3.4}$$

where $\theta < 1 < \Theta$.

REMARK 3.1. Because the convergence rate of $|y_p - y|$ can be arbitrarily slow, it is impossible to establish any rate for the convergence of $\|EF_p - F_y\|$ if we know nothing about the convergence rate of $|y_p - y|$. Conversely, if we know the convergence rate of $|y_p - y|$, then from (3.3) or (3.4), we can easily derive a convergence rate for $\|EF_p - F_y\|$. This is the reason why F_{y_p}, instead of the limit distribution F_y, is used in Theorem 3.1.

For simplicity, we shall drop the index p from y_p. We need only to prove the theorems for the cases where $\theta \le y \le \Theta < 1$ and $\Theta \le y \le 1$. For the case of $y > 1$, since the first n largest eigenvalues of the matrix $X_p X_p'$ are identical with those of $X_p' X_p$, the theorem then follows by considering the analogous result for the matrix $(1/p)X_p' X_p$. For simplicity, these two cases are referred to as $y \le \Theta$ and $y \le 1$, respectively.

To begin with, we shall establish two propositions applicable to both cases.

PROPOSITION 3.3. *Under condition* (3.1) *and the following additional assumption*:

$$|x_{ij}| < \sqrt{n}\,\eta, \quad \text{with } \eta \to 0, \tag{3.5}$$

we have

$$\int_B^\infty |EF_p(x) - F_y(x)|\,dx = o(n^{-t}), \tag{3.6}$$

where $B = b + 1$, $b = b(y)$ *is defined below* (3.2) *and the constant* $t > 0$ *is fixed but can be arbitrarily large.*

REMARK 3.2. For any rate of $\eta = \eta_p \to 0$, Proposition 3.3 is true. In application of this proposition, the choice of η in (3.5) involves a truncation and normalization procedure described at the end of this section [see (3.62)].

PROOF. In Yin, Bai and Krishnaiah (1988), after truncation and normalization, it is actually proved (their arguments still work even without the assumption of identical distributions of the entries of X_p) that,

$$E(\lambda_p)^m \le (b + \zeta)^m, \tag{3.7}$$

for some sequence of constants $\zeta = \zeta_p \to 0$ and some sequence of integers $m = m_p$ such that $\log n/m \to 0$, under condition (i) of (3.1), (3.5) and

$$E|x_{ij}|^k \le dn^{(k-3)/2}, \quad \text{for all } k \ge 3, \tag{3.8}$$

where λ_p denotes the largest eigenvalue of $n^{-1}X_p X_p'$ and b is defined in (3.2). Note that (3.8) is implied by (3.5) and (ii) of (3.1).

Note that

$$1 - F_p(x) \le I_{[\lambda_p \ge x]}, \quad \text{for } x \ge 0. \tag{3.9}$$

We have

$$\int_B^\infty |EF_p(x) - F(x)|\,dx \le \int_B^\infty P(\lambda_p \ge x)\,dx$$

$$(3.10)\qquad \le \int_B^\infty \left(\frac{b+\zeta}{x}\right)^m dx$$

$$= O\left(\left(\frac{b+\zeta}{B}\right)^{m-1}\right) = o(n^{-t}),$$

for any fixed $t > 0$. Proposition 3.3 is proved. □

Theorem 3.1 can be proved via Theorem 2.2. From the above results and the fact that $F_p(x) = F(x) = 0$ for all $x < 0$, we need only estimate $s_p(z) - s_y(z)$ for $z = u + iv$, $v > 0$, $|u| < A$, where A is a constant chosen according to (2.2).

From (2.3), the Stieltjes transform of the limiting spectral distribution F_y is given by

$$(3.11)\qquad s_y(z) = -\frac{1}{2yz}\left\{z + y - 1 - \sqrt{(z+y-1)^2 - 4yz}\right\}.$$

Set

$$(3.12)\qquad s_p(z) = \frac{1}{p} E\,\mathrm{tr}(W_p - zI_p)^{-1}.$$

PROPOSITION 3.4. *Choose $v = (10C_0(A+1)/n)^{1/6}$, where C_0 is a constant which will be specified in* (3.32). *Then, if* (3.1) *and* (3.5) *hold, we have that*

$$(3.13)\qquad \int_{-A}^{A} |s_p(z) - s_y(z)|\,du \le Cv,$$

where C is a positive constant.

PROOF. By the inverse matrix formula [see (3.8) in Part I], we have

$$s_p(z) = \frac{1}{p}\sum_{k=1}^{p} E\frac{1}{w_{kk} - z - \alpha'(k)(W_p(k) - zI_{n-1})^{-1}\alpha(k)}$$

$$(3.14)\qquad = \frac{1}{p}\sum_{k=1}^{p} E\frac{1}{\varepsilon_k + 1 - y - z - yzs_p(z)}$$

$$= -\frac{1}{z + y - 1 + yzs_p(z)} + \delta,$$

where $W_p(k)$ is the matrix obtained from W_p by deleting the kth row and kth column, $\alpha(k)$ denotes the vector obtained from the kth column of W_p by removing the kth element, and

$$\varepsilon_k = \frac{1}{n}\sum_{j=1}^{n}\left(x_{kj}^2 - 1\right) + y + yzs_p(z) - \alpha'(k)\left(W_p(k) - zI_{p-1}\right)^{-1}\alpha(k),$$

$$\delta = \delta_p = -\frac{1}{p}\sum_{k=1}^{p} E\frac{\varepsilon_k}{\left(z + y - 1 + yzs_p(z)\right)\left(z + y - 1 + yzs_p(z) - \varepsilon_k\right)}$$

$$(3.15)\qquad = -\frac{1}{p}\sum_{k=1}^{p} E\frac{\varepsilon_k}{\left(z + y - 1 + yzs_p(z)\right)^2}$$

$$-\frac{1}{p}\sum_{k=1}^{p} E\frac{\varepsilon_k^2}{\left(z + y - 1 + yzs_p(z)\right)^2\left(z + y - 1 + yzs_p(z) - \varepsilon_k\right)}.$$

Solving $s_p(z)$ from (3.14), we get two roots

$$(3.16)\qquad s_{(1),(2)}(z) = -\frac{1}{2yz}\left(z + y - 1 - yz\delta \pm \sqrt{(z + y - 1 + yz\delta)^2 - 4yz}\right).$$

Comparing (3.16) with (3.11), it seems that the Stieltjes transform $s_p(z)$ should be the solution $s_{(2)}(z)$, for all values of z with $v > 0$, that is,

$$(3.17)\qquad s_p(z) = -\frac{1}{2yz}\left(z + y - 1 - yz\delta - \sqrt{(z + y - 1 + yz\delta)^2 - 4yz}\right).$$

Now, we prove (3.17). First, we note that

$$(3.18)\qquad \begin{aligned} \operatorname{Im}\left(z + y - 1 + yzs_p(z)\right) &= \operatorname{Im}\left(z - 1 + y\int_a^b \frac{x\,dF_p(x)}{x - z}\right) \\ &= v\left(1 + y\int_a^b \frac{x\,dF_p(x)}{(x - u)^2 + v^2}\right) > v. \end{aligned}$$

It follows immediately from (3.18) that

$$(3.19)\qquad \frac{1}{\left|z + y - 1 + yzs_p(z)\right|} \le v^{-1}.$$

It is obvious that $|s_p(z)| \le v^{-1}$. Therefore,

$$(3.20)\qquad |\delta| \le 2/v.$$

For any fixed u, when $v \to \infty$, we have $s_p(z) \to 0$, $s_{(2)}(z) \to 0$ and $s_{(1)}(z) \to -1/y$. This shows that (3.17) is true for all large v.

As in Part I, one may easily see that both $s_{(1)}(z)$ and $s_{(2)}(z)$ are continuous functions on the upper-half plane. Thus, to prove that (3.17) is true for all z on the upper-half plane, it is sufficient to show that $s_{(1)}(z) \ne s_{(2)}(z)$ for all z on the upper-half plane. Otherwise, there would be some z on the upper-half plane such that $s_p(z) = s_{(1)}(z) = s_{(2)}(z)$. Then the square root term in (3.16) would be zero. Hence, we have

$$s_p(z) = -\frac{y + z - 1 - yz\delta}{2yz}.$$

Substituting the expression derived for δ from (3.14) into the above expression, we obtain

$$s_p(z) = \frac{1 - y - z}{yz} + \frac{1}{y + z - 1 + yzs_p(z)}.$$

However, this is impossible, since the imaginary parts of the two terms are obviously negative [for the second term, see (3.18)]. This contradiction proves that (3.17) is true for all z on the upper-half plane.

Comparing (3.17) with (3.11), we need to show that both δ and the integral of the absolute value of δ with respect to u on a finite interval are "small." Then, we begin to find a bound for $|s_p(z) - s_y(z)|$ in terms of δ.

We now proceed to estimate $|\delta|$. First, we note that, by (3.11) of Part I,

$$\frac{1}{\left|z + y - 1 + yzs_p(z) - \varepsilon_k\right|} \le v^{-1}. \tag{3.21}$$

Then, by (3.15) and (3.21), we have

$$|\delta| \le \frac{1}{p} \sum_{k=1}^{p} \left(|E\varepsilon_k| + v^{-1}E|\varepsilon_k^2|\right)\left(\left|z + y - 1 + yzs_p(z)\right|\right)^{-2}. \tag{3.22}$$

Denote by $X_p(k)$ the $(p-1) \times n$ matrix obtained from X_p by eliminating the kth row, and denote by $x(k)$ the n vector of the kth row of X_p. Then, $\alpha(k) = (1/n)X_p(k)x(k)$ and $W_p(k) = (1/n)X_p(k)X_p'(k)$. Recalling the definition of ε_k, one finds that

$$\begin{aligned} &E^{(k)}\alpha'(k)\left(W_p(k) - zI_{p-1}\right)^{-1}\alpha(k) \\ &= n^{-2}E^{(k)}x'(k)X_p'(k)\left(W_p(k) - zI_{p-1}\right)^{-1}X_p(k)x(k) \\ &= n^{-2}\operatorname{tr} X_p'(k)\left(W_p(k) - zI_{p-1}\right)^{-1}X_p(k) \\ &= n^{-1}\operatorname{tr}\left(W_p(k) - zI_{p-1}\right)^{-1}W_p(k) \\ &= y - \frac{1}{n} + zn^{-1}\operatorname{tr}\left(W_p(k) - zI_{p-1}\right)^{-1}, \end{aligned} \tag{3.23}$$

where $E^{(k)}$ denotes the conditional expectation given $\{x_{ij},\ i \ne k\}$. Then by (3.10) of Part I, for all z with $|u| \le A$ we have

$$\begin{aligned} |E\varepsilon_k| &= \frac{|z|}{n}\left|E\left[\operatorname{tr}\left(W_p(k) - zI_{p-1}\right)^{-1} - \operatorname{tr}\left(W_p - zI_p\right)^{-1}\right]\right| + \frac{1}{n} \\ &\le \frac{C}{nv}, \end{aligned} \tag{3.24}$$

where the constant C may take the value $A + 1$.

Next, we proceed to estimate $E|\varepsilon_k|^2$. We have

$$E|\varepsilon_k^2| \le \frac{M}{n} + R_1 + R_2 + |E(\varepsilon_k)|^2, \tag{3.25}$$

where

$$R_1 = E\left|\alpha'(k)\left(W_p(k) - zI_{p-1}\right)^{-1}\alpha(k) - E^{(k)}\left(\alpha'(k)\left(W_p(k) - zI_{p-1}\right)^{-1}\alpha(k)\right)\right|^2$$

$$R_2 = \frac{|z|^2}{n}E\left|\operatorname{tr}\left(W_p(k) - zI_{p-1}\right)^{-1} - E\operatorname{tr}\left(W_p(k) - zI_{p-1}\right)^{-1}\right|^2$$

and M is the constant in (3.1).

Let

$$\Gamma_k = \left(r_{ij}(k)\right) = \frac{1}{n}X_p'(k)\left(W_p(k) - zI_{p-1}\right)^{-1}X_p(k).$$

Then, we have

$$\begin{aligned}
R_1 &\le \frac{2M}{n^2}\sum_{i,j}E\left|r_{ij}^2(k)\right| = \frac{2M}{n^2}E\operatorname{tr}\left(\Gamma_k\bar{\Gamma}_k\right)\\
&= \frac{2M}{n^2}E\operatorname{tr}\Big[\left(W_p(k) - zI_{p-1}\right)^{-1}W_p(k)\\
&\qquad\times\left(W_p(k) - \bar{z}I_{p-1}\right)^{-1}W_p(k)\Big]\\
&= \frac{2M}{n^2}E\operatorname{tr}W_p^2(k)\left(\left(W_p(k) - uI_{p-1}\right)^2 + v^2I_{p-1}\right)^{-1}\\
&\le \frac{4M}{n^2}E\operatorname{tr}\left[\left(W_p(k) - uI_{p-1}\right)^2 + u^2I_{p-1}\right]\\
&\qquad\times\left(\left(W_p(k) - uI_{p-1}\right)^2 + v^2I_{p-1}\right)^{-1}\\
&\le \frac{4M}{n^2}\Big[p - 1 + |u|^2E\operatorname{tr}\Big(\left(W_p(k) - uI_{p-1}\right)^2 \qquad (3.26)\\
&\qquad\qquad + v^2I_{p-1}\Big)^{-1}\Big]\\
&\le 4Mn^{-1} + 4MA^2n^{-1}v^{-2} \le Cn^{-1}v^{-2}. \qquad (3.27)
\end{aligned}$$

Here, the constant C can be taken as $4M(A^2 + 1)$.

Define $\gamma_d(d) = 0$, and define for $d \ne k$,

$$\begin{aligned}
\gamma_d(k) &= E_{d-1}\operatorname{tr}\left(W_p(k) - zI_{p-1}\right)^{-1} - E_d\operatorname{tr}\left(W_p(k) - zI_{p-1}\right)^{-1}\\
&= E_{d-1}\sigma_d(k) - E_d\sigma_d(k), \qquad d = 1, 2, \ldots, p,
\end{aligned} \tag{3.28}$$

where

$$\sigma_d(k) = \operatorname{tr}\left(W_p(k) - zI_{p-1}\right)^{-1} - \operatorname{tr}\left(W(d,k) - zI_{p-2}\right)^{-1},$$

$W(d, k)$ is the matrix obtained from W_p by deleting the dth and kth rows and the dth and kth columns, $\alpha(d, k)$ is the vector obtained from the dth column vector of W_p by deleting the dth and kth elements and E_d denotes the conditional expectation given $\{x_{ij}, d+1 \le i \le p, 1 \le j \le n\}$.

Again, by (3.11) of Part I, we have

$$|\sigma_d(k)| \le v^{-1}. \tag{3.29}$$

Therefore, we obtain

$$R_2 \le \frac{(A+1)^2}{n^2} \sum_{d=1}^{p} E|\gamma_d^2(k)| \le Cn^{-1}v^{-2}. \tag{3.30}$$

Then by the definition of ε_k and (3.24)–(3.30), we obtain that for some positive constant C,

$$E|\varepsilon_k|^2 \le \frac{C}{nv^2}. \tag{3.31}$$

Throughout the paper, the letter C denotes a generic positive constant which may take different values at different places. From (3.19), (3.22), (3.24) and (3.31), it follows that for some positive constant C_0,

$$|\delta| \le \frac{C_0}{nv^5}. \tag{3.32}$$

Choose $v = (10C_0(A+1)/n)^{1/6}$. By (3.31), we know that

$$|\delta| \le \frac{v}{10(A+1)^2}. \tag{3.33}$$

By (3.22) and (3.33), for large n, we have

$$\begin{aligned}
\int_{-A}^{A} |\delta|\, du &\le \frac{C}{nv^3} \int_{-A}^{A} |z + y - 1 + yzs_n(z)|^{-2}\, du \\
&\le \frac{C}{nv^3} \left[\int_{A}^{A} |s_n(z)|^2\, du + \int_{-A}^{A} |\delta|^2\, du \right] \\
&\le \frac{C}{nv^3} \left[\int_{A}^{A} |s_n(z)|^2\, du + v \int_{-A}^{A} |\delta|\, du \right] \\
&\le \frac{C}{nv^3} \int_{-A}^{A} |s_n(z)|^2\, du \\
&\le \frac{C}{nv^3} \int_{-A}^{A} \left[\int_{-\infty}^{\infty} \frac{1}{(x-u)^2 + v^2}\, du \right] dF_n(x) \\
&\le \frac{C}{nv^4} \le Cv^2.
\end{aligned} \tag{3.34}$$

Here, in the derivation of the fourth inequality, we have used the fact that

$$a \le c + ba \Rightarrow c + ba \le c/(1-b) \tag{3.35}$$

for any positive a, c and $b < 1$.

Now, we are in position to estimate $|s_p(z) - s_y(z)|$. By (3.11) and (3.17), we have

$$|s_p(z) - s_y(z)| \le \left|\frac{\delta}{2}\right|\left[1 + \frac{|2(z+y-1) - yz\delta|}{\left|\sqrt{(z+y-1)^2 - 4yz} + \sqrt{(z+y-1+yz\delta)^2 - 4yz}\right|}\right]. \tag{3.36}$$

As done in Part I, we need to find the condition guaranteeing that the real parts of $\sqrt{(z+y-1)^2 - 4yz}$ and $\sqrt{(z+y-1+yz\delta)^2 - 4yz}$ have the same sign. We claim this is true for $|u - y - 1| > y/[2(A+1)]$. In fact, if $|u - y - 1| > y/[2(A+1)]$, then (3.33) implies that

$$v + \mathrm{Im}(yz\delta) > 0,$$
$$|u - y - 1| > |\mathrm{Re}(yz\delta)|$$

and

$$\begin{aligned}&|(u - y - 1 - \mathrm{Re}(yz\delta))(v + \mathrm{Im}(yz\delta))| - 2y^2|\mathrm{Im}(z\delta)| \\ &\quad > \left(\frac{y}{2(A+1)} - \frac{vy}{10(A+1)}\right)\left(v - \frac{v}{10(A+1)}\right) - \frac{vy^2}{10(A+1)} > 0.\end{aligned}$$

From the above estimates it follows that the sign of the real part of $\sqrt{(z+y-1+yz\delta)^2 - 4yz}$ is

$$\begin{aligned}&\mathrm{sign}[2((u+y-1+\mathrm{Re}(yz\delta))(v+\mathrm{Im}(yz\delta)) - 4yv] \\ &\quad = \mathrm{sign}((u-y-1-\mathrm{Re}(yz\delta))(v+\mathrm{Im}(yz\delta)) + 2y\,\mathrm{Im}(yz\delta)) \\ &\quad = \mathrm{sign}(u-y-1).\end{aligned} \tag{3.37}$$

Since the sign of the real part of $\sqrt{(z+y-1)^2 - 4yz}$ is

$$\mathrm{sign}(2v(u-y-1)) = \mathrm{sign}(u-y-1),$$

for $|u - y - 1| > y/[2(A+1)]$, by (3.37), the real parts of both $\sqrt{(z+y-1)^2 - 4yz}$ and $\sqrt{(z+y-1+yz\delta)^2 - 4yz}$ have a common sign. Hence for large p, (3.36) implies that

$$|s_p(z) - s_y(z)| \le \frac{1}{2}|\delta|\left[1 + \frac{2A+2}{\sqrt{|(u-y-1)^2 - v^2 - 4y|}}\right]. \tag{3.38}$$

If $|u - y - 1| \le y/[2(A+1)]$, then for all large p, we have

$$\begin{aligned}\left|\sqrt{(z-y-1)^2 - 4y} - 2i\sqrt{y}\right| &= \frac{|z-y-1|^2}{\left|\sqrt{(z-y-1)^2 - 4y} + 2i\sqrt{y}\right|} \\ &\le \frac{1}{2\sqrt{y}}\left[\frac{y^2}{4(A+1)^2} + v^2\right] \\ &\le \frac{1}{2}y\end{aligned}$$

and

$$
\begin{aligned}
&\left|\sqrt{(z+y-1+yz\delta)^2-4yz}\ -2i\sqrt{y}\right| \\
&\quad = \frac{\left|(z-y-1+yz\delta)^2+4y^2z\delta\right|}{\left|\sqrt{(z+y-1+yz\delta)^2-4yz}\ +2i\sqrt{y}\right|} \\
&\quad \le \frac{1}{\sqrt{y}}\left[(u-y-1)^2+|v+yz\delta|^2+4y^2|z\delta|\right] \\
&\quad \le \frac{1}{\sqrt{y}}\left[\frac{y^2}{4(A+1)^2}+v^2\left(1+\frac{1}{10(A+1)}\right)^2+\frac{4y^2v}{10(A+1)}\right] \\
&\quad \le \frac{1}{2}y.
\end{aligned}
$$

Therefore, for $|u-y-1| \le y/[2(A+1)]$, we have

$$\left|\sqrt{(z+y-1)^2-4yz}+\sqrt{(z+y-1+yz\delta)^2-4yz}\right| \ge 4\sqrt{y}-y > 2\sqrt{y}.$$

Combining the above and (3.38) together, for $|u| \le A$, we have

$$
(3.39)\qquad |s_n(z)-s_y(z)| \le \begin{cases} \dfrac{1}{2}|\delta|\left[1+\dfrac{2A+2}{\sqrt{|(u+y-1)^2-v^2-4yu|}}\right], & \\ \qquad\qquad \text{if } |u-y-1| > \dfrac{y}{2(A+1)}, & \\ C_1|\delta|, \qquad \text{if } |u-y-1| \le \dfrac{y}{2(A+1)}, & \end{cases}
$$

where $C_1 = C_1(y)$ is a positive constant depending upon y; for example, here we may take $C = (A+2)/2\sqrt{y}$.

By (3.34) and (3.39), one finally gets

$$
\begin{aligned}
&\int_{-A}^{A}|s_n(z)-s_y(z)|\,du \\
&\quad = \left\{\int_{[|u-y-1|>y/[2(A+1)],\,|u|\le A]} \right. \\
(3.40)\qquad &\qquad \left. +\int_{[|u-y-1|\le y/[2(A+1)],\,|u|\le A]}\right\}|s_n(z)-s_y(z)|\,du \\
&\quad \le Cv\int_{-A}^{A}\frac{1}{\sqrt{|(u+y-1)^2-v^2-4yu|}}\,du + C\int_{-A}^{A}|\delta|\,du \\
&\quad \le \eta v + Cv^2,
\end{aligned}
$$

where

$$\eta = C \sup_{v<1} \int_{-A}^{A} \frac{du}{\sqrt{\left|(u+y-1)^2 - v^2 - 4yu\right|}}.$$

The proof of Proposition 3.4 is complete. □

PROOF OF THEOREM 3.1. Since the density function of the Marchenko–Pastur distribution is bounded when $\theta \le y \le \Theta < 1$, applying Theorem 2.2 and Proposition 3.4, we obtain a preliminary bound

$$\|EF_p - F_y\| = O(n^{-1/6}), \tag{3.41}$$

under the additional restriction of (3.5). Next, we shall improve the result as we did in Part I.

Assume that $\|EF_p - F_y\| \le \Delta_1 = \eta n^{-1/6}$, for some $\eta > 1$. We shall refine the estimates of $\Sigma_{ij} E|r_{ij}^2(k)|$ and $\Sigma_d E|\gamma_d^2(k)|$. Assume that $\Delta_1 \ge v \ge n^{-1/4}$. The exact value of v will be chosen later. Noticing that $y \le \Theta$, by (3.26), we get

$$\begin{aligned}
R_1 &\le \frac{4M(p-1)}{n^2}\left[1 + E\int_0^\infty \frac{(u^2 - v^2)\, dF_{p-1}^{(k)}(x)}{(x-u)^2 + v^2}\right] \\
&\le \frac{4MA^2}{n^2}\left[ny + E\int_0^\infty \frac{(p-1)\, dF_{p-1}^{(k)}(x)}{(x-u)^2 + v^2}\right] \\
&= \frac{4MA^2}{n^2}\left[ny + \int_0^\infty \frac{p\, dF_y(x)}{(x-u)^2 + v^2} + E\int_0^\infty \frac{p\, d\big(F_p(x) - F_y(x)\big)}{(x-u)^2 + v^2}\right. \\
&\qquad \left. + E\int_0^\infty \frac{d\big((p-1)F_{p-1}^{(k)}(x) - pF_p(x)\big)}{(x-u)^2 + v^2}\right] \qquad (3.42) \\
&\le \frac{4MA^2}{n^2}\left[ny + \frac{n\sqrt{y}}{v(1-y)} + E\int_0^\infty \frac{2p(x-u)\big(F_p(x) - F_y(x)\big)\, dx}{\big((x-u)^2 + v^2\big)^2}\right. \\
&\qquad \left. + E\int_0^\infty \frac{2(x-u)\big((p-1)F_{p-1}^{(k)}(x) - pF_p(x)\big)\, dx}{\big((x-u)^2 + v^2\big)^2}\right] \\
&\le \frac{4MA^2}{n^2}\left[ny + \frac{n\sqrt{y}}{v(1-y)} + 2ny\,\Delta_1 v^{-2} + 2v^{-2}\right] \\
&\le Cn^{-1}\Delta_1 v^{-2}. \qquad (3.43)
\end{aligned}$$

Here, in the derivation of the third inequality, the first integral in (3.42) was estimated by the upper bound $1/(\pi\sqrt{y}(1-y))$ of the density of F_y, the second by $\|F_p - F_y\| \le \Delta_1$ and the third by the fact that $|(p-1)F_{p-1}^{(k)}(x) - pF_p(x)| \le 1$, established in Lemma 3.3 in Part I.

Now, we estimate $E|\gamma_d(k)|^2$. Rewrite $\sigma_d(k)$ as

$$\begin{aligned}\sigma_d(k) &= \frac{1+\alpha'(d,k)\big(W(d,k)-zI_{p-2}\big)^{-2}\alpha(d,k)}{w_{dd}-z-\alpha'(d,k)\big(W(d,k)-zI_{p-2}\big)^{-1}\alpha(d,k)}\\
&= \frac{1+(1/n)\operatorname{tr}\Gamma^{(2)}}{1-z-(1/n)\operatorname{tr}\Gamma^{(1)}} + \frac{(1/n)x_d'\Gamma^{(2)}x_d-(1/n)\operatorname{tr}\Gamma^{(2)}}{1-z-(1/n)\operatorname{tr}\Gamma^{(1)}}\\
&\quad+\frac{\big(1+(1/n)x_d'\Gamma^{(2)}x_d\big)\big((1/n)x_d'\Gamma^{(1)}x_d-(1/n)\operatorname{tr}\Gamma^{(1)}-w_{dd}+1\big)}{\big(1-z-(1/n)\operatorname{tr}\Gamma^{(1)}\big)\big(w_{dd}-z-(1/n)x_d'\Gamma^{(1)}x_d\big)}\\
&:= \sigma_d'(k)+\sigma_d''(k)+\sigma_d'''(k),\end{aligned}$$

where

$$\Gamma^{(1)} = \frac{1}{n}X'(d,k)\big(W(d,k)-zI_{p-2}\big)^{-1}X(d,k),$$

$$\Gamma^{(2)} = \frac{1}{n}X'(d,k)\big(W(d,k)-zI_{p-2}\big)^{-2}X(d,k),$$

$X(d,k)$ is the $(p-2)\times n$ matrix obtained from X_p by eliminating its dth and kth rows, and x_d is the n vector of the dth row of X_p. It is easy to see that

$$E_{d-1}\sigma_d'(k) - E_d\sigma_d'(k) = 0. \tag{3.44}$$

Similarly to the proof of (3.18), we may prove that

$$\left|\operatorname{Im}\left(1-z-\frac{1}{n}\operatorname{tr}\Gamma^{(1)}\right)\right| > v.$$

We may also derive that

$$\begin{aligned}\operatorname{tr}\Gamma^{(1)}\bar\Gamma^{(1)} &= \operatorname{tr}W^2(d,k)\Big(\big(W(d,k)-uI_{p-2}\big)^2+v^2I_{p-2}\Big)^{-1}\\
&\le 2\operatorname{tr}\Big(\big(W(d,k)-uI_{p-2}\big)^2+u^2I_{p-2}\Big)\\
&\qquad\times\Big(\big(W(d,k)-uI_{p-2}\big)^2+v^2I_{p-2}\Big)^{-1}\\
&\le 2\operatorname{tr}\Big[I_{p-2}+A^2\Big(\big(W(d,k)-uI_{p-2}\big)^2+v^2I_{p-2}\Big)^{-1}\Big]\\
&\le C(p-2)v^{-2} < Cnv^{-2}\end{aligned}$$

and

$$\begin{aligned}\operatorname{tr}\Gamma^{(2)}\bar\Gamma^{(2)} &= \operatorname{tr}W^2(d,k)\Big(\big(W(d,k)-uI_{p-2}\big)^2+v^2I_{p-2}\Big)^{-2}\\
&\le v^{-2}\operatorname{tr}W^2(d,k)\Big(\big(W(d,k)-uI_{p-2}\big)^2+v^2I_{p-2}\Big)^{-1}\\
&\le Cnv^{-4}.\end{aligned}$$

Thus, we have

$$E\left|E_{d-1}\sigma_d''(k)-E_d\sigma_d''(k)\right|^2 \le E\left|1-z-\frac{1}{n}\operatorname{tr}\Gamma^{(1)}\right|^{-2}\left|\frac{1}{n}\left(x_d'\Gamma^{(2)}x_d-\operatorname{tr}\Gamma^{(2)}\right)\right|^2$$

$$\le Cn^{-2}E\left|1-z-\frac{1}{n}\operatorname{tr}\Gamma^{(1)}\right|^{-2}\operatorname{tr}\Gamma^{(2)}\overline{\Gamma}^{(2)} \tag{3.45}$$

$$\le Cn^{-1}v^{-4}E\left|1-z-\frac{1}{n}\operatorname{tr}\Gamma^{(1)}\right|^{-2}$$

and by (3.29),

$$E\left|E_{d-1}\sigma_d'''(k)-E_d\sigma_d'''(k)\right|^2 \le v^{-2}E\left|1-z-\frac{1}{n}\operatorname{tr}\Gamma^{(1)}\right|^{-2}\left|\frac{1}{n}x_d'\Gamma^{(1)}x_d-\operatorname{tr}\Gamma^{(1)}\right|^2$$

$$\le n^{-2}v^{-2}E\left|1-z-\frac{1}{n}\operatorname{tr}\Gamma^{(1)}\right|^{-2}\operatorname{tr}\Gamma^{(1)}\overline{\Gamma}^{(1)} \tag{3.46}$$

$$\le Cn^{-1}v^{-4}E\left|1-z-\frac{1}{n}\operatorname{tr}\Gamma^{(1)}\right|^{-2}.$$

Summing up (3.44), (3.45) and (3.46), we obtain

$$E\left|\gamma_d(k)\right|^2 \le Cn^{-1}v^{-4}E\left|1-z-\frac{1}{n}\operatorname{tr}\Gamma^{(1)}\right|^{-2}. \tag{3.47}$$

By (3.10) of Part I, we have

$$\left|\operatorname{tr}\left(W_p(k)-zI_{p-1}\right)^{-1}-\operatorname{tr}\left(W_p-zI_p\right)^{-1}\right|<v^{-1} \tag{3.48}$$

and

$$\left|\operatorname{tr}\left(W_p(k)-zI_{p-1}\right)^{-1}-\operatorname{tr}\left(W(d,k)-zI_{p-2}\right)^{-1}\right|<v^{-1}. \tag{3.49}$$

Thus, by (3.48) and (3.49), we have

$$\begin{aligned}\operatorname{tr}(\Gamma^{(1)}) &= p-2+z\operatorname{tr}\left(\left(W(d,k)-zI_{p-2}\right)^{-1}\right.\\ &= ny+z\operatorname{tr}\left(\left(W_p-zI_p\right)^{-1}\right)+zR(d,k),\end{aligned} \tag{3.50}$$

with $|R(d,k)|<4/v$. Consequently, we obtain that

$$E\left|\gamma_d(k)\right|^2 \le Cn^{-1}v^{-4}E\left|1-z-y-zn^{-1}\operatorname{tr}\left(W_p-zI_p\right)^{-1}+zn^{-1}R(d,k)\right|^{-2}$$

$$= Cn^{-1}v^{-4}E\left|1-z-y-zn^{-1}\operatorname{tr}\left(W_p-zI_p\right)^{-1}\right|^{-2}$$

$$\times\left|1+\frac{zn^{-1}R(d,k)}{1-z-(1/n)\operatorname{tr}\Gamma^{(1)}}\right|^2 \tag{3.51}$$

$$\le Cn^{-1}v^{-4}E\left|1-z-y-zn^{-1}\operatorname{tr}\left(W_p-zI_p\right)^{-1}\right|^{-2}\left|1+\frac{4(A+1)}{nv^2}\right|^2$$

$$\le Cn^{-1}v^{-4}E\left|1-z-y-zn^{-1}\operatorname{tr}\left(W_p-zI_p\right)^{-1}\right|^{-2},$$

for some positive constant C. Therefore, by (3.30) we obtain

$$R_2 \le Cn^{-2}v^{-4}E\left|1 - z - y - zn^{-1}\operatorname{tr}(W_p - zI_p)^{-1}\right|^{-2}. \tag{3.52}$$

Similarly to the proof of (3.30), we may prove that

$$E\left|\frac{1}{n}\operatorname{tr}(W_p - zI_p)^{-1} - ys_p(z)\right|^2 \le \frac{1}{n^2}\sum_{k=1}^{p} E|\gamma_k|^2 \le \frac{1}{nv^2}, \tag{3.53}$$

where

$$\begin{aligned}\gamma_k &= E_{k-1}\operatorname{tr}(W_p - zI_p)^{-1} - E_k\operatorname{tr}(W_p - zI_p)^{-1}\\ &= E_{k-1}\operatorname{tr}(\sigma_k) - E_k\operatorname{tr}(\sigma_k)\end{aligned}$$

and

$$\sigma_k = \frac{1 + \alpha_k'\big(W_p(k) - zI_{p-1}\big)^{-2}\alpha_k}{\varepsilon_k + 1 - y - z - yzs_p(z)}$$

with $|\sigma_k| \le v^{-1}$. Therefore, from (3.51)–(3.53), we get

$$\begin{aligned}R_2 &\le Cn^{-2}v^{-4}\left|1 - z - y - yzs_p(z)\right|^{-2}\\ &\quad\times E\left|1 + \frac{zn^{-1}\operatorname{tr}(W_p - zI_p)^{-1} - yzs_p(z)}{1 - y - z - zn^{-1}\operatorname{tr}(W_p - zI_p)^{-1}}\right|^2\\ &\le Cn^{-2}v^{-4}\left|1 - z - y - yzs_p(z)\right|^{-2}\\ &\quad\times\left(1 + v^{-2}E\left|zn^{-1}\operatorname{tr}(W_p - zI_p)^{-1} - yzs_p(z)\right|^2\right)\\ &\le Cn^{-2}v^{-4}\left|1 - z - y - yzs_p(z)\right|^{-2}(1 + n^{-1}v^{-4})\\ &\le Cn^{-2}v^{-4}\left|1 - z - y - yzs_p(z)\right|^{-2},\end{aligned} \tag{3.54}$$

for some positive constant C.

By (3.14), we have

$$\left|1 - z - y - yzs_p(z)\right|^{-2} \le 3\left(|\delta|^2 + \left|s_p(z) - s_y(z)\right|^2 + \left|s_y(z)\right|^2\right).$$

Noting the bounds of $s_y(z)$ established in (2.8), we have

$$\begin{aligned}\left|1 - z - y - yzs_p(z)\right|^{-2} &\le C\left(|\delta|^2 + \frac{\pi^2\Delta_1^2}{v^2} + \frac{\left(1 + 3\sqrt{y}\right)^2}{y(1 - y)^2}\right)\\ &\le C\left(|\delta|^2 + \frac{\Delta_1^2}{v^2}\right).\end{aligned} \tag{3.55}$$

Here we used the fact that $|s_p(z) - s_y(z)| \le \pi\Delta_1/v$, which can be easily obtained by integration by parts.

Substitute (3.43), (3.54) and (3.55) into (3.25) first and then substitute the result into (3.22). We obtain

$$|\delta| \le C\left[\frac{1}{nv} + \frac{\Delta_1}{nv^3} + \frac{1}{n^2v^5}\left|1 - y - z - yzs_p(z)\right|^{-2} + \frac{1}{n^2v^2}\right]$$

$$\times\left|1 - y - z - yzs_p(z)\right|^{-2}$$

$$\text{(3.56)}\qquad \le C\left[\frac{\Delta_1}{nv^3} + \frac{1}{n^2v^5}\left|1 - y - z - yzs_p(z)\right|^{-2}\right]\left|1 - y - z - yzs_p(z)\right|^{-2}$$

$$\text{(3.57)}\qquad \le C_0\left[\frac{\Delta_1}{nv^3} + \frac{1}{n^2v^5}|\delta|^2\right]\left[|\delta|^2 + \frac{\Delta_1^2}{v^2}\right].$$

Choose $v = (40C_0\eta^3(A+1)^2)^{1/6}n^{-1/4}$. Then, by (3.57) and the fact that $|\delta| \le 2/v$ [see (3.20)], we obtain, for all large p,

$$|\delta| \le C_0\left[\frac{8}{n^2v^8} + \frac{2\Delta_1}{nv^4} + \frac{2\Delta_1^2}{n^2v^8}\right]|\delta| + \frac{C_0\Delta_1^3}{nv^5}$$

$$\text{(3.58)}\qquad \le \frac{1}{2}|\delta| + \frac{C_0\Delta_1^3}{nv^5} \le \frac{2C_0\Delta_1^3}{nv^5}$$

$$\le \frac{v}{10(A+1)^2}.$$

Then, by (3.55), (3.56) and (3.58), we have

$$|\delta| \le C\left(\frac{\Delta_1}{nv^3} + \frac{1}{n^2v^5}\left(|\delta|^2 + \frac{\Delta^2}{v^2}\right)\right)\left|1 - y - z - yzs_p(z)\right|^{-2}$$

$$\le C\left(\frac{\Delta_1}{nv^3}\left|1 - y - z - yzs_p(z)\right|^{-2}\right)$$

$$\text{(3.59)}\qquad \le \frac{C\Delta_1}{nv^3}\left[|\delta|^2 + \left|s_p(z) - s_y(z)\right|^2 + \left|s_y(z)\right|^2\right]$$

$$\le \frac{C\Delta_1}{nv^3}\left[\left|s_p(z) - s_y(z)\right|^2 + \left|s_y(z)\right|^2\right]$$

$$\le C\frac{\Delta_1}{nv^3}\left[\left|s_p(z) - s_y(z)\right|^2 + \left|s_y(z)\right|^2\right].$$

Hence, by Lemma 2.2 and (2.10),

$$\int_{-\infty}^{\infty}|\delta|\,du \le \frac{C\Delta_1}{nv^3}\left[\int_{-\infty}^{\infty}\left|s_p(z) - s_y(z)\right|^2 du + \int_{-\infty}^{\infty}\left|s_y(z)\right|^2 du\right]$$

$$\text{(3.60)}\qquad \le \frac{C\Delta_1}{nv^3}\left[\frac{\Delta_1}{v} + 1\right] \le Cv.$$

Recall that (3.39) holds for $|u| \le A$ provided that (3.33) is true. Therefore, by (3.58) we conclude that (3.39) is true for the newly chosen v and $|u| \le A$.

By (3.60), repeating the procedure of (3.40), we may prove that

$$\|EF_p - F_y\| = O(n^{-1/4}),$$

under the additional assumption (3.5).

To finish the proof of Theorem 3.1, we drop the restriction (3.5). Define

$$\hat{x}_{ij} = \sigma_{ij}^{-1}\left[x_{ij}I_{[|x_{ij}| \le \sqrt{n}\eta]} - Ex_{ij}I_{[|x_{ij}| \le \sqrt{n}\eta]}\right], \tag{3.61}$$

where σ_{ij}^2 is the variance of $x_{ij}I_{[|x_{ij}| \le \eta\sqrt{n}]}$, η is chosen such that $\eta \to 0$ and

$$\sup_{ij} E|x_{ij}^4| I_{[|x_{ij}| \ge \eta\sqrt{n}]} = o(\eta^2). \tag{3.62}$$

By the second condition in (3.1), it is easy to select η fulfilling the above condition.

Let $\hat{F}_p$ denote the spectral distribution of $(1/n)\hat{X}_p\hat{X}_p'$ with $\hat{X}_p = (\hat{x}_{ij})$. Then, by what we have proved under the restriction (3.5), we have

$$\|E\hat{F}_p - F_y\| = O(n^{-1/4}). \tag{3.63}$$

Note that, when $\theta \le y \le \Theta$, the density function of the Marchenko–Pastur distribution has an upper bound $D = 1/(\sqrt{y}(1-y))$. Applying Lemma 2.4 and the triangular inequality, we have

$$\begin{aligned}
\|EF_p - F_y\| &\le (D+1)L(EF_p, F_y) \\
&\le (D+1)\left[L\left(EF_p, E\hat{F}_p\right) + L\left(E\hat{F}_p, F_y\right)\right] \\
&\le (D+1)\left[L\left(EF_p, E\hat{F}_p\right) + \|E\hat{F}_p - F_y\|\right],
\end{aligned} \tag{3.64}$$

where $L(\cdot,\cdot)$ denotes the Lévy distance between distribution functions.

Denote the eigenvalues of the matrices $(1/n)X_pX_p'$ and $(1/n)\hat{X}_p\hat{X}_p'$ by $\lambda_1 \le \cdots \le \lambda_p$ and $\hat{\lambda}_1 \le \cdots \le \hat{\lambda}_p$, respectively. By Lemma 2.3, we have

$$L^2\left(EF_p, E\hat{F}_p\right) \le E\int\left|F_p(x) - \hat{F}_p(x)\right|dx = \frac{1}{p}\sum_{s=1}^{p} E|\lambda_s - \hat{\lambda}_s|. \tag{3.65}$$

Following the approach of Yin (1986), we prove that

$$\begin{aligned}
L^2\left(EF_p, E\hat{F}_p\right) &\le \frac{1}{p}\sum_{s=1}^{p} E|\lambda_s - \hat{\lambda}_s| \\
&\le \left(\frac{1}{np}E\operatorname{tr}\left(X_p + \hat{X}_p\right)\left(X_p + \hat{X}_p\right)'\right)^{1/2} \\
&\quad \times \left(\frac{1}{np}E\operatorname{tr}\left(X_p - \hat{X}_p\right)\left(X_p - \hat{X}_p\right)'\right)^{1/2} \\
&\le C\left(\sup_{ij} Ex_{ij}^2 I_{[|x_{ij}| \ge \eta\sqrt{n}]}\right)^{1/2} = o(n^{-1/2}).
\end{aligned} \tag{3.66}$$

Then (3.3) follows from (3.63), (3.64) and (3.66), and the proof of Theorem 3.1 is complete. □

PROOF OF THEOREM 3.2. Now, we consider the case of $0 < \Theta < y \le 1$. When applying Theorem 2.2 of Part I to this case, the reader should note that the density function is no longer bounded and hence the third term on the right-hand side of (2.1) does not have the order of $O(v)$. Therefore, we cannot get a preliminary estimate as good as (3.41), although the estimate of Proposition 3.4 is still true. However, we may obtain an estimate as follows:

$$\sup_x \int_{|u|\le 2va} |F_y(x+u) - F_y(x)|\, du \le Cv^{3/2}, \tag{3.67}$$

where C may be chosen as $\sqrt{2a}(1+\sqrt{y})/(\pi y)$ and the constant a is defined in Theorem 2.2 of Part I. By this and Proposition 3.4, applying Theorem 2.2 of Part I, we obtain the following preliminary estimate:

$$\|EF_p - F_y\| = O(n^{-1/12}). \tag{3.68}$$

Now, based on (3.68), we get an improved estimate by refining the estimates of $\sum_{ij} E|r_{ij}^2(k)|$ and $\sum_d E|\gamma_d^2(k)|$. Assume that $\|EF_p - F_y\| \le \Delta_1 = \eta_0 n^{-1/12}$ for some $\eta_0 > 1$ and assume that $\Delta_1 \ge v > n^{-5/24}$.

Corresponding to (3.43), applying Lemma 2.2 to the first integral in (3.42) [note that (3.42) is true for both the two cases], we find that (3.43) is still true for the newly defined Δ_1 and v, that is,

$$\begin{aligned} \sum_{i,j} E\left|r_{ij}^2(k)\right| &\le 2A^2\left[ny + \frac{\sqrt{2}n(1+\sqrt{y})}{\sqrt{v^3}} + 2ny\,\Delta_1 v^{-2} + 2v^{-2}\right] \\ &\le Cn\,\Delta_1 v^{-2}. \end{aligned} \tag{3.69}$$

We now refine the estimate of $\sum_d E|\gamma_d^2(k)|$. Using the same notation defined in the proof of Theorem 3.1 and checking the proof of formulae (3.44)–(3.54), we find that they are still true for the present case. Corresponding to (3.55), applying (2.9), we obtain

$$\left|1 - z - y - yzs_p(z)\right|^{-2} \le C\left(|\delta|^2 + \frac{\pi^2\Delta_1^2}{v^2} + \frac{4}{vy}\right) \le C\left(|\delta|^2 + \frac{\Delta_1^2}{v^2}\right). \tag{3.70}$$

This means that (3.55) is still formally true for the newly defined Δ_1 and v. Consequently, the expressions (3.56) and (3.57) are still formally true.

Choose $v = (40C_0\eta_0^3(A+1)^2)^{1/6} n^{-5/24}$. Corresponding to (3.58), for all large p, we may directly obtain from (3.57) that

$$\begin{aligned} |\delta| &\le 2C_0 n^{-1} v^{-3}\Delta_1\left[2v^{-1}|\delta| + v^{-2}\Delta_1^2\right] \\ &\le 4C_0 n^{-1} v^{-5}\Delta_1^3 = \frac{v}{10(A+1)^2}. \end{aligned} \tag{3.71}$$

By (3.56) we may similarly prove that (3.59) is true for the present case. Hence, by Lemma 2.2 and (2.11),

$$\begin{aligned} \int_{-\infty}^{\infty} |\delta|\, du &\le \frac{C\Delta_1}{nv^3}\left[\int_{-\infty}^{\infty} \left|s_p(z) - s_y(z)\right|^2 du + \int_{-\infty}^{\infty} \left|s_y(z)\right|^2 du\right] \\ &\le \frac{C\Delta_1}{nv^3}\left[\frac{\Delta_1}{v} + \frac{1}{\sqrt{v}}\right] \le Cv. \end{aligned} \tag{3.72}$$

By (3.71) and (3.72), repeating the procedure of (3.40), one may prove that

$$\int_{-A}^{A} |s_p(z) - s_y(z)| \, du \le Cv.$$

Then applying Theorem 2.2 of Part I and (3.67), we obtain that

$$\|EF_p - F_y\| = O(n^{-5/48}),$$

under the additional assumption (3.5).

As done in the proof of Theorem 3.1, make the truncation and normalization for the entries of X_p in the same way. Use the same notation defined in the proof of Theorem 3.1. By what we have proved, we have

$$\|E\hat{F}_p - F_y\| = O(n^{-5/48}). \tag{3.73}$$

In the proof of Theorem 3.1 [see (3.65)], we have proved that

$$\int |EF_p(x) - E\hat{F}_p(x)| \, dx = o(n^{-1/2}). \tag{3.74}$$

Note that F_y satisfies the condition of Lemma 2.5 with $\beta = 1/2$ and $D = (1 + \sqrt{y})/(\pi y)$. Applying Lemma 2.5 and by (3.73) and (3.74), we obtain

$$\|EF_p - F_y\|^3 \le O(n^{-5/48}) \|EF_p - F_y\|^2 + o(n^{-1/2}), \tag{3.75}$$

which implies (3.4). The proof of Theorem 3.2 is complete. □

Acknowledgment. The author would like to thank Professor J. W. Silverstein again for pointing out that the density of the Marchenko–Pastur distribution when $y = 1$ is unbounded, which led to the establishment of Theorem 3.2.

REFERENCES

Bai, Z. D. (1993). Convergence rate of expected spectral distributions of large-dimensional random matrices: Part I. Wigner matrices. *Ann. Probab.* **21** 625–648.

Marchenko, V. A. and Pastur, L. A. (1967). The distribution of eigenvalues in certain sets of random matrices. *Mat. Sb.* **72** 507–536 (*Math. USSR-Sb.* **1** 457–483).

Wachter, K. W. (1978). The strong limits of random matrix spectra for sample matrices of independent elements. *Ann. Probab.* **6** 1–18.

Yin, Y. Q. (1986). Limiting spectral distribution for a class of random matrices. *J. Multivariate Anal.* **20** 50–68.

Yin, Y. Q., Bai, Z. D. and Krishnaiah, P. R. (1988). On the limit of the largest eigenvalue of the large dimensional sample covariance matrix. *Probab. Theory Related Fields* **78** 509–521.

Department of Statistics
341 Speakman Hall
Temple University
Philadelphia, Pennsylvania 19122

The Annals of Probability
1993, Vol. 21, No. 3, 1275–1294

LIMIT OF THE SMALLEST EIGENVALUE OF A LARGE DIMENSIONAL SAMPLE COVARIANCE MATRIX

BY Z. D. BAI AND Y. Q. YIN

Temple University and University of Massachusetts, Lowell

In this paper, the authors show that the smallest (if $p \leq n$) or the $(p-n+1)$-th smallest (if $p > n$) eigenvalue of a sample covariance matrix of the form $(1/n)XX'$ tends almost surely to the limit $(1-\sqrt{y})^2$ as $n \to \infty$ and $p/n \to y \in (0, \infty)$, where X is a $p \times n$ matrix with iid entries with mean zero, variance 1 and fourth moment finite. Also, as a by-product, it is shown that the almost sure limit of the largest eigenvalue is $(1+\sqrt{y})^2$, a known result obtained by Yin, Bai and Krishnaiah. The present approach gives a unified treatment for both the extreme eigenvalues of large sample covariance matrices.

1. Introduction. Suppose A is a $p \times p$ matrix with real eigenvalues $\lambda_1, \lambda_2, \ldots, \lambda_p$. Then the spectral distribution of the matrix A is defined by

$$F^A(x) = \frac{1}{p}\#\{i \leq p : \lambda_i \leq x\}.$$

We are especially interested in the matrix of the form $S = S_n = (1/n)XX'$, where $X = X_n = (X_{ij})$, and where X_{ij}, $i = 1, \ldots, p$; $j = 1, \ldots, n$, are iid random variables with zero mean and variance σ^2. We will call it a sample covariance matrix.

There are many studies on the limiting behavior of the spectral distributions of sample covariance matrices. For example, under various conditions, Grenander and Silverstein (1977), Jonsson (1982) and Wachter (1978) prove that the spectral distribution $F^S(x)$ converges to

$$F_y(x) = (1 - y^{-1} \wedge 1)\delta(x) + \int_{-\infty}^{x} f_y(u)\, du,$$

where $\delta(x)$ is the distribution function with mass 1 at 0, and

$$f_y(x) = \begin{cases} \dfrac{1}{2\pi y x}\sqrt{(x-a)(b-x)}, & \text{if } a < x < b, \\ 0, & \text{otherwise,} \end{cases}$$

as $p = p(n) \to \infty$, $n \to \infty$ and $p/n \to y \in (0, \infty)$. Here

$$a = a(y) = \left(1 - \sqrt{y}\right)^2 \sigma^2, \qquad b = b(y) = \left(1 + \sqrt{y}\right)^2 \sigma^2.$$

As a consequence of Yin (1986), if the second moment of X_{11} is finite, the above convergence holds with probability 1. Note that σ^2 appears in the

Received April 1990; revised June 1992.
AMS 1991 *subject classifications*. Primary 60F15; secondary 62H99.
Key words and phrases. Random matrix, sample covariance matrix, smallest eigenvalue of a random matrix, spectral radius.

definition of $F_y(x)$. Thus, the condition on the existence of the second moment of X_{11} is also necessary.

It is not hard to see that if $F^S(x)$ converges to $F_y(x)$ a.s., then

$$\liminf \max_{1\le i\le p} \lambda_i \ge b \quad \text{a.s.}$$

However, the converse assertion

$$\limsup \max_{1\le i\le p} \lambda_i \le b \quad \text{a.s.}$$

is not trivial.

The first success in establishing the last relation ($\le$) was made by Geman (1980), who did it under the condition that

$$E|X_{11}|^k \le Mk^{\alpha k}, \quad \text{for some } M > 0,\ \alpha > 0, \text{ and all } k \ge 3.$$

Yin, Bai and Krishnaiah (1988) established the same conclusion under the condition that

$$E|X_{11}|^4 < \infty,$$

which is further proved to be necessary in Bai, Silverstein and Yin (1988) by showing that

$$E|X_{11}|^4 = \infty \quad \Rightarrow \quad \limsup \max_{1\le i\le n} \lambda_i = \infty \quad \text{a.s.}$$

It is much harder to study the convergence of the smallest eigenvalue of a sample covariance matrix. The first breakthrough was given by Silverstein (1985), who proved that if $X_{11} \sim N(0,1)$, then

$$\min_{1\le i\le p} \lambda_i \to a(y) \quad \text{a.s.},$$

as $p \to \infty$, $p/n \to y < 1$. However, it is hard to use his method to get the general result, since his proof depends heavily on the normality hypothesis.

In this paper, we shall prove the following theorems.

THEOREM 1. *Let* $[X_{uv};\ u,v = 1,2,\dots]$ *be a double array of independent and identically distributed (iid) random variables with zero mean and unit variance, and let*

$$X = [X_{uv}\colon u = 1,\dots,p;\ v = 1,\dots,n], \qquad S = (1/n)XX'.$$

Then, if $E|X_{11}|^4 < \infty$, *as* $n \to \infty$, $p \to \infty$, $p/n \to y \in (0,1)$,

$$-2\sqrt{y} \le \liminf \lambda_{\min}(S - (1+y)I) \le \limsup \lambda_{\max}(S - (1+y)I) \le 2\sqrt{y} \quad a.s.$$

As an easy corollary of Theorem 1, we have the following.

THEOREM 2. *Under the conditions of Theorem* 1, *as* $n \to \infty$, $p \to \infty$, $p/n \to y \in (0,1)$,

$$\lim \lambda_{\min} = \left(1 - \sqrt{y}\right)^2 \quad a.s. \tag{1.1}$$

$$\lim \lambda_{\max} = \left(1 + \sqrt{y}\right)^2 \quad a.s. \tag{1.2}$$

REMARK 1. The assertion (1.1) is trivially true for $y = 1$. If $y > 1$, then $p > n$ for all large p, and the $p - n$ smallest eigenvalues of S must be 0. In this case, (1.1) is no longer true as it stands. However, if we redefine $\lambda_{\min}$ to be the $(p - n + 1)$-th smallest eigenvalue of S, then (1.1) is still true. In fact, for the case of $y > 1$, define $S^* = (1/p)X'X$ and $y^* = 1/y \in (0, 1)$. By Theorem 2, we have

$$\lambda_{\min}(S^*) \to \left(1 - \sqrt{y^*}\right)^2 = \frac{1}{y}\left(1 - \sqrt{y}\right)^2 \quad \text{a.s.},$$

Therefore,

$$\lambda_{\min}(S) = \frac{p}{n}\lambda_{\min}(S^*) \to \left(1 - \sqrt{y}\right)^2 \quad \text{a.s.}$$

By a similar argument, one may easily show that the conclusion (1.2) is also true for $y \geq 1$.

REMARK 2. The conclusion (1.2) has already been proved in Yin, Bai and Krishnaiah (1988). Here, we prove it by an alternative approach as a by-product of our Theorem 1, which is the key step for the proof of the limit of the smallest eigenvalue.

REMARK 3. From the proof given later, one can see that if the condition $EX_{11}^4 < \infty$ is weakened to

$$n^2 P\left(|X_{11}| > \sqrt{n}\right) \to 0, \tag{1.3}$$

then the two limit relations (1.1) and (1.2) hold in probability.

In fact, if (1.3) is true, then for each $\varepsilon > 0$,

$$E|X_{11}|^{4-\varepsilon} < \infty$$

and there exists a sequence of positive constants $\delta = \delta_n \to 0$ such that

$$n^2 P\left(|X_{11}| > \delta\sqrt{n}\right) \to 0.$$

Here, we may assume that the rate of $\delta \to 0$ is sufficiently slow.

As done in Silverstein (1989) for the largest eigenvalue, one may prove that the probability of the event that the smallest eigenvalue of the sample covariance matrix constructed by the truncated variables at $\delta\sqrt{n}$ differs from the original by a quantity controlled by

$$n^2 P\left(|X_{11}| > \delta\sqrt{n}\right).$$

Also, employing von Neumann's inequality, one may conclude that the difference between the square root of the smallest eigenvalue of the truncated sample covariance matrix and that of the truncated and then centralized sample covariance matrix is controlled by

$$\sqrt{p}\,\left|EX_{11} I_{[|X_{11}| > \delta\sqrt{n}]}\right| \to 0$$

(For details of the application of von Neumann's inequality, see the beginning of Section 2.) Then the truncated and then centralized variables satisfy the conditions given in (2.1), and the desired result can be proved by the same lines of the proof of the main result.

REMARK 4. In Girko (1989), an attempt is made to prove the weak convergence of the smallest eigenvalue under a stronger condition. However, this proof contains some serious errors. Regardless, the result we get here is strong convergence under much weaker conditions.

2. Some lemmas. In this section we prove several lemmas. By the truncation lemma proved in Yin, Bai and Krishnaiah (1988), one may assume that the entries of X_n have already been truncated at $\delta\sqrt{n}$ for some slowly varying $\delta = \delta_n \to 0$. Let

$$V_{uv} = X_{uv}I(|X_{uv}| \le \delta\sqrt{n}) - EX_{uv}I(|X_{uv}| \le \delta\sqrt{n}).$$

In 1937, von Neumann proved that

$$\sum_{i=1}^{p} \lambda_i \tau_i \ge \operatorname{tr}(A'B),$$

if A and B are $p \times n$ matrices with singular values $\lambda_1 \ge \cdots \ge \lambda_p$ and $\tau_1 \ge \cdots \ge \tau_p$, respectively. Then, using von Neumann's inequality, we have

$$\begin{aligned} &\left|\lambda_{\min}^{1/2}\left(n^{-1}\hat{X}_n\hat{X}_n'\right) - \lambda_{\min}^{1/2}\left(n^{-1}V_nV_n'\right)\right|^2 \\ &\le \sum_{k=1}^{p}\left(\lambda_k^{1/2}\left(n^{-1}\hat{X}_n\hat{X}_n'\right) - \lambda_k^{1/2}\left(n^{-1}V_nV_n'\right)\right)^2 \\ &\le \frac{1}{n}\operatorname{tr}\left(\hat{X}_n - V_n\right)\left(\hat{X}_n - V_n\right)' \\ &\le pE^2|X_{11}|I_{[|X_{11}|>\delta\sqrt{n}]} \to 0, \end{aligned}$$

where $\hat{X}_n$ and V are $n \times p$ matrices with (u,v)-th entries $X_{u,v}I_{[|X_{uv}|\le\delta\sqrt{n}]}$ and V_{uv}, respectively. In fact, the above convergence is true provided $n\delta^3 \to 0$. Therefore, we can assume that for each n the entries $X_{uv} = X_{uv}(n)$ of the matrix X_n are iid and satisfy

$$\begin{aligned} &EX_{uv} = 0, \quad EX_{uv}^2 \le 1 \quad \text{and} \quad EX_{uv}^2 \to 1 \quad \text{as } n \to \infty, \\ &E|X_{uv}|^l \le \left(\delta\sqrt{n}\right)^{l-3}, \quad \text{for } l \ge 3, \end{aligned} \tag{2.1}$$

where $\delta = \delta_n \to 0$ is nonrandom and sufficiently slow.

Replacing all diagonal elements of S by 0, we get a matrix T. Define $T(0) = I_p$ and $T(1) = T$, and, for each $l \ge 2$, let $T(l) = (T_{ab}(l))$ be a $p \times p$ matrix with

$$T_{ab}(l) = n^{-l}\sum{}' X_{av_1}X_{u_1v_1}X_{u_1v_2}X_{u_2v_2}\cdots X_{u_{l-1}v_l}X_{bv_l}, \tag{2.2}$$

where the summation Σ' runs over all integers $u_1, \ldots, u_{l-1}$ from the set

$\{1,\dots,p\}$ and $v_1,\dots,v_l$ from $\{1,\dots,n\}$, subject to the conditions that

$$a \neq u_1, \qquad u_1 \neq u_2,\dots,u_{l-1} \neq b, \qquad (2.3)$$
$$v_1 \neq v_2, \qquad v_2 \neq v_3,\dots,v_{l-1} \neq v_l.$$

LEMMA 1.

$$\limsup_{n\to\infty} \|T(l)\| \le (2l+1)(l+1)y^{(l-1)/2} \quad a.s.,$$

where the matrix norm used here and throughout the paper is the operator norm, that is, the largest singular value of the matrix.

PROOF. For integers $u_0,\dots,u_r$ from $\{1,\dots,p\}$ and integers $v_1,\dots,v_r$ from $\{1,\dots,n\}$, define a graph

$$G\{u_0, v_1, u_1,\dots,v_r,u_r\}$$

as follows. Let $u_0,\dots,u_r$ be plotted on a straight line, and let $v_1,\dots,v_r$ be plotted on another straight line. The two lines are supposed to be parallel. $u_0,\dots,u_r,v_1,\dots,v_r$ are vertices. The graph has $2r$ edges: $e_1,\dots,e_{2r}$. The two ends of e_{2i-1} are u_{i-1}, v_i and those of e_{2i} are v_i, u_i. Two edges are said to be coincident if they have the same set of end vertices.

An edge e_i is said to be single up to e_j, $j \ge i$, if it does not coincide with any $e_1,\dots,e_j$ other than itself.

If $e_{2i-1} = u_{i-1}v_i$ ($e_{2i} = v_i u_i$) is such that

$$v_i \notin \{v_1,\dots,v_{i-1}\} \ (u_i \notin \{u_1,\dots,u_{i-1}\}),$$

then e_{2i-1} (e_{2i}) is called a column (row) innovation. T_1 denotes the set of all innovations. If e_j is such that there is an innovation e_i, $i<j$, and e_j is the first one to coincide with e_i, then we say $e_j \in T_3$. Other edges constitute the set T_4. Thus, edges are partitioned into three disjoint classes: T_1, T_3, T_4. Edges which are not innovations and single up to themselves form a set T_2. It is obvious that $T_2 \subset T_4$.

If e_j is a T_3 edge and there is more than one single (up to e_{j-1}) innovation among $e_1,\dots,e_{j-1}$ which is adjacent to e_j, we say that e_j is a regular T_3 edge. We can prove that for a regular T_3 edge, the number of such innovations is bounded by $t+1$, where t is the maximum number of noncoincident T_4 edges [see Yin, Bai and Krishnaiah (1988)], and the number of regular T_3 edges is bounded by twice the number of T_4 edges [see Yin, Bai and Krishnaiah (1988)].

In order to establish Lemma 1, we estimate $E \operatorname{tr} T^{2m}(l)$. By (2.2),

$$\begin{aligned}
\operatorname{tr} T^{2m}(l) &= \sum T_{b_1b_2}(l)T_{b_2b_3}(l)\cdots T_{b_{2m}b_1}(l)\\
&= n^{-2ml} \sum_{(b_i)\sum_1'\sum_2'\cdots\sum_{2m}'} X_{b_1v_1'}X_{u_1'v_1'}X_{u_1'v_2'}X_{u_2'v_2'}\\
&\qquad \cdots X_{u_{l-1}'v_l'}X_{b_2v_l'}X_{b_2v_1''}\cdots X_{b_{2m}v_l^{(2m-1)}}X_{b_{2m}v_1^{(2m)}}X_{u_1^{(2m)}v_2^{(2m)}}\\
&\qquad \cdots X_{u_{l-1}^{(2m)}v_l^{(2m)}}X_{b_1v_l^{(2m)}}.
\end{aligned} \qquad (2.4)$$

Here the summation Σ_i' is taken with respect to $u_1^{(i)},\dots,u_{l-1}^{(i)}$ running over $\{1,\dots,p\}$ and $v_1^{(i)},\dots,v_l^{(i)}$ running over $\{1,\dots,n\}$ subject to the condition that

$$b_i \neq u_1^{(i)}, \qquad u_1^{(i)} \neq u_2^{(i)},\dots,u_{l-1}^{(i)} \neq b_{i+1};$$
$$v_1^{(i)} \neq v_2^{(i)},\dots,v_{l-1}^{(i)} \neq v_l^{(i)},$$

for each $i = 1,2,\dots,2m$; and $\Sigma_{(b_i)}$ is the summation for b_i, $i = 1,\dots,2m$, running over $\{1,\dots,p\}$.

Now we can consider the sum (2.4) as a sum over all graphs G of the form

$$\begin{aligned} G = G\big[&b_1, v_1', u_1', v_2',\dots,u_{l-1}', v_l', b_2, v_1'', u_1'',\dots,v_l'',\\ &b_3,\dots,b_{2m}, v_1^{(2m)}, u_1^{(2m)},\dots,u_{l-1}^{(2m)}, v_{l-1}^{(2m)}, b_1\big]. \end{aligned} \tag{2.5}$$

At first we partition all these graphs into isomorphism classes. We take the sum within each isomorphism class, and then take the sum of all such sums over all isomorphism classes. (Here we say that two graphs are isomorphic, if equality of two vertex indices in one graph implies the equality of the corresponding vertex indices in the other graph.)

Within each isomorphism class, the ways of arranging the three different types of edges are all the same. In other words, if two graphs of the form (2.5) are isomorphic, the corresponding edges must have the same type. However, two graphs with the same arrangements of types are not necessarily isomorphic.

We claim that

$$\begin{aligned} &E\big[\operatorname{tr} T^{2m}(l)\big]\\ &\quad \le n^{-2ml}\sum{}^{*}\left(\prod_{i=1}^{2m}\binom{2l+1}{2a_i+1}\right)\binom{4ml-k}{k}\binom{4ml}{t}\\ &\qquad \times t^{4ml-2k}(t+1)^{8lm-4k} n^{k+1}\left(\frac{p}{n}\right)^{(k-t-2m)/2}\left(\sqrt{n}\,\delta\right)^{4ml-2k-t}, \end{aligned} \tag{2.6}$$

where the summation Σ^* is taken with respect to k, t and a_i, $i = 1,\dots,2m$, under some restrictions to be specified. Here:

(i) k $(= 1,\dots,2ml)$ is the total number of innovations in G.

(ii) t $(= 0,\dots,4ml - 2k)$ is the number of noncoincident T_4 edges in G.

(iii) a_i $(= 0,\dots,l)$ is the number of pairs of consecutive edges (e, e') in the graph

$$G_i = G\big[b_i, v_1^{(i)}, u_1^{(i)},\dots,u_{l-1}^{(i)}, v_l^{(i)}, b_{i+1}\big] \tag{2.7}$$

in which e is an innovation but e' is not.

Now we explain the reasons why (2.6) is true:

(i) The factor n^{-2ml} is obvious.

(ii) If there is an overall single edge in a graph G, then the mean of the product of X_{ij} corresponding to this graph [denoted by $EX(G)$] is zero. Thus, in any graph corresponding to a nonzero term, we have $k \le 2ml$.

(iii) The number of T_3 edges is also k. Hence the number of T_4 edges is $4ml - 2k$, and $t \leq 4ml - 2k$.

(iv) The graph G is split into $2m$ subgraphs $G_1, \dots, G_{2m}$ defined in (2.6). Obviously, $0 \leq a_i \leq l$.

(v) The number of sequences of consecutive innovations in G_i is either a_i or $a_i + 1$ (the latter happens when the last edge in G_i is an innovation). Hence the number of ways of arranging these consecutive sequences in G_i is at most

$$\binom{2l}{2a_i} + \binom{2l}{2a_i + 1} = \binom{2l+1}{2a_i+1}.$$

(vi) Given the position of innovations, there are at most $\binom{4ml-k}{k}$ ways to arrange T_3 edges.

(vii) Given the positions of innovations and T_3 edges, there are at most $\binom{4ml}{t}$ ways to choose t distinguishable positions for the t noncoincident T_4 edges. When these positions have been chosen, there are at most t^{4ml-2k} ways to distribute the $4ml - k$ T_4 edges into these t positions.

(viii) Yin, Bai and Krishnaiah (1988) proved that a regular T_3 edge e has at most $t + 1$ choices and that the number of regular T_3 edges is dominated by $2(4ml - 2k)$. Therefore, there are at most $(t+1)^{8ml-4k}$ different ways to arrange the T_3 edges.

(ix) Let r and c be the number of row and column innovations. Then $r + c = k$, and the number of graphs G within the isomorphism class is bounded by $n^c p^{r+1} = n^{k+1}(p/n)^{r+1}$.

Suppose that in the pair (e, e'), e is an innovation in G_i and e' is not an innovation in G_i. Then it is easy to see that e' is of type T_4 and is single up to itself. Therefore,

$$t \geq \sum_{i=1}^{2m} a_i. \tag{2.8}$$

In each G_i, there are at most $a_i + 1$ sequences of consecutive innovations. Therefore,

$$|r - c| \leq \sum_{i=1}^{2m} a_i + 2m. \tag{2.9}$$

Since $r + c = k$, by (2.8) and (2.9) we obtain

$$r \geq \tfrac{1}{2}(k - t) - m,$$

by which we conclude that (by noticing that we can assume $p/n < 1$)

$$n^{k+1}\left(\frac{p}{n}\right)^{r+1} \leq n^{k+1}\left(\frac{p}{n}\right)^{(k-t-2m)/2} \tag{2.10}$$

(x) By the same argument as in Yin, Bai and Krishnaiah (1988), we have

$$|EX(G)| \leq \left(\sqrt{n}\,\delta\right)^{4ml-2k-t} \tag{2.11}$$

The above 10 points are discussed for $t > 0$, but they are still valid when $t = 0$, if we take the convention that $0^0 = 1$ for the term t^{4ml-2k}. Thus we have established (2.6). Now we begin to simplify the estimate (2.6). Note that

$$\binom{2l+1}{2a_i+1} \le (2l+1)^{2a_i+1}.$$

By (2.8), we have

$$\prod_{i=1}^{2m} \binom{2l+1}{2a_i+1} \le (2l+1)^{2\Sigma a_i + 2m} \le (2l+1)^{2t+2m} \tag{2.12}$$

The number of choices of the a_i's does not exceed $(l+1)^{2m}$. Therefore, by the elementary inequality $a^{-t}t^b \le (b/\log a)^b$, for all $a > 1$, $b > 0$, $t \ge 1$, and letting m be such that $m/\log n \to \infty$, $m\delta^{1/3}/\log n \to 0$ and $m/(\delta\sqrt{n}) \to 0$, we obtain from (2.6), for sufficiently large n,

$$\begin{aligned}
E[\operatorname{tr} T^{2m}(l)] &\le \sum_{k=1}^{2ml}\sum_{t=0}^{4ml-2k} n(2l+1)^{2m}(l+1)^{2m} \\
&\quad\times \binom{4ml-k}{k}\left(\frac{4ml(2l+1)^2}{\sqrt{p}\,\delta}\right)^t \\
&\quad\times (t+1)^{12ml-6k}\left(\frac{p}{n}\right)^{(k-2m)/2}\delta^{4ml-2k} \\
&\le n^2(2l+1)^{2m}(l+1)^{2m}\left(\frac{p}{n}\right)^{-m} \\
&\quad\times \sum_{k=1}^{2ml}\binom{4ml-k}{k}\left(\frac{12ml-6k}{\left|\log[36ml^3/(\sqrt{p}\,\delta)]\right|}\right)^{12ml-6k} \\
&\quad\times \left(\frac{p}{n}\right)^{k/2}\delta^{4ml-2k} \\
&\le n^2((2l+1)(l+1))^{2m}\left(\frac{p}{n}\right)^{-m} \\
&\quad\times \sum_{k=1}^{2ml}\binom{4ml}{2k}\left(\frac{24ml\delta^{1/3}}{\frac{1}{2}\log n}\right)^{12ml-6k}\left(\frac{p}{n}\right)^{k/2} \\
&\le n^2((2l+1)(l+1))^{2m}\left(\frac{p}{n}\right)^{-m} \\
&\quad\times \left[\left(\frac{p}{n}\right)^{1/4} + \left(\frac{24ml\delta^{1/3}}{\frac{1}{2}\log n}\right)^3\right]^{4ml}.
\end{aligned} \tag{2.13}$$

Here, in the proof of the second inequality, we have used the fact that

$$\left(\frac{4ml(2l+1)}{\sqrt{p}\,\delta}\right)^{t} \le n\left(\frac{4ml(2l+1)}{\sqrt{p}\,\delta}\right)^{t+1}.$$

If we choose z to be

$$z = (2l+1)(l+1)y^{(l-1)/2}(1+\varepsilon),$$

where $\varepsilon > 0$, then

$$\sum z^{-2m} E \operatorname{tr} T^{2m}(l) < \infty.$$

Thus the lemma is proved. □

LEMMA 2. *Let $\{X_{ij},\ i, j, = 1, 2, \ldots,\}$ be a double array of iid random variables and let $\alpha > \frac{1}{2}$, $\beta \ge 0$ and $M > 0$ be constants. Then as $n \to \infty$,*

$$\max_{j \le Mn^{\beta}} \left| n^{-\alpha} \sum_{i=1}^{n} (X_{ij} - c) \right| \to 0 \quad a.s., \tag{2.14}$$

if and only if the following hold:

(i) $$E|X_{11}|^{(1+\beta)/\alpha} < \infty;$$

(ii) $$c = \begin{cases} EX_{11}, & \text{if } \alpha \le 1, \\ \text{any number}, & \text{if } \alpha > 1. \end{cases}$$

The proof of Lemma 2 is given in the Appendix.

LEMMA 3. *If $f > 0$ is an integer and $X^{(f)}$ is the $p \times n$ matrix $[X_{uv}^{f}]$, then*

$$\limsup \lambda_{\max}\{n^{-f} X^{(f)} X^{(f)\prime}\} \le 7 \quad a.s.$$

PROOF. When $f = 1$, we have

$$\left\| \frac{1}{n} X^{(1)} X^{(1)\prime} \right\| \le \|T(1)\| + \left\| \operatorname{diag}\left[\frac{1}{n} \sum_{j=1}^{n} X_{ij}^{2},\ i = 1, \ldots, p \right] \right\|$$

$$\le \|T(1)\| + \frac{1}{n} \max_{i \le p} \sum_{j=1}^{n} X_{ij}^{2}.$$

So, by Lemmas 1 and 2, we get

$$\limsup_{n} \|n^{-1} X^{(1)} X^{(1)\prime}\| \le 7 \quad \text{a.s.}$$

For $f = 2$, by the Gerŝgorin theorem and Lemma 2, we have

$$\begin{aligned}\lambda_{\max}\{n^{-2}X^{(2)}X^{(2)\prime}\} &\le \max_i n^{-2}\sum_{j=1}^{n} X_{ij}^4 + \max_i n^{-2}\sum_{k\ne i}\sum_{j=1}^{n} X_{ij}^2X_{kj}^2 \\ &\le \max_i n^{-2}\sum_{j=1}^{n} X_{ij}^4 + \left(\max_i n^{-1}\sum_{j=1}^{n} X_{ij}^2\right)\left(\max_j n^{-1}\sum_{k=1}^{p} X_{kj}^2\right) \\ &\to y \quad \text{a.s.}\end{aligned}$$

For $f > 2$, the conclusion of Lemma 4 is stronger than that of this lemma. □

REMARK 5. For the case of $f = 1$, the result can easily follow from a result in Yin, Bai and Krishnaiah (1988) with a bound of 4. Here, we prove this lemma by using our Lemma 1 to avoid the use of the result of Yin, Bai and Krishnaiah (1988), since we want to get a unified treatment for limits of both the largest and the smallest eigenvalues of large covariance matrices, as we remarked after the statement of Theorem 2.

In the following, we say that a matrix is $o(1)$ if its operator norm tends to 0.

LEMMA 4. *Let $f > 2$ be an integer, and let $X^{(f)}$ be as defined in Lemma 3. Then*

$$\|n^{-f/2}X^{(f)}\| = o(1) \quad a.s.$$

PROOF. Note that, by Lemma 2, we have

$$\|n^{-f/2}X^{(f)}\|^2 \le n^{-f}\sum_{u,v} X_{uv}^{2f} \to 0 \quad \text{a.s.},$$

since $E|X_{11}^{2f}|^{2/f} = EX_{11}^4 < \infty$. The proof is complete. □

LEMMA 5. *Let H be a $p \times n$ matrix. If $\|H\|$ is bounded a.s. and $f > 2$, or $H = o(1)$ a.s. and $f \ge 1$, then the following matrices are $o(1)$ a.s.:*

$$A(k,f) = \left(n^{-k+1-f/2}\sum_{\substack{a\ne u_1\ne\cdots\ne u_{k-1}\ne b\\ v_1\ne\cdots\ne v_k}} H_{av_1}X_{u_1v_1}^f X_{u_1v_2}\cdots X_{u_{k-1},v_k}X_{bv_k}\right);$$

$$B(k,f) = \left(n^{-k+1-f/2}\sum_{\substack{u_1\ne\cdots\ne u_{k-1}\ne b\\ v_1\ne\cdots\ne v_k}} H_{av_1}X_{u_1v_1}^f X_{u_1v_2}X_{u_2v_2}\cdots X_{bv_k}\right).$$

PROOF. For the case of $k = 1$, by Lemma 3, we have

$$B(1,f) = \left(n^{-f/2}\sum_{v_1} H_{av_1}X_{bv_1}^f\right) = n^{-f/2}HX^{(f)\prime} = Hn^{-f/2}X^{(f)\prime} = o(1)$$

and

$$A(1, f) = B(1, f) - n^{-f/2}\,\mathrm{diag}\left[\sum_{v_1} H_{av_1} X_{av_1}^f\right]$$

$$= B(1, f) - \mathrm{diag}(B(1, f)) = o(1),$$

where diag(B) denotes the diagonal matrix whose diagonal elements are the same as the matrix B. Here, in the proof of diag(B(1, f)) $= o(1)$, we have used the fact that $\|\mathrm{diag}(B)\| \le \|B\|$.

For the general case $k > 1$, by Lemma 1 and the assumptions, we have

$$(B_{ab}) = \left(n^{-k+1-f/2} \sum_{\substack{u_1 \neq \cdots \neq u_{k-1} \neq b \\ v_1 \neq \cdots \neq v_k}} H_{av_1} X_{u_1v_1}^f X_{u_1v_2} X_{u_2v_2} \cdots X_{bv_k}\right)$$

$$= \left(n^{-k+1-f/2} \sum_{\substack{u_1 \neq \cdots \neq u_{k-1} \neq b \\ v_2 \neq \cdots \neq v_k}} \left(\sum_{v_1} H_{av_1} X_{u_1v_1}^f\right) X_{u_1v_2} X_{u_2v_2} \cdots X_{bv_k}\right)$$

$$- \left(n^{-k+1-f/2} \sum_{\substack{u_1 \neq \cdots \neq u_{k-1} \neq b \\ v_2 \neq \cdots \neq v_k}} H_{av_2} X_{u_1v_2}^{f+1} X_{u_2v_2} \cdots X_{bv_k}\right)$$

$$= n^{-f/2} H X^{(f)\prime} T(k-1) - C = o(1) - C.$$

However, the entries of the matrix C satisfy

$$C_{ab} = n^{-k+1-f/2} \sum_{\substack{u_2 \neq \cdots \neq u_{k-1} \neq b \\ v_2 \neq \cdots \neq v_k}} H_{av_2} \sum_{u_1} X_{u_1v_2}^{f+1} X_{u_2v_2} \cdots X_{bv_k}$$

$$- n^{-k+1-f/2} \sum_{\substack{u_2 \neq \cdots \neq u_{k-1} \neq b \\ v_2 \neq \cdots \neq v_k}} H_{av_2} X_{u_2v_2}^{f+2} X_{u_2v_3} X_{u_3v_3} \cdots X_{bv_k}$$

$$= D_{ab} - E_{ab}.$$

Note that the matrix E is of the form of B with a smaller k-index. Thus, by the induction hypothesis, we have $E = o(1)$ a.s. The matrix D also has the same form as B with

$$1, k-1, H^* = \left(n^{-(f+1)/2} H_{av} \sum_u X_{uv}^{f+1}\right)$$

in place of f, k, H_{av}. Evidently, by Lemma 2, we have

$$H^* = H\,\mathrm{diag}\left\{n^{-(f+1)/2} \sum_u X_{uv}^{f+1};\ v = 1, \ldots, n\right\} = o(1).$$

Thus, $D = o(1)$ and hence $B(k, f) = o(1)$.

For matrices A, it is seen that

$$A_{ab} = B_{ab} - n^{-k+1-f/2} \sum_{\substack{a \neq u_2 \neq \cdots \neq u_{k-1} \neq b \\ v_2 \neq \cdots \neq v_k}} \Big(\sum_{v_1} H_{av_1} X_{av_1}^f \Big) X_{av_2} X_{u_2 v_2} \cdots X_{bv_k}$$

$$+ n^{-k+1-f/2} \sum_{\substack{a \neq u_2 \neq \cdots \neq u_{k-1} \neq b \\ v_2 \neq \cdots \neq v_k}} H_{av_2} X_{av_2}^{f+1} X_{u_2 v_2} \cdots X_{bv_k}$$

$$= B_{ab} - F_{ab} + K_{ab}.$$

Note that

$$\|F\| = \big\| \big[\operatorname{diag}(Hn^{-f/2} X^{(f)'})\big] T(k-1) \big\| \leq \big\| H(n^{-f/2} X^{(f)'}) \big\| \, \|T(k-1)\| = o(1).$$

It is easy to see that the matrix K is of the form of A with

$$1, k-1, \hat{H} = \big(n^{-(f+1)/2} H_{uv} X_{uv}^{f+1}\big)$$

in places of f, k and H. Note that $\hat{H} = H \circ (n^{(f+1)/2} X^{(f+1)'})$, where $A \circ B = (A_{uv} B_{uv})$ denotes the Hadamard product of the matrices A and B. By the fact that $\|A \circ B\| \leq \|A\| \, \|B\|$ [when A and B are Hermitian and positive definite, this inequality can be found in Marcus (1964); a simple proof for the general case is given in the Appendix], we have $\hat{H} = o(1)$. Hence, by the induction hypothesis, $K = o(1)$. Thus, we conclude that $A = o(1)$ and the proof of this lemma is complete. □

LEMMA 6. *The following matrices are* $o(1)$ *a.s.*:

$$A_1 = \Bigg(n^{-k-1} \sum_{\substack{u_1 \neq \cdots \neq u_{k-1} \neq b \\ v_1 \neq \cdots \neq v_k}} X_{av_1}^3 X_{u_1 v_1} X_{u_1 v_2} \cdots X_{u_{k-1} v_k} X_{bv_k} \Bigg),$$

$$A_2 = \Bigg(n^{-k-1} \sum_{\substack{a \neq u_2 \neq \cdots \neq u_{k-1} \neq b \\ v_1 \neq \cdots \neq v_k}} X_{av_1}^4 X_{av_2} X_{u_2 v_2} \cdots X_{bv_k} \Bigg),$$

$$A_3 = \Bigg(n^{-k-1} \sum_{\substack{u_1 \neq \cdots \neq b \\ v_1 \neq \cdots \neq v_k}} X_{av_1} X_{u_1 v_1}^3 X_{u_1 v_2} X_{u_2 v_2} \cdots X_{bv_k} \Bigg),$$

$$A_4 = \Bigg(n^{-k} \sum_{\substack{u_1 \neq \cdots \neq b \\ v_1 \neq \cdots \neq v_k}} X_{av_1} Z_{v_1} X_{u_1 v_1} X_{u_1 v_2} X_{u_2 v_2} \cdots X_{bv_k} \Bigg),$$

$$A_5 = \Bigg(n^{-k} \sum_{\substack{a \neq u_2 \neq \cdots \neq b \\ v_1 \neq \cdots \neq v_k}} W_a X_{av_1} X_{u_1 v_1} X_{u_2 v_2} \cdots X_{bv_k} \Bigg),$$

where $Z = \operatorname{diag}[Z_1, \ldots, Z_n] = o(1)$ *and* $W = \operatorname{diag}[W_1, \ldots, W_p] = o(1)$.

PROOF. All are simple consequences of Lemmas 2–5. For instance, A_1 is a matrix of type B as in Lemma 5, with $f = 1$ and $H = n^{-3/2}X^{(3)} = o(1)$ a.s.

LEMMA 7. *For all $k \geq 1$,*

$$TT(k) = T(k+1) + yT(k) + yT(k-1) + o(1) \quad a.s.,$$

where $T = T(1)$ and $T(k)$ are defined in (2.2).

PROOF. Recall that $T(0) = I$ and $T(1) = T$. We have

$$\begin{aligned}
TT(k) &= \left[n^{-k-1} \sum_{\substack{a \neq u_0 \neq \cdots \neq u_{k-1} \neq b \\ v_0, v_1 \neq \cdots \neq v_k}} X_{av_0} X_{u_0 v_0} X_{u_0 v_1} X_{u_1 v_1} \cdots X_{u_{k-1} u_k} X_{b v_k} \right] \\
&= T(k+1) \\
&\quad + \left[n^{-k-1} \sum_{\substack{a \neq u_0 \neq \cdots \neq u_{k-1} \neq b \\ v_1 \neq \cdots \neq v_k}} X_{av_1} X_{u_0 v_1}^2 X_{u_1 v_1} \cdots X_{u_{k-1} v_k} X_{b v_k} \right] \\
&= T(k+1) + n^{-k-1} \\
&\quad \times \left[\sum_{\substack{u_1 \neq \cdots \neq u_{k-1} \neq b \\ v_1 \neq \cdots \neq v_k}} X_{av_1} \sum_{u_0} X_{u_0 v_1}^2 (1 - \delta_{u_0 a})(1 - \delta_{u_0 u_1}) X_{u_1 v_1} \cdots X_{b v_k} \right] \\
&= T(k+1) + n^{-k-1} \left[\sum\nolimits^* X_{av_1} \sum_{u} X_{u v_1}^2 X_{u_1 v_1} \cdots X_{b v_k} \right] \\
&\quad - n^{-k-1} \left[\sum\nolimits^* X_{av_1}^3 X_{u_1 v_1} \cdots X_{b v_k} \right] \\
&\quad - n^{-k-1} \left[\sum\nolimits^* X_{av_1} X_{u_1 v_1}^3 X_{u_1 v_2} X_{u_2 v_2} \cdots X_{b v_k} \right] \\
&\quad + n^{-k-1} \left[\sum_{\substack{a \neq u_2 \neq \cdots \neq u_{k-1} \neq b \\ v_1 \neq \cdots \neq v_k}} X_{av_1}^4 X_{av_2} X_{u_2 v_2} \cdots X_{b v_k} \right] \\
&= T(k+1) + R_1 - R_2 - R_3 + R_4,
\end{aligned}$$

where δ_{uv} is the Kronecker delta and

$$\sum\nolimits^* \quad \text{stands for} \quad \sum_{\substack{u_1 \neq \cdots \neq u_{k-1} \neq b \\ v_1 \neq \cdots \neq v_k}}.$$

By Lemmas 1 and 6, and the fact that $EX_{uv}^2 \to 1$, we obtain

$$R_1 = \Big[n^{-k-1}p\sum{}^* X_{av_1}\frac{1}{p}\sum_u\big(X_{uv_1}^2 - 1\big)X_{u_1v_1}X_{u_1v_2}X_{u_2v_2}\cdots X_{bv_k}$$

$$+n^{-k-1}p\sum{}^* X_{av_1}X_{u_1v_1}\cdots X_{bv_k}\Big]$$

$$= o(1) + yT(k) + n^{-k-1}p\Bigg[\sum_{\substack{a\neq u_2\neq\cdots\neq u_{k-1}\neq b\\ v_1\neq\cdots\neq v_k}} X_{av_1}^2 X_{av_2}X_{u_2v_2}\cdots X_{bv_k}\Bigg]$$

$$= o(1) + yT(k) + n^{-k}p$$

$$\times\Bigg[\sum_{\substack{a\neq u_2\neq\cdots\neq u_{k-1}\neq b\\ v_2\neq\cdots\neq v_k}} n^{-1}\sum_{v_1}\big(X_{av_1}^2 - 1\big)X_{av_2}X_{u_2v_2}\cdots X_{bv_k}\Bigg]$$

$$- n^{-k-1}p\Bigg[\sum_{\substack{a\neq u_2\neq\cdots\neq b\\ v_2\neq\cdots\neq v_k}} X_{av_2}^3 X_{u_2v_2}\cdots X_{bv_k}\Bigg] + yT(k-1) + o(1)$$

$$= yT(k) + yT(k-1) + o(1)$$

and

$$R_2 = o(1), \qquad R_3 = o(1), \qquad R_4 = o(1).$$

LEMMA 8. $(T - yI)^k = \sum_{r=0}^k (-1)^{r+1}T(r)\sum_{i=0}^{[(k-r)/2]}C_i(k, r)y^{k-r-i} + o(1)$ *a.s., where*

$$|C_k(k,r)| \le 2^k. \tag{2.15}$$

PROOF. We proceed with our proof by induction on k. When $k = 1$, it is trivial. Now suppose the lemma is true for k. By Lemma 7, we have

$$(T - yI)^{k+1} = \sum_{r=1}^{k}(-1)^{r+1}\{T(r+1) + yT(r) + yT(r-1)\}$$

$$\times\sum_{i=0}^{[(k-r)/2]} C_i(k,r)y^{k-r-i} - T\sum_{i=0}^{[k/2]} C_i(k,0)y^{k-i}$$

$$-\sum_{r=0}^{k}(-1)^{r+1}T(r)\sum_{i=0}^{[(k-r)/2]} C_i(k,r)y^{k-r-i+1} + o(1)$$

$$= \sum_{r=1}^{k+1} (-1)^{r+1} T(r) \sum_{i=0}^{[(k+1-r)/2]} \{-C_i(k, r-1)\} y^{k+1-r-i}$$
$$+ \sum_{r=0}^{k-1} (-1)^{r+1} T(r) \sum_{i=1}^{[(k+1-r)/2]} \{-C_{i-1}(k, r+1)\} y^{k+1-r-i}$$
$$+ I \sum_{i=1}^{[k/2]} C_i(k, 0) y^{k-i} + o(1)$$
$$= \sum_{r=0}^{k+1} (-1)^{r+1} T(r)$$
$$\times \sum_{i=0}^{[(k+1-r)/2]} C_i(k+1, r) y^{k+1-r-i} + o(1) \quad \text{a.s.}$$

Here $C_i(k+1, r)$ is a sum of one or two terms of the form $-C_i(k, r+1)$ and $-C_i(k, r-1)$, which are also quantities satisfying (2.15). By induction, we conclude that (2.15) is true for all fixed k. Thus, the proof of this lemma is complete. □

3. Proof of Theorem 1. By Lemma 2, with $\alpha = \beta = 1$, we have

$$\|S - I - T\| = \max_{i \le p} \left| n^{-1} \sum_{i=1}^{n} \left(X_{ij}^2 - 1 \right) \right| \to 0 \quad \text{a.s.} \tag{3.1}$$

Therefore, to prove Theorem 1, we need only to show that

$$\limsup \|T - yI\| \le 2\sqrt{y} \quad \text{a.s.} \tag{3.2}$$

By Lemmas 1 and 8, we have, for any fixed even integer k,

$$\limsup_{n \to \infty} \|T - yI\|^k \le \sum_{r=0}^{k} Ck^2 y^{r/2} [(k-r)/2] 2^k y^{(k-r)/2}$$
$$\le Ck^4 2^k y^{k/2} \quad \text{a.s.}$$

Taking the k-th root on both sides of this inequality and then letting $k \to \infty$, we obtain (3.2). The proof of Theorem 1 is complete. □

APPENDIX

PROOF OF LEMMA 2 (*Sufficiency*). Without loss of generality, assume that $c = 0$. Since, for $\varepsilon > 0$ and $N \ge 1$,

$$P\left\{ \max_{j \le Mn^\beta} \left| \frac{1}{n^\alpha} \sum_{i=1}^{n} X_{ij} \right| \ge \varepsilon, \text{i.o.} \right\} \le \sum_{k \ge N} P\left\{ \max_{2^{k-1} < n \le 2^k} \max_{j \le M2^{k\beta}} \left| \sum_{i=1}^{n} X_{ij} \right| \ge \varepsilon' 2^{k\alpha} \right\}$$
$$\le \sum_{k \ge N} M 2^{k\beta} P\left\{ \max_{2^{k-1} < n \le 2^k} \left| \sum_{i=1}^{n} X_{i1} \right| \ge 2^{k\alpha} \varepsilon' \right\},$$

where $\varepsilon' = 2^{-\alpha}\varepsilon$, to conclude that the probability on the left-hand side of this inequality is equal to zero, it is sufficient to show that

$$\text{(A.1)} \qquad \sum_{k=1}^{\infty} 2^{k\beta} P\left\{ \max_{n \le 2^k} \left| \sum_{i=1}^{n} X_{i1} \right| \ge 2^{k\alpha}\varepsilon \right\} < \infty.$$

Let $Y_{ik} = X_{i1} I\ (|X_{i1}| < 2^{k\alpha})$ and $Z_{ik} = Y_{ik} - EY_{ik}$. Then $|Z_{ik}| \le 2^{k\alpha+1}$ and $EZ_{ik} = 0$.

Let g be an even integer such that $g(\alpha - \frac{1}{2}) > \beta + 2\alpha$. Then, by the submartingale inequality, we have

$$\text{(A.2)} \qquad \begin{aligned} P\left\{ \max_{n \le 2^k} \left| \sum_{i=1}^{n} Z_{ik} \right| \ge \varepsilon 2^{k\alpha} \right\} &\le C 2^{-kg\alpha} E \left| \sum_{i=1}^{2^k} Z_{ik} \right|^g \\ &\le C 2^{-kg\alpha} \left\{ 2^k E|Z_{1k}^g| + 2^{kg/2} \left(EZ_{1k}^2 \right)^{g/2} \right\}, \end{aligned}$$

where the last inequality follows from Lemma A.1, which will be given later. Hence

$$\text{(A.3)} \qquad \sum_{k=N}^{\infty} 2^{k\beta} P\left\{ \max_{n \le 2^k} \left| \sum_{i=1}^{n} Z_{ik} \right| \ge \varepsilon 2^{k\alpha} \right\} < \infty,$$

which follows from the following bounds:

$$\begin{aligned} \sum_{k=1}^{\infty} 2^{k\beta - kg\alpha + k} E|Z_{1k}^g| &\le C \sum_{k=1}^{\infty} 2^{k(\beta - g\alpha + 1)} E|X_{11}^g| I\left[|X_{11}| < 2^{k\alpha} \right] \\ &\le C \sum_{k=1}^{\infty} 2^{k(\beta - g\alpha + 1)} \\ &\quad \times \left[\sum_{l=1}^{k} E|X_{11}|^g I\left[2^{\alpha(l-1)} \le |X_{11}| < 2^{\alpha l} \right] + 1 \right] \\ &\le C \sum_{l=1}^{\infty} E|X_{11}|^{(\beta+1)/\alpha} I\left[2^{\alpha(-1)} \le |X_{11}| < 2^{\alpha l} \right] + C_1 < \infty \end{aligned}$$

[note that $g\alpha - \beta - 1 > g(\alpha - \frac{1}{2}) - (\beta + 2\alpha) > 0$] and, when $(1 + \beta)/\alpha \ge 2$ (hence $EZ_{1k}^2 \le EX_{11}^2 < \infty$),

$$\sum_{k=1}^{\infty} 2^{k\beta - kg\alpha + kg/2} \left(EX_{1k}^2 \right)^{g/2} \le C \sum_{k=1}^{\infty} 2^{k(\beta + 2\alpha - g(\alpha - 1/2))} < \infty.$$

If $(1 + \beta)/\alpha < 2$, we have

$$\sum_{k=1}^{\infty} 2^{k\beta - kg\alpha + gk/2}\left(EZ_{1k}^2\right)^{g/2}$$

$$\le \sum_{k=1}^{\infty} 2^{k\beta - kg\alpha + kg/2}\left(EZ_{1k}^{(1+\beta)/\alpha + 2 - (1+\beta)/\alpha}\right)^{g/2-1}\left(EZ_{1k}^2\right)$$

$$\le C\sum_{k=1}^{\infty} 2^{k\beta - kg\alpha + kg/2} 2^{k\alpha(2-(1+\beta)/\alpha)(g/2-1)} E\left[X_{11}^2 I(|X_{11}| < 2^{k\alpha})\right]$$

$$\le C\sum_{k=1}^{\infty} 2^{k(\beta - g\alpha + g/2 + g\alpha - 2\alpha - (1+\beta)g/2 + 1 + \beta)}$$

$$\times\left[\sum_{l=1}^{k} E\left(X_{11}^2 I\left(2^{(l-1)\alpha} \le |X_{11}| < 2^{l\alpha}\right)\right) + 1\right]$$

$$\le C\sum_{k=1}^{\infty} 2^{k(\beta - 2\alpha - \beta g/2 + 1 + \beta)} \sum_{l=1}^{k} E\left(|X_{11}|^2 I\left(2^{(l-1)\alpha} \le |X_{11}| < 2^{l\alpha}\right)\right) + C_1$$

$$= C\sum_{l=1}^{\infty} 2^{l(\beta - 2\alpha - \beta g/2 + 1 + \beta)} E\left(|X_{11}|^2 I\left(2^{(l-1)\alpha} \le |X_{11}| < 2^{l\alpha}\right)\right) + C_1$$

$$\le C\sum_{l=1}^{\infty} 2^{l(\beta - 2\alpha - \beta g/2 + 1 + \beta)} E|X_{11}|^{(1+\beta)/\alpha + 2 - (1+\beta)/\alpha}$$

$$\times I\left(2^{(l-1)\alpha} \le |X_{11}| < 2^{l\alpha}\right) + C_1$$

$$\le C\sum_{l=1}^{\infty} 2^{l(\beta - 2\alpha - \beta g/2 + 1 + \beta)} 2^{l\alpha(2-(1+\beta)/\alpha)} E|X_{11}|^{(1+\beta)/\alpha}$$

$$\times I\left(2^{(l-1)\alpha} \le |X_{11}| < 2^{l\alpha}\right) + C_1$$

$$< \infty.$$

Now we estimate EY_{ik} for large k. We have

$$\max_{n \le 2^k}\left|\sum_{i=1}^{n} EY_{ik}\right| \le 2^k |EY_{1k}|$$

$$\text{(A.4)}\qquad \le \begin{cases} 2^k E|X_{11}| I\left[|X_{11}| \ge 2^{k\alpha}\right] \\ \quad \le 2^{k(\alpha-\beta)} E|X_{11}|^{(\beta+1)/\alpha} I\left[|X_{11}| \ge 2^{k\alpha}\right], & \text{if } \alpha \le 1, \\ 2^k \log k + 2^{k(\alpha-\beta)} E|X_{11}|^{(\beta+1)/\alpha} I\left[|X_{11}| > \log k\right], & \text{if } \alpha > 1, \end{cases}$$

$$\le 2^{-1}\varepsilon 2^{k\alpha}.$$

Because (A.3) and (A.4) are true for all $\varepsilon > 0$, the inequality (A.3) is still true if the Z_{ik}'s are replaced by Y_{ik}'s.

Finally, since $E|X_{11}|^{(\beta+1)/\alpha} < \infty$, we have

$$\text{(A.5)} \qquad \sum_{k=1}^{\infty} 2^{k\beta} P\left[\bigcup_{i=1}^{2^k} \{|X_{i1}| \geq 2^{k\alpha}\} \right] \leq \sum_{k=1}^{\infty} 2^{k(\beta+1)} P\left[|X_{11}| \geq 2^{k\alpha}\right] < \infty.$$

Then, (A.1) follows from (A.3)–(A.5).

(*Necessity*.) If $\beta = 0$, then it reduces to the Marcinkiewicz's strong law of large numbers, which is well known. We only need to prove the case of $\beta > 0$. By (2.14) we have

$$\max_{j \leq M(n-1)^\beta} n^{-\alpha} \left| \sum_{i=1}^{n} X_{ij} \right| \to 0 \quad \text{a.s.}$$

and

$$\max_{j \leq M(n-1)^\beta} n^{-\alpha} \left| \sum_{i=1}^{n-1} X_{ij} \right| \to 0 \quad \text{a.s.}$$

By changing to a smaller M, we may change $(n-1)^\beta$ to n^β for simplicity. Thus, we obtain

$$\max_{j \leq Mn^\beta} n^{-\alpha} |X_{nj}| \to 0 \quad \text{a.s.},$$

which, together with the Borel–Cantelli lemma, implies

$$\sum_n P\left[\max_{j \leq Mn^\beta} |X_{nj}| \geq n^\alpha \right] < \infty.$$

By the convergence theorem for an infinite product, this above inequality is equivalent to

$$\begin{aligned} \prod_{n=1}^{\infty} P\left[\max_{j \leq Mn^\beta} |X_{nj}| < n^\alpha \right] &= \prod_{n=1}^{\infty} \left(P(|X_{11}| < n^\alpha) \right)^{[Mn^\beta]} \\ &= \prod_{n=1}^{\infty} \left(1 - P(|X_{11}| \geq n^\alpha) \right)^{[Mn^\beta]} > 0, \end{aligned}$$

which, by using the same theorem again, implies that

$$\sum_{n=1}^{\infty} Mn^\beta P(|X_{11}| \geq n^\alpha) < \infty.$$

This routinely implies $E|X_{11}|^{(\beta+1)/\alpha} < \infty$. Then, applying the sufficiency part, condition (ii) in Lemma 2 follows. □

LEMMA A.1. *Suppose $X_1, \ldots, X_n$ are iid random variables with mean 0 and finite g-th moment, where $g \geq 2$ is an even integer. Then, for some constant $C = C(g)$,*

$$\text{(A.6)} \qquad E\left| \sum_{i=1}^{n} X_1 \right|^g \leq C\left[nEX_1^g + n^{g/2}(EX_1^2)^{g/2} \right].$$

PROOF. We need only to show (A.6) for $g > 2$. We have

$$E\left|\sum_{i=1}^{n} X_i\right|^g \le \sum_{l=1}^{g/2} \sum_{\substack{i_1+\cdots+i_l=g\\ i_1\ge 2,\ldots,i_l\ge 2}} \frac{n!g!E|X_1|^{i_1}\cdots E|X_1|^{i_l}}{l!(n-l)!i_1!\cdots i_l!}. \tag{A.7}$$

By Hölder's inequality, we have

$$E|X_1^{i_t}| \le (EX_1^g)^{(i_t-1)/(g-2)}(EX_1^2)^{(g-i_t)/(g-2)},$$

which, together with (A.7), implies that

$$\begin{aligned} E\left|\sum_{i=1}^{n} X_i\right|^g &\le C\sum_{l=1}^{g/2}(nEX_1^g)^{(g-2l)/(g-2)}(nEX_1^2)^{g(l-1)/(g-2)} \\ &\le \begin{cases} C(nEX_1^2)^{g/2}, & \text{if } (nEX_1^2)^{g/(g-2)} \ge (nEX_1^g)^{2/(g-2)}, \\ CnEX_1^g, & \text{otherwise.} \end{cases} \end{aligned}$$

This implies (A.6), and the proof is finished. □

LEMMA A.2. *Let A and B be two $n \times p$ matrices with entries A_{uv} and B_{uv}, respectively. Denote by $A \circ B$ the Hadamard product of the matrices A and B. Then*

$$\|A \circ B\| \le \|A\|\,\|B\|.$$

PROOF. Let $\mathbf{x} = (x_1, \ldots, x_p)'$ be a unit p-vector. Then the lemma follows from

$$\begin{aligned} \|A \circ B\mathbf{x}\|^2 &= \sum_{u=1}^{n}\left(\sum_{v=1}^{p} A_{uv}B_{uv}x_v\right)^2 \\ &\le \sum_{k=1}^{n}\sum_{u=1}^{n}\left(\sum_{v=1}^{p} A_{kv}B_{uv}x_v\right)^2 \\ &= \operatorname{tr}(BXA'AXB') \\ &= \operatorname{tr}(XA'AXB'B) \le \|A\|^2\|B\|^2 \operatorname{tr}(X^2) = \|A\|^2\|B\|^2, \end{aligned}$$

where $X = \operatorname{diag}(\mathbf{x})$. □

Recently, it was found that this result was proved in Horn and Johnson [(1991), page 332]. Because the proof is very simple, we still keep it here.

REFERENCES

BAI, Z. D., SILVERSTEIN, J. W., and YIN, Y. Q. (1988). A note on the largest eigenvalue of a large dimensional sample covariance matrix. *J. Multivariate Anal.* **26** 166–168.

GEMAN, S. (1980). A limit theorem for the norm of random matrices. *Ann. Probab.* **8** 252–261.

GIRKO, V. L. (1989). Asymptotics of the distribution of the spectrum of random matrices. *Russian Math. Surveys* **44** 3–36.

GRENANDER, U. and SILVERSTEIN, J. (1977). Spectral analysis of networks with random topologies. *SIAM J. Appl. Math.* **32** 499–519.

HORN, R. A. and JOHNSON, C. R. (1991). *Topics in Matrix Analysis*. Cambridge Univ. Press.

JONSSON, D. (1982). Some limit theorems for the eigenvalues of a sample covariance matrix. *J. Multivariate Anal.* **12** 1–38.

MARCUS, M. (1964). *A Survey of Matrix Theory and Matrix Inequalities*. Allyn and Bacon, Boston.

SILVERSTEIN, J. W. (1985). The smallest eigenvalue of a large dimensional Wishart matrix. *Ann. Probab.* **13** 1364–1368.

SILVERSTEIN, J. W. (1989). On the weak limit of the largest eigenvalue of a large dimensional sample covariance matrix. *J. Multivariate Anal.* **30** 307–311.

VON NEUMANN, J. (1937). Some matrix inequalities and metrization of matric space. *Tomsk Univ. Rev.* **1** 286–300.

WACHTER, K. W. (1978). The strong limits of random matrix spectra for sample matrices of independent elements. *Ann. Probab.* **6** 1–18.

YIN, Y. Q. (1986). Limiting spectral distribution for a class of random matrices. *J. Multivariate Anal.* **20** 50–68.

YIN, Y. Q., BAI, Z. D. and KRISHNAIAH, P. R. (1988). On the limit of the largest eigenvalue of the large dimensional sample covariance matrix. *Probab. Theory Related Fields* **78** 509–521.

DEPARTMENT OF STATISTICS
TEMPLE UNIVERSITY
PHILADELPHIA, PENNSYLVANIA 19122

DEPARTMENT OF MATHEMATICS
UNIVERSITY OF MASSACHUSETTS, LOWELL
LOWELL, MASSACHUSETTS 01854

The Annals of Probability
1997, Vol. 25, No. 1, 494–529

CIRCULAR LAW

BY Z. D. BAI[1]

National Sun Yat-sen University

It was conjectured in the early 1950's that the empirical spectral distribution of an $n \times n$ matrix, of iid entries, normalized by a factor of $1/\sqrt{n}$, converges to the uniform distribution over the unit disc on the complex plane, which is called the circular law. Only a special case of the conjecture, where the entries of the matrix are standard complex Gaussian, is known. In this paper, this conjecture is proved under the existence of the sixth moment and some smoothness conditions. Some extensions and discussions are also presented.

1. Introduction. Suppose that Ξ_n is an $n \times n$ matrix with entries $\xi_{kj} = (1/\sqrt{n})x_{kj}$ and $\{x_{kj}, k, j = 1, 2, \dots,\}$ forms an infinite double array of iid complex random variables of mean zero and variance one. Using the complex eigenvalues $\lambda_1, \lambda_2, \dots, \lambda_n$ of Ξ_n, we can construct a two-dimensional empirical distribution by

$$\mu_n(x, y) = \frac{1}{n}\#\{i \le n: \mathrm{Re}(\lambda_k) \le x, \mathrm{Im}(\lambda_k) \le y\},$$

which is called the empirical spectral distribution of the matrix Ξ_n.

The motivation for the study of spectral analysis of large-dimensional random matrices comes from quantum mechanics. The energy level of a quantum is not directly observable and it is known that the energy levels of quantums can be described by the eigenvalues of a matrix of observations. Since the 1960's, the spectral analysis of large-dimensional random matrices has attracted considerable interest from probabilists, mathematicians and statisticians. For a general review, the reader is referred to, among others, Bai (1993a, b), Bai and Yin (1993, 1988a, b, 1986), Geman (1980, 1986), Silverstein and Bai (1995), Wachter (1978, 1980) and Yin, Bai and Krishnaiah (1988).

Most of the important existing results are on symmetric large-dimensional random matrices. Basically, two powerful tools are used in this area. The first is the moment approach which was successfully used in finding the limiting spectral distributions of large-dimensional random matrices and in establishing the strong convergence of extreme eigenvalues. See, for example, Bai and Yin (1993, 1988a, b, 1986), Geman (1980, 1986), Jonsson (1982) and Yin, Bai

Received March 1996.
[1]Supported by ROC Grant NSC84-2816-M110-009L.
AMS 1991 *subject classifications.* Primary 60F15; secondary 62H99.
Key words and phrases. Circular law, complex random matrix, noncentral Hermitian matrix, largest and smallest eigenvalue of random matrix, spectral radius, spectral analysis of large-dimensional random matrices.

and Krishnaiah (1988). The second is the Stieltjes transform which was used in Bai (1993a, b), Bai and Silverstein (1995), Marčenko and Pastur (1967), Pastur (1972, 1973), Silverstein and Choi (1995) and Wachter (1978, 1980). Unfortunately, these two approaches are not suitable for dealing with nonsymmetric random matrices. Due to lack of appropriate methodologies, very few results were known about nonsymmetric random matrices. The only known result is about the spectral radius of the matrix Ξ_n. Bai and Yin [(1986), under the fourth moment] and Geman [(1986), under some growth restrictions on all moments], independently proved that with probability 1, the upper limit of the spectral radius of Ξ_n is not greater than 1.

Since the early 1950's, it has been conjectured that the distribution $\mu_n(x, y)$ converges to the so-called circular law, that is, the uniform distribution over the unit disk in the complex plane. This problem has been unsolved, except where the entries are complex normal variables [given in an unpublished paper of Silverstein in 1984 but reported in Hwang (1986)]. Silverstein's proof relies on the explicit expression of the joint distribution density of the eigenvalues of Ξ_n [see, e.g., Ginibre (1965)]. Hence his approach cannot be extended to the general case. Girko presented (1984a, b) a proof of this conjecture under some conditions. However, the paper contained too many mathematical gaps, leaving the problem still open. After Girko's flaw was found, "many have tried to understand Girko's 'proofs' without success," [Edelman (1995)]. When the entries are iid real normal random variables, Edelman (1995) found the conditional joint distribution of the complex eigenvalues when the number of real eigenvalues are given and showed that the expected empirical spectral distribution of Ξ_n tends to the circular law.

In spite of mathematical gaps in his arguments, Girko had come up with an important idea (his Lemma 1), which established a relation between the characteristic function of the empirical spectral distribution of Ξ_n and an integral involving the empirical spectral distribution of a Hermitian matrix. Girko's Lemma 1 is presented below for easy reference.

GIRKO'S LEMMA 1. *For any $uv \neq 0$, we have*

$$
\begin{aligned}
m_n(u, v) &= \iint_2 \exp(iux + ivy)\mu_n(dx, dy) \\
&= \frac{u^2 + v^2}{4iu\pi} \iint \frac{\partial}{\partial s}\left[\int_0^\infty \ln x \nu_n(dx, z)\right] \exp(ius + ivt)\, dt\, ds,
\end{aligned} \tag{1.1}
$$

where $z = s + it$, $i = \sqrt{-1}$ and $\nu_n(x, z)$ is the empirical spectral distribution of the nonnegative definite Hermitian matrix $\mathbf{H}_n = \mathbf{H}_n(z) = (\Xi_n - z\mathbf{I})^*(\Xi_n - z\mathbf{I})$. Here and throughout this paper, Ξ^* denotes the complex conjugate and transpose of the matrix Ξ.

It is easy to see that $m_n(u, v)$ is an entire function in both u and v. By Bai and Yin (1986) or Geman (1986), the family of distributions $\mu_n(x, y)$ is tight. And hence, every subsequence of $\mu_n(x, y)$ contains a completely convergent subsequence and the characteristic function $m(u, v)$ of the limit must be also entire. Therefore, to prove the circular law, applying Girko's Lemma 1, one needs only show that the right-hand side of (1.1) converges to its counterpart generated by the circular law. Note that the function $\ln x$ is not bounded at both infinity and zero. Therefore, the convergence of the right hand side of (1.1) cannot be simply reduced to the convergence of ν_n. In view of the results of Yin, Bai and Krishnaiah (1988), there would not be a serious problem for the upper limit of the inner integral, since the support of ν_n is a.s. eventually bounded from the right by $(2 + \varepsilon + |z|)^2$ for any positive ε. In his 1984 papers, Girko failed only in dealing with the lower limit of the integral.

In this paper, making use of Girko's lemma, we shall provide a proof of the famous circular law.

THEOREM 1.1 (Circular law). *Suppose that the entries of* $\mathbf{X}$ *have finite sixth moment and that the joint distribution of the real and imaginary part of the entries has a bounded density. Then, with probability* 1, *the empirical distribution* $\mu_n(x, y)$ *tends to the uniform distribution over the unit disc in two-dimensional space.*

The proof of the theorem will be rather tedious. Thus, for ease of understanding, an outline of the proof is provided first.

The proof of the theorem will be presented by showing that with probability 1, $m_n(u, v) \to m(u, v)$ for every (u, v) such that $uv \neq 0$. To this end, we need the following steps.

1. Reduce the range of integration. First we need to reduce the range of integration to a finite rectangle, so that the dominated convergence theorem is applicable. As will be seen, proof of the circular law reduces to showing that for every large $A > 0$ and small $\varepsilon > 0$,

$$\int\int_T \left[\frac{\partial}{\partial s}\int_0^\infty \ln x\nu_n(dx, z)\right]\exp(ius + ivt)\,ds\,dt$$
$$\to \iint_T \left[\frac{\partial}{\partial s}\int_0^\infty \ln x\nu(dx, z)\right]\exp(ius + ivt)\,ds\,dt, \tag{1.2}$$

   where $T = \{(s, t);\ |s| \le A,\ |t| \le A^2,\ |\sqrt{s^2 + t^2} - 1| \ge \varepsilon\}$ and $\nu(x, z)$ is the limiting spectral distribution of the sequence of matrices $\mathbf{H}_n$ which determines the circular law.
2. Find the limiting spectrum $\nu(\cdot, z)$ of $\nu_n(\cdot, z)$ and show that it determines the circular law.
3. Find a convergence rate of $\nu_n(x, z)$ to $\nu(x, z)$ uniformly in every bounded region of z. Then, we will be able to apply the convergence rate to establish (1.2). As argued earlier, it is sufficient to show the following.

4. Show that for suitably defined sequence ε_n, with probability 1:

$$\limsup_{n\to\infty} \iint_T \left| \int_{\varepsilon_n}^{\infty} \ln x(\nu_n(dx, z) - \nu(dx, z)) \right| = 0, \tag{1.3}$$

and

$$\limsup_{n\to\infty} \left| \iint_T \int_0^{\varepsilon_n} \ln x \nu_n(dx, z)\, ds\, dt \right| = 0 \tag{1.4}$$

The convergence rate of $\nu_n(\cdot, z)$ will be used in proving (1.3). The proof of (1.4) will be specifically treated. The proofs of the above four steps are rather long and thus the paper is organized into several sections. For convenience, a list of symbols and their definitions are given in Section 2. Section 3 is devoted to the reduction of the integral range. In Section 4, we shall present some lemmas discussing the properties of the limiting spectrum ν and its Stieltjes transform, and some lemmas establishing a convergence rate of ν_n. The most difficult part of this work, namely, the proof of (1.4), is given in Section 5 and the proof of Theorem 1.1 is present in Section 6. Some discussions and extensions are given in Section 7. Some technical lemmas are presented in the Appendix.

2. List of notations. The definitions of the notations presented below will be given again when the notations appear.

$\{x_{kj}\}$: a double array of iid complex random variables with $\mathrm{E}(x_{kj}) = 0$, $\mathrm{E}|x_{kj}|^2 = 1$ and $\mathrm{E}|x_{kj}|^6 < \infty$;

$\mathbf{X}_n = (x_{kj})_{k,j=1,2,\dots,n}$. Its kth column vector is denoted by $\mathbf{x}_k$.

$\Xi_n = (1/\sqrt{n})\mathbf{X}_n = (\xi_{jk}) = (\boldsymbol{\xi}_k)$.

$\mathbf{R}(z) = \Xi_n - z\mathbf{I}_n$ with $z = s + it$ and $i = \sqrt{-1}$. Its kth column vector is denoted by $\mathbf{r}_k$.

$\mathbf{H}_n = \mathbf{R}^*(z)\mathbf{R}(z)$.

$\mathbf{A}^*$ denotes the complex conjugate and transpose of the matrix $\mathbf{A}$.

$m_n(u, v)$ and $m(u, v)$ denote the characteristic functions of the distributions μ_n and the circular law μ.

$F^{\mathbf{X}}$ denotes the empirical spectral distribution of $\mathbf{X}$ if $\mathbf{X}$ is a matrix. However, we do not use this notation for the matrix Ξ_n since it is traditionally and simply denoted as F_n.

$\alpha = x + iy$. In most cases, $y = y_n = n^{-1/60} \ln^{-1} n$. But in some places, y denotes a fixed positive number.

$\nu_n(x, z)$ denotes the empirical spectral distribution of $\mathbf{H}_n$ and $\nu(x, z)$ denotes its limiting spectral distribution.

$\Delta_n(\alpha)$ and $\Delta(\alpha)$ are the Stieltjes transforms of $\nu_n(x, z)$ and $\nu(x, z)$ respectively.

Boldface capitals will be used to denote matrices and boldface lower case used for vectors.

The symbol K_d denotes the upper bound of the joint density of its real and imaginary parts of the entries x_{kj}. In Section 7, it is also used for the upper bound of the conditional density of the real part of the entry of $\mathbf{X}$ when its imaginary part is given.

$\varepsilon_n = \exp(-n^{1/120})$, a constant.

Re(·) and Im(·) denote the real and imaginary parts of a complex number.

$I(\cdot)$ denotes the indicator function of the set in parentheses.

$\|f\|$ denotes the uniform norm of the function f, that is, $\|f\| = \sup_{\mathbf{x}}|f(\mathbf{x})|$.

$\|\mathbf{A}\|$ denotes the operation norm of the matrix $\mathbf{A}$, that is, its largest singular value.

3. Integral range reduction. Let $\mu_n(x, y)$ denote the empirical spectral distribution of the matrix $\Xi_n = (1/\sqrt{n})X_n$ and $\nu_n(x, z)$ denote the empirical distribution of the Hermitian matrix $\mathbf{H} = \mathbf{H}_n = (\Xi_n - z\mathbf{I})^*(\Xi_n - z\mathbf{I})$, for each fixed $z = s + it \in \mathcal{C}$. The following lemma is the same as Girko's Lemma 1. We present a proof here for completeness; this proof is easier to understand than that provided by Girko (1984a, b).

LEMMA 3.1. *For all $u \neq 0$ and $v \neq 0$, we have*

$$
\begin{aligned}
m_n(u, v) &= \iint \exp(ius + ivy)\,\mu_n(dx, dy) \\
&= \frac{u^2 + v^2}{4iu\pi} \iint g_n(s, t) \exp(ius + ivt)\, dt\, ds
\end{aligned} \tag{3.1}
$$

where $\iint \cdots dt\,ds$ denotes the iterated integral $\int[\int \cdots dt]\,ds$ and

$$
g_n(s, t) = \frac{1}{n}\sum_{k=1}^{n} \frac{2(s - \mathrm{Re}(\lambda_k))}{(s - \mathrm{Re}(\lambda_k))^2 + (t - \mathrm{Im}(\lambda_k))^2} = \frac{\partial}{\partial s}\int_0^\infty \ln x \nu_n(dx, z).
$$

REMARK 3.1. When $z = \lambda_k$ for some $k \leq n$, $\nu_n(x, z)$ will have a positive measure of $1/n$ at $x = 0$ and hence the inner integral of $\ln x$ is not well defined. Therefore, the iterated integral in (3.1) should be understood as the generalized integral. That is, we cut off the n discs with centers $[\mathrm{Re}(\lambda_k), \mathrm{Im}(\lambda_k)]$ and radius ε from the s, t plane. Take the integral outside the n discs in the s, t plane and then take $\varepsilon \to 0$. Then, the outer integral in (3.1) is defined to be the limit [w.r.t. (with respect to) $\varepsilon \to 0$]) of the integral over the reduced integration range.

REMARK 3.2. Note that $g_n(s, t)$ is twice the real part of the Stieltjes transform of the two-dimensional empirical distribution μ_n, that is,

$$
\iint \frac{\mu_n(dx, dy)}{x + iy - z} = \frac{1}{n}\sum_{k=1}^{n} \frac{1}{\lambda_k - z},
$$

which has exactly n simple poles at the n eigenvalues of Ξ_n. The function $g_n(s, t)$ uniquely determines the n eigenvalues of the matrix Ξ_n. On the other hand, $g_n(s, t)$ can also be regarded as the derivative (w.r.t. s) of the logarithm of the determinant of $\mathbf{H}$ which can be expressed as an integral w.r.t. the empirical spectral distribution of **H**, as given in the second equality in the definition of $g_n(s,t)$. In this way, the problem of the spectrum of a non-Hermitian matrix is transformed as one of the spectrum of a Hermitian matrix, so that the approach via Stieltjes transforms can be applied to this problem.

PROOF. Note that for all $uv \neq 0$,

$$\begin{aligned}
&\frac{u^2+v^2}{2iu\pi}\iint \frac{s}{s^2+t^2}\exp(ius+ivt)\,dt\,ds\\
&\quad=\frac{u^2+v^2}{2iu\pi}\iint \frac{\operatorname{sign}(s)}{1+t^2}\exp(ius+iv|s|t)\,dt\,ds\\
&\quad=\frac{u^2+v^2}{2iu}\int \operatorname{sign}(s)\exp(ius-|vs|)\,ds\\
&\quad=\frac{u^2+v^2}{2|u|}\int \sin|us|\exp(-|vs|)\,ds=1.
\end{aligned}$$

Therefore,

$$\begin{aligned}
&\iint \exp(iux+ivy)\mu_n(dx,dy)\\
&\quad=\frac{u^2+v^2}{2iu\pi}\iint \frac{1}{n}\sum_{k=1}^{n}\frac{s}{s^2+t^2}\\
&\qquad\qquad\times\exp(ius+ivt+iu\operatorname{Re}(\lambda_k)+iv\operatorname{Im}(\lambda_k))\,dt\,ds\\
&\quad=\frac{u^2+v^2}{4iu\pi}\iint \frac{1}{n}\sum_{k=1}^{n}\frac{2(s-\operatorname{Re}(\lambda_k))}{(s-\operatorname{Re}(\lambda_k))^2+(t-\operatorname{Im}(\lambda_k))^2}\\
&\qquad\qquad\times\exp(ius+ivt)\,dt\,ds\\
&\quad=\frac{u^2+v^2}{4iu\pi}\iint\left[\frac{\partial}{\partial s}\int_0^\infty \ln x\,\nu_n(dx,z)\right]\exp(ius+ivt)\,dt\,ds.
\end{aligned}$$

The proof of Lemma 3.1 is complete. □

LEMMA 3.2. *For all $uv \neq 0$, we have*

$$\begin{aligned}
m(u,v)&=\frac{1}{\pi}\iint_{x^2+y^2\le 1}\exp(iux+ivy)\,dx\,dy\\
&=\frac{u^2+v^2}{4iu\pi}\iint g(s,t)\exp(ius+ivt)\,dt\,ds,
\end{aligned}\tag{3.2}$$

where

$$g(s,t)=\begin{cases}\dfrac{2s}{s^2+t^2}, & \text{if } s^2+t^2>1,\\ 2s, & \text{otherwise.}\end{cases}$$

PROOF. As in the proof of Lemma 3.1, we have, for all $uv \neq 0$,

$$\text{(3.3)}\quad m(u,v)=\frac{u^2+v^2}{4iu\pi}\iint\left[\frac{1}{\pi}\iint_{x^2+y^2\le 1}\frac{2(s-x)}{(s-x)^2+(t-y)^2}\,dx\,dy\right]\times\exp(ius+ivt)\,ds\,dt.$$

Then, the lemma follows from the fact that the inner integral on the right-hand side of (3.3) equals $g(s,t)$, using Green's formula. □

LEMMA 3.3. *For any $uv \neq 0$ and $A > 2$, with probability* 1, *when n is large, we have*

$$\text{(3.4)}\quad \left|\int_{|s|\ge A}\int_{-\infty}^{\infty} g_n(s,t)\exp(ius+ivt)\,ds\,dt\right| \le \frac{4\pi}{|v|}\exp\left(-\frac{1}{2}|v|A\right)$$

and

$$\text{(3.5)}\quad \left|\int_{|s|\le A}\int_{|t|\ge A^2} g_n(s,t)\exp(ius+ivt)\,ds\,dt\right| \le \frac{8(A+1+\varepsilon)}{A^2}.$$

Furthermore, the two inequalities above hold if the function $g_n(s,t)$ is replaced by $g(s,t)$.

PROOF. From Bai and Yin (1986), it follows that with probability 1, when n is large, we have $\max_k\{|\lambda_k|\} \le 1+\varepsilon$. Hence,

$$\begin{aligned}
&\left|\int_{|s|\ge A}\int_{-\infty}^{\infty} g_n(s,t)\exp(ius+ivt)\,ds\,dt\right|\\
&=\left|\int_{|s|\ge A}\int_{-\infty}^{\infty}\frac{1}{n}\sum_{k=1}^{n}\frac{2(s-\operatorname{Re}(\lambda_k))}{(s-\operatorname{Re}(\lambda_k))^2+(t-\operatorname{Im}(\lambda_k))^2}\right.\\
&\qquad\qquad\left.\times\exp(ius+ivt)\,ds\,dt\right| \qquad \text{(3.6)}\\
&=\left|\frac{\pi}{n}\sum_{k=1}^{n}\int_{|s|\ge A}\operatorname{sign}(s-\operatorname{Re}(\lambda_k))\exp(ius-|v(s-\operatorname{Re}(\lambda_k))|)\,ds\right|\\
&\le\frac{4\pi}{|v|}\exp\left(-\frac{1}{2}|v|A\right),
\end{aligned}$$

and

$$(3.7)\qquad \left|\int_{|s|\le A}\int_{|t|\ge A^2} g_n(s,t)\exp(ius+ivt)\,ds\,dt\right| \le \frac{8(A+1+\varepsilon)}{A^2}.$$

Similarly, one can prove the above two inequalities for $g(s,t)$. The proof of Lemma 3.3 is complete. □

From Lemma 3.3, one can see that the right-hand sides of (3.4) and (3.5) can be made arbitrarily small by making A large enough. The same is true when $g_n(s,t)$ is replaced by $g(s,t)$. Therefore, the proof of the circular law is reduced to showing

$$(3.8)\qquad \int_{|s|\le A}\int_{|t|\le A^2}[g_n(s,t)-g(s,t)]\exp(ius-ivt)\,ds\,dt \to 0.$$

Finally, define sets

$$T=\{(s,t):|s|\le A,\ |t|\le A^2 \text{ and } ||z|-1|\ge\varepsilon\}$$

and

$$T_1=\{(s,t):||z|-1|<\varepsilon\},$$

where $z=s+it$.

LEMMA 3.4. *For all fixed A and $0<\varepsilon<1$, for all n,*

$$(3.9)\qquad \left|\iint_{T_1} g_n(s,t)\,ds\,dt\right| \le 24\pi\sqrt{\varepsilon}.$$

Furthermore, when $g_n(s,t)$ in (3.9) *is replaced by $g(s,t)$, the estimation* (3.9) *remains true.*

PROOF. For any fixed u and v, by a polar transformation, we obtain

$$\left|\iint_{T_1}\frac{(s-u)\,dt\,ds}{(s-u)^2+(t-v)^2}\right| = \left|\int_0^{2\pi}2D(\theta)\cos\theta\,d\theta\right| \le 24\pi\sqrt{\varepsilon},$$

where $D(\theta)$ is the sum of lengths of at most two segments which are the intersection of the ring T_1 and the straight line $(s-u)\cos\theta+(t-v)\sin\theta=0$. In the above, we have used the fact that $\max_\theta D(\theta)\le 2\sqrt{4\varepsilon+2\varepsilon^2}\le 6\sqrt{\varepsilon}$.

This completes the proof of (3.9) for $g_n(s,t)$. The proof of (3.9) for $g(s,t)$ is similar and thus omitted. The proof of Lemma 3.4 is complete. □

Note that the right-hand side of (3.9) can be made arbitrarily small by choosing ε small. Thus, by Lemmas 3.3 and 3.4, to prove the circular law, one needs only to show that, for each fixed $A>0$ and $\varepsilon\in(0,1)$,

$$(3.10)\qquad \iint_T (g_n(s,t)-g(s,t))\,ds\,dt \to 0 \quad \text{a.s.}$$

4. Convergence of $\nu_n(x, z)$ and the limiting spectrum $\nu(x, z)$. In this section, we shall establish a convergence rate of $\nu_n(x, z)$ and discuss properties of the limiting distribution $\nu(x, z)$ of $\nu_n(x, z)$. Throughout the remainder of this paper, we shall use the notations $o(1)$ and $O(1)$ in the sense of "almost surely." Furthermore, if the quantities represented by the symbols $o(1)$ or $O(1)$ are involved with indices j, l or k, or variables α or z, then the orders are uniform about these indices and variables.

Suppose that $\nu(\cdot, z)$ is the limiting spectral distribution of some convergent subsequence of $\nu_n(\cdot, z)$. Denote by $\Delta_n(\alpha, z)$ and $\Delta(\alpha, z)$, $\alpha = x + iy$, $y > 0$, the Stieltjes transforms of $\nu_n(\cdot, z)$ and $\nu(\cdot, z)$, respectively, that is,

$$\Delta_n(\alpha, z) = \int \frac{1}{x - \alpha} \nu_n(dx, z) = \frac{1}{n} \operatorname{tr}(\mathbf{H} - \alpha \mathbf{I})^{-1}$$

and

$$\Delta(\alpha, z) = \int \frac{1}{x - \alpha} \nu(dx, z),$$

where α is a complex number with positive imaginary part. The variable z in these symbols will be omitted when there is no confusion. We will prove the following lemmas.

LEMMA 4.1. *Suppose that the conditions of Theorem* 1.1 *are true. Write*

$$\Delta_n(\alpha)^3 + 2\Delta_n(\alpha)^2 + \frac{\alpha + 1 - |z|^2}{\alpha} \Delta_n(\alpha) + \frac{1}{\alpha} = r_n, \tag{4.1}$$

where $r_n = r_n(\alpha, z)$. *Then, we have*

$$\Delta^3 + 2\Delta^2 + \frac{\alpha + 1 - |z|^2}{\alpha} \Delta + \frac{1}{\alpha} = 0 \tag{4.2}$$

the remainder term r_n *satisfies*

$$\begin{gathered} \sup\{|r_n|: \alpha = x + iy \text{ with } -\infty < x < \infty,\, u \geq y_n, |z| \leq M\} = o(\delta_n), \\ y_n = n^{-1/60} \ln^{-1} n \quad \text{and} \quad \delta_n = n^{-1/60}. \end{gathered} \tag{4.3}$$

LEMMA 4.2. *The limiting distribution function* $\nu(x, z)$ *satisfies*

$$|\nu(x + u, z) - \nu(x, z)| \leq \pi^{-1} \sqrt{2} \max\{\sqrt{|u|}, |u|\} \quad \text{for all } z. \tag{4.4}$$

Also, the limiting distribution function $\nu(x, z)$ *is supported by the interval* (x_1, x_2) *when* $|z| > 1$ *and by* $(0, x_2)$ *when* $|z| \leq 1$, *where*

$$x_1 = \frac{1}{8|z|^2}\left[-1 + 20|z|^2 + 8|z|^4 - \sqrt{(1 + 8|z|^2)^3}\right],$$

$$\begin{aligned} x_2 &= \frac{1}{8|z|^2}\left[\sqrt{(1 + 8|z|^2)^3} - 1 + 20|z|^2 + 8|z|^4\right] \quad \text{when } z \neq 0, \\ &= 4 \quad \text{when } z = 0. \end{aligned}$$

LEMMA 4.3. *Let $m_2(\alpha)$ and $m_3(\alpha)$ denote the two solutions of* (4.2) *other than $\Delta(\alpha)$. For any given constants $N > 0$, $A > 0$ and $\varepsilon \in (0, 1)$, there exist positive constants ε_0 and ε_1 such that for all large n, $|\alpha| \le N$, $y \ge 0$ and $z \in T$, we have the following*:

(i)

$$\max_{j=2,3} |\Delta(\alpha) - m_j(\alpha)| \ge \varepsilon_0, \tag{4.5}$$

(ii) *for* $|\alpha - x_2| \ge \varepsilon_1$, (*and* $|\alpha - x_1| \ge \varepsilon_1$, *if* $|z| > 1$),

$$\min_{j=2,3} |\Delta(\alpha) - m_j(\alpha)| \ge \varepsilon_0, \tag{4.6}$$

(iii) *for* $|\alpha - x_2| < \varepsilon_1$,

$$\min_{j=2,3} |\Delta(\alpha) - m_j(\alpha)| \ge \varepsilon_0 \sqrt{|\alpha - x_2|}, \tag{4.7}$$

(iv) *for* $|z| > 1 + \varepsilon$, *and* $|\alpha - x_1| < \varepsilon_1$,

$$|\Delta(\alpha) - m_j(\alpha)| \ge \varepsilon_0 \sqrt{|\alpha - x_1|}. \tag{4.8}$$

REMARK 4.1. This lemma basically says that the Stieltjes transform of the limiting spectral distribution $\nu(\cdot, z)$ is distinguishable from the other two solutions of the equation (4.2). Here, we give a more explicit estimate of the distance of $\Delta(\alpha)$ from the other two solutions. This lemma simply implies that the limiting spectral distribution of the sequence of matrices $\mathbf{H}_n$ is unique and nonrandom since the variation from ν_n to ν_{n+1} is of order $O(1/n)$ and hence the variation from $\Delta_n(\alpha)$ to $\Delta_{n+1}(\alpha)$ is $O(1/ny)$.

LEMMA 4.4. *We have*

$$\frac{\partial}{\partial s} \int_0^\infty \ln x \nu(dx, z) = g(s, t). \tag{4.9}$$

LEMMA 4.5. *Under the conditions of Theorem* 1.1, *for any* $M_2 > M_1 > 0$,

$$\begin{aligned} \sup_{M_1 \le |z| \le M_2} \|\nu_n(\cdot, z) - \nu(\cdot, z)\| &:= \sup_{x,\, M_1 \le |z| \le M_2} |\nu_n(x, z) - \nu(x, z)| \\ &= o(n^{-1/120}). \end{aligned} \tag{4.10}$$

REMARK 4.2. Lemma 4.5 is used only in proving (1.3) for a suitably chosen ε_n. From the proof of the lemma and comparing with the results in Bai (1993a, b) one can see that a better rate of convergence can be obtained by considering more terms in the expansion. As the rate given in (4.10) is enough for our purposes, we restrict ourselves to the weaker result (4.10) by a simpler proof, rather than trying to get a better rate by long and tedious arguments.

PROOF OF LEMMA 4.1. This lemma plays a key role in establishing a convergence rate of the empirical spectral distribution $\nu_n(\cdot, z)$ of $\mathbf{H}$. The approach used in the proof of this lemma is in a manner typical in the application of Stieltjes transforms to the spectral analysis of large-dimensional random matrices. The basic idea of the proof relies on the following two facts: (1) the n diagonal elements of $(\mathbf{H} - \alpha\mathbf{I})^{-1}$ are identically distributed and asymptotically the same as their average, the Stieltjes transform of the empirical spectral distribution of $\mathbf{H}$; (2) for all $k \le n$, $(1/n)\mathrm{tr}((\tilde{\mathbf{H}}_k - \alpha\mathbf{I}_{n-1})^{-1})$ are identically distributed and asymptotically equivalent to $(1/n)\mathrm{tr}((\mathbf{H} - \alpha\mathbf{I}_n)^{-1})$, where the matrix $\tilde{\mathbf{H}}_k$ is defined similarly as $\mathbf{H}$ by Ξ with the kth column and row removed. By certain steps of expansion, one can obtain the equation (4.1) which determines the Stieltjes transform $\Delta_n(\alpha)$ of $\mathbf{H}$.

Since $\Delta(\alpha)$ is the limit of some convergent subsequence of $\Delta_n(\alpha)$ and hence (4.2) is a consequence of (4.3), only (4.3) need be shown.

To begin, we need to reduce the uniform convergence of (4.3) over an uncountable set to that over a finite set. Yin, Bai and Krishnaiah (1988), proved that $\|\Xi_n\| \to 2$, a.s., where $\|\Xi_n\|$ denotes the operator norm, that is, the largest singular value, of the matrix Ξ_n when the entries of $\mathbf{X}$ are all real. Their proofs can be translated to the complex case word for word so that the above result is still true when the entries are complex. Therefore, with probability 1, when n is large enough, for all $|z| \le M$,

$$\Lambda_{\max}(\mathbf{H}_n) \le (\|\Xi_n\| + |z|)^2 \le (3 + M)^2. \tag{4.11}$$

Hence, when $|\alpha| \ge n^{1/60} \ln n$ and (4.11) is true, we have for all large n

$$|\Delta_n(\alpha)| \le \frac{1}{|\alpha| - (3 + M)^2}$$

and consequently,

$$\begin{aligned} |r_n| &= \left| \Delta_n^2 + 2\Delta_n^2 + \frac{\alpha + 1 - |z|^2}{\alpha}\Delta_n + \frac{1}{\alpha} \right| \\ &\le 4Mn^{-1/60} \ln^{-1} n = o(\delta_n). \end{aligned} \tag{4.12}$$

If $\max(|\alpha|, |\alpha'|) < n^{1/60} \ln n$ and $|\alpha - \alpha'| \le n^{-1/7}$, then

$$|\Delta_n(\alpha) - \Delta_n(\alpha')| \le [\min(y, y')]^{-2} |\alpha - \alpha'| \le y_n^{-2} n^{-1/7},$$

which implies that

$$|r_n(\alpha) - r_n(\alpha')| \le My_n^{-4} n^{-1/7} \le Mn^{-1/14} \tag{4.13}$$

for some positive constant M.

Suppose that $|z - z'| \le n^{-1/4}$. Let $\Lambda_k(z)$ and $\Lambda_k(z')$ (arranged in increasing order) be eigenvalues of the matrices $\mathbf{H}(z) = (\Xi_n - z\mathbf{I})^*(\Xi_n - z\mathbf{I})$ and

$\mathbf{H}(z') = (\Xi_n - z'\mathbf{I})^*(\Xi_n - z'\mathbf{I})$, respectively. Then for any fixed α, by Lemma A.5, we have

$$
(4.14)\quad \begin{aligned}
&|\Delta_n(\alpha, z) - \Delta_n(\alpha, z')| \\
&\leq \frac{1}{n}\sum_{k=1}^{n}\frac{|\Lambda_k(z) - \Lambda_k(z')|}{|\Lambda_k(z) - \alpha||\Lambda_k(z') - \alpha|} \\
&\leq y^{-2}|z - z'|\left(\frac{1}{n}\mathrm{tr}(2\Xi_n - (z + z')\mathbf{I})^*(2\Xi_n - (z + z')\mathbf{I})\right)^{1/2} \\
&\leq y^{-2}|z - z'|(3 + 2M) \leq Mn^{-1/6}.
\end{aligned}
$$

This, together with (4.12) and (4.13), shows that to finish the proof of (4.3), it is sufficient to show that

$$
(4.15)\quad \max_{l, j \leq n}\{|r_n(\alpha_l, z_j)|\} = o(\delta_n),
$$

where $\alpha_l = x(l) + iy(l)$, $l = 1, 2, \ldots, [n^{1/6}]$ and z_j, $j = 1, 2, \ldots, [n^{1/3}]$ are selected so that $|x(l)| \leq n^{1/60}\ln n$, $y_n \leq y(l) \leq n^{1/60}\ln n$ and for each $|\alpha| \leq n^{1/60}\ln n$ with $y \geq y_n$, there is an l such that $|\alpha - \alpha_l| < n^{-1/7}$; and for each $|z| \leq M$, there is a j such that $|z - z_j| \leq n^{-1/4}$.

In the rest of the proof of this lemma, we shall suppress the indices l and j from the variables α_l and z_j. The reader should remember that we shall only consider those α_l and z_j which are selected to satisfy the properties described in the last paragraph.

Let $\mathbf{R} = \mathbf{R}_n(z) = (r_{kj})$, where $r_{kj} = \xi_{kj}$ for $j \neq k$ and $r_{kk} = \xi_{kk} - z$. Then $\mathbf{H} = \mathbf{R}^*\mathbf{R}$. We have

$$
(4.16)\quad \begin{aligned}
\Delta_n(\alpha) &= \frac{1}{n}\mathrm{tr}(\mathbf{H} - \alpha\mathbf{I})^{-1} \\
&= \frac{1}{n}\sum_{k=1}^{n}\frac{1}{|\mathbf{r}_k|^2 - \alpha - \mathbf{r}_k^*\mathbf{R}_k(\mathbf{H}_k - \alpha\mathbf{I}_{n-1})^{-1}\mathbf{R}_k^*\mathbf{r}_k},
\end{aligned}
$$

where $\mathbf{r}_k$ denotes the kth column vector of $\mathbf{R}$, $\mathbf{R}_k$ consists of the remaining $n-1$ columns of $\mathbf{R}$ when $\mathbf{r}_k$ is removed and $\mathbf{H}_k = \mathbf{R}_k^*\mathbf{R}_k$.

First, notice that

$$
(4.17)\quad \begin{aligned}
&\left||\mathbf{r}_k|^2 - \alpha - \mathbf{r}_k^*\mathbf{R}_k(\mathbf{H}_k - \alpha\mathbf{I}_{n-1})^{-1}\mathbf{R}_k^*\mathbf{r}_k\right| \\
&\geq \left|\mathrm{Im}\left(|\mathbf{r}_k|^2 - \alpha - \mathbf{r}_k^*\mathbf{R}_k(\mathbf{H}_k - \alpha\mathbf{I}_{n-1})^{-1}\mathbf{R}_k^*\mathbf{r}_k\right)\right| \geq y.
\end{aligned}
$$

By Lemma A.4, we conclude that

$$
(4.18)\quad \max_{j, l, k \leq n}\left||\mathbf{r}_k|^2 - (1 + |z|^2)\right| = o(n^{-5/36}\ln^2 n).
$$

As mentioned earlier, with probability 1 for all large n, the norm of $\mathbf{R}$ is not greater than $3 + M$. We conclude that with probability 1, for all large n, the eigenvalues and hence the entries of $\mathbf{R}_k(\mathbf{H}_k - \alpha\mathbf{I}_{n-1})^{-1}\mathbf{R}_k^*$ are bounded by $(3 + M)^2/y \leq (3 + M)^2/y_n$. Therefore, the sum of squares of absolute values of entries of any row of $\mathbf{R}_k(\mathbf{H}_k - \alpha \mathbf{I}_{n-1})^{-1}\mathbf{R}_k^*$ is not greater than $(3+M)^4/y^2 \le (3+M)^4/y_n^2$. By applying Lemma A.4 and noticing that $\mathbf{r}_k = (1/\sqrt{n})\mathbf{x}_k - z\mathbf{e}_k$, where $\mathbf{e}_k$ is the vector whose elements are all zero except the kth element which is 1, we obtain

$$\begin{aligned}\max_{j,l,k\le n}\Bigg|\mathbf{r}_k^*\mathbf{R}_k(\mathbf{H}_k - \alpha\mathbf{I}_{n-1})^{-1}\mathbf{R}_k^*\mathbf{r}_k \\ -\bigg(\frac{1}{n}\operatorname{tr}\Big(\mathbf{R}_k(\mathbf{H}_k - \alpha\mathbf{I}_{n-1})^{-1}\mathbf{R}_k^*\Big) \\ +|z|^2\Big[\mathbf{R}_k(\mathbf{H}_k - \alpha\mathbf{I}_{n-1})^{-1}\mathbf{R}_k^*\Big]_{kk}\bigg)\Bigg| \\ = O\big(y_n^{-1}n^{-5/36}\ln^2 n\big),\end{aligned} \tag{4.19}$$

where $[\mathbf{A}]_{kk}$ denote the (k, k)th element of the matrix $\mathbf{A}$.

Now, denote by $\Lambda_1 \le \cdots \le \Lambda_n$ and $\Lambda_{k,1} \le \cdots \le \Lambda_{k(n-1)}$ the eigenvalues of $\mathbf{H}$ and those of $\mathbf{H}_k$, respectively. Then by the relation $0 \le \Lambda_l - \Lambda_{k,l-1} \le \Lambda_l - \Lambda_{l-1}$, and by the fact that with probability 1

$$\Lambda_n \le (2+|z|)^2 + \varepsilon \quad \text{for all large } n,$$

$$\frac{1}{n}\operatorname{tr}\Big(\mathbf{R}_k(\mathbf{H}_k - \alpha\mathbf{I}_{n-1})^{-1}\mathbf{R}_k^*\Big) = 1 - \frac{1}{n} + \frac{\alpha}{n}\operatorname{tr}\Big((\mathbf{H}_k - \alpha\mathbf{I}_{n-1})^{-1}\Big),$$

and

$$\begin{aligned}\Big|\operatorname{tr}\big((\mathbf{H} - \alpha\mathbf{I})^{-1}\big) - \operatorname{tr}\big((\mathbf{H}_k - \alpha\mathbf{I}_{n-1})^{-1}\big)\Big| \\ = \left|\sum_{l=2}^{n}\frac{\Lambda_l - \lambda_{k,l-1}}{(\Lambda_l - \alpha)(\Lambda_{k,l-1} - \alpha)} + \frac{1}{\Lambda_1 - \alpha}\right| \\ \le \Lambda_n/y^2 + 1/y,\end{aligned} \tag{4.20}$$

we conclude that

$$\begin{aligned}\max_{j,l,k\le n}\left|\frac{1}{n}\operatorname{tr}\Big(\mathbf{R}_k(\mathbf{H}_k - \alpha\mathbf{I}_{n-1})^{-1}\mathbf{R}_k^*\Big) - 1 - \alpha\Delta_n(\alpha)\right| \\ \le \frac{1}{n} + \frac{|\alpha|}{n}\big(\Lambda_n/y^2 + 1/y\big) = o(n^{-4/5}).\end{aligned} \tag{4.21}$$

We now estimate $[\mathbf{R}_k(\mathbf{H}_k - \alpha\mathbf{I}_{n-1})^{-1}\mathbf{R}_k^*]_{kk}$. Let β_k' denote the kth row of $\mathbf{R}_k$, and $\tilde{\mathbf{R}}_k$ denote the matrix of the remaining $n-1$ rows of $\mathbf{R}_k$ when β_k' is removed. Also, write $\tilde{\mathbf{H}}_k = \tilde{\mathbf{R}}_k^*\tilde{\mathbf{R}}_k$. Note that β_k' is just the kth row of $\mathbf{\Xi}_n$ with the kth element removed. Then we have

$$\begin{aligned}\Big[\mathbf{R}_k(\mathbf{H}_k - \alpha\mathbf{I}_{n-1})^{-1}\mathbf{R}_k^*\Big]_{kk} = \beta_k'\Big(\tilde{\mathbf{H}}_k - \alpha\mathbf{I}_{n-1} + \bar{\beta}_k\beta_k'\Big)^{-1}\bar{\beta}_k \\ = \frac{\beta_k'\big(\tilde{\mathbf{H}}_k - \alpha\mathbf{I}_{n-1}\big)^{-1}\bar{\beta}_k}{1 + \beta_k'\big(\tilde{\mathbf{H}}_k - \alpha\mathbf{I}_{n-1}\big)^{-1}\bar{\beta}_k}.\end{aligned} \tag{4.22}$$

Applying Lemma A.4 with $K_a = y_n^{-1}$, we obtain

$$(4.23)\qquad \max_{j,l,k\le n}\left|\beta_k'(\tilde{\mathbf{H}}_k-\alpha\mathbf{I}_{n-1})^{-1}\bar{\beta}_k-\frac{1}{n}\mathrm{tr}\left((\tilde{\mathbf{H}}_k-\alpha\mathbf{I}_{n-1})^{-1}\right)\right| = O(y_n^{-1}n^{-5/36}\ln^2 n).$$

By elementary knowledge of matrix theory, $\Lambda_j(\mathbf{H})) \le \Lambda_j(\tilde{\mathbf{H}}_k) \le \Lambda_{j+2}(\mathbf{H})$, and we have

$$(4.24)\qquad \left|\frac{1}{n}\mathrm{tr}\left((\tilde{\mathbf{H}}_k-\alpha\mathbf{I}_{n-1})^{-1}\right)-\Delta_n(\alpha)\right| \le 4y_n^{-2}n^{-1} = o(n^{-9/10}).$$

By noticing

$$|\alpha(1+\Delta_n(\alpha))| \ge |\mathrm{Im}(\alpha(1+\Delta_n(\alpha)))| \ge y,$$
$$\left|\alpha\left(1+\beta_k'(\tilde{\mathbf{H}}_k-\alpha\mathbf{I}_{n-1})^{-1}\tilde{\beta}_k\right)\right| \ge y$$

and by (4.23) and (4.24), one obtains

$$(4.25)\qquad \begin{aligned}&\max_{j,l,k\le n}\left|\left[\mathbf{R}_k(\mathbf{H}_k-\alpha\mathbf{I}_{n-1})^{-1}\mathbf{R}_k^*\right]_{kk}-\frac{\Delta_n(\alpha)}{1+\Delta_n(\alpha)}\right|\\ &\quad\le \max_{j,l,k\le n}|\alpha|^2y^{-2}\left|\beta_k'(\tilde{\mathbf{H}}_k-\alpha\mathbf{I}_{n-1})^{-1}\bar{\beta}_k-\Delta_n(\alpha)\right|\\ &\quad= o(\alpha^2y_n^{-3}n^{-5/36}\ln n).\end{aligned}$$

Combining estimates (4.16)–(4.25), we conclude that

$$(4.26)\qquad \max_{j,l\le n}\left|\Delta_n(\alpha)-\frac{1+\Delta_n(\alpha)}{|z|^2-\alpha(1+\Delta_n(\alpha))^2}\right| = o(\alpha^2y_n^{-3}n^{-5/36}\ln^2 n).$$

From this, one can see that r_n is controlled by $o(\alpha^2y_n^{-5}n^{-5/36}\ln^2 n)) = o(\delta_n)$ and thus the error estimate (4.1) follows. The proof of Lemma 4.1 is complete. □

Proof of Lemma 4.2. Note that the Stieltjes transform $\Delta(\alpha)$ of the limiting spectral distribution $\nu(\cdot, z)$ is an analytic solution in α on the upper half plane $y > 0$ to the equation (4.2). It can be continuously extended to the "closed" upper plane $y \ge 0$ (but $\alpha \ne 0$). By way of the Stieltjes transform [see Bai (1993a) or Silverstein and Choi (1995)], it can be shown that $\nu(\cdot, z)$ has a continuous density (probably excluding $x = 0$ when $|z| \le 1$), say $p(\cdot, z)$, such that

$$\nu(x, z) = \int_0^x p(u, z)\, du$$

and $p(x, z) = \pi^{-1}\,\mathrm{Im}(\Delta(x))$. Since $p(x, z)$ is the density of the limiting spectral distribution $\nu(\cdot, z)$, $p(x, z) = 0$ for all $x < 0$ and $x > (2 + |z|)^2$. Let $x > 0$ be an inner point of the support of $\nu(\cdot, z)$. Write $\Delta(x) = g(x) + ih(x)$. Then, to prove (4.4), it suffices to show

$$h(x) \le \max\{\sqrt{2/x}, 1\}.$$

Rewrite (4.2) for $\alpha = x$ as

$$\Delta^2 + 2\Delta + 1 + \frac{1 - |z|^2}{x} + \frac{1}{x\Delta} = 0.$$

Comparing the imaginary and real parts of both sides of the above equation, we obtain

$$2(g(x) + 1) = \frac{1}{x(g^2(x) + h^2(x))}. \tag{4.27}$$

and

$$\begin{aligned} h^2(x) &= \frac{1 - |z|^2}{x} + (g(x) + 1)^2 + \frac{g(x)}{x(g^2(x) + h^2(x))} \\ &\leq \frac{1}{x} + \frac{1}{4x^2(g^2(x) + h^2(x))^2} + \frac{1}{2xh(x)} \\ &\leq \frac{1}{x} + \frac{1}{4x^2h^4(x)} + \frac{1}{2xh(x)}. \end{aligned} \tag{4.28}$$

This implies $h(x) \leq \max\{\sqrt{2/x}\ 1\}$, because substituting the reverse inequality $h(x) > \sqrt{2/x}$ (or $h(x) > 1$) will lead to a contradiction if $0 < x < 2$ (or $x \geq 2$, correspondingly). Thus, (4.4) is established.

Now, we proceed to find the boundary of the support of $\nu(\cdot, z)$. Since $\nu(\cdot, z)$ has no mass on the negative half line, we need only consider $x > 0$. Suppose $h(x) > 0$. Comparing the real and imaginary parts for both sides of (4.2) and then making x approach the boundary [namely, $h(x) \to 0$], we obtain

$$x(g^3 + 2g^2 + g) + (1 - |z|^2)g + 1 = 0$$

and

$$x(3g^2 + 4g + 1) + 1 - |z|^2 = 0. \tag{4.29}$$

Thus,

$$\left[(1 - |z|^2)g + 1\right](3g + 1) = (1 - |z|^2)g(g + 1).$$

For $|z| \neq 1$, the solution to this quadratic equation in g is

$$g = \frac{-3 \pm \sqrt{1 + 8|z|^2}}{4 - 4|z|^2} \quad \left(g = -\frac{1}{3} \text{ if } |z| = 1\right) \tag{4.30}$$

which, together with (4.29), implies that, for $|z| \neq 1$,

$$\begin{aligned} x_{1,2} &= -\frac{1 - |z|^2}{(g + 1)(3g + 1)} \\ &= \begin{cases} -\dfrac{1}{8|z|^2}\left\{1 - 20|z|^2 - 8|z|^4 \pm \sqrt{(1 + 8|z|^2)^3}\right\}, & \text{if } z \neq 0, \\ x_1 = -\infty \text{ and } x_2 = 4, & \text{if } z = 0. \end{cases} \end{aligned} \tag{4.31}$$

Note that $0 < x_1 < x_2$ when $|z| > 1$. Hence, the interval (x_1, x_2) is the support of $\nu(\cdot, z)$ since $p(x, z) = 0$ when x is very large. When $|z| < 1$, $x_1 < 0 < x_2$. Note that for the case $|z| < 1$, $g(x_1) < 0$ which contradicts the fact that $\Delta(x) > 0$ for all $x < 0$ and hence x_1 is not a solution of the boundary. Thus, the support of $\nu(\cdot, z)$ is the interval $(0, x_2)$. For $|z| = 1$, there is only one solution $x_2 = -1/[g(g+1)^2] = 27/4$, which can also be expressed by (4.31). In this case, the support of $\nu(\cdot, z)$ is $(0, x_2)$. The proof of Lemma 4.2 is complete. □

PROOF OF LEMMA 4.3. We first prove that $\Delta(\alpha)$ does not coincide with other roots of the equation (4.2) for $y \ge 0$ and $\alpha \ne x_{1,2}$. Otherwise, if for some α, $\Delta(\alpha)$ is a multiple root of (4.2), then it must be also a root of the derivative of the equation (4.2), that is,

$$3\Delta^2 + 4\Delta + \frac{\alpha + 1 - |z|^2}{\alpha} = 0. \tag{4.32}$$

Similar to the proof of Lemma 4.2, solving equations (4.2) and (4.32), one obtains $\alpha = x_1$ or x_2 and Δ is the same as g given in (4.30). Our assertion is proved.

We now prove (4.7). Let $\Delta + \rho$ be either m_2 or m_3. Since both Δ and $\Delta + \rho$ satisfy (4.2), we obtain

$$\rho = -\frac{3\Delta^2(\alpha) + 4\Delta(\alpha) + 1 + (1 - |z|^2)/\alpha}{3\Delta(\alpha) + 2 + \rho}. \tag{4.33}$$

Write $\hat{\rho} = \Delta(\alpha) - \Delta(x_2)$. By (4.29), we have

$$\begin{aligned} &3\Delta^2(\alpha) + 4\Delta(\alpha) + 1 + (1 - |z|^2)/\alpha \\ &\quad = 3\Delta^2(\alpha) + 4\Delta(\alpha) + 1 + (1 - |z|^2)/\alpha \\ &\qquad - \left[3\Delta^2(x_2) + 4\Delta(x_2) + 1 + (1 - |z|^2)/x_2\right] \\ &\quad = \hat{\rho}[6\Delta(x_2) + 4 + 3\hat{\rho}] + \frac{(1 - |z|^2)(x_2 - \alpha)}{x_2 \alpha}. \end{aligned} \tag{4.34}$$

From (4.2) and (4.29), it follows that

$$\begin{aligned} 0 &= \left[3\Delta^2(x_2) + 4\Delta(x_2) + 1 + (1 - |z|^2)/\alpha\right]\hat{\rho} \\ &\quad + [3\Delta(x_2) + 2]\,\hat{\rho}^2 + \hat{\rho}^3 \\ &\quad + \frac{(x_2 - \alpha)\big(\Delta(x_2)(1 - |z|^2) + 1\big)}{x_2 \alpha} \\ &= [3\Delta(x_2) + 2 + \hat{\rho}]\,\hat{\rho}^2 \\ &\quad + \frac{(x_2 - \alpha)\left[(\Delta(x_2) + \hat{\rho})(1 - |z|^2) + 1\right]}{x_2 \alpha}. \end{aligned} \tag{4.35}$$

Note that $\Delta(x_2)(1-|z|^2)+1/4 = 1/4(1+\sqrt{1+8|z|^2}) \geq 1/2$. Equation (4.35) implies that

$$(4.36)\qquad |\hat{\rho}| \geq \min\left(\frac{1}{4M^2}, \sqrt{\frac{|x_2-\alpha|}{6(|\Delta(x_2)|+1)}}\right) \geq c_1\sqrt{|x_2-\alpha|},$$

for some positive constant c_1. Note that Δ is continuous in the rectangle $\{(\alpha, z);\ z \in T,\ x_{2,\min} - \varepsilon_1 \leq x \leq x_{2,\max},\ 0 \leq y \leq N\}$, where $x_{2,\min} = 4$ (corresponding to $z = 0$) and $x_{2,\max} = (1/8M^2)[(1+8M^2)^{3/2} - 1 + 20M^2 + 8M^4]$ (corresponding to $|z| = M = \sqrt{A^2 + A^4}$). Therefore, we may select a positive constant ε_1 such that for all $|z| \leq M$ and $|\alpha - x_2| \leq \varepsilon_1$, $|\hat{\rho}| \leq \min(\frac{1}{8}, c_1^2/M^4)$. Then, from (4.33) and (4.34) and the fact that when $|\rho(\alpha)| \leq \frac{1}{8}$,

$$|3\Delta(\alpha) + 2 + \rho(\alpha)| \leq 4,$$

we conclude that

$$(4.37)\qquad \begin{aligned} |\rho(\alpha)| &\geq \min\left(\frac{1}{8}, \frac{1}{4}\left|\hat{\rho}[6\Delta(x_2) + 4 + 3\hat{\rho}] + \frac{(1-|z|^2)(x_2-\alpha)}{x_2\alpha}\right|\right) \\ &\geq \min\left(\frac{1}{8}, \frac{1}{8}c_1\sqrt{|x_2-\alpha|} - \frac{1}{36}M^2|x_2-\alpha|\right) \\ &\geq c_2\sqrt{|x_2-\alpha|}. \end{aligned}$$

This concludes the proof of (4.7).

The proof of (4.8) is similar to that of (4.7). Checking the proof of (4.7), one finds that equations (4.33)–(4.35) are still true if x_2 is replaced by x_1. The rest of the proof depends on the fact that for all $z \in T$, $|z| \geq 1 + \varepsilon$ and $|\alpha - x_1| \leq \varepsilon_1$, $|3\Delta(\alpha) + 2 + \rho(\alpha)|$ has a uniform upper bound and $\hat{\rho}$ can be made as small as desired provided ε_1 is small enough. Indeed, this can be done because x_1 has a strictly positive minimum $x_{1,\min}$ at $|z| = 1 + \varepsilon$, and hence, $\Delta(\alpha)$ is uniformly continuous in the rectangle $\{(\alpha, z);\ z \in T,\ x_{1,\min} - \varepsilon_1 \leq x \leq x_{1,\max},\ 0 \leq y \leq N\}$, provided ε_1 is chosen so that $x_{1,\min} - \varepsilon_1 > 0$.

We claim that (4.6) is true. If not, then for each k, there exist α_k and z_k with $z_k \in T$ and $|\alpha_k - x_2| \geq \varepsilon_1$ (and $|\alpha_k - x_1| \geq \varepsilon_1$ if $|z_k| \geq 1 + \varepsilon$), such that

$$\min_{j=2,3} |\Delta(\alpha_k) - m_j(\alpha_k)| < \frac{1}{k}.$$

Then, we may select a subsequence $\{k'\}$ such that $\alpha_{k'} \to \alpha_0$ and $z_{k'} \to z_0 \in T$ and $|\alpha_0 - x_2| \geq \varepsilon_1$. If $|z_0| \geq 1 + \varepsilon$, we also have $|\alpha_0 - x_1| \geq \varepsilon_1$. For at least one of $j = 2$ or 3, say $j = 2$,

$$|\Delta(\alpha_{k'}) - m_2(\alpha_{k'})| < \frac{1}{k'}.$$

If $\alpha_0 \neq 0$, by continuity of $\Delta(\alpha)$ and $m_2(\alpha)$, we shall have $\Delta(\alpha_0) = m_2(\alpha_0)$ which contradicts the fact that $\Delta(\alpha)$ does not coincide with $m_2(\alpha)$ except $\alpha = x_2$ or $\alpha = x_1$ when $|z| > 1$. It is impossible that $\alpha_0 = 0$ and $|z_0| \geq 1 + \varepsilon$, since $\Delta(\alpha_{k'}) \to 1/(|z_0|^2 - 1)$ while $\min_{j=2,3}|m_j(\alpha_{k'})| \to \infty$. It is also impossible that $\alpha_0 = 0$ and $|z_0| \le 1 - \varepsilon$, since in this case, we should have $\mathrm{Re}(\Delta(\alpha_{k'})) \to +\infty$, $m_2(\alpha_{k'}) \to 1/(|z_0|^2 - 1)$ and $\mathrm{Re}(m_3(\alpha_{k'})) \to -\infty$ which follows from $\Delta(\alpha_{k'}) + m_2(\alpha_{k'}) + m_3(\alpha_{k'}) = -2$. This concludes the proof of (4.6).

The assertion (4.5) follows from the fact that equation (4.2) has no three identical roots for any α and z, since the second derivative of (4.2) gives $\Delta(\alpha) = -2/3$ equals neither $\Delta(x_2)$ nor $\Delta(x_1)$. The proof of Lemma 4.3 is then complete. □

PROOF OF LEMMA 4.4. For $x < 0$, we have: (1) $\Delta(x) > 0$ (real); (2) $\Delta(x) \to 0$ as $x \to -\infty$ and (3) from (4.2), as $x \uparrow 0$,

$$
\begin{cases} \sqrt{|x|}\,\Delta(x) \uparrow \sqrt{1 - |z|^2}, & \text{if } |z| < 1, \\ \sqrt[3]{|x|}\,\Delta(x) \uparrow 1, & \text{if } |z| = 1, \\ \Delta(x) \uparrow (|z|^2 - 1)^{-1}, & \text{if } |z| > 1. \end{cases} \tag{4.38}
$$

Thus, for any $C > 0$, the integral $\int_{-C}^{0} \Delta(x)\,dx$ exists. We have by exchanging the integration order,

$$
\begin{aligned}
\int_{-C}^{0} \Delta(x)\,dx &= \int_0^C \Delta(-x)\,dx = \int_0^C \int_0^\infty \frac{1}{u + x}\nu(du, z)\,dx \\
&= \int_0^\infty [\ln(C + u) - \ln u]\,\nu(du, z) \\
&= \ln C + \int_0^\infty \ln(1 + u/C)\,\nu(du, z) - \int_0^\infty \ln u\,\nu(du, z).
\end{aligned} \tag{4.39}
$$

Differentiating both sides with respect to s, we get

$$
\begin{aligned}
\frac{\partial}{\partial s}\int_0^\infty \ln u\,\nu(du, z) &= \frac{\partial}{\partial s}\int_0^\infty \ln(1 + u/C)\,\nu(du, z) \\
&\quad - \int_{-C}^{0} \frac{\partial}{\partial s}\Delta(x)\,dx.
\end{aligned} \tag{4.40}
$$

[The reasons for the exchangability of the order of the integral and derivative are given after (4.47).]

Differentiating both sides of (4.2) with respect to s and x, we obtain

$$
\frac{\partial}{\partial s}\Delta(x)\left[3\Delta^2(x) + 4\Delta(x) + \frac{x + 1 - |z|^2}{x}\right] = \frac{2s\Delta(x)}{x}, \tag{4.41}
$$

and

$$
\frac{\partial}{\partial x}\Delta(x)\left[3\Delta^2(x) + 4\Delta(x) + \frac{x + 1 - |z|^2}{x}\right] = \frac{\Delta(x)(1 - |z|^2) + 1}{x^2}.
$$

Comparing the two equations, we get

$$
\frac{\partial}{\partial s}\Delta(x) = \frac{2sx\Delta(x)}{1 + \Delta(x)(1 - |z|^2)}\frac{\partial}{\partial x}\Delta(x) = -\frac{2s}{(1 + \Delta(x))^2}\frac{\partial}{\partial x}\Delta(x), \tag{4.42}
$$

where the last equality follows from the fact that

$$x = \frac{|z|^2}{(1+\Delta(x))^2} - \frac{1}{\Delta(x)(1+\Delta(x))} = -\frac{1+\Delta(x)(1-|z|^2)}{\Delta(x)(1+\Delta(x))^2}, \tag{4.43}$$

which is a solution of (4.2).

By (4.42), we obtain

$$\begin{aligned} \int_{-C}^{0} \frac{\partial}{\partial s}\Delta(x)\,dx &= -\int_{-C}^{0} \frac{\partial}{\partial x}\Delta(x)\frac{2s}{(1+\Delta(x))^2}\,dx \\ &= -2s\int_{\Delta(-C)}^{\Delta(0_-)} \frac{1}{(1+\Delta)^2}\,d\Delta \\ &= \frac{2s}{1+\Delta(0_-)} - \frac{2s}{1+\Delta(-C)}. \end{aligned} \tag{4.44}$$

Letting $x \uparrow 0$ in (4.2), we get

$$\Delta(0_-) = \begin{cases} \infty, & \text{if } |z|^2 \le 1, \\ \dfrac{1}{|z|^2-1}, & \text{if } |z|^2 > 1. \end{cases} \tag{4.45}$$

We also have $\Delta(-C) \to 0$ as $C \to \infty$. Thus, we get

$$\int_{-C}^{0} \frac{\partial}{\partial s}\Delta(x)\,dx \to -g(s,t). \tag{4.46}$$

Note that (4.42) is still true for $x > 0$. Therefore, by noticing

$$\nu(dx,z)/dx = \pi^{-1}\operatorname{Im}(\Delta(x)) = p(x,z),$$

we have

$$\begin{aligned} &\left|\frac{\partial}{\partial s}\int_0^\infty \ln(1+u/C)\nu(du,z)\right| \\ &\quad= \left|\frac{1}{\pi}\operatorname{Im}\left(\int_0^\infty \ln(1+u/C)\frac{\partial}{\partial s}(\Delta(u))\,du\right)\right| \\ &\quad= \left|\frac{1}{\pi}\operatorname{Im}\left(\int_0^{(2+|z|)^2} \ln(1+u/C)\frac{2s}{(1+\Delta(u))^2}\frac{\partial}{\partial u}\Delta(u)\,du\right)\right| \\ &\quad\le \left|\frac{2|s|(2+|z|)^2}{C\pi}\operatorname{Im}\left(\int_0^{(2+|z|)^2} \frac{1}{(1+\Delta(u))^2}\frac{\partial}{\partial u}\Delta(u)\,du\right)\right| \\ &\quad\le \frac{2|s|(2+|z|)^2}{C\pi}\int_0^\infty \frac{1}{(1+\Delta)^2}\,d\Delta \\ &\quad\le \frac{|s|(2+|z|)^2}{C} \to 0 \quad \text{as } C \to \infty. \end{aligned} \tag{4.47}$$

In the first equality above and in (4.40), the justification of the exchangability of the order of the integral and derivative follows from the dominated convergence theorem and the following facts:

(i) When $|z| > 1$, $\mathrm{Im}((\partial/\partial s)(\Delta(u))$ is continuous in u and vanishes when $u > x_2$ and $u < x_1$.

(ii) When $|z| < 1$, for $u > 0$ $\mathrm{Im}((\partial/\partial s)(\Delta(u))$ is continuous in u and vanishes when $u > x_2$, and for small u, by $u\Delta^2(u) \to -1 + |z|^2$ [see (4.2) and (4.41)],

$$\left|\mathrm{Im}\left(\frac{\partial}{\partial s}(\Delta(u))\right)\right| \le \frac{2|s\Delta(u)|}{|3u\Delta^2(u) + 4u\Delta(u) + u + 1 - |z|^2|} \le \frac{4|s|}{\sqrt{xu(1-|z|^2)^3}}$$

which is integrable w.r.t. u.

(iii) When $|z| = 1$ and u small, by $u\Delta^3(u) \to -1$,

$$\left|\mathrm{Im}\left(\frac{\partial}{\partial s}(\Delta(u))\right)\right| \le 4|s|u^{-2/3}$$

which is also integrable w.r.t. u.

The assertion (4.9) then follows from (4.40), (4.46) and (4.47) and Lemma 4.4 is proved. □

PROOF OF LEMMA 4.5. We shall prove (4.10) by employing Corollary 2.3 of Bai (1993a). For all $z \in T$, the supports of $\nu(\cdot, z)$ are commonly bounded. Therefore, we may select a constant N such that, for some absolute constant C,

$$\begin{aligned} &\|\nu_n(\cdot, z) - \nu\cdot, z)\| \\ &\quad \le C\left(\int_{|x|\le N} |\Delta_n(\alpha) - \Delta(\alpha)|\, dx \right. \\ &\qquad \left. + y_n^{-1} \sup_x \int_{|y|\le 2y_n} |\nu(x+y, z) - \nu(x, z)|\, dy\right) \\ &\quad \le C\left(\int_{|x|\le N} |\Delta_n(\alpha) - \Delta(\alpha)|\, dx + y_n^{1/2}\right), \end{aligned} \tag{4.48}$$

where the last step follows from (4.4).

Denoted by $m_1(\alpha) = \Delta(\alpha)$, $m_2(\alpha)$ and $m_3(\alpha)$ the three solutions of the equation (4.2). Note that $\Delta(\alpha)$ is analytic in α for $\mathrm{Im}(\alpha) > 0$. By a suitable selection, the three solutions are all analytic in α on the upper half complex plane.

By Lemma 4.3, there are constant ε_0 and ε_1 such that (4.5)–(4.8) hold. By Lemma 4.1, there is an n_0 such that for all $n \ge n_0$,

$$|(\Delta_n - m_1)(\Delta_n - m_2)(\Delta_n - m_3)| = o(\delta_n) \le \tfrac{4}{27}\varepsilon_0^3\delta_n. \tag{4.49}$$

Now, choose an $\alpha_0 = x_0 + iy_0$ with $|x_0| \le N$, $y_0 > 0$ and $\min_{k=1,2}(|x_0 - x_k|) \ge \varepsilon_1$. For a fixed $z \in T$, as argued earlier, $\Delta_n(\alpha_0)$ converges to $\Delta(\alpha_0)$ when n goes to infinity along some subsequence. Then, for infinitely many $n > n_0$, $|\Delta_n(\alpha_0) - \Delta(\alpha_0)| < \varepsilon_0/3$. Hence,

$$\min_{k=2,3}(|\Delta_n(\alpha_0) - m_k(\alpha_0)|)$$
$$\ge \min_{k=2,3}(|\Delta(\alpha_0) - m_k(\alpha_0)|) - |\Delta_n(\alpha_0) - \Delta(\alpha_0)|) > \tfrac{2}{3}\varepsilon_0.$$

This and (4.49) imply, for infinitely many n,

$$|\Delta_n(\alpha_0) - \Delta(\alpha_0)| = o(\delta_n) \le \tfrac{1}{3}\varepsilon_0\delta_n. \tag{4.50}$$

Let n_0 be also such that $2/(y_n^2 n_0) + \frac{1}{3}\varepsilon_0 n_0^{-1/120} < \varepsilon_0/3$. We claim that (4.50) is true for all $n \ge n_0$. In fact, if (4.50) is true for some $n > n_0$, then

$$|\Delta_{n-1}(\alpha_0) - \Delta(\alpha_0)| \le |\Delta_{n-1}(\alpha_0) - \Delta_n(\alpha_0)| + |\Delta_n(\alpha_0) - \Delta(\alpha_0)|$$
$$< 2/(y_n n) + \tfrac{1}{3}\varepsilon_0 n^{-1/120} < \varepsilon_0/3.$$

Here we have used the trivial fact that $\|\nu_n(\cdot, z) - \nu_{n-1}(\cdot, z)\| \le 2/n$ which implies $|\Delta_{n-1}(\alpha_0) - \Delta_n(\alpha_0)| \le 2/(y_n n)$. This shows that

$$\min_{k=2,3}(|\Delta_{n-1}(\alpha_0) - m_k(\alpha_0)|) > \tfrac{2}{3}\varepsilon_0,$$

which implies that (4.50) is true for $n-1$. This completes the proof of our assertion.

Now, we claim that (4.50) is true for all $n > n_0$ and $|\alpha| \le N$, $\min_{k=1,2}(|x - x_k|) \ge \varepsilon_1$, that is,

$$|\Delta_n(\alpha) - \Delta(\alpha)| \le o(\delta_n) \le \tfrac{1}{3}\varepsilon_0\delta_n. \tag{4.51}$$

By (4.6) and (4.49), we conclude that (4.51) is equivalent to

$$\min_{k=2,3}(|\Delta_n(\alpha) - m_k(\alpha)|) > \tfrac{2}{3}\varepsilon_0. \tag{4.52}$$

Note that both Δ_n and $m_j(\alpha)$, $j = 1, 2, 3$, are continuous functions in both α and z. Therefore, on the boundary of the set of points (α, z) at which (4.51) does not hold, we should have $|\Delta_n(\alpha) - \Delta(\alpha)| = \frac{1}{3}\varepsilon_0\delta_n$ and $\min_{k=2,3}(|\Delta_n(\alpha) - m_k(\alpha)|) = \frac{2}{3}\varepsilon_0$. This is impossible because these two equalities contradict (4.6).

For $|\alpha - x_k| \le \varepsilon_1$, $k = 1$ or 2, (4.5), (4.7) and (4.8) imply that

$$|\Delta_n(\alpha) - \Delta(\alpha)| \le o\left(\delta_n/\sqrt{|\alpha - x_k|}\right). \tag{4.53}$$

This, together with (4.48) and (4.51), implies (4.10). The proof of Lemma 4.5 is complete. □

5. Proof of (1.4). In this section, we shall show that probability 1,

$$\int_{z\in T}\left|\int_0^{\varepsilon_n} \ln x\nu_n(dx, z)\right| dt\, ds \to 0, \tag{5.1}$$

where $\varepsilon_n = \exp(-n^{1/120})$.

Denote by $\mathbf{Z}_1$ and $\mathbf{Z}$ the matrix of the first two columns of $\mathbf{R}$ and that formed by the last $n-2$ columns. Let $\Lambda_1 \le \cdots \le \Lambda_n$ denote the eigenvalues of the matrix $\mathbf{R}^*\mathbf{R}$ and let $\eta_1 \le \cdots \le \eta_{n-2}$ denote the eigenvalues of $\mathbf{Z}^*\mathbf{Z}$. Then, for any $k \le n-2$, we have $\Lambda_k \le \eta_k \le \Lambda_{k+2}$ and $\det(\mathbf{R}^*\mathbf{R}) = \det(\mathbf{Z}^*\mathbf{Z})\det(\mathbf{Z}_1^*\mathbf{Q}\mathbf{Z}_1)$, where $\mathbf{Q} = \mathbf{I} - \mathbf{Z}(\mathbf{Z}^*\mathbf{Z})^{-1}\mathbf{Z}^*$. This identity can be written as

$$\sum_{k=1}^{n} \ln(\Lambda_k) = \ln(\det(\mathbf{Z}_1^*\mathbf{Q}\mathbf{Z}_1)) + \sum_{k=1}^{n-2} \ln(\eta_k).$$

If l is the smallest integer such that $\eta_l \ge \varepsilon_n$, then $\Lambda_{l-1} < \varepsilon_n$ and $\Lambda_{l+2} > \varepsilon_n$. Therefore, we have

$$\begin{aligned} 0 > \int_0^{\varepsilon_n} \ln x \nu_n(dx, z) &= \frac{1}{n} \sum_{\Lambda_k < \varepsilon_n} \ln \Lambda_k \\ &\ge \frac{1}{n}\min\{\ln(\det(\mathbf{Z}_1^*\mathbf{Q}\mathbf{Z}_1)), 0\} + \frac{1}{n}\sum_{\eta_k < \varepsilon_n} \ln \eta_k - \frac{2}{n}\ln(\max(\Lambda_n, 1)). \end{aligned} \tag{5.2}$$

To prove (5.1), we first estimate the integral of $(1/n)\ln(\det(\mathbf{Z}_1^*\mathbf{Q}\mathbf{Z}_1))$ with respect to s and t. Note that with probability one, the rank of the matrix $\mathbf{Q}$ is 2. Hence, there are two orthogonal complex unit vectors γ_1 and γ_2 such that $\mathbf{Q} = \gamma_1\gamma_1^* + \gamma_2\gamma_2^*$. Denote the two column vectors of $\mathbf{Z}_1$ by $\mathbf{r}_1$ and $\mathbf{r}_2$. Then we have

$$\frac{1}{n}\ln(\det(\mathbf{Z}_1^*\mathbf{Q}\mathbf{Z}_1)) = \frac{1}{n}\ln\left(|\gamma_1^*\mathbf{r}_1\gamma_2^*\mathbf{r}_2 - \gamma_2^*\mathbf{r}_1\gamma_1^*\mathbf{r}_2|^2\right).$$

Define the random sets

$$E = \left\{(s,t) : |\gamma_1^*\mathbf{r}_1\gamma_2^*\mathbf{r}_2 - \gamma_2^*\mathbf{r}_1\gamma_1^*\mathbf{r}_2| \ge n^{-14}, |\xi_1| \le n, |\xi_2| \le n\right\}$$

and

$$F = \left\{(s,t) : |\gamma_1^*\mathbf{r}_1\gamma_2^*\mathbf{r}_2 - \gamma_2^*\mathbf{r}_1\gamma_1^*\mathbf{r}_2| < n^{-14}, |\xi_1| \le n, |\xi_2| \le n\right\}.$$

It is trivial to see that

$$\mathrm{P}(|\xi_1| > n \quad \text{or} \quad |\xi_2| > n) < 2n^{-2}. \tag{5.3}$$

When $|\xi_1| \le n$ and $|\xi_2| \le n$, we have $\det(\mathbf{Z}_1^*\mathbf{Q}\mathbf{Z}_1) = |\gamma_1^*\mathbf{r}_1\gamma_2^*\mathbf{r}_2 - \gamma_2^*\mathbf{r}_1\gamma_1^*\mathbf{r}_2|^2 \le 4(n+M)^4$. Thus,

$$\frac{1}{n}\int_{z\in T} |I_E \ln(\det(\mathbf{Z}_1^*\mathbf{Q}\mathbf{Z}_1))|\, dt\, ds \le Cn^{-1}\ln n \to 0. \tag{5.4}$$

On the other hand, for any $\varepsilon > 0$, we have

$$\begin{aligned} &\mathrm{P}\left(\frac{1}{n}\int_{z\in T} |I_F \ln(\det(\mathbf{Z}_1^*\mathbf{Q}\mathbf{Z}_1))|\, dt\, ds \ge \varepsilon\right) \\ &\quad\le \frac{1}{\varepsilon n}\int_{z\in T} \mathrm{E}\left| I_F \ln\left(|\gamma_1^*\mathbf{r}_1\gamma_2^*\mathbf{r}_2 - \gamma_2^*\mathbf{r}_1\gamma_1^*\mathbf{r}_2|^2\right)\right| dt\, ds. \end{aligned} \tag{5.5}$$

Note that the elements of $\sqrt{n}\mathbf{r}_1$ and $\sqrt{n}\mathbf{r}_2$ are independent of each other and the joint densities of their real and imaginary parts have a common upper bound K_d. Also, they are independent of γ_1 and γ_2. Therefore, by Corollary A.2, the conditional joint density of the real and imaginary parts of $\sqrt{n}\gamma_1^*\mathbf{r}_1$, $\sqrt{n}\gamma_2^*\mathbf{r}_2$ $\sqrt{n}\gamma_2^*\mathbf{r}_1$ and $\sqrt{n}\gamma_1^*\mathbf{r}_2$, when γ_1 and γ_2 are given, is bounded by $(2K_d n)^4$. Hence, the conditional joint density of the real and imaginary parts of $\gamma_1^*\mathbf{r}_1$, $\gamma_2^*\mathbf{r}_2$, $\gamma_2^*\mathbf{r}_1$ and $\gamma_1^*\mathbf{r}_2$, when γ_1 and γ_2 are given, is bounded by $K_d^4 2^4 n^8$. Set $\mathbf{x} = (\gamma_1^*\mathbf{r}_1, \gamma_2^*\mathbf{r}_1)'$ and $\mathbf{y} = (\mathbf{r}_2^*\gamma_2, -\mathbf{r}_2^*\gamma_1)'$. Note that by Corollary A.2, the joint density of $\mathbf{x}$ and $\mathbf{y}$ is bounded by $K_d^4 2^r n^8$.

If $|\xi_1| \le n$, $|\xi_2| \le n$, then $\max(|\mathbf{x}|, |\mathbf{y}|) \le n + |z| \le n + M$. Applying Lemma A.3 with $f(t) = \ln t$, $M = \mu = 1$, we obtain

$$\frac{2}{n}\int_{z\in T}\left|\mathrm{E}\left(I_{(|\mathbf{x}^*\mathbf{y}| < n^{-14},\, |\xi_1|\le n,\, |\xi_2|\le n)} \ln(|\mathbf{x}^*\mathbf{y}|)\big|\gamma_1, \gamma_2\right)\right| dt\,ds \le Cn^{12}n^{-14} \le Cn^{-2}, \tag{5.6}$$

for some positive constant C.

From (5.3), (5.5) and (5.6), it follows that

$$\frac{1}{n}\int_{z\in T} |I_F \ln(\det(\mathbf{Z}_1^*\mathbf{Q}\mathbf{Z}_1))|\, dt\,ds \to 0 \quad \text{a.s.} \tag{5.7}$$

Next, we estimate the second term in (5.2). We have

$$\begin{aligned}\frac{1}{n}\left|\sum_{\eta_k<\varepsilon_n} \ln(\eta_k)\right| &\le n^{-119/120}\varepsilon_n \sum_{k=1}^{n-2}\frac{1}{\eta_k}\\ &= n^{-119/120}\varepsilon_n \operatorname{tr}\left((\mathbf{Z}^*\mathbf{Z})^{-1}\right)\\ &= n^{-119/120}\varepsilon_n \sum_{k=3}^{n}\frac{1}{\mathbf{r}_k^*\mathbf{Q}_k\mathbf{r}_k}\\ &= n^{-119/120}\varepsilon_n \sum_{k=3}^{n}\frac{1}{|\mathbf{r}_k^*\gamma_{k1}|^2 + |\mathbf{r}_k^*\gamma_{k2}|^2 + |\mathbf{r}_k^*\gamma_{k3}|^2},\end{aligned} \tag{5.8}$$

where for each k, γ_{kj}, $j = 1, 2, 3$, are orthonormal complex vectors such that $\mathbf{Q}_k = \gamma_{k1}\gamma_{k1}^* + \gamma_{k2}\gamma_{k2}^* + \gamma_{k3}\gamma_{k3}^*$ which is the projection matrix onto the orthogonal complement of the space spanned by the $3, \dots, k-1, k+1, \dots, n$ columns of $\mathbf{R}(z)$.

As in the proof of (5.7), one can show that the conditional joint density of the real and imaginary parts of $\mathbf{r}_k^*\gamma_{k1}$, $\mathbf{r}_k^*\gamma_{k2}$ and $\mathbf{r}_k^*\gamma_{k3}$ when γ_{kj}, $j = 1, 2, 3$ are given, is bounded by $CK_d n^{12}$. Therefore, we have

$$\begin{aligned}&n^{-119/120}\varepsilon_n \sum_{k=3}^{n}\int_{z\in T}\mathrm{E}\left(\frac{dt\,ds}{|\mathbf{r}_k^*\gamma_{k1}|^2 + |\mathbf{r}_k^*\gamma_{k2}|^2 + |\mathbf{r}_k^*\gamma_{k3}|^2}\right)\\ &\le Cn^{-119/120}\varepsilon_n\left(K_d^3 n^{13}\int\cdots\int_{u_1^2+\cdots+u_6^2<1}\frac{du_1\cdots du_6}{u_1^2+\cdots+u_6^2} + n\right)\\ &\le Cn^{13}\varepsilon_n \quad \text{by a polar transformation}\\ &\le Cn^{-2}.\end{aligned} \tag{5.9}$$

Therefore, by the Borel–Cantelli lemma,

$$n^{-119/120}\varepsilon_n \sum_{k=3}^{n} \int_{z\in T}\left(\frac{dt\,ds}{|\mathbf{r}_k^*\gamma_{k1}|^2+|\mathbf{r}_k^*\gamma_{k2}|^2+|\mathbf{r}_k^*\gamma_{k3}|^2}\right)\to 0 \quad \text{a.s.}$$

and hence, with probability 1,

$$\int_{z\in T}\frac{1}{n}\sum_{\eta_k<\varepsilon_n}\ln(\eta_k)\,dt\,ds \le n^{-119/120}\varepsilon_n\sum_{k=1}^{n-2}\int_{z\in T}\frac{dt\,ds}{\eta_k}\to 0. \tag{5.10}$$

Finally, we estimate the integral of the third term in (5.2). By Yin, Bai and Krishnaiah (1988), we have $\Lambda_n \le (\|\Xi_n\| + |z|)^2 \to (2+|z|)^2$, a.s. We conclude that

$$\frac{2}{n}\int_{z\in T}\ln(\max(\Lambda_n, 1))\,dt\,ds\to 0 \quad \text{a.s.} \tag{5.11}$$

Hence, (5.1) follows from (5.7), (5.10) and (5.11).

6. Proof of Theorem 1.1. In Section 3, the problem is reduced to showing (3.10). Recalling the definitions of $g_n(s,t)$ and $g(s,t)$, we have by integration by parts,

$$\begin{aligned}
&\left|\int_{z\in T}(g_n(s,t)-g(s,t))\exp(ius+itv)\,dt\,ds\right|\\
&\quad=\Bigg|-\int_{z\in T} iu\tau(s,t)\,dt\,ds\\
&\qquad+\int_{|t|\le A^2}[\tau(A,t)\,dt-\tau(-A,t)]\,dt\\
&\qquad+\int_{|t|\le 1+\varepsilon}\Big[\tau\Big(\sqrt{(1+\varepsilon)^2-t^2},t\Big)\,dt\\
&\qquad\qquad-\tau\Big(-\sqrt{(1+\varepsilon)^2-t^2},t\Big)\Big]\,dt\\
&\qquad+\int_{|t|\le 1-\varepsilon}\Big[\tau\Big(\sqrt{(1-\varepsilon)^2-t^2},t\Big)\,dt\\
&\qquad\qquad-\tau\Big(-\sqrt{(1-\varepsilon)^2-t^21},t\Big)\Big]\,dt\Bigg|,
\end{aligned} \tag{6.1}$$

where

$$\tau(s,t)=\exp(ius+itv)\int_0^\infty \ln x(\nu_n(dx,z)-\nu(dx,z)).$$

When A is large enough, with probability 1, for all large n, the support of $\nu_n(\cdot,\pm A+it)$ is uniformly bounded by $(A-3)^2>1$ from the left and by

$(A + A^2 + 3)^2$ from the right. By Lemma 4.5, we have

$$\left|\int_{|t|\le A^2} \tau(\pm A, t)\, dt\right|$$

$$\le \int_{|t|\le A^2}\left|\int_{(A-3)^2}^{(A+A^2+3)^2} \ln x(\nu_n(dx, \pm A + it) - \nu(dx, \pm A + it))\right| dt$$

$$\to 0 \quad \text{a.s.}$$

Let $\varepsilon_n = \exp(-n^{1/120})$. In Section 5, we proved that

$$\int_{z\in T}\left|\int_0^{\varepsilon_n} \ln x\nu_n(dx, z)\right| dt\, ds \to 0, \quad \text{a.s.}$$

By (4.4), we have

$$\int_{z\in T}\left|\int_0^{\varepsilon_n} \ln x\nu(dx, z)\right| dt\, ds \to 0.$$

By Lemma 4.5, we have

$$\int_{z\in T}\left|\int_{\varepsilon_n}^{(A+A^2+3)^2} \ln x(\nu_n(dx, z) - \nu(dx, z))\right| dt\, ds$$

$$\le 4CA^3|\ln(\varepsilon_n)|\max_{z\in T}\|\nu_n(\cdot, z) - \nu(\cdot, z)\| \to 0.$$

This proves that

$$iu\int_{z\in T} \tau(s, t)\, dt\, ds \to 0.$$

Similarly, we can prove that

$$\int_{|t|\le 1\pm\varepsilon}\left[\pm\tau\left(\sqrt{(1\pm\varepsilon)^2 - t^2}, t\right)\right] dt \to 0.$$

The proof of Theorem 1.1 is complete. □

7. Comments and extensions.

7.1. *Relaxation of conditions assumed in Theorem* 1.1.

7.1.1. *On the moment of the underlying distribution.* Reviewing the definition of ε_n and checking the proofs given in Sections 5 and 6, one finds that $|\ln(\varepsilon_n)| = \sqrt{1/\delta_n}$ and $\max_{z\in T}\|\nu_n(\cdot, z) - \nu(\cdot, z)\| = o(\sqrt{\delta_n})$. Hence, (1.3) is always true for any choice of $\delta_n (\to 0)$. The rate of ε_n is required to be $o(n^{-M})$ for some large M, for the proof of (5.9). Reexaming the proofs of Lemmas 4.1, 4.5 and A.4, one may find that if $\mathrm{E}|x_{11}|^{4+\varepsilon} < \infty$, then $\max_{z\in T}\|\nu_n(\cdot, z) - \nu(\cdot, z)\| = o(n^{-\beta})$ for some $\beta > 0$. Therefore, the circular law is true when the moment condition in Theorem 1.1 is reduced to the existence of the $4 + \varepsilon$th moment. The details of the proof are omitted.

7.1.2. *On the smoothness of the underlying distribution.* The purpose of this subsection is to consider the circular law for real random matrices whose entries have a bounded density. The circular law for this case does not follow from Theorem 1.1 since the joint distribution of the real and imaginary parts of the entries does not have a joint two-dimensional density. In the following, we shall consider a more general case where the conditional density of one linear combination of the real and imaginary parts of the entry when another is given is uniformly bounded. Without loss of generality, we assume that the two linear combinations are $\mathrm{Re}(x_{11})\cos(\theta) + \mathrm{Im}(x_{11})\sin(\theta)$ and $\mathrm{Re}(x_{11})\sin(\theta) - \mathrm{Im}(x_{11})\cos(\theta)$. Note that the proof of the circular law for the matrix $\mathbf{X}$ is equivalent to that for the matrix $e^{i\theta}\mathbf{X}$ under the condition that the conditional density of the real part when the imaginary part is given is uniformly bounded. We shall establish the following theorem.

THEOREM 7.1. *Assume that the conditional density of the real part of the entries of* $\mathbf{X}$ *when given the imaginary part is uniformly bounded and assume that the entries have finite* $4 + \varepsilon$ *moment. Then the circular law holds.*

SKETCH OF THE PROOF. A review of the proof of Theorem 1.1 reveals that it is sufficient to prove the inequalities (5.7) and (5.10) under the conditions of Theorem 7.1. We start the proof of (5.7) from (5.5). Rewrite

$$\ln(|\mathbf{y}^*\mathbf{x}|^2) = \ln(|\mathbf{y}|^2) + \ln(|\tilde{\mathbf{y}}^*\mathbf{x}|^2),$$

where $\tilde{\mathbf{y}} = \mathbf{y}/|\mathbf{y}|$.

Denote by $\mathbf{x}_{jr}$ and $\mathbf{x}_{ji}$ the real and imaginary parts of the vector $\mathbf{x}_j$. Without loss of generality, we assume that $|\gamma_{ir}| \ge 1/\sqrt{2}$. Then, we have

$$\begin{aligned}|\mathbf{y}|^2 &= \mathbf{r}_2^*(\gamma_1\gamma_1^* + \gamma_2\gamma_2^*)\mathbf{r}_2\\ &\ge (\gamma_{1r}'\mathbf{r}_{2r} + \gamma_{1i}'\mathbf{r}_{2i})^2.\end{aligned}$$

Applying Lemma A.1, we find that the conditional density of $\gamma_{1r}'\mathbf{r}_{2r} + \gamma_{1i}'\mathbf{r}_{2i}$ when γ_1, γ_2 and $\mathbf{r}_{2i}$ are given is bounded by $CK_d n$. Therefore, by Lemma A.3,

$$\begin{aligned}&\frac{1}{n}\int_{z\in T}\left|\mathrm{E}\left(I_{(|\mathbf{y}|^2<n^{-14},\,|\xi_2|\le n)}\ln(|\mathbf{y}|^2)\right)\right|dt\,ds\\ &\quad\le \frac{1}{n}\int_{z\in T}\mathrm{E}\Big|\mathrm{E}\Big(I_{(|\gamma_{1r}'\mathbf{r}_{2r}+\gamma_{1i}'\mathbf{r}_{2i}|^2<n^{-14},\,|\xi_2|\le n)}\\ &\qquad\qquad\times\ln(|\gamma_{1r}'\mathbf{r}_{2r}+\gamma_{1i}'\mathbf{r}_{2i}|^2)\Big|\gamma_1,\gamma_2,\mathbf{r}_{2i}\Big)\Big|\,dt\,ds\\ &\quad\le CK_d\int_0^{n^{-7}}\ln x\,dx \le Cn^{-7}\ln n,\end{aligned}\tag{7.1}$$

for some positive constant C.

Rewrite

$$|\tilde{\mathbf{y}}^*\mathbf{x}|^2 = (\beta_1'\mathbf{r}_{1r} + \zeta_1'\mathbf{r}_{1i})^2 + (\beta_2'\mathbf{r}_{1r} + \zeta_2'\mathbf{r}_{1i})^2,$$

where

$$\beta_1 = (\tilde{\mathbf{y}}_r', \mathbf{y}_i')(\gamma_{1r}, \gamma_{2r}, -\gamma_{1i}, \gamma_{2i})', \qquad \beta_2 = (\tilde{\mathbf{y}}_r', \tilde{\mathbf{y}}_i')(-\gamma_{1i}, -\gamma_{2i}, \gamma_{1r}, -\gamma_{2r})'$$
$$\zeta_1 = (\tilde{\mathbf{y}}_r', \tilde{\mathbf{y}}_i')(\gamma_{1i}, \gamma_{2i}, \gamma_{1r}, -\gamma_{2r})', \qquad \zeta_2 = (\tilde{\mathbf{y}}_r', \tilde{\mathbf{y}}_i')(\gamma_{1r}, -\gamma_{2r}, -\gamma_{1i}, -\gamma_{2i})'.$$

It can be verified that $|\beta_1|^2 + |\beta_2|^2 = 1$. Thus, we may assume that $|\beta_1| \geq 1/\sqrt{2}$. By Lemma A.1, the conditional density of $\beta_1'\mathbf{r}_{1r} + \zeta_1'\mathbf{r}_{1i}$ when $\gamma_1, \gamma_2, \mathbf{y}$ and $\mathbf{r}_{1i}$ are given is bounded by $2K_d n$. Consequently, we can prove that

$$\begin{aligned}
&\frac{1}{n}\int_{z\in T} \mathrm{E}\left|\mathrm{E}\left(I_{(|\tilde{\mathbf{y}}^*\mathbf{x}|<n^{-7})}\ln(|\tilde{\mathbf{y}}^*\mathbf{x}|^2)\,\middle|\,\gamma_1, \gamma_2, \mathbf{y}, \mathbf{r}_{1i}\right)\right| dt\,ds \\
&\quad \leq \frac{1}{n}\int_{z\in T} \mathrm{E}\left|\mathrm{E}\left(I_{(|\beta_1'\mathbf{r}_{1r}+\zeta_1'\mathbf{r}_{1i}|^2|<n^{-7})}\ln(|\beta_1'\mathbf{r}_{1r}+\zeta_1'\mathbf{r}_{1i}|^2)\,\middle|\,\gamma_1, \gamma_2, \mathbf{y}, \mathbf{r}_{1i}\right)\right| dt\,ds \\
&\quad \leq CK_d\int_0^{n^{-7}} \ln x\,dx \leq Cn^{-7}\ln n.
\end{aligned}$$

This, together with (7.1), completes the proof of (5.7).

Now, we prove (5.10). For each k, consider the $2n \times 6$ matrix $\mathbf{A}$ whose first three columns are $(\gamma_{jkr}', -\gamma_{jki}')'$, $j = 1, 2, 3$, and other three columns are $(\gamma_{jki}', \gamma_{jkr}')'$. Since γ_{kj} are orthonormal, we have $\mathbf{A}'\mathbf{A} = \mathbf{I}_6$. Using the same approach as the proof of Lemma A.1, one may select a 6×6 submatrix $\mathbf{A}_1$ of $\mathbf{A}$ such that $|\det(\mathbf{A}_1)| \geq n^{-3}$. Within the six rows of $\mathbf{A}_1$, either three rows come from the first n rows of $\mathbf{A}$ or three come from the last n rows. Without loss of generality, assume that $\mathbf{A}_1$ has three rows coming from the first n rows of $\mathbf{A}$. Then, consider the Laplace expansion of the determinant of $\mathbf{A}_1$ with respect to the first three rows. Within the 20 terms, we may select one whose absolute value is not less than $\frac{1}{20}n^{-3}$. This term is the product of a minor from the first three rows of $\mathbf{A}_1$ and its cofactor. Since the absolute value of the entries of $\mathbf{A}$ is not greater than 1, the absolute value of the cofactor is not greater than 6. Therefore, the absolute value of the minor is not less than $\frac{1}{20}n^{-3}$. Suppose the three columns of the minor come from the first, second and fourth columns of $\mathbf{A}$, that is, from γ_{1kr}, γ_{2kr} and γ_{1ki} (the proof of the other 19 cases is similar). Then, as in the proof of Lemma A.1, one can prove that the conditional joint density of $\gamma_{1kr}'\mathbf{r}_{kr}$, $\gamma_{2kr}'\mathbf{r}_{kr}$ and $\gamma_{1ki}'\mathbf{r}_{kr}$ when γ_{jk} and $\mathbf{r}_{ki}$ are given is uniformly bounded by $120K_d n^{4.5}$. Finally, from (5.8), we have

$$\begin{aligned}
&\varepsilon_n \sum_{k=3}^{n} \frac{1}{|\mathbf{r}_k^*\gamma_{k1}|^2 + |\mathbf{r}_k^*\gamma_{k2}|^2 + |\mathbf{r}_k^*\gamma_{k3}|^2} \\
&\quad \leq \varepsilon_n \sum_{k=3}^{n} \frac{1}{(\mathbf{r}_{kr}'\gamma_{k1r} + \mathbf{r}_{ki}'\gamma_{k1i})^2 + (\mathbf{r}_{kr}'\gamma_{k2r} + \mathbf{r}_{ki}'\gamma_{k2i})^2 + (\mathbf{r}_{kr}'\gamma_{k1i} - \mathbf{r}_{ki}'\gamma_{k1r})^2}.
\end{aligned}$$

Using this and the same approach as in Section 5, one may prove that the right-hand side of the above tends to zero almost surely. Thus, (5.10) is proved and consequently, Theorem 7.1 follows. □

7.1.3. *Extension to the nonidentical case.* Reviewing the proofs of Theorem 1.1, one finds that the moment condition and the distributional identity of the entries of the random matrix were used only in Lemma A.4, for establishing the uniform convergence rate of certain quadratic forms. One requirement for this purpose is that the variables can be truncated at $n^{1/3}$ (actually, $n^{1/2-\varepsilon}$ is good enough as discussed in subsection 7.1.1). Two other requirements are $\max_{j_1, j_2}|\mathrm{E}(X_{m, j_1, j_2})| = o(n^{-1})$ and $\max_{j_1, j_2}|\mathrm{E}(|X_{m, j_1, j_2}|^2) - 1| = o(1)$. Therefore, we have the following theorem.

THEOREM 7.2. *In additional to the smoothness condition assumed in Theorem* 1.1, *we further assume that*

$$\max_{j_1, j_2}|\mathrm{E}(X_{j,k}) I_{(|X_{j,k}| \le n^{1/2-\varepsilon})}| = o(n^{-1}),$$
$$\max_{j_1, j_2}\left|\mathrm{E}\left(|X_{j_1, j_2}|^2 I_{(|X_{j,k}| \le n^{1/2-\varepsilon})}\right) - 1\right| = o(1) \tag{7.2}$$

and

$$\limsup_{n\to\infty} \mathrm{P}\left(\bigcup_{k, j \le n} \left(|x_{nkj}| > n^{1/2-\varepsilon}\right)\right) = 0. \tag{7.3}$$

Then the circular law is true.

A sufficient condition for (7.2) and (7.3) is the following: in addition to $\mathrm{E}(X_{jk}) = 0$, $\mathrm{E}(|X_{jk}|^2) = 1$,

$$\max_{k, j} \mathrm{E}|x_{kj}|^{4+\varepsilon} < \infty \quad \text{if all } x_{kj} \text{ come from a double array,}$$

or

$$\max_{n, k, j} \mathrm{E}|x_{nkj}|^{6+\varepsilon} < \infty \quad \text{if } x_{nkj} \text{ depends on } n.$$

7.2. *Spectral radius.* As mentioned earlier, Bai and Yin (1986) and Geman (1986), proved that with probability 1, the upper limit of the spectral radius of Ξ_n is not greater than 1. Combining this result together with Theorem 1.1, it follows immediately that with probability 1, the spectral radius of Ξ_n converges to 1. In fact, we can get more, that is, under the conditions of Theorem 1.1, we have, with probability 1,

$$\lim_{n\to\infty} \inf_{a^2+b^2=1} \max_{k\le n}(a\,\mathrm{Re}(\lambda_k) + b\,\mathrm{Im}(\lambda_k))$$
$$= \lim_{n\to\infty} \sup_{a^2+b^2=1} \max_{k\le n}(a\,\mathrm{Re}(\lambda_k) + b\,\mathrm{Im}(\lambda_k)) = 1.$$

APPENDIX

Elementary lemmas.

A1. Lemmas on densities or expectations of functions of random variables.

LEMMA A.1. *Let* $\mathbf{X} = (\mathbf{x}_1, \ldots, \mathbf{x}_n)$ *be a* $p \times n$ *real random matrix of* n *independent column vectors whose probability densities have a common bound* K_d *and let* $\boldsymbol{\alpha}_1, \ldots, \boldsymbol{\alpha}_k$, $(k < n)$ *be* k *orthogonal real unit* n-*vectors. Then, the joint density of the random* p-*vectors* $\mathbf{y}_j = \mathbf{X}\boldsymbol{\alpha}_j$, $j = 1, \ldots, k$, *is bounded by* $K_d^k n^{kp/2}$.

PROOF. Write $\mathbf{C} = (\boldsymbol{\alpha}_1 \cdots \boldsymbol{\alpha}_k)'$ and let $\mathbf{C}(j_1, \ldots, j_k)$ denote the $k \times k$ submatrix formed by the $j_1 \cdots j_k$th columns of C. By Bennett's formula, we have

$$\sum_{1 \le j_1 < \cdots < j_k \le n} \det{}^2(\mathbf{C}(j_1, \ldots, j_k)) = \det(\mathbf{C}'\mathbf{C}) = 1.$$

Thus, we may select $1 \le j_1 < \cdots < j_k \le n$, say, $1, 2, \ldots, k$ for simplicity, such that $|\det(\mathbf{C}(1, \ldots, k))| \ge n^{-k/2}$. Let $\mathbf{C}_1$ and $\mathbf{C}_2$ denote the submatrices of the first k and last $n-k$ columns of $\mathbf{C}$ and let $\mathbf{X}_1$ and $\mathbf{X}_2$ denote the submatrices of the first k and last $n-k$ columns of $\mathbf{X}$, respectively. Furthermore, denote by $\mathbf{c}_1, \ldots, \mathbf{c}_k$, the row vectors of the matrix $\mathbf{C}^{-1}(1, 2, \ldots, k)$. Then, the joint density of $\mathbf{y}_1, \ldots, \mathbf{y}_k$ is given by

$$p(\mathbf{y}_1, \ldots, \mathbf{y}_k) = |\det{}^{-p}(C(1, 2, \ldots, k))| \mathrm{E}\left(\prod_{i=1}^{k} f_i((\mathbf{y} - \mathbf{X}_2\mathbf{C}_2')\mathbf{c}_i) \right) \le K_d^k n^{kp/2},$$

where $\mathbf{Y} = (\mathbf{y}_1, \ldots, \mathbf{y}_k)'$. The proof of the lemma is complete. □

For the complex case, we have the following corollary.

COROLLARY A.2. *If the vectors and matrices in Lemma* A.1 *are assumed to be complex and the joint density of the real and imaginary parts of* $\mathbf{x}_j$ *are uniformly bounded by* K_d, *then the joint density of the real and imaginary parts of* $\mathbf{y}_1, \ldots, \mathbf{y}_k$ *is bounded by* $K_d^{2k} n^{2kp}$.

LEMMA A.3. *Suppose that* $f(t)$ *is a function such that* $\int_0^\delta t|f(t)|\,dt \le M\delta^\mu$, *for some* $\mu > 0$ *and all small* δ. *Let* $\mathbf{x}$ *and* $\mathbf{y}$ *be two complex random* k-*vectors* $(k > 1)$ *whose joint density of the real and imaginary parts of* $\mathbf{x}$ *and* $\mathbf{y}$ *is bounded by* K_d. *Then,*

$$\mathrm{E}(f(|\mathbf{x}^*\mathbf{y}|) I(|\mathbf{x}^*\mathbf{y}| < \delta, |\mathbf{x}| \le K_e, |\mathbf{y}| \le K_e) \le C_k M\delta^\mu K_d K_e^{4k-4}, \tag{A.1}$$

where C_k *is a positive constant depending on* k *only.*

PROOF. Note that the measure of $\mathbf{x} = 0$ is zero. For each $\mathbf{x} \ne 0$, define a unitary $k \times k$ matrix U with $\mathbf{x}/|\mathbf{x}|$ as its first column. Now, make a change of variables $\mathbf{u} = \mathbf{x}$ and $\mathbf{v} = \mathbf{U}^* \mathbf{y}$. It is known that the Jacobian of this variable transformation is 1. This leads to $|\mathbf{x}^* \mathbf{y}| = |\mathbf{u}||v_1|$. Thus,

$$\mathrm{E}(f(|\mathbf{x}^* \mathbf{y}|) I(|\mathbf{x}^* \mathbf{y}| < \delta, |\mathbf{x}| \le K_e, |\mathbf{y}| \le K_e)$$

$$(\mathrm{A.2}) \qquad = \int \cdots \int f(|\mathbf{u}||v_1|) I(|\mathbf{u}||v_1| < \delta, |\mathbf{u}| \le K_e, |\mathbf{v}| \le K_e) \times p(\mathbf{u}, U\mathbf{v})\, d\mathbf{u}\, d\mathbf{v}$$

$$\le K_d s_{2k} 2\pi (2K_e)^{2k-2} \int_0^{\sqrt{k} K_e} \rho_1^{2k-1}\, d\rho_1 \int_0^{\delta/\rho_1} \rho_2 f(\rho_1 \rho_2)\, d\rho_2$$

$$(\mathrm{A.3}) \qquad \le K_d s_{2k} 2\pi (2K_e)^{2k-2} (2k-2)^{-1} (\sqrt{k} K_e)^{2k-2} \int_0^{\delta} t f(t)\, dt,$$

where s_{2k} denotes the Euclidean area of the $2k$-dimensional unit sphere. Here, the inequality (A.2) follows from a polar transformation for the real and imaginary parts of $\mathbf{u}$ (dimension $= 2k$) and from a polar transformation for the real and imaginary parts (dimension $= 2$) of v_1. The lemma now follows from (A.3). □

LEMMA A.4. *Let $\{a_{nlkj_1, j_2}\}$, $l \le n^d$, $k, j_1, j_2 \le n$, be complex random variables satisfying* $\max_{n, l, k, j_2} \sum_{j_1} |a_{nlkj_1, j_2}|^2 < K^2$, $\max_{n, l, k, j_1} \sum_{j_2} |a_{nlkj_1, j_2}|^2 < K^2$, *and $|z_l| \le n^{1/36}$ are complex constants. Suppose that $\{X_{kj},\ k, j = 1, 2, \dots\}$ is a double array of iid complex random variables with mean zero and finite sixth moment. Assume for each fixed k, $\{X_{kj},\ j = 1, 2, \dots\}$ is independent of $\{a_{nlkj_1, j_2}\}$, $l \le n^d$, $j_1, j_2 \le n$. Then,*

$$(\mathrm{A.4}) \qquad \sup_{l \le n^d,\, k \le n} \left| \sum_{j_1, j_2} a_{nlkj_1, j_2} \left[\left(\frac{1}{\sqrt{n}} X_{kj_1} - z_l \delta_{k, j_1} \right) \left(\frac{1}{\sqrt{n}} \bar{X}_{kj_2} - \bar{z}_l \delta_{k, j_2} \right) - \delta_{j_1, j_2} \left(\frac{1}{n} \mathrm{E}|X_{11}|^2 + \delta_{k, j_1} |z_l|^2 \right) \right] \right|$$

$$= o(n^{-5/36} K \ln^2 n),$$

where $d > 0$ is a positive constant and δ_{kj} is the Kronecker delta, that is, $= 1$ *or* 0 *corresponding to $k = j$ or not.*

PROOF. Without loss of generality, we may assume that $K = 1$, $\mathrm{E}(|X_{11}|^2) = 1$, and that a_{nlkj_1, j_2} are real nonrandom constants and $\{z_l\}$ and $X_{k,j}$ are real constants and random variables, respectively.

Now, let m be a positive integer. For k, j, defined $X_{mkj} = X_{kj}$ or zero according to $|X_{kj}| \le 2^{m/3}$ or not, respectively. Note that

$$\mathrm{P}\left(\bigcup_{m=1}^{\infty} \bigcup_{k, j \le 2^m} (X_{mkj} \ne X_{kj}) \right) \le \sum_{m=1}^{\infty} 2^{2m} \mathrm{P}(|X_{11}| > 2^{m/3}) < \infty$$

by the finiteness of the sixth moment of X_{11}. Therefore, by the Borel–Cantelli lemma, the variables X_{kj} in (A.4) can be replaced by X_{mkj}, for all $n \in (2^m, 2^{m+1}]$. In other words, we may assume that for each n, $|X_{mkj}| \le n^{1/3}$.

In the rest of the proof of this lemma, all probabilities and expectations are conditional probabilities and expectations for the a-variables given, namely, we treat the a-variables as nonrandom. For fixed $\varepsilon > 0$, by Bernstein's inequality, we have

$$\sum_n \sum_{l,k} \mathrm{P}\left(\frac{1}{n} \left| \sum_{j=1}^{n} a_{nlkjj}\left[X_{mkj}^2 - \mathrm{E} X_{m11}^2 \right] \right| \ge \varepsilon n^{-1/3} \ln^2 n \right)$$
$$\le M \sum_n n^{d+1} \exp\left(-\varepsilon^2 \ln^4 n / \left[\mathrm{E}|X_{m11}|^4 + \varepsilon \ln^2 n \right] \right) < \infty.$$

which, together with Borel–Cantelli, implies that

$$\text{(A.5)} \qquad \max_{l \le n^d,\, k \le n} \left| \frac{1}{n} \sum_{j=1}^{n} a_{nlkjj}\left[X_{mjj}^2 - \mathrm{E} X_{m11}^2 \right] \right| = o(n^{-1/3} \ln^2 n).$$

Because of the truncation, we have

$$\text{(A.6)} \qquad \max_{l \le n^d,\, k \le n} \left| \frac{1}{\sqrt{n}} a_{nlkkk} z_l X_{m,kk} \right| < M n^{-5/36} = o(n^{-5/36} \ln^2 n).$$

By (A.5) and (A.6) and the fact that

$$\max_{l \le n^d,\, k \le n} \left| \frac{1}{n} \sum_{j=1}^{n} a_{nlkjj}\left[\mathrm{E}(X_{11}^2) - \mathrm{E}(X_{m11}^2) \right] \right| \le M \max_{l \le n^d,\, k \le n} n^{-7/3} \sum_{j=1}^{n} |a_{nlkjj}|$$
$$\le M n^{-4/3} = o(n^{-1/2} \ln^2 n),$$

to finish the proof of the lemma, one need only show that

$$\text{(A.7)} \qquad \max_{l \le n^d,\, k \le n} \left| \frac{1}{n} \sum_{j_1 \ne j_2} a_{nlkj_1,j_2} X_{mkj_1} X_{mkj_2} \right| = o(n^{-5/36} \ln^2 n),$$

and

$$\text{(A.8)} \qquad \max_{l \le n^d,\, k \le n} |z_l| \left| \frac{1}{\sqrt{n}} \sum_{j=1 (j \ne k)}^{n} a_{nlkk,j} X_{mkj} \right| = o(n^{-5/36} \ln^2 n).$$

Note that (A.8) is implied by

$$\text{(A.9)} \qquad \max_{l \le n^d,\, k \le n} \left| \frac{1}{\sqrt{n}} \sum_{j=1 (j \ne k)}^{n} a_{nlkk,j} X_{mkj} \right| = o(n^{-5/36} \ln^2 n),$$

which can be proved by the same lines in the proof of (A.5).

In the proof of (A.7), for convenience of notation, we shall omit the subscripts n, k and l from a_{nlk,j_1,j_2} and rewrite j_1, j_2 as k and j. Also, we simplify X_{mkj} as X_j and assume that $a_{j_1 j_2} = 0$ if $j_1 \ge j_2$. We will finish the proof of (A.7) by establishing that the probability when the left-hand side of (A.7) is greater than $\varepsilon n^{-5/36}\ln^2 n$ can be smaller than any a fixed negative power of n.

Define $b_{nkj} = n^{-31/36} a_{kj}$, and for $2 \le h < n$ and $1 \le k < j \le h$, $b_{hkj} = b_{h+1,kj} + 2b_{h+1,j,h+1} b_{h+1,k,h+1}$. By induction and the condition that $\sum_{1 \le k < j \le n} |b_{nkj}^2| \le n^{-13/18}$, one can prove that for any $2 \le h < n$ and $n > 60$,

$$(A.10)\quad \begin{aligned}
&\sum_{1 \le k < j \le h} b_{hkj}^2 \\
&= \sum_{1 \le k < j \le h} \Big[b_{h+1,k,j}^2 + 4 b_{h+1,k,j} b_{h+1,k,h+1} b_{h+1,j,h+1} \\
&\qquad\qquad + 4 b_{h+1,k,h+1}^2 b_{h+1,j,h+1}^2 \Big] \\
&\le \sum_{1 \le k < j \le h} b_{h+1,k,j}^2 + 4\Bigg(\sum_{1 \le k < j \le h} b_{h+1,k,j}^2 \Bigg)^{1/2} \\
&\quad \times \sum_{k=1}^{h} b_{h+1,k,h+1}^2 + 4\Bigg(\sum_{k=1}^{h} b_{h+1,k,h+1}^2 \Bigg)^2 \\
&\le \sum_{1 \le k < j \le h+1} b_{h+1,k,j}^2 + (4n^{-13/36} + 4n^{-13/18}) \sum_{k=1}^{h} b_{h+1,k,h+1}^2 \\
&\le \sum_{1 \le k < j \le h+1} b_{h+1,k,j}^2 \le n^{-13/18}.
\end{aligned}$$

Let $q = \sum_{h=2}^{n} \sum_{k=1}^{h-1} b_{hkh}^2$. Then, by definition, we have

$$(A.11)\quad \begin{aligned}
q &= \sum_{k=1}^{n-1} n^{-31/18} a_{n-1,n}^2 \\
&\quad + \sum_{h=2}^{n-1} \sum_{k=1}^{h-1} \big(b_{h+1,kh}^2 + 4 b_{h+1,kh} b_{h+1,k,h+1} b_{h+1,h,h+1} \\
&\qquad\qquad + 4 b_{h+1,k,h+1}^2 b_{h+1,h,h+1}^2 \big) \\
&= \sum_{k<h} n^{-31/18} a_{k,h}^2 \\
&\quad + 4 \sum_{h=2}^{n-1} \sum_{k=1}^{h-1} \sum_{s=h+1}^{n} \big(b_{s,kh} b_{s,k,s} b_{s,h,s} + 4 b_{s,k,s}^2 b_{s,h,s}^2 \big) \\
&\le n^{-13/18} + 4 \sum_{s=3}^{n} \Bigg(\sum_{k<h<s} b_{s,kh}^2 \Bigg)^{1/2} \Bigg(\sum_{k<s} b_{sks}^2 \Bigg)^{1/2} \Bigg(\sum_{h<s} b_{shs}^2 \Bigg)^{1/2} \\
&\quad + 4 \sum_{s=3}^{n} \Bigg(\sum_{k<s} b_{sks}^2 \Bigg)^2 \\
&\le n^{-13/18} + 4(n^{-13/36} + n^{-13/18}) q \quad \text{[by (A.10)]} \\
&\le 2n^{-13/18}.
\end{aligned}$$

Write $\tilde{b}_j = \sum_{k=j+1}^{n} b_{jkj}^2$. The estimate (A.11) implies that $\tilde{b}_j < 2n^{-13/18}$. Notice that

$$\sum_{j=1} \tilde{b}_j \mathrm{E}(X_j^2) = O(n^{-13/18}).$$

Then, by (A.11) and applying Bernstein's inequality, we obtain

$$\text{(A.12)} \qquad \begin{aligned} \mathrm{P}\Bigl(\sum_{j=1} \tilde{b}_j X_j^2 \geq 2\Bigr) &\leq \mathrm{P}\Bigl(\sum_{j=1} \tilde{b}_j\bigl(X_j^2 - \mathrm{E}(X_j^2)\bigr) \geq 1\Bigr) \\ &\leq \exp(-cn^{1/18}) \end{aligned}$$

for some positive constant c.

Now, we shall find the bound for the b_{khj}'s. By the definition of b_{hkj} and the estimation (A.11), we have

$$\text{(A.13)} \qquad \begin{aligned} |b_{khj}| &\leq |b_{h+1,kj}| + |b_{h+1,k,h+1}||b_{h+1,j,h+1}| \\ &\leq |b_{n,kj}| + \sum_{s=h+1}^{n} |b_{sks}||b_{sjs}| \\ &\leq n^{-31/36} + q \leq 3n^{-13/18}. \end{aligned}$$

Define

$$E_h = \left\{\left|\sum_{k=1}^{h-1} b_{hkh} X_k\right| < \tfrac{1}{2} n^{-13/36}\right\},$$

$$E_0 = \left\{\sum_{j=1} \tilde{b}_j X_j^2 < 2\right\} = \left\{\sum_{h=2}^{n} \sum_{j=1}^{h-1} b_{hjh}^2 X_j^2 < 2\right\}$$

and

$$F_h = \bigcap_{l=1}^{h} E_l.$$

Then, by (A.13) and applying Bernstein's inequality or Kolmogorov's inequality [Loève (1977), page 266], we have

$$\mathrm{P}(E_h^c) \leq C\exp(-cn^{1/36}),$$

for some positive constants C and c. Consequently,

$$\begin{aligned} &\mathrm{P}\left(n^{-1}\left|\sum_{k<j} a_{kj} X_k X_j\right| \geq \varepsilon n^{-5/36} \ln^2 n\right) \\ &\leq \mathrm{P}(E_0) + \sum_{h=3}^{n} \mathrm{P}(E_m^c) \\ &\qquad + \exp(-\varepsilon \ln^2 n) \mathrm{E}\, I(F_n)\, I(E_0) \exp\left\{\sum_{j=2}^{n} \sum_{k=1}^{j-1} b_{nkj} X_k X_j\right\}. \end{aligned}$$

Thus, to complete the proof of the lemma, it suffices to show that

$$\text{(A.14)}\qquad \mathrm{E}\, I(F_n)\, I(E_0) \exp\left\{ \sum_{j=2}^{n} \sum_{k=1}^{j-1} b_{nkj} X_k X_j \right\} = O(1),$$

which is obviously [see (A.12)] implied by

$$\text{(A.15)}\qquad \mathrm{E}\, I(F_n) \exp\left\{ -\sum_{h=2}^{n} \sum_{j=1}^{h-1} b_{hjh}^2 X_j^2 + \sum_{j=2}^{n} \sum_{k=1}^{j-1} b_{nkj} X_k X_j \right\} = O(1).$$

In fact, by induction, we have

$$\begin{aligned}
&\mathrm{E}\big(I(F_n) \exp\left\{ -\sum_{h=2}^{n} \sum_{j=1}^{h-1} b_{hjh}^2 X_j^2 + \sum_{j=2}^{n} \sum_{k=1}^{j-1} b_{nkj} X_k X_j \right\} \\
&\quad \le \mathrm{E}\big(I(F_n) \exp\left\{ -\sum_{h=2}^{n} \sum_{j=1}^{h-1} b_{hjh}^2 X_j^2 + \sum_{j=2}^{n-1} \sum_{k=1}^{j-1} b_{nkj} X_k X_j \right\} \\
&\qquad \times \left[1 + \sum_{k=1}^{n-1} b_{nkj} X_k X_n + \left(\sum_{k=1}^{n-1} b_{nkj} X_k X_n \right)^2 \right] \quad \left(\text{by } \left| \sum_{k=1}^{n-1} b_{nkj} X_k X_n \right| \le \tfrac{1}{2} \right) \\
&\quad \le \mathrm{E}\big(I(F_n) \exp\left\{ \sum_{j=2}^{n-1} \sum_{k=1}^{j-1} b_{nkj} X_k X_j \right\} \left[1 + n^{-5/3} + \left(\sum_{k=1}^{n-1} b_{nkj} X_k \right)^2 \right] \\
&\qquad\qquad [\, X_n \text{ is independent of } F_n \text{ and } |\mathrm{E}(X_n)| \le C n^{-5/3},] \\
&\quad \le \mathrm{E}\big(I(F_{n-1}) \exp\left\{ -\sum_{h=2}^{n-1} \sum_{j=1}^{h-1} b_{hjh}^2 X_j^2 + \sum_{j=2}^{n-1} \sum_{k=1}^{j-1} b_{n-1,kj} X_k X_j + n^{-5/3} \right\} \cdots \\
&\quad \le \mathrm{E} \exp\{ b_{212} X_1 X_2 + n^{-2/3} \} \le \exp\{ n^{-1/18} + n^{-2/3} \} \to 1.
\end{aligned}$$

This establishes (A.15) and consequently the proof of the lemma is complete. □

A2. A result known in the literature.

LEMMA A.5. *Let* **A** *and* **B** *be two* $m \times n$ *complex matrices and denote by* ν_a *and* ν_b *the empirical spectral distributions of* $\mathbf{A}^*\mathbf{A}$ *and* $\mathbf{B}^*\mathbf{B}$, *respectively. Then, we have*

$$\text{(A.16)}\qquad \left(\int |\nu_a(u) - \nu_b(u)|\, du \right)^2 \le 2n^{-2} \operatorname{tr}(\mathbf{A}^*\mathbf{A} + \mathbf{B}^*\mathbf{B}) \operatorname{tr}((\mathbf{A} - \mathbf{B})^* (\mathbf{A} - \mathbf{B})).$$

Similar versions of this lemma were used by Bai and Silverstein (1995), Wachter (1978) and Yin (1986). The exact version of this lemma (but for real

A and **B**) was used in Bai (1993b) but it was not stated as a lemma. An outline of the proof of the lemma is given below:

$$\begin{aligned}\left(\int|\nu_a(u)-\nu_b(u)|\,du\right)^2 &\le \left(\frac{1}{n}\sum_{k=1}^{n}|\lambda_k-\eta_k|\right)^2\\ &\le \left(\frac{1}{n}\sum_{k=1}^{n}\left|\sqrt{\lambda_k}+\sqrt{\eta_k}\right|^2\right)\left(\frac{1}{n}\sum_{k=1}^{n}\left|\sqrt{\lambda_k}-\sqrt{\eta_k}\right|^2\right)\\ &\le 2n^{-2}\operatorname{tr}(\mathbf{AA}^*+\mathbf{BB}^*)\operatorname{tr}((\mathbf{A}-\mathbf{B})^*(\mathbf{A}-\mathbf{B})).\end{aligned}$$

The last step follows from the von Neumann inequality.

Acknowledgments. The author thanks Professor J. W. Silverstein and the referee for their careful review of the paper and their helpful comments and suggestions in improving the paper.

REFERENCES

BAI, Z. D. (1993a). Convergence rate of expected spectral distributions of large random matrices. I. Wigner matrices. *Ann. Probab.* **21** 625–648.

BAI, Z. D. (1993b). Convergence rate of expected spectral distributions of large random matrices. II. Sample covariance matrices. *Ann. Probab.* **21** 649–672.

BAI, Z. D. and YIN, Y. Q. (1986). Limiting behavior of the norm of products of random matrices and two problems of Geman-Hwang. *Probab. Theory Related Fields* **73** 555–569.

BAI, Z. D. and YIN, Y. Q. (1988a). A convergence to the semicircle law. *Ann. Probab.* **16** 863–875.

BAI, Z. D. and YIN, Y. Q. (1988b). Necessary and sufficient conditions for the almost sure convergence of the largest eigenvalue of Wigner matrices. *Ann. Probab.* **16** 1729–1741.

BAI, Z. D. and YIN, Y. Q. (1993). Limit of the smallest eigenvalue of large dimensional covariance matrix. *Ann. Probab.* **21** 1275–1294.

EDELMAN, A. (1995). The circular law and the probability that a random matrix has k real eigenvalues. Unpublished manuscript.

GEMAN, S. (1980). A limit theorem for the norm of random matrices. *Ann. Probab.* **8** 252–261.

GEMAN, S. (1986). The spectral radius of large random matrices. *Ann. Probab.* **14** 1318–1328.

GINIBRE, J. (1965). Statistical ensembles of complex, quaterion and real matrices. *J. Math. Phys.* 440–449.

GIRKO, V. L. (1984a). Circle law. *Theory Probab. Appl.* 694–706.

GIRKO, V. L. (1984b). On the circle law. *Theory Probab. Math. Statist.* 15–23.

HWANG, C. R. (1986). A brief survey on the spectral radius and the spectral distribution of large dimensional random matrices with iid entries. In *Random Matrices and Their Applications* (M. L. Mehta, ed.) 145–152. Amer. Math. Soc., Providence, RI.

JONSSON, D. (1982). Some limit theorems for the eigenvalues of a sample covariance matrix. *J. Multivariate Anal.* **12** 1–38.

LOÈVE, M. (1977). *Probability Theory*, 4th ed. Springer, New York.

MARČENKO, V. A. and PASTUR, L. A. (1967). Distribution for some sets of random matrices. *Math. USSR-Sb.* **1** 457–483.

PASTUR, L. A. (1972). On the spectrum of random matrices. *Teoret. Mat. Fiz.* **10** 102–112. (English translation in *Theoret. and Math. Phys.* **10** 67–74.)

PASTUR, L. A. (1973). Spectra of random self-adjoint operators. *Uspekhi Mat. Nauk* **28** 4–63. (English translation in *Russian Math. Surveys* **28** 1–67.)

SILVERSTEIN, J. W. and BAI, Z. D. (1995). On the empirical distribution of eigenvalues of a class of large dimensional random matrices. *J. Multivariate Anal.* **54** 175–192.

SILVERSTEIN, W. J. and CHOI, S. I. (1995). Analysis of the limiting spectral distribution of large dimensional random matrices. *J. Multivariate Anal.* **54** 295–309.

WACHTER, K. W. (1978). The strong limits of random matrix spectra for sample matrices of independent elements. *Ann. Probab.* **6** 1–18.

WACHTER, K. W. (1980). The limiting empirical measure of multiple discriminant ratios. *Ann. Statist.* **8** 937–957.

YIN, Y. Q., BAI, Z. D. and KRISHNAIAH, P. R. (1988). On the limit of the largest eigenvalue of the large dimensional sample covariance matrix. *Probab. Theory Related Fields* **78** 509–521.

DEPARTMENT OF APPLIED MATHEMATICS
NATIONAL SUN YAT-SEN UNIVERSITY
KAOHSIUNG, TAIWAN 80424
E-MAIL: baiz@math.nsysu.edu.tw

The Annals of Applied Probability
1998, Vol. 8, No. 3, 886–895

ON THE VARIANCE OF THE NUMBER OF MAXIMA IN RANDOM VECTORS AND ITS APPLICATIONS

BY ZHI-DONG BAI, CHERN-CHING CHAO,[1] HSIEN-KUEI HWANG AND WEN-QI LIANG[1]

National Sun Yat-Sen University, Academia Sinica, Academia Sinica and Academia Sinica

We derive a general asymptotic formula for the variance of the number of maxima in a set of independent and identically distributed random vectors in $\mathbb{R}^d$, where the components of each vector are independently and continuously distributed. Applications of the results to algorithmic analysis are also indicated.

1. Introduction. Let $X = \{\mathbf{x}_1, \mathbf{x}_2, \ldots, \mathbf{x}_n\}$ be a set of independent and identically distributed (iid) random vectors in $\mathbb{R}^d$. A point $\mathbf{x}_i = (x_{i1}, \ldots, x_{id})$ is said to be *dominated* by $\mathbf{x}_j$ if $x_{ik} < x_{jk}$ for all $k = 1, \ldots, d$, and a point $\mathbf{x}_i$ is called a *maximum* of X if none of the other points dominates it. This paper is concerned with the number of maxima, denoted by $K_{n,d}$, of X.

The study of the number of maxima of a set of points was initiated by Barndorff-Nielsen and Sobel (1966) as an attempt to describe the boundary of a set of random points in $\mathbb{R}^d$. Due to its close relationships to convex hull, this problem has been developed to be one of the core problems in computational geometry, with many applications in diverse disciplines such as pattern classification, graphics, economics, data analysis, etc. The reader is referred to Preparata and Shamos (1985), Bentley, Kung, Schkolnick and Thompson (1978), Becker, Denby, McGill and Wilks (1987) and Bentley, Clarkson and Levine (1993), Golin (1993) for more information. This problem also arose in the multicriterial choice problem in operations research. Let x_{ij} represent a utility of variant (alternative, plan) i according to criterion j, $i = 1, \ldots, n$, $j = 1, \ldots, d$. If there is no relation of criteria according to importance, the choice is often made by relying on the partial order relation $\mathbf{x}_i \succ \mathbf{x}_j$ if $x_{ik} \geq x_{jk}$ for all k and $x_{il} > x_{jl}$ for some l. Then the optimal variants constitute the so-called *Pareto* set of X, that is, the set of all $\mathbf{x}_i$ which are not "$\prec$" by others. The Pareto set has been actively investigated since the seventies, notably in Russia; see the survey paper by Sholomov (1983). Under the assumptions that $\mathbf{x}_1, \ldots, \mathbf{x}_n$ are iid and that the components of each vector are identically and continuously distributed, the Pareto set is identical to the set of maxima. In the sequel, all (with only one exception) results concerning the random variables $K_{n,d}$ mentioned in this paper are under the above assumptions.

Received September 1997; revised November 1997.
[1]Supported in part by NSC Grant NSC-86-2115-M001-022.
AMS 1991 *subject classifications.* Primary 60D05; secondary 68Q25, 65Y25.
Key words and phrases. Maximal points, multicriterial optimization, Eulerian sums.

Dominance is clearly one of the natural order relations in multivariate observations. Thus, the random variables $K_{n,d}$ play a fundamental role in diverse fields, and some of their probabilistic properties have been rediscovered in the literature. Barndorff-Nielsen and Sobel (1966) first showed, as a special case of their general results, that

$$
\begin{aligned}
\mu_{n,d} := \mathrm{E}(K_{n,d}) &= \sum_{1\le k\le n} \binom{n}{k}(-1)^{k-1}k^{1-d} \\
&= \frac{(\log n)^{d-1}}{(d-1)!}(1+O((\log n)^{-1})), \qquad n\to\infty,
\end{aligned}
\tag{1.1}
$$

for $d \ge 2$. This problem is the object of many papers (some simplifying the proofs of the others) by Berezovskii and Travkin (1975), Ivanin (1975b), Bentley, Kung, Schkolnick and Thompson (1978), Devroye (1980), O'Neill (1980), Buchta (1989). The only exception mentioned above is that Ivanin (1975a) dropped the assumption of independence of components and derived an asymptotic formula for $\mathrm{E}(K_{n,d})$ for multivariate normal random variables $\mathbf{x}_i$.

Finding the distribution of $K_{n,d}$ for general $d \ge 3$ is definitely more difficult (see discussions in the Bulletin Board of the newly established WEB site: `http://www-rocq.inria.fr/algo/AofA/index.html`). However, for $d = 2$, if we arrange x_{i1}, $i = 1, \dots, n$, in decreasing order, then it is easily seen that $K_{n,2}$ is essentially identical to the number of record values in a set of n iid random variables with a common continuous distribution. Thus, the exact distribution is nothing but the Stirling numbers of the first kind given by [see Barndorff-Nielsen and Sobel (1966)],

$$
\mathrm{E}\big(z^{K_{n,2}}\big) = \frac{z(z+1)\cdots(z+n-1)}{n!}, \qquad n\ge 1,
$$

and its asymptotic normality is also implied. In addition, Barndorff-Nielsen and Sobel put forth methods for calculating the distribution of $K_{n,d}$ for (1) small d and general n, and (2) small n and general d, and carried out the computations for $n = 2, 3, 4, 5$.

For the variance, it is known that

$$
\mathrm{Var}(K_{n,2}) = H_n - H_n^{(2)} = \log n + \gamma - \frac{\pi^2}{6} + O(n^{-1}), \qquad n\to\infty,
$$

where $H_n^{(j)} = \sum_{1\le k\le n} k^{-j}$ denotes the harmonic numbers and γ is Euler's constant. Barndorff-Nielsen and Sobel (1966) also showed that

$$
\begin{aligned}
\mathrm{Var}(K_{n,3}) &= 6\sum_{1\le i\le j<k\le l\le n}\frac{1}{ijkl} + \sum_{1\le i\le j\le n}\frac{1}{ij} - \big(\mathrm{E}(K_{n,3})\big)^2 \\
&= \left(\frac{1}{2}+\frac{\pi^2}{6}\right)(\log n)^2(1+O((\log n)^{-1})), \qquad n\to\infty.
\end{aligned}
$$

For general d, Devroye (1997) derived the general estimate

$$
\mathrm{Var}(K_{n,d}) = O\big(\mathrm{E}(K_{n,d})\big).
$$

This implies, by Chebyshev's inequality, that

$$\frac{K_{n,d}}{(\log n)^{d-1}/(d-1)!} \to 1 \quad \text{in probability,} \qquad n \to \infty.$$

On the other hand, Ivanin [(1976), page 99] derived an exact formula for the second moment of $K_{n,d}$:

$$\text{(1.2)} \qquad \mathrm{E}(K_{n,d}^2) = \mu_{n,d} + \sum_{1 \le t < d} \binom{d}{t} \sum_{l=1}^{n-1} \frac{1}{l} \sum{}^{(1)} \frac{1}{i_1 \cdots i_{d-2} j_1 \cdots j_{d-1}},$$

where the summation $\sum^{(1)}$ runs over all indices satisfying the inequalities

$$1 \le i_1 \le \cdots \le i_{t-1} \le l, \qquad 1 \le i_t \le \cdots \le i_{d-2} \le l$$

and

$$l < j_1 \le \cdots \le j_{d-1} \le n.$$

From this expression, the asymptotics of the variances for $d = 2, 3, 4$ were further simplified:

$$\text{(1.3)} \quad \mathrm{Var}(K_{n,4}) \sim \biggl(\sum_{1 \le j \le n} j^{-3} + \sum_{2 \le j \le n} j^{-2} \sum_{1 \le i < j} j^{-1} + \tfrac{1}{6} \biggr) (\log n)^3, \qquad n \to \infty.$$

As the main result of this paper, we establish the following theorem.

THEOREM. *For* $d \ge 2$,

$$\text{(1.4)} \quad \mathrm{Var}(K_{n,d}) = \biggl(\frac{1}{(d-1)!} + c_d \biggr) (\log n)^{d-1} \bigl(1 + O((\log n)^{-1})\bigr), \qquad n \to \infty,$$

where

$$\text{(1.5)} \qquad \begin{aligned} c_d = \frac{1}{(d-1)!} \sum_{l \ge 1} \frac{1}{l^2} \sum_{1 \le p, q \le l} \binom{l}{p} \binom{l}{q} (-1)^{p+q} pq \\ \times \bigl((p^{-1} + q^{-1})^{d-1} - p^{1-d} - q^{1-d} \bigr). \end{aligned}$$

Thus asymptotically $\mathrm{Var}(K_{n,d}) \ge \mathrm{E}(K_{n,d})$ as n becomes large.

The proof will be presented in Section 2. We first give an alternative derivation of (1.2) and then consider $\mathrm{E}(K_{n,d}^2) - \mu_{n,d}^2$. Comparing (1.1) and (1.4), we see that the major task in proving (1.4) is to cancel the first $d - 1$ terms in the asymptotic expansion of $\mu_{n,d}^2$ and to identify the dth term.

For constants c_d, we have, in particular,

$$c_2 = 0 \quad \text{and} \quad c_3 = \zeta(2) = \sum_{l \ge 1} l^{-2} = \frac{\pi^2}{6},$$

where $\zeta(s)$ denotes Riemann's zeta function $\zeta(s) = \sum_{k\ge 1} k^{-s}$ for $\Re s > 1$. While a general reduction of c_d to Riemann's zeta function at integer arguments may seem impossible in view of current knowledge on multiple harmonic sums [cf. Bailey, Borwein and Girgensohn (1994), Flajolet and Salvy (1996), Hoffman (1992) and Zagier (1992)], we have, by a well-known formula due to Euler,

$$c_4 = \sum_{l\ge 1} \frac{H_l}{l^2} = 2\zeta(3)$$

[simplifying (1.3)], and, by reductions to suitable Euler sums,

$$c_5 = \frac{5}{12}\sum_{l\ge 1}\frac{H_l^2}{l^2} + \frac{1}{6}\sum_{l\ge 1}\frac{H_l^{(2)}}{l^2} = \frac{33}{16}\zeta(4),$$

$$c_6 = \frac{7}{72}\sum_{l\ge 1}\frac{H_l^3}{l^2} + \frac{1}{8}\sum_{l\ge 1}\frac{H_l H_l^{(2)}}{l^2} + \frac{1}{36}\sum_{l\ge 1}\frac{H_l^{(3)}}{l^2} = \frac{5}{4}\zeta(5) + \frac{1}{6}\zeta(2)\zeta(3).$$

These identities can be derived by the results in Flajolet and Salvy (1996); details are omitted here. (They are numerically easy to check by using symbolic computation packages like MAPLE or MATHEMATICA.) Note that

$$c_d = \sum_{l\ge 1}\frac{1}{l^2}\sum_{j=1}^{d-2}\frac{\mu_{l,j}\mu_{l,d-1-j}}{j!(d-1-j)!}.$$

Applications of our theorem to algorithmic analysis will be briefly discussed in Section 3. Based on numerical simulations, we predict that the asymptotic distribution of $K_{n,d}$ would be Gaussian. However, we have not found any proof for $d \ge 3$.

2. Proof of the theorem. Without loss of generality, we assume that n iid random vectors $\mathbf{x}_1, \dots, \mathbf{x}_n$ are uniformly distributed over $(0, 1)^d$. Denote by G_k the event (as well as the indicator of the event) that $\mathbf{x}_k$ is a maximum in $\mathbf{x}_1, \dots, \mathbf{x}_n$. Then

$$K_{n,d} = \sum_{k=1}^{n} G_k.$$

If there are exactly $r-1$ points dominating $\mathbf{x}_k$, then $\mathbf{x}_k$ is called an rth layer maximum. Denote this event by $G_k(r)$. Thus, the total number of rth layer maxima can be expressed by

$$K_{n,d}(r) = \sum_{k=1}^{n} G_k(r).$$

To prove the theorem, we first derive a lemma and the mean of $K_{n,d}(r)$.

LEMMA 1. *Let* $0 \le t < d$. *Then,*

$$\int_{(0,1)^d} \left(1 - \prod_{i=1}^{d} x_i\right)^n \left(\prod_{i=t+1}^{d} x_i\right)^l d\mathbf{x} = \sum_{1 \le i_1 \le \cdots \le i_{d-1} \le n+1} \frac{n!(i_t + l - 1)!}{(n+l+1)!(i_1 \cdots i_{t-1}) i_t!(i_{t+1} + l) \cdots (i_{d-1} + l)}. \tag{2.1}$$

PROOF. Rewrite

$$1 - \prod_{i=1}^{d} x_i = (1 - x_1)x_2 \cdots x_d + (1 - x_2)x_3 \cdots x_d + \cdots + (1 - x_d).$$

Then

$$\left(1 - \prod_{i=1}^{d} x_i\right)^n \left(\prod_{i=t+1}^{d} x_i\right)^l = \sum_{\substack{i_1 + \cdots + i_d = n \\ i_1, \ldots, i_d \ge 0}} \frac{n!}{i_1! \cdots i_d!} (1 - x_1)^{i_1} x_2^{i_1} (1 - x_2)^{i_2} \cdots$$
$$\times x_t^{i_1 + \cdots + i_{t-1}} (1 - x_t)^{i_t} x_{t+1}^{i_1 + \cdots + i_t + l} (1 - x_{t+1})^{i_{t+1}} \cdots$$
$$\times x_d^{i_1 + \cdots + i_{d-1} + l} (1 - x_d)^{i_d},$$

and (2.1) follows. □

In particular, for $t = 0$, we have

$$\int_{(0,1)^d} \left(1 - \prod_{i=1}^{d} x_i\right)^n \left(\prod_{i=1}^{d} x_i\right)^l d\mathbf{x} = \frac{n!l!}{(n+l+1)!} \sum_{l+1 \le i_1 \le \cdots \le i_{d-1} \le n+l+1} \frac{1}{i_1 \cdots i_{d-1}}. \tag{2.2}$$

COROLLARY 1 [Barndorff-Nielsen and Sobel (1966)]. *The mean number of* r*th layer maxima is given by*

$$\mu_{n,d}(r) := \mathrm{E}(K_{n,d}(r)) = \sum_{r \le i_1 \le \cdots \le i_{d-1} \le n} \frac{1}{i_1 \cdots i_{d-1}}. \tag{2.3}$$

PROOF. The result follows from

$$\mu_{n,d}(r) = n\,\mathrm{E}(G_1(r)) = n\binom{n-1}{r-1} \int_{(0,1)^d} \left(1 - \prod_{i=1}^{d} x_i\right)^{n-r} \left(\prod_{i=1}^{d} x_i\right)^{r-1} d\mathbf{x}$$

and (2.2). □

REMARK. It is interesting to note that the probability that a point, say $\mathbf{x}_i$, is a maximal point satisfies

$$\frac{\mu_{n,d}}{n} = \mathrm{P}(Y_2 + \cdots + Y_n < d),$$

where each Y_j is geometrically distributed:

$$\mathrm{E}(z^{Y_j}) = \frac{1-1/j}{1-z/j}, \qquad 2 \le j \le n.$$

This follows from the fact that $\mu_{n,d}/n$ equals the coefficient of z^{d-1} in

$$\frac{1}{n}\prod_{1\le j\le n}\frac{1}{1-z/j} = \frac{1}{1-z}\prod_{2\le j\le n}\frac{1-1/j}{1-z/j}.$$

Also from a computational point of view, it is useful to use the recurrence $\mu_{n,1} = 1$ for $n \ge 1$ and for $d \ge 2$,

$$\mu_{n,d} = \frac{1}{d-1}\sum_{1\le j\le d-1} H_n^{(d-j)}\mu_{n,j},$$

by taking derivatives on both sides of $\sum_{d=1}^{\infty}\mu_{n,d}z^{d-1} = \prod_{j=1}^{n} 1/(1-z/j)$ and by equating coefficients of the same powers.

Next, we derive the second moment of $K_{n,d}$. Let $(t) = \{(\mathbf{x},\mathbf{y});\, x_1 > y_1, \ldots, x_t > y_t, x_{t+1} < y_{t+1}, \ldots, x_d < y_d\}$. We have

$$\begin{aligned}
\mathrm{E}(K_{n,d}^2) &= \mu_{n,d} + n(n-1)\mathrm{P}(G_1G_2)\\
&= \mu_{n,d} + n(n-1)\sum_{t=1}^{d-1}\binom{d}{t}\int_{(t)}\Bigg(1-\prod_{i=1}^{d}(1-x_i)-\prod_{i=1}^{d}(1-y_i)\\
&\qquad\qquad + \prod_{i=1}^{t}(1-x_i)\prod_{i=t+1}^{d}(1-y_i)\Bigg)^{n-2} d\mathbf{x}\,d\mathbf{y}\\
&= \mu_{n,d} + n(n-1)\sum_{t=1}^{d-1}\binom{d}{t}\\
&\quad\times\int_{(0,1)^d}\int_{(0,1)^d}\Bigg(1-\prod_{i=1}^{d}x_i\prod_{i=1}^{t}y_i - \prod_{i=t+1}^{d}x_i\prod_{i=1}^{d}y_i + \prod_{i=1}^{d}x_i\prod_{i=1}^{d}y_i\Bigg)^{n-2}\\
&\qquad\times\prod_{i=t+1}^{d}x_i\prod_{i=1}^{t}y_i\,d\mathbf{x}\,d\mathbf{y}\\
&= \mu_{n,d} + n(n-1)\sum_{t=1}^{d-1}\binom{d}{t}\\
&\quad\times\int_{(0,1)^d}\int_{(0,1)^d}\Bigg(1-\prod_{i=1}^{t}x_i\prod_{i=1}^{d}y_i - \prod_{i=t+1}^{d}x_i\prod_{i=1}^{d}y_i + \prod_{i=1}^{d}x_i\prod_{i=1}^{d}y_i\Bigg)^{n-2}\\
&\qquad\times\prod_{i=1}^{d}y_i\,d\mathbf{x}\,d\mathbf{y}\\
&:= \mu_{n,d} + \sum_{t=1}^{d-1}\binom{d}{t}J_t.
\end{aligned}$$

Noting that

$$1-\prod_{i=1}^{t}x_i\prod_{i=1}^{d}y_i-\prod_{i=t+1}^{d}x_i\prod_{i=1}^{d}y_i+\prod_{i=1}^{d}x_i\prod_{i=1}^{d}y_i$$
$$=\left(1-\prod_{i=1}^{d}y_i\right)+\prod_{i=1}^{d}y_i\left(1-\prod_{i=1}^{t}x_i\right)\left(1-\prod_{i=t+1}^{d}x_i\right),$$

we have, by Lemma 1,

$$J_t=\sum_{l=0}^{n-2}\frac{n!}{l!(n-2-l)!}\int_{(0,1)^d}\int_{(0,1)^d}\left(1-\prod_{i=1}^{d}y_i\right)^{n-2-l}\left(\prod_{i=1}^{d}y_i\right)^{l+1}$$
$$\times\left(1-\prod_{i=1}^{t}x_i\right)^{l}\left(1-\prod_{i=t+1}^{d}x_i\right)^{l}d\mathbf{x}\,d\mathbf{y}$$
$$=\sum_{l=0}^{n-2}\frac{n!}{l!(n-2-l)!}\frac{(l+1)!(n-2-l)!\mu_{n,d}(l+2)}{n!}\frac{\mu_{l+1,t}}{l+1}\frac{\mu_{l+1,d-t}}{l+1}$$
$$=\sum_{l=1}^{n-1}\frac{1}{l}\mu_{n,d}(l+1)\mu_{l,t}\mu_{l,d-t}.$$

Therefore,

$$\mathrm{E}(K_{n,d}^2)=\mu_{n,d}+\sum_{t=1}^{d-1}\binom{d}{t}\sum_{l=1}^{n-1}\frac{1}{l}\mu_{n,d}(l+1)\mu_{l,t}\mu_{l,d-t},$$

and we finally obtain (1.2).

Noting that in (1.2) the sum of those terms with at least two identical j indices in $\sum^{(1)}$ is at most $O((\log n)^{d-3})$, we further have

$$\mathrm{E}(K_{n,d}^2)=\mu_{n,d}+\sum_{t=1}^{d-1}\binom{d}{t}\sum_{l=1}^{n-1}\frac{1}{l}\sum^{(*)}\frac{1}{i_1\cdots i_{d-2}j_1\cdots j_{d-1}}+O\big((\log n)^{d-3}\big),\tag{2.4}$$

where the last summation $\sum^{(*)}$ is extended over all indices satisfying the inequalities

$$1\le i_1\le\cdots\le i_{t-1}\le l,\qquad 1\le i_t\le\cdots\le i_{d-2}\le l$$

and

$$l<j_1<\cdots<j_{d-1}\le n.$$

Now, let us compare the second term in the above expression (2.4) with $\mu_{n,d}^2$. By (2.3),

$$\mu_{n,d}^2=\sum^{(2)}\frac{1}{i_1\cdots i_{d-1}j_1\cdots j_{d-1}},$$

where the summation $\sum^{(2)}$ runs over all combinations

$$1 \le i_1 \le \cdots \le i_{d-1} \le n \quad \text{and} \quad 1 \le j_1 \le \cdots \le j_{d-1} \le n.$$

Write the $(d-1)$st largest index among $\{i_1, \dots, i_{d-1}, j_1, \dots, j_{d-1}\}$ as l and the $d-1$ indices greater than or equal to l as $k_1, \dots, k_{d-1}$. If the k_j's are not all distinct or if there is one $k_h = l$ $(1 \le h \le d-1)$, then the sum of all those terms is $O((\log n)^{d-2})$. Now, consider the sum of those terms for which the $d-1$ k-indices are distinct and not equal to l. Rearrange the k-indices as $l < k_1 < \cdots < k_{d-1}$. Suppose that there are t $(1 \le t \le d-1)$ indices (among $l, k_1, \dots, k_{d-1}$) from the j-indices. There are $\binom{d}{t}$ such possibilities. Write the other $t-1$ i-indices as $1 \le i_1 \le \cdots \le i_{t-1} \le l$ and the remaining $d-1-t$ j-indices as $1 \le j_1 \le \cdots \le j_{d-1-t} \le l$. Note that the reindexing is unique if $i_{t-1} < l$ and $j_{d-1-t} < l$. However, ambiguity arises when $i_{t-1} = l$ or $j_{d-1-t} = l$. For, if there is a term with $i_{t-1} = l$ and $j_{d-t-1} = l$, then this term is counted once in the case when l is an i-index and the number of j-indices in $\{l, k_1, \dots, k_{d-1}\}$ is t as well as once in the case when l is a j-index and the number of j-indices in $\{l, k_1, \dots, k_{d-1}\}$ is $t+1$. Thus,

$$\begin{aligned}
\mu_{n,d}^2 &= \sum_{t=1}^{d-1} \binom{d}{t} \sum_{l=1}^{n-1} \frac{1}{l} \sum{}^{(*)} \frac{1}{i_1 \cdots i_{d-2} j_1 \cdots j_{d-1}} \\
&\quad - \sum_{t=1}^{d-2} \binom{d-1}{t} \sum_{l=1}^{n-1} \frac{1}{l^2} \sum{}^{(**)} \frac{1}{i_1 \cdots i_{t-1} j_1 \cdots j_{d-2-t} k_1 \cdots k_{d-1}} \\
&\quad + O\big((\log n)^{d-2}\big),
\end{aligned}$$

where the summation $\sum^{(**)}$ is extended over all indices satisfying

$$1 \le i_1 \le \cdots \le i_{t-1} \le l, \qquad 1 \le j_1 \le \cdots \le j_{d-2-t} \le l$$

and

$$l < k_1 < \cdots < k_{d-1} \le n.$$

Therefore, we finally obtain

$$\begin{aligned}
\operatorname{Var}(K_{n,d}) &= \mu_{n,d} + \sum_{t=1}^{d-2} \binom{d-1}{t} \sum_{l=1}^{n-1} \frac{1}{l^2} \sum{}^{(**)} \frac{1}{i_1 \cdots i_{t-1} j_1 \cdots j_{d-2-t} k_1 \cdots k_{d-1}} \\
&\quad + O\big((\log n)^{d-2}\big) \\
&= \left(\frac{1}{(d-1)!} + c_d \right) (\log n)^{d-1} \big(1 + O\big((\log n)^{-1}\big)\big),
\end{aligned}$$

where

$$c_d = \sum_{t=1}^{d-2} \frac{1}{t!(d-1-t)!} \sum_{l \ge 1} \frac{1}{l^2} \sum_{\substack{1 \le i_1 \le \cdots \le i_{t-1} \le l \\ 1 \le j_1 \le \cdots \le j_{d-2-t} \le l}} \frac{1}{i_1 \cdots i_{t-1} j_1 \cdots j_{d-2-t}}.$$

Using the finite difference formula

$$\mu_{n,h} = \sum_{1\le i_1\le i_2\le\cdots\le i_h\le n} \frac{1}{i_1 i_2\cdots i_h} = \sum_{1\le j\le n} \binom{n}{j}\frac{(-1)^{j-1}}{j^h}, \qquad h=1,2,\ldots,$$

we obtain (1.5) and complete the proof of the theorem. □

3. Algorithmic applications. In this section we briefly discuss an implication of our main result: the asymptotic linearity of the variance of the cost of maxima-finding algorithms using divide-and-conquer paradigm.

There exists a large number of algorithms for finding the maxima in a given set of points [cf. Preparata and Shamos (1985), Bentley, Clarkson and Levine (1993), Devroye (1997)]. A naive divide-and-conquer algorithm runs as follows [cf. Devroye (1983)]. Divide the points $\{\mathbf{x}_1,\ldots,\mathbf{x}_n\}$ into two groups $\{\mathbf{x}_1,\ldots,\mathbf{x}_{\lfloor n/2\rfloor}\}$ and $\{\mathbf{x}_{\lfloor n/2\rfloor+1},\ldots,\mathbf{x}_n\}$, where $\lfloor y\rfloor$ denotes the largest integer less than or equal to y. Find (recursively) the (set of) maxima of each group, denoted by M_1 and M_2, respectively. Then find, by pairwise comparisons, the maxima of M_1 and M_2. Note that the randomness is preserved in the process. The worst case behavior of this algorithm is obviously quadratic in n. But the expected number of comparisons as well as the variance are both linear under our uniform distribution assumption. This is seen by noting that both quantities satisfy recurrences of the form

$$f_n = f_{\lfloor n/2\rfloor} + f_{\lfloor (n+1)/2\rfloor} + g_n,$$

for $n \ge n_0 \ge 1$ with suitable initial conditions and that $g_n = O((\log n)^{2d-2})$ for mean and $g_n = O((\log n)^{3d-3})$ for variance. It follows that $f_n = O_d(n)$, where the implied constant depends on d. The linear terms are oscillating in nature; see Flajolet and Golin (1993).

Other divide-and-conquer algorithms, such as that in Bentley and Shamos (1978), can also be shown to have linear variance for its cost.

Acknowledgments. The authors thank Luc Devroye for making his preprint available to them. They also thank the referee for pointing out an error in an earlier version of the paper.

REFERENCES

BAILEY, D. D., BORWEIN, J. M. and GIRGENSOHN, R. (1994). Experimental evaluation of Euler sums. *Experiment. Math.* **3** 17–30.

BARNDORFF-NIELSEN, O. and SOBEL, M. (1966). On the distribution of the number of admissible points in a vector random sample. *Theory Probab. Appl.* **11** 249–269.

BECKER, R. A., DENBY, L., MCGILL, R. and WILKS, A. R. (1987). Analysis of data from Places Rated Almanac. *Amer. Statist.* **41** 169–186.

BENTLEY, J. L., CLARKSON, K. L. and LEVINE, D. B. (1993). Fast linear expected-time algorithms for computing maxima and convex hulls. *Algorithmica* **9** 168–183.

BENTLEY, J. L., KUNG, H. T., SCHKOLNICK, M. and THOMPSON, C. D. (1978). On the average number of maxima in a set of vectors and applications. *J. Assoc. Comput. Mach.* **25** 536–543.

BENTLEY, J. L. and SHAMOS, M. I. (1978). Divide and conquer for linear expected time. *Inform. Process. Lett.* **7** 87–91.

BEREZOVSKII, B. A. and TRAVKIN, S. I. (1975). Supervision of queues of requests in computer systems. *Automat. Remote Control* **36** 1719–1725.

BUCHTA, C. (1989). On the average number of maxima in a set of vectors. *Inform. Process. Lett.* **33** 63–65.

DEVROYE, L. (1980). A note on finding convex hulls via maximal vectors. *Inform. Process. Lett.* **11** 53–56.

DEVROYE, L. (1983). Moment inequalities for random variables in computational geometry. *Computing* **30** 111–119.

DEVROYE, L. (1997). A note on the expected time for finding maxima by list algorithms. *Algorithmica*. To appear.

FLAJOLET, P. and GOLIN, M. (1993). Exact asymptotics of divide-and-conquer recurrences. *Lecture Notes in Comput. Sci.* **700** 137–149. Springer, Berlin.

FLAJOLET, P. and SALVY, B. (1996). Euler sums and contour integral representations. *Experiment. Math.* To appear.

GOLIN, M. J. (1993). How many maxima can there be? *Comput. Geom.* **2** 335–353.

HOFFMAN, M. E. (1992). Multiple harmonic sums. *Pacific J. Math.* **152** 275–290.

IVANIN, V. M. (1975a). Asymptotic estimate for the mathematical expectation of the number of elements in the Pareto set. *Cybernetics* **11** 108–113.

IVANIN, V. M. (1975b). Estimate of the mathematical expectation of the number of elements in a Pareto set. *Cybernetics* **11** 506–507.

IVANIN, V. M. (1976). Calculation of the dispersion of the number of elements of the Pareto set for the choice of independent vectors with independent components. In *Theory of Optimal Decisions* 90–100. Akad. Nauk. Ukrain. SSR Inst. Kibernet., Kiev. (In Russian.)

O'NEILL, B. (1980). The number of outcomes in the Pareto-optimal set of discrete bargaining games. *Math. Oper. Res.* **6** 571–578.

PREPARATA, F. P. and SHAMOS, M. I. (1985). *Computational Geometry: An Introduction*. Springer, New York.

SHOLOMOV, L. A. (1983). Survey of estimational results in choice problems. *Engrg. Cybernetics* **21** 51–75.

ZAGIER, D. (1992). Values of zeta functions and their applications. In *First European Congress of Mathematics* **2** 497–512. Birkhäuser, Berlin.

Z.-D. BAI
DEPARTMENT OF MATHEMATICS
NATIONAL UNIVERSITY OF SINGAPORE
KENT RIDGE, SINGAPORE 0511
SINGAPORE

C.-C. CHAO
H.-K. HWANG
W.-Q. LIANG
INSTITUTE OF STATISTICAL SCIENCE
ACADEMIA SINICA
TAIPEI, 115
TAIWAN
E-MAIL: ccchao@stat.sinica.edu.tw
hkwang@stat.sinica.edu.tw
wqliang@stat.sinica.edu.tw

Statistica Sinica **9**(1999), 611-677

METHODOLOGIES IN SPECTRAL ANALYSIS OF LARGE DIMENSIONAL RANDOM MATRICES, A REVIEW

Z. D. Bai

National University of Singapore

Abstract: In this paper, we give a brief review of the theory of spectral analysis of large dimensional random matrices. Most of the existing work in the literature has been stated for real matrices but the corresponding results for the complex case are also of interest, especially for researchers in Electrical and Electronic Engineering. Thus, we convert almost all results to the complex case, whenever possible. Only the latest results, including some new ones, are stated as theorems here. The main purpose of the paper is to show how important methodologies, or mathematical tools, have helped to develop the theory. Some unsolved problems are also stated.

Key words and phrases: Circular law, complex random matrix, largest and smallest eigenvalues of a random matrix, noncentral Hermitian matrix, spectral analysis of large dimensional random matrices, spectral radius.

1. Introduction

The necessity of studying the spectra of LDRM (Large Dimensional Random Matrices), especially the Wigner matrices, arose in nuclear physics during the 1950's. In quantum mechanics, the energy levels of quantums are not directly observable, but can be characterized by the eigenvalues of a matrix of observations. However, the ESD (Empirical Spectral Distribution, the empirical distribution of the eigenvalues) of a random matrix has a very complicated form when the order of the matrix is high. Many conjectures, e.g., the famous circular law, were made through numerical computation.

Since then, research on the LSA (Limiting Spectral Analysis) of LDRM has attracted considerable interest among mathematicians, probabilitists and statisticians. One pioneering work is Wigner's semicircular law for a Gaussian (or Wigner) matrix (Wigner (1955, 1958)). He proved that the expected ESD of a large dimensional Wigner matrix tends to the so-called semicircular law. This work was generalized by Arnold (1967, 1971) and Grenander (1963) in various aspects. Bai and Yin (1988a) proved that the ESD of a suitably normalized sample covariance matrix tends to the semicircular law when the dimension is relatively smaller than the sample size. Following the work by Marčenko and Pastur (1967) and Pastur (1972, 1973), the LSA of large dimensional sample covariance matrices was developed by many researchers, including Bai, Yin and

Krishnaiah (1986), Grenander and Silverstein (1977), Jonsson (1982), Wachter (1978), Yin (1986) and Yin and Krishnaiah (1983). Also, Bai, Yin and Krishnaiah (1986, 1987), Silverstein (1985a), Wachter (1980), Yin (1986) and Yin and Krishnaiah (1983) investigated the LSD (Limiting Spectral Distribution) of the multivariate F-matrix, or more generally, of products of random matrices.

Two important problems arose after the LSD was found: bounds on extreme eigenvalues and the convergence rate of the ESD. The literature on the first problem is extensive. The first success was due to Geman (1980), who proved that the largest eigenvalue of a sample covariance matrix converges almost surely to a limit under a growth condition on all the moments of the underlying distribution. Yin, Bai and Krishnaiah (1988) proved the same result under the existence of the fourth moment, and Bai, Silverstein and Yin (1988) proved that the existence of the fourth moment is also necessary for the existence of the limit. Bai and Yin (1988b) found necessary and sufficient conditions for the almost sure convergence of the largest eigenvalue of a Wigner matrix. Bai and Yin (1993), Silverstein (1985b) and Yin, Bai and Krishnaiah (1983) considered the almost sure limit of the smallest eigenvalue of a sample covariance matrix. Some related work can be found in Geman (1986) and Bai and Yin (1986).

The second problem, the convergence rate of the ESD's of LDRM, is of practical interest, but had been open for decades. The first success was made in Bai (1993a,b), in which convergence rates for the expected ESD of a large Wigner matrix and sample covariance matrix were established. Further extensions of this work can be found in Bai, Miao and Tsay (1996a,b, 1997).

The most perplexing problem is the so-called circular law that conjectures that the ESD of a non-symmetric random matrix, after suitable normalization, tends to the uniform distribution over the unit disc in the complex plane. The difficulty lies in the fact that two most important tools for symmetric matrices do not apply to non-symmetric matrices. Furthermore, certain truncation and centralization techniques cannot be used. The first known result, a partial solution for matrices whose entries are i.i.d. standard complex normal (whose real and imaginary parts are i.i.d. real normal with mean zero and variance $1/2$), was given in Mehta (1991) and an unpublished result of Silverstein reported in Hwang (1986). They used the explicit expression of the joint density of the complex eigenvalues of a matrix with independent standard complex Gaussian entries, found by Ginibre (1965). The first attempt to prove this conjecture under some general conditions was made in Girko (1984a,b). However, his proofs have puzzled many who attempted to understand the arguments. Recently, Edelman (1997) found the joint distribution of the eigenvalues of a matrix whose entries are real normal $N(0,1)$ and proved that the expected ESD of a matrix of i.i.d. real Gaussian entries tends to the circular law. Under the existence of the $(4+\varepsilon)$th moment and some smoothness conditions, Bai (1997) proved the strong version of the circular law.

In this paper, we give a brief review of the main results in this area and some of their applications. We convert most known results for real matrices to complex matrices. Since some of the extensions are non-trivial, we give brief outlines of their proofs. The review will be given in accordance with methodologies by which the results were obtained, rather than in a chronological order. The organization of the paper is as follows. In Section 2, we give results obtained by the moment approach. In Section 3, the Stieltjes transform technique is introduced. Recent achievements on the circular law are presented in Section 4. Some applications are mentioned in Section 5, and some open problems or conjectures are presented in Section 6.

2. Moment Approach

Throughout this section, we consider only Hermitian matrices which include real symmetric matrices as a special case. Let $\mathbf{A}$ be a $p \times p$ Hermitian matrix and denote its eigenvalues by $\lambda_1 \leq \cdots \leq \lambda_p$. The ESD of $\mathbf{A}$ is defined by $F^{\mathbf{A}}(x) = p^{-1}\#\{k \leq p,\ \lambda_k \leq x\}$, where $\#\{.\}$ denotes the number of elements in the set indicated. A simple fact is that the hth moment of $F^{\mathbf{A}}$ can be written as

$$\beta_h(\mathbf{A}) = \int x^h F^{\mathbf{A}}(dx) = p^{-1}\mathrm{tr}(\mathbf{A}^h). \tag{2.1}$$

This formula plays a fundamental role in the theory of LDRM. Most of the results were obtained by estimating the mean, variance or higher moments of $p^{-1}\mathrm{tr}(\mathbf{A}^h)$.

2.1. Limiting spectral distributions

To show that $F^{\mathbf{A}}$ tends to a limit, say F, usually the Moment Convergence Theorem (MCT) is employed, i.e., verifying

$$\beta_h(\mathbf{A}) \to \beta_h = \int x^h F(dx)$$

in some sense (e.g., almost surely (a.s.) or in probability) and the Carleman's condition $\sum_{h=1}^{\infty} \beta_{2h}^{-1/(2h)} = \infty$. Note that Carleman's condition is slightly weaker than requiring that the characteristic function of the LSD be analytic near 0. For most cases, the LSD has bounded support and its characteristic function is analytic. Thus, the MCT can be applied to show the existence of the LSD of a sequence of random matrices.

2.1.1. Wigner matrix

In this subsection, we first consider the famous semicircular law. A Wigner matrix $\mathbf{W}$ of order n is defined as an $n \times n$ Hermitian matrix whose entries above the diagonal are i.i.d. complex random variables with variance σ^2, and whose diagonal elements are i.i.d. real random variables (without any moment requirement). We have the following theorem.

Theorem 2.1. *Under the conditions described above, as $n \to \infty$, with probability 1, the ESD $F^{(n^{-1/2}\mathbf{W})}$ tends to the semicircular law with scale-parameter σ, whose density is given by*

$$p_\sigma(x) = \begin{cases} \frac{1}{2\pi\sigma^2}\sqrt{4\sigma^2 - x^2}, & \text{if } |x| \le 2\sigma, \\ 0, & \text{otherwise.} \end{cases} \tag{2.2}$$

The proof of the theorem consists of centralization, truncation and convergence of moments. First, we present two lemmas. Throughout the paper, $\|f\| = \sup_x |f(x)|$.

Lemma 2.2 (Rank Inequality). *Let $\mathbf{A}$ and $\mathbf{B}$ be two $n \times n$ Hermitian matrices. Then*

$$\|F^{\mathbf{A}} - F^{\mathbf{B}}\| \le \frac{1}{n}\text{rank}(\mathbf{A} - \mathbf{B}). \tag{2.3}$$

Suppose that $\text{rank}(\mathbf{A} - \mathbf{B}) = k$. To prove (2.3), we may assume that

$$\mathbf{A} = \begin{bmatrix} \mathbf{A}_{11} & \mathbf{A}_{12} \\ \mathbf{A}_{21} & \mathbf{A}_{22} \end{bmatrix} \quad \text{and} \quad \mathbf{B} = \begin{bmatrix} \mathbf{B}_{11} & \mathbf{A}_{12} \\ \mathbf{A}_{21} & \mathbf{A}_{22} \end{bmatrix},$$

where the order of $\mathbf{A}_{22}$ is $(n-k) \times (n-k)$, since (2.3) is invariant under unitary similarity. Denote the eigenvalues of $\mathbf{A}$, $\mathbf{B}$ and $\mathbf{A}_{22}$ by $\lambda_1 \le \cdots \le \lambda_n$, $\eta_1 \le \cdots \le \eta_n$ and $\tilde\lambda_1 \le \cdots \le \tilde\lambda_{(n-k)}$, respectively. By the interlacing inequality $\max(\lambda_j, \eta_j) \le \tilde\lambda_j \le \min(\lambda_{(j+k)}, \eta_{(j+k)})$ (see Rao (1976, p.64), referred to as *Poincare Separation Theorem*), we conclude that for any $x \in (\tilde\lambda_{(j-1)}, \tilde\lambda_j)$, $\frac{j-1}{n} \le F^{\mathbf{A}}(x),\ F^{\mathbf{B}}(x) < \frac{j+k}{n}$, which implies (2.3).

Lemma 2.3 (Difference Inequality). *Let $\mathbf{A}$ and $\mathbf{B}$ be two $n \times n$ complex normal matrices with complex eigenvalues $\lambda_1, \ldots, \lambda_n$ and $\eta_1, \ldots, \eta_n$, respectively. Then*

$$\min_\pi \sum_{k=1}^n |\lambda_k - \eta_{\pi_k}|^2 \le \text{tr}(\mathbf{A} - \mathbf{B})(\mathbf{A} - \mathbf{B})^*, \tag{2.4}$$

where π is a permutation of the set $\{1, \ldots, n\}$. Consequently, if both $\mathbf{A}$ and $\mathbf{B}$ are Hermitian, then

$$L^3(F^{\mathbf{A}}, F^{\mathbf{B}}) \le \frac{1}{n}\text{tr}(\mathbf{A} - \mathbf{B})^2 \tag{2.5}$$

where $L(F, G)$ denotes the Levy distance between distribution functions F and G. This lemma is an improvement to Lemma 2.3 of Bai (1993b), *where the power of the Levy distance is* 4.

To prove (2.4), we may assume, without loss of generality, that $\mathbf{A} = \mathrm{diag}(\lambda_1, \ldots, \lambda_n)$ and $\mathbf{B} = \mathbf{U}^*\mathrm{diag}(\eta_1, \ldots, \eta_n)\mathbf{U}$, where $\mathbf{U} = (u_{jk})$ is a unitary matrix. Then we have $\mathrm{tr}(\mathbf{AA}^*) = \sum_{k=1}^n |\lambda_k^2|$, $\mathrm{tr}(\mathbf{BB}^*) = \sum_{k=1}^n |\eta_k^2|$ and for some permutation π, $\mathrm{Re}(\mathrm{tr}(\mathbf{AB}^*)) = \sum_{jk} \mathrm{Re}(\lambda_j \bar{\eta}_k)|u_{jk}^2| \leq \sum_{k=1}^n \mathrm{Re}(\lambda_k \bar{\eta}_{\pi_k})$. The last inequality holds since the maximum of $\sum_{jk} \mathrm{Re}(\lambda_j \bar{\eta}_k) a_{jk}$, subject to the constraints $a_{jk} \geq 0$ and $\sum_j a_{jk} = \sum_k a_{jk} = 1$, is equal to $\sum_{k=1}^n \mathrm{Re}(\lambda_k \bar{\eta}_{\pi_k})$ for some permutation π. Then (2.4) follows.

If $\mathbf{A}$ and $\mathbf{B}$ are both Hermitian, their eigenvalues λ_k's and η_k's are real and hence can be arranged in increasing order. In this case, the solution of the minimization on the left hand side of (2.4) is $\pi_k = k$. One can show that

$$L^3(F^{\mathbf{A}}, F^{\mathbf{B}}) \leq \frac{1}{n} \sum_{k=1}^n |\lambda_k - \eta_k|^2,$$

which, together with (2.4), implies (2.5). The proof of the lemma is then complete.

Now, we outline a proof of Theorem 2.1. Note that $\mathrm{E}(\mathrm{Re}(w_{jk})) = \mathrm{E}(\mathrm{Re}(w_{kj}))$. Applying Lemma 2.2, we conclude that the LSD's of $n^{-1/2}\mathbf{W}$ and $n^{-1/2}(\mathbf{W} - \mathrm{E}(\mathrm{Re}(w_{12}))\mathbf{11}')$ exist and are the same if either one of them exists. Thus, we can remove the real parts of the expectation of off-diagonal entries (i.e., to replace them by 0). Further, note that the eigenvalues of $i\mathrm{E}(\mathrm{Im}(n^{-1/2}\mathbf{W}))$ are given by $n^{-1/2}\mathrm{E}(\mathrm{Im}(w_{12}))\cot(\pi(2k-1)/2n)$, $k = 1, \ldots, n$. Thus, applying Lemma 2.2, we can remove the eigenvalues of $i\mathrm{E}(\mathrm{Im}(n^{-1/2}\mathbf{W}))$ whose absolute values are greater than $n^{-1/4}$ (the number of such eigenvalues is less than $4n^{3/4}|E\mathrm{Im}(w_{12})|$), and by Lemma 2.3, we can also remove the remaining eigenvalues. Therefore, we may assume that the means of the off-diagonal entries of $n^{-1/2}\mathbf{W}$ are zero. Now we remove the diagonal elements of $\mathbf{W}$ by employing Hoeffding's (1963) inequality, which states that for any $\varepsilon > 0$,

$$\mathrm{P}(|\xi_n - \eta| \geq \varepsilon) \leq 2\exp(-\varepsilon^2/(2\eta + \varepsilon)),$$

where $\eta = \mathrm{E}(\xi_n)$ and ξ_n is the sum of n independent random variables taking values 0 or 1 only.

By this inequality and the fact that $\mathrm{P}(|w_{11}| \geq \varepsilon\sqrt{n}) = o(1)$, we have

$$\mathrm{P}(\sum_{k=1}^n I_{\{|n^{-1/2}w_{kk}| \geq \varepsilon\}} \geq \varepsilon n) \leq \mathrm{P}(\sum_{k=1}^n \left[I_{\{|n^{-1/2}w_{kk}| \geq \varepsilon\}} - \mathrm{P}(|n^{-1/2}w_{11}| \geq \varepsilon)\right] \geq \frac{1}{2}\varepsilon n)$$
$$\leq 2\exp\{-n^2/(2n\mathrm{P}(|w_{11}| \geq \varepsilon\sqrt{n}) + \varepsilon n)\} \leq 2e^{-bn}$$

for some $b > 0$ and all large n. Applying Lemma 2.2, one can remove the diagonal entries greater than ε and, by Lemma 2.3, can also remove those smaller than ε without altering the limiting distribution of $n^{-1/2}\mathbf{W}$.

Now we use the truncation and centralization techniques. Let $\widetilde{\mathbf{W}}$ be the matrix with zero diagonal entries and off-diagonal entries $w_{jk}I_{(|w_{jk}|\le C)} - \mathrm{E}(w_{jk} I_{(|w_{jk}|\le C)})$. By Lemma 2.3 and the Law of Large Numbers, with probability 1,

$$\begin{aligned} &L^4(F^{(n^{-1/2}\mathbf{W})}, F^{(n^{-1/2}\widetilde{\mathbf{W}})}) \\ &\le n^{-2}\sum_{j,k} |w_{jk}I_{(|w_{jk}|>C)} - \mathrm{E}(w_{jk}I_{(|w_{jk}|>C)})|^2 \\ &\to \mathrm{E}|w_{12}I_{(|w_{12}|>C)} - \mathrm{E}(w_{12}I_{(|w_{12}|>C)})|^2 \le \mathrm{E}|w_{12}I_{(|w_{12}|>C)}|^2. \qquad (2.6) \end{aligned}$$

Note that $\mathrm{E}|w_{12}I_{(|w_{12}|>C)}|^2$ can be arbitrarily small if C is large enough. Thus, in the proof of Theorem 2.1, we may assume that the entries of $\mathbf{W}$ are bounded by C.

Next we establish the convergence of moments of the ESD of $n^{-1/2}\mathbf{W}$, see (2.1). For given integers $j_1, \ldots, j_h (\le n)$, construct a W-graph G by plotting $j_1, \ldots, j_h$ on a straight line as vertices and drawing h edges from j_r to j_{r+1} (with $j_{h+1} = j_1$), $r = 1, \ldots, h$. An example of a W-graph is given in Figure 1, in which there are 10 vertices (i_1-i_{10}), 4 non-coincident vertices (v_1-v_4), 9 edges, 4 non-coincident edges and 1 single edge (v_4, v_3). For this graph, we say the edge (v_1, v_2) has multiplicity 2 and the edges (v_2, v_3) and (v_2, v_4) have multiplicities 3. An edge (u, v) corresponds to the variable $w_{u,v}$ and a W-graph corresponds to the product of variables corresponding to the edges making up this W-graph.

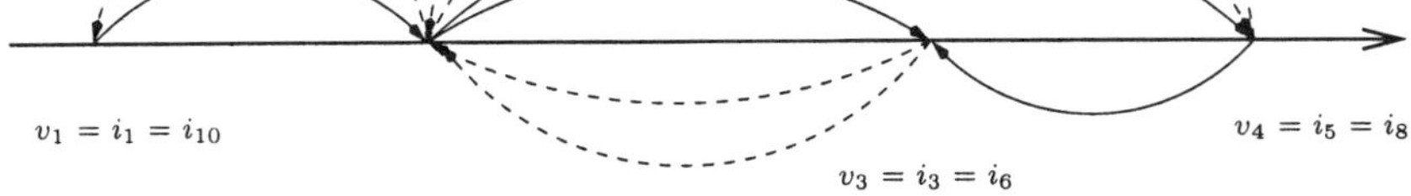

Figure 1. Types of edges in a W-graph.

Note that

$$\frac{1}{n}\mathrm{tr}((n^{-1/2}\mathbf{W})^h) = n^{-1-h/2} \sum_{j_1,\ldots,j_h} w_{j_1j_2}w_{j_2,j_3}\cdots w_{j_hj_1} := n^{-1-h/2}\sum_G w_G.$$

Then Theorem 2.1 follows by showing that

$$\frac{1}{n}\mathrm{E}(\mathrm{tr}((n^{-1/2}\mathbf{W})^h)) = n^{-1-h/2}\sum_G \mathrm{E}(w_G) = \begin{cases} \frac{(2s)!\sigma^h}{s!(s+1)!} + O(n^{-1}), & \text{if } h = 2s \\ O(n^{-1/2}), & \text{if } h = 2s+1 \end{cases} \tag{2.7}$$

and

$$\mathrm{Var}(\frac{1}{n}\mathrm{tr}((n^{-1/2}\mathbf{W})^h)) = n^{-2-h}\sum_{G_1,G_2}[\mathrm{E}(w_{G_1}w_{G_2}) - \mathrm{E}(w_{G_1})\mathrm{E}(w_{G_2})] = O(n^{-2}), \tag{2.8}$$

through the following arguments.

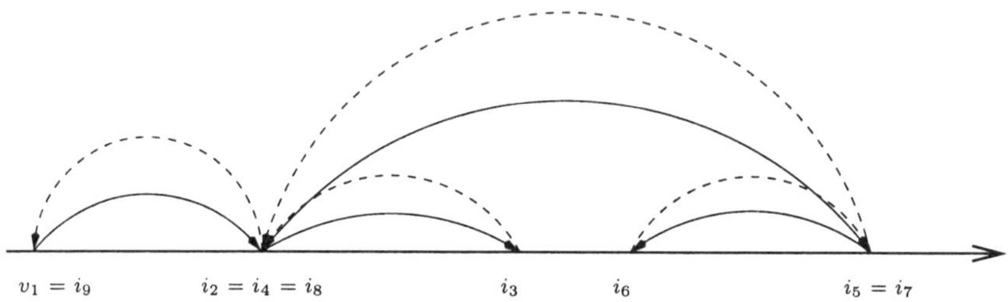

Figure 2. A W-graph with 8 edges, 4 non-coincident edges and 5 vertices.

To prove (2.7), we note that if there is a single edge in the W-graph, the corresponding expectation is zero. When $h = 2s+1$, there are at most s non-coincident edges and hence at most $s+1$ non-coincident vertices. This shows that there are at most n^{s+1} graphs (or non-zero terms in the expansion). Then the second conclusion in (2.7) follows since the denominator is $n^{s+3/2}$ and the absolute value of the expectation of each term is not larger than C^h. When $h = 2s$, classify the graphs into two types. The first type, consisting of all graphs which have at most s non-coincident vertices, gives the estimation of the remainder term $O(n^{-1})$. The second type consists of all graphs which have exactly $s+1$ non-coincident vertices and s non-coincident edges. There are no cycles of non-coincident edges in such graphs and each edge (u, v) must coincide with and only with the edge (v, u) which corresponds to $\mathrm{E}|w_{uv}|^2 = \sigma^2$. Thus, each term corresponding to a second type W-graph is σ^h. To complete the proof of the first conclusion of (2.7), we only need to count the number of second type W-graphs. We say that two W-graphs are isomorphic if one can be coverted to the other by a permutation of $\{1, \ldots, n\}$ on the straight line. We first compute the number of isomorphic classes. If an edge (u, v) is single in the subgraph $[(i_1, i_2), \ldots, (i_t, i_{t+1})]$, we say that this edge is single up to the edge (i_t, i_{t+1}) or the vertex i_{t+1}. In a second type W-graph, there are s edges which are single up to themselves, and the other s edges coincide with existing edges when they are drawn. Define a CS (Characteristic Sequence) for a W-graph by $u_i = 1$ (or -1) if the ith edge is single up to its end vertex (or coinciding with an existing edge, respectively). For example, for the graph in Figure 2, its CS is $1, 1, -1, 1, 1, -1, -1, -1$. The sequence $u_1, \ldots, u_{2s}$ should satisfy $\sum_{i=1}^{j} u_i \geq 0$, for all $j = 1, \ldots, 2s$.

The number of all arrangements of the ± 1's is $\frac{(2s)!}{s!s!}$. By the reflection principle (see Figure 3), the number of arrangements such that at least one of the requirements $\sum_{i=1}^{j} u_i \geq 0$ is violated is $\frac{(2s)!}{(s-1)!(s+1)!}$ (see the broken curve which reaches the line $y = -1$; reflecting the rear part of the curve across the axis $y = -1$ results in the dotted curve which ends at $y = -2$ and consists of $s - 1$ ones and $s + 1$ minus ones). It follows that the number of isomorphic classes is $\frac{(2s)!}{s!s!} - \frac{(2s)!}{(s-1)!(s+1)!} = \frac{(2s)!}{s!(s+1)!}$. The number of graphs in each isomorphic class is $n(n-1)\cdots(n-s) = n^{1+s}(1 + O(n^{-1}))$. Then the first conclusion in (2.7) follows. The proof of (2.8) follows from the following observation. When G_1 has no edges coincident with any G_2-edges, the corresponding term is zero since $\mathrm{E}(w_{G_1} w_{G_2}) = \mathrm{E}(w_{G_1})\,\mathrm{E}(w_{G_2})$, due to independence. If there is a single edge in $G = G_1 \bigcup G_2$, the corresponding term is also zero. There are two cases in which the terms in (2.8) may be non-zero. In the first, both G_1 and G_2 have no single edges in themselves and G_1 has at least one edge coincident with an edge of G_2. In the second, there is at least one cycle in both G_1 and G_2. In both cases the number of non-coincident vertices of G is at most h.

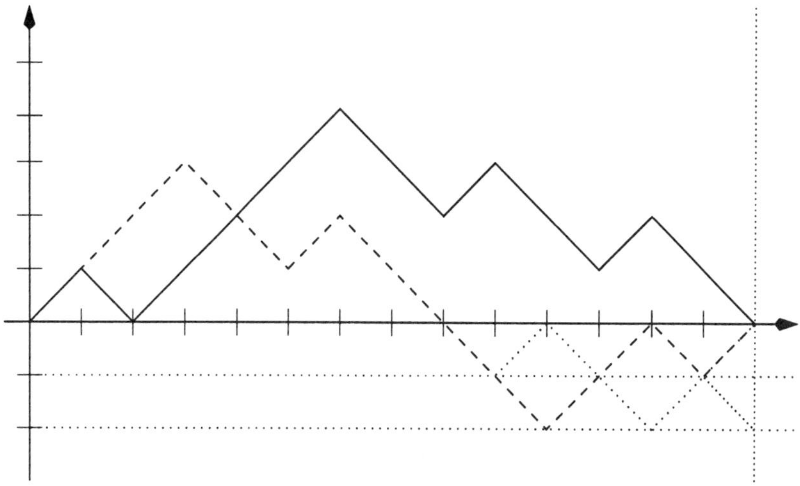

Figure 3. The solid curve represents a CS, the broken curve represents a non-CS and the dotted curve is the reflection of the broken curve.

Remark 2.1. The existence of the second moment of the off-diagonal entries is obviously a necessary and sufficient condition for the semicircular law since the LSD involves the parameter σ^2. It is interesting that there is no moment requirement on the diagonal elements. This fact makes the proof of Theorem 2.12 much easier than exists in the literature.

Sometimes it is of practical interest to consider the case where, for each n, the entries above the diagonal of $\mathbf{W}$ are independent complex random variables with mean zero and variance σ^2, but which may not be identically distributed and may depend on n. We have the following result.

Theorem 2.4. *If* $\mathrm{E}(w_{jk}^{(n)}) = 0$, $\mathrm{E}|w_{jk}^{(n)}|^2 = \sigma^2$ *and for any* $\delta > 0$

$$\lim_{n\to\infty} \frac{1}{\delta^2 n^2} \sum_{jk} \mathrm{E}|w_{jk}^{(n)}|^2 I_{(|w_{jk}^{(n)}| \geq \delta\sqrt{n})} = 0, \tag{2.9}$$

then the conclusion of Theorem 2.1 *holds.*

The proof of this theorem is basically the same as that of Theorem 2.1. At first, we note that one can select a sequence $\delta_n \downarrow 0$ such that (2.9) is still true with δ replaced by δ_n. Then one may truncate the variables at $C = \delta_n\sqrt{n}$. For brevity, in the proof of Theorem 2.4, we suppress the dependence on n from entries of $\mathbf{W}_n$. By Lemma 2.2, we have

$$\|F^{\mathbf{W}_n} - F^{\widetilde{\mathbf{W}}_n}\| \leq \frac{1}{n}\sum_{jk} I_{\{|w_{jk}| \geq \delta_n\sqrt{n}\}},$$

where $\widetilde{\mathbf{W}}_n$ is the matrix of truncated variables. By Condition (2.9),

$$\sum_{jk} \mathrm{P}\{|w_{jk}| \geq \delta_n\sqrt{n}\} \leq \frac{1}{\delta_n^2 n}\sum_{jk} \mathrm{E}|w_{jk}|^2 I_{(|w_{jk}| \geq \delta_n\sqrt{n})} = o(n).$$

Applying Hoeffding's inequality to the sum of the $n(n+1)/2$ independent terms of $I_{(|w_{jk}| \geq \delta_n\sqrt{n})}$, we have

$$\begin{aligned}
&\mathrm{P}\Big(\sum_{j\leq k} I_{\{|w_{jk}|\geq\delta_n\sqrt{n}\}} \geq \varepsilon n\Big) \leq \mathrm{P}\Big(\Big|\sum_{j\leq k}[I_{\{|w_{jk}|\geq\delta_n\sqrt{n}\}} - \mathrm{P}(|w_{jk}| \geq \delta_n\sqrt{n})]\Big| \geq \frac{1}{3}\varepsilon n\Big) \\
&\leq 2\exp\Big(-\frac{9^{-1}\varepsilon^2 n^2}{2[\sum_{jk}\mathrm{P}\{|w_{jk}| \geq \delta_n\sqrt{n}\} + 3^{-1}\varepsilon n]}\Big) \leq 2\exp(-bn),
\end{aligned}$$

for some positive constant b. By the Borel-Cantelli Lemma, with probability 1, the truncation does not affect the LSD of $\mathbf{W}_n$. Then, applying Lemma 2.3, one can re-centralize the truncated variable and replace the diagonal entries by zero without changing the LSD.

Then for the truncated and re-centralized matrix (still denoted by $\mathbf{W}_n$), it can be shown that, by estimates similar to those given in the proof of Theorem 2.1 and corresponding to (2.7),

$$\frac{1}{n}\mathrm{E}(\mathrm{tr}((n^{-1/2}\mathbf{W}_n)^h)) = \begin{cases} \frac{(2s)!\sigma^h}{s!(s+1)!} + o(1), & \text{if } h = 2s \\ o(1), & \text{if } h = 2s+1. \end{cases} \tag{2.10}$$

However, we cannot prove the counterpart for $\mathrm{Var}(\frac{1}{n}\mathrm{tr}((n^{-1/2}\mathbf{W}_n)^h))$ since its order is only $O(\frac{\delta_n^2}{n})$, which implies convergence "in probability", but not "almost surely". In this case, we can consider the fourth moment and prove

$$\begin{aligned} &\mathrm{E}\Big|\frac{1}{n}\mathrm{tr}((n^{-1/2}\mathbf{W}_n)^h) - \mathrm{E}(\frac{1}{n}\mathrm{tr}((n^{-1/2}\mathbf{W}_n)^h))\Big|^4 \\ &= n^{-4-2h} \sum_{G_1,\ldots,G_4} \mathrm{E}\Big[\prod_{i=1}^{4}(w_{G_i} - \mathrm{E}(w_{G_i}))\Big] = O(n^{-2}). \end{aligned} \tag{2.11}$$

In fact, if there is one subgraph among $G_1, \ldots, G_4$ which has no edge coincident with any edges of the other three, the corresponding term is zero. Thus, we only need to estimate those terms for the graphs whose every subgraph has at least one edge coincident with an edge of other subgraphs. Then (2.11) can be proved by analyzing various cases. The details are omitted.

Remark 2.2. In Girko's book (1990), it is stated that condition (2.9) is necessary and sufficient for the conclusion of Theorem 2.4.

2.1.2. Sample covariance matrix

Suppose that $\{x_{jk},\ j,k = 1,2,\ldots\}$ is a double array of i.i.d. complex random variables with mean zero and variance σ^2. Write $\mathbf{x}_k = (x_{1k}, \ldots, x_{pk})'$ and $\mathbf{X} = (\mathbf{x}_1, \ldots, \mathbf{x}_n)$. The sample covariance matrix is usually defined by $\mathbf{S} = \frac{1}{n-1}\sum_{k=1}^{n}(\mathbf{x}_k - \bar{\mathbf{x}})(\mathbf{x}_k - \bar{\mathbf{x}})^*$. However, in spectral analysis of LDRM, the sample covariance matrix is simply defined as $\mathbf{S} = \frac{1}{n}\sum_{k=1}^{n}\mathbf{x}_k\mathbf{x}_k^* = \frac{1}{n}\mathbf{X}\mathbf{X}^*$.

The first success in finding the LSD of $\mathbf{S}$ is due to Marčenko and Pastur (1967). Subsequent work was done in Bai and Yin (1988a), Grenander and Silverstein (1977), Jonsson (1982), Wachter (1978) and Yin (1986). When the entries of $\mathbf{X}$ are not independent, Yin and Krishnaiah (1985) investigated the LSD of $\mathbf{S}$ when the underlying distribution is isotropic. The next theorem is a consequence of a result in Yin (1986), where the real case is considered. Here we state it in the complex case.

Theorem 2.5. *Suppose that $p/n \to y \in (0, \infty)$. Under the assumptions stated at the beginning of this section, the ESD of* $\mathbf{S}$ *tends to a limiting distribution with* *density*

$$p_y(x) = \begin{cases} \frac{1}{2\pi x y \sigma^2}\sqrt{(b-x)(x-a)}, & \text{if } a \le x \le b \\ 0, & \text{otherwise}, \end{cases} \tag{2.12}$$

and a point mass $1 - 1/y$ *at the origin if* $y > 1$, *where* $a = a(y) = \sigma^2(1 - y^{1/2})^2$ *and* $b = b(y) = \sigma^2(1 + y^{1/2})^2$.

The limit distribution of Theorem 2.5 is called the Marčenko — Pastur law with ratio index y and scale index σ^2. The proof relies on the following lemmas.

Lemma 2.6 (Rank Inequality). *Let* $\mathbf{A}$ *and* $\mathbf{B}$ *be two* $p \times n$ *complex matrices. Then*

$$||F^{\mathbf{AA}^*} - F^{\mathbf{BB}^*}|| \le \frac{1}{p}\text{rank}(\mathbf{A} - \mathbf{B}). \tag{2.13}$$

From Lemma 2.2, one easily derives a weaker result (but enough for applications to LSA of large sample covariance matrices) that $||F^{\mathbf{AA}^*} - F^{\mathbf{BB}^*}|| \le \frac{2}{p}\text{rank}(\mathbf{A} - \mathbf{B})$.

To prove Lemma 2.6, one may assume that $\mathbf{A}' = (\mathbf{A}_1' \vdots \mathbf{A}_2')'$ and $\mathbf{B}' = (\mathbf{B}_1' \vdots \mathbf{A}_2')'$, where the number of rows of $\mathbf{A}_1$ (also $\mathbf{B}_1$) is k =rank$(\mathbf{A}-\mathbf{B})$. Then, as in the proof of Lemma 2.2, Lemma 2.6 can be proven by applying the interlacing inequality to the matrices $\mathbf{A}_2\mathbf{A}_2^*$, $\mathbf{AA}^*$ and $\mathbf{BB}^*$.

Lemma 2.7 (Difference Inequality). *Let* $\mathbf{A}$ *and* $\mathbf{B}$ *be two* $p \times n$ *complex matrices. Then*

$$L^4(F^{(\mathbf{AA}^*)}, F^{(\mathbf{BB}^*)}) \le \frac{2}{p^2}\text{tr}((\mathbf{A}-\mathbf{B})(\mathbf{A}-\mathbf{B})^*)\text{tr}(\mathbf{AA}^* + \mathbf{BB}^*). \tag{2.14}$$

This lemma relies on the following:

$$L^4(F^{(\mathbf{AA}^*)}, F^{(\mathbf{BB}^*)}) \le \Big(\frac{1}{p}\sum_{k=1}^{p}|\lambda_k - \eta_k|\Big)^2 \le \frac{2}{p^2}\Big(\sum_{k=1}^{p}(\sqrt{\lambda_k}-\sqrt{\eta_k})^2\Big)\Big(\sum_{k=1}^{p}(\lambda_k+\eta_k)\Big),$$

$$\text{tr}(\mathbf{AA}^*) = \sum_{k=1}^{p}\lambda_k, \qquad \text{tr}(\mathbf{BB}^*) = \sum_{k=1}^{p}\eta_k,$$

and for some unitary matrices $\mathbf{U} = (u_{jk})$ and $\mathbf{V} = (v_{jk})$,

$$\begin{aligned} \text{Re}(\text{tr}(\mathbf{AB}^*)) &= \sum_{j,k}\sqrt{\lambda_j\eta_k}\text{Re}(u_{jk}\bar{v}_{jk}) \\ &\le \Big(\sum_{j,k}\sqrt{\lambda_j\eta_k}|u_{jk}|^2\sum_{j,k}\sqrt{\lambda_j\eta_k}|v_{jk}|^2\Big)^{1/2} \le \sum_{k=1}^{p}\sqrt{\lambda_k\eta_k}. \end{aligned}$$

Now we are in a position to sketch a proof of Theorem 2.5. Define $\tilde{x}_{jk} = x_{jk}I_{(|x_{jk}|<C)} - \text{E}(x_{jk}I_{(|x_{jk}|<C)})$ and denote by $\widetilde{\mathbf{S}}$ the sample covariance matrix

constructed of the $\tilde{x}_{jk}$. Similar to the proof of (2.6), employing Lemma 2.7, we can show that with probability 1

$$\limsup_{n\to\infty} L^4(F^{\mathbf{S}}, F^{\widetilde{\mathbf{S}}}) \leq 4\sigma^2 \mathrm{E}|x_{11}^2| I_{(|x_{11}|\geq C)}.$$

Also, $\mathrm{E}(|\tilde{x}_{jk}|^2) \to \sigma^2$ as $C \to \infty$. Therefore, in the proof of Theorem 2.5, we may assume that the variables x_{jk} are uniformly bounded, since the right hand side in the above inequality can be arbitrarily small if C is chosen large enough. Then we use the expression

$$\begin{aligned} p^{-1}\mathrm{tr}(\mathbf{S}^h) &= p^{-1}n^{-h} \sum_{i_1,\ldots,i_h} \sum_{j_1,\ldots,j_h} x_{i_1j_1}\bar{x}_{i_2j_1}x_{i_2j_2}\bar{x}_{i_3j_2}\cdots \bar{x}_{i_hj_{h-1}}x_{i_hj_h}\bar{x}_{i_1j_h} \\ &:= p^{-1}n^{-h}\sum_G x_G, \end{aligned}$$

where in the summations, the indices $i_1,\ldots,i_h$ run over $1,\ldots,p$, the indices $j_1,\ldots,j_h$ run over $1,\ldots,n$. An S-graph G is constructed by plotting the i's and j's on two parallel straight lines respectively, drawing h (down) edges from i_v to j_v and another h (up) edges from j_v to i_{v+1} (with the convention $i_{h+1} = i_1$), $v = 1,\ldots,h$. Finally, we show that

$$\mathrm{E}(p^{-1}\mathrm{tr}(\mathbf{S}^h)) = p^{-1}n^{-h}\sum_G \mathrm{E}(x_G) = \sigma^{2h}\sum_{r=0}^{h-1}\frac{y_n^r}{r+1}\binom{h}{r}\binom{h-1}{r} + O(n^{-1}) \quad (2.15)$$

where $y_n = p/n$, and

$$\mathrm{Var}(p^{-1}\mathrm{tr}(\mathbf{S}^h)) = p^{-2}n^{-2h}\sum_{G_1,G_2}[\mathrm{E}(x_{G_1}x_{G_2}) - \mathrm{E}(x_{G_1})\mathrm{E}(x_{G_2})] = O(n^{-2}). \quad (2.16)$$

Similar to the proof of (2.7), the proof of (2.15) reduces to, for each $r = 0,\ldots,h-1$, the calculation of the number of graphs which have no single edges, $r+1$ non-coincident i-vertices and $h-r$ noncoincident j-vertices. In such graphs, each down edge (a,b) must coincide with and only with the up-edge (b,a) which contributes a factor $\mathrm{E}|x_{ab}|^2 \to \sigma^2$ (as $C \to \infty$). We say that two S-graphs are *isomorphic* if one can be coverted to the other through a permutation of $\{1,\ldots,p\}$ on the i-line and a permutation of $\{1,\ldots,n\}$ on the j-line. To compute the number of isomorphic classes, define $d_\ell = -1$ if the path of the graph ultimately leaves an i-vertex (other than the initial i-vertex) after the ℓth down edge and define $u_\ell = 1$ if the ℓth up-edge leads to a new i-vertex. For other cases, define d_ℓ and u_ℓ as 0. Note that we always have $d_1 = 0$. It is obvious that $u_1 + \cdots + u_{\ell-1} + d_1 + \cdots + d_\ell \geq 0$. Ignoring this restriction, we have $\binom{h}{r}\binom{h-1}{r}$ ways to arrange r ones into the h positions of up-edges and r minus ones into the

$h-1$ positions of down-edges (except the first). If ℓ is the first integer such that $u_1+\cdots+u_{\ell-1}+d_1+\cdots+d_\ell<0$, then $u_{\ell-1}=0$ and $d_\ell=-1$. By changing $u_{\ell-1}$ to 1 and d_ℓ to 0, we get a d-sequence with $r-1$ minus ones and a u-sequence with $r+1$ ones. Thus, the number of isomorphic classes is $\binom{h}{r}\binom{h-1}{r}-\binom{h}{r+1}\binom{h-1}{r-1}=\frac{1}{r+1}\binom{h}{r}\binom{h-1}{r}$. The number of graphs in each isomorphic class is obviously $p(p-1)\cdots(p-r)n(n-1)\cdots(n-h+r-1)=pn^hy_n^r(1+O(n^{-1}))$. Then (2.15) follows. The proof of (2.16) is similar to that of (2.8). This completes the proof of the theorem.

Remark 2.3. The existence of the second moment of the entries is obviously necessary and sufficient for the Marčenko-Pastur Law since the LSD involves the parameter σ^2. The condition of zero mean can be relaxed to having a common mean, since the means of the entries form a rank-one matrix which can be removed by applying Lemma 2.6.

Remark 2.4. As argued before the statement of Theorem 2.4, sometimes it is of practical interest to consider the case where the entries of $\mathbf{X}$ depend on n, and for each n they are independent but not identically distributed. Similar to Theorem 2.4, truncating the variables at $\delta_n\sqrt{n}$ for some $\delta_n\downarrow 0$ by using Lemma 2.6, and recentralizing by using Lemma 2.7, one can prove the following generalization.

Theorem 2.8. *Suppose that for each n, the entries of $\mathbf{X}_n$ are independent complex variables, with a common mean and variance σ^2. Assume that $p/n\to y\in(0,\infty)$ and that for any $\delta>0$,*

$$\frac{1}{\delta^2 np}\sum_{j,k}\mathrm{E}(|x_{jk}^{(n)}|^2 I_{(|x_{jk}^{(n)}|\ge\delta\sqrt{n})})\to 0. \tag{2.17}$$

Then $F^{\mathbf{S}}$ tends almost surely to the Marčenko-Pastur law with ratio index y and scale index σ^2.

Now we consider the case $p\to\infty$ but $p/n\to 0$ as $p\to\infty$. It is conceivable that almost all eigenvalues tend to 1 and hence the ESD of $\mathbf{S}$ tends to a degenerate one. In turn, to investigate the behavior of the eigenvalues of the sample covariance matrix $\mathbf{S}$, we consider the ESD of the matrix $\mathbf{W}=\sqrt{n/p}(\mathbf{S}-\sigma^2\mathbf{I}_p)=\frac{1}{\sqrt{np}}(\mathbf{XX}^*-n\sigma^2\mathbf{I}_p)$. When the entries of $\mathbf{X}$ are real, under the existence of the fourth moment, Bai and Yin (1988a) showed that its ESD tends to the semicircular law almost surely as $p\to\infty$. Now we give a generalization of this result.

Theorem 2.9. *Suppose that for each n the entries of the matrix $\mathbf{X}_n$ are independent complex random variables with a common mean and variance σ^2. Assume that for any constant $\delta>0$, as $p\to\infty$ with $p/n\to 0$,*

$$\frac{1}{p\delta^2\sqrt{np}}\sum_{jk}\mathrm{E}(|x_{jk}^{(n)}|^2 I_{(|x_{jk}^{(n)}|\ge\delta\sqrt[4]{np})})=o(1) \tag{2.18}$$

and

$$\frac{1}{np^2}\sum_{jk}\mathrm{E}(|x_{jk}^{(n)}|^4 I_{(|x_{jk}^{(n)}|\le\delta\sqrt[4]{np})}) = o(1). \tag{2.19}$$

Then with probability 1 *the ESD of* $\mathbf{W}$ *tends to the semi-circular law with scale index* σ^2.

Remark 2.5. Conditions (2.18) and (2.19) hold if the entries of $\mathbf{X}$ have bounded fourth moments. This is the condition assumed in Bai and Yin (1988a).

The proof of Theorem 2.9 consists of the following steps. Applying Lemma 2.6, we may assume that the common mean is zero. Truncate the entries of $\mathbf{X}$ at $\delta_p\sqrt[4]{np}$, where $\delta_p \to 0$ such that (2.18) and (2.19) hold with δ replaced by δ_p. By Condition (2.18), $\sum_{jk}\mathrm{P}(|x_{jk}^{(n)}|\ge\delta_p\sqrt[4]{np}) = o(p)$. From this and applying Hoeffding's inequality, one can prove that the probability that the number of truncated elements of $\mathbf{X}$ is greater than εp is less than Ce^{-bp} for some $b>0$.

One needs to recentralize the truncated entries of $\mathbf{X}$. The application of Lemma 2.7 requires

$$\frac{1}{p}\sum_{jk}|\mathrm{E}(x_{jk}^{(n)})I_{[|x_{jk}^{(n)}|\ge\delta_p\sqrt[4]{np}]}|^2 = o(1)$$

and

$$\frac{1}{np^2}\sum_{jk}|x_{jk}^{(n)}|^2 I_{[|x_{jk}^{(n)}|\le\delta_p\sqrt[4]{np}]} = o(1), \text{a.s.}$$

Here, the first assertion is an easy consequence of (2.18). The second can be proved by applying Bernstein's inequality (see Prokhorov (1968)).

The next step is to remove the diagonal elements of $\mathbf{W}$. Write $y_\ell = I_{\{\frac{1}{\sqrt{np}}\left|\sum_{j=1}^n(|x_{\ell j}|^2-\sigma^2)\right|\ge\varepsilon\}}$. Note that by Condition (2.19),

$$\sum_{\ell=1}^{p}\mathrm{E}(y_\ell)\le\frac{1}{\varepsilon^2\sqrt{np}}\sum_{\ell,j}\mathrm{E}(|x_{\ell j}|^4 I_{|x_{\ell j}|\le\delta_p\sqrt[4]{np}}) = o(p).$$

Applying Hoeffding's inequality, we have

$$\mathrm{P}(\sum_{\ell=1}^{p} y_\ell\ge\varepsilon p)\le 2e^{-bp} \tag{2.20}$$

for some $b>0$. Then applying Lemma 2.2, we can replace the diagonal elements of $\mathbf{W}$ which are greater than ε by zero, since the number of such elements is $o(p)$ by (2.20). By Lemma 2.3, we can also replace those smaller than ε by zero.

In the remainder of the proof, we assume that $\mathbf{W} = (\frac{1}{\sqrt{np}}\sum_{j=1}^n x_{i_1 j}\overline{x}_{i_2 j}(1-\delta_{i_1,i_2}))$, where δ_{jk} is the Kronecker delta . Then we need to prove that

$$\mathrm{E}(\mathrm{tr}(\mathbf{W}^h)) = \begin{cases} \frac{(2s)!\sigma^{2h}}{s!(s+1)!} + o(1), & \text{if } h = 2s \\ o(1), & \text{if } h = 2s+1 \end{cases}$$

$$\mathrm{E}|\mathrm{tr}(\mathbf{W}^h) - \mathrm{E}(\mathrm{tr}(\mathbf{W}^h))|^4 = O(\frac{1}{p^2}).$$

Similar to the proof of Theorem 2.5, construct graphs for estimating $\mathrm{E}(\mathrm{tr}(\mathbf{W}^h))$. Denote by r and s the numbers of i and j vertices. Note that the number of non-coincident edges is not less than twice the number of non-coincident j vertices, since consecutive i vertices are not equal. It is obvious that the number of non-coincident edges is not less than $r+s-1$. Therefore, the contribution of each isomorphic class to the sum is not more than

$$\begin{cases} p^{-1}(np)^{-h/2}n^s p^r(\delta_p \sqrt[4]{np})^{2h-4s}\sigma^{4s} = \delta_p^{2h-4s}p^{r-s-1}\sigma^{4s} & \text{if } s+1 \geq r, \\ p^{-1}(np)^{-h/2}n^s p^r(\delta_p \sqrt[4]{np})^{2h-2s-2r+2}\sigma^{2s+2r} & \\ \quad = \delta_p^{2h-2s-2r+2}(p/n)^{r-s-1}\sigma^{2s+2r} & \text{if } s+1 < r. \end{cases}$$

The quantities on the right hand sides of the above estimations are $o(1)$ unless $h = 2s = 2r-2$. When $h = 2s = 2r-2$, the contribution of each isomorphic class is $\sigma^{2h}(1+O(p^{-1}))$ and the number of non-isomorphic graphs is $(2s)!/[s!(s+1)!]$. The rest of the proof is similar to that of Theorem 2.4 and hence omitted.

2.1.3. Product of two random matrices

The motivation for studying products of random matrices originates from the investigation of the spectral theory of large sample covariance matrices when the population covariance matrix is not a multiple of an identity matrix, and that of *multivariate* $\mathbf{F} = \mathbf{S}_1\mathbf{S}_2^{-1}$ *matrices*. When $\mathbf{S}_1$ and $\mathbf{S}_2$ are independent Wishart, the LSD of $\mathbf{F}$ follows from Wachter (1980) and its explicit forms can be found in Bai, Yin and Krishnaiah (1987), Silverstein (1985a) and Yin, Bai and Krishnaiah (1983). Relaxation of the Wishart assumption on $\mathbf{S}_1$ and $\mathbf{S}_2$ relies on the investigation of the strong limit of the smallest eigenvalue of a sample covariance matrix. Based on the results in Bai and Yin (1993) and Yin (1986), and using the approach in Bai, Yin and Krishnaiah (1985), one can show that the LSD of $\mathbf{F}$ is the same as if both $\mathbf{S}_1$ and $\mathbf{S}_2$ were Wishart when the underlying distribution of $\mathbf{S}_1$ has finite second moment and that of $\mathbf{S}_2$ has finite fourth moment. Yin and Krishnaiah (1983) investigated the limiting distribution of a product of a Wishart matrix $\mathbf{S}$ and a positive definite matrix $\mathbf{T}$. Later work was done in Bai, Yin and Krishnaiah (1986), Silverstein (1995), Silverstein and Bai (1995) and Yin (1986). Here we present the following result.

Theorem 2.10. *Suppose that the entries of* $\mathbf{X}$ *are independent complex random variables satisfying* (2.17), *and assume that* $\mathbf{T}(=\mathbf{T}_n)$ *is a sequence of* $p\times p$ *Hermitian matrices independent of* $\mathbf{X}$ *such that its ESD tends to a non-random and non-degenerate distribution* H *in probability* (or almost surely). *Further assume that* $p/n\to y\in(0,\infty)$. *Then the ESD of the product* $\mathbf{ST}$ *tends to a non-random limit in probability (or almost surely, respectively).*

This theorem contains Yin (1986) as a special case. In Yin (1986), the entries of $\mathbf{X}$ are assumed to be real and i.i.d. with mean zero and variance 1, the matrix $\mathbf{T}$ is assumed to be real and positive definite and to satisfy, for each fixed h,

$$\frac{1}{p}\mathrm{tr}(\mathbf{T}^h)\to\alpha_h,\quad \text{(in pr. or a.s.,)} \tag{2.21}$$

where the sequence $\{\alpha_h\}$ satisfies Carleman's condition.

There are two directions to generalize Theorem 2.10. One is to relax the independence assumption on the entries of $\mathbf{S}$. Bai, Yin and Krishnaiah (1986) assume the columns of $\mathbf{X}$ are i.i.d. and each column is isotropically distributed with certain moment conditions, for example. It could be that Theorems 2.1, 2.4, 2.5, 2.8 and 2.10 are still true when the underlying variables defining the Wigner or sample covariance matrices are weakly dependent, say ϕ-mixing, although I have not found any such results yet. It may be more interesting to investigate the case where the entries are dependent, say the columns of $\mathbf{X}$ are i.i.d. and the entries of each column are uncorrelated but not independent.

Another direction is to generalize the structure of the setup. An example is given in Theorem 3.4. Since the original proof employs the Stieltjes transformation technique, we postpone its statement and proof to Section 3.1.2.

To sketch the proof of Theorem 2.10, we need the following lemma.

Lemma 2.11. *Let* G^0 *be a connected graph with* m *vertices and* h *edges. To each vertex* $v(=1,\ldots,m)$ *there corresponds a positive integer* n_v, *and to each edge* $e_j=(v_1,v_2)$ *there corresponds a matrix* $\mathbf{T}_j=(t^{(j)}_{\eta,\tau})$ *of order* $n_{v_1}\times n_{v_2}$. *Let* E_c *and* E_{nc} *denote the sets of cutting edges (those edges whose removal causes the graph disconnected) and non-cutting edges, respectively. Then there is a constant* C, *depending upon* m *and* h *only, such that*

$$\Big|\sum_{i_1,\ldots,i_m}\prod_{j=1}^{h}t^{(j)}_{i_{fini(e_j)},i_{fend(e_j)}}\Big|\le Cn\prod_{e_j\in E_{nc}}\|\mathbf{T}_j\|\prod_{e_j\in E_c}\|\mathbf{T}_j\|_0,$$

where $n=\max(n_1,\ldots,n_m)$, $\|\mathbf{T}_j\|$ *denotes the maximum singular value, and* $\|\mathbf{T}_j\|_0$ *equals the product of the maximum dimension and the maximum absolute value of the entries of* $\mathbf{T}_j$*; in the summation* i_v *runs over* $\{1, \ldots, n_v\}$, $f_{ini}(e_j)$ *and* $f_{end}(e_j)$ *denote the initial and end vertices of the edge* e_j.

If there are no cutting edges in G^0, then the lemma follows from the norm inequality $\|\mathbf{AB}\| \le \|\mathbf{A}\|\|\mathbf{B}\|$. For the general case, the lemma can be proved by induction with respect to the number of cutting edges. The details are omitted.

In the proof of Theorem 2.10, we only consider a.s. convergence, since the case of convergence in probability can be reduced to the a.s. case by using the strong representation theorem (see Yin (1986) for details).

For given $\tau_0 > 0$, define a matrix $\widetilde{\mathbf{T}}$ by replacing, in the spectral decomposition of $\mathbf{T}$, the eigenvalues of $\mathbf{T}$ whose absolute values are greater than τ_0 by zero. Then the ESD of $\widetilde{\mathbf{T}}$ converges to

$$H_{\tau_0}(x) = \int_{-\infty}^{x} I_{[-\tau_0,\tau_0]}(u)H(du) + (H(-\tau_0) + 1 - H(\tau_0))I_{[0,\infty)}(x), \quad \text{a.s.}$$

and (2.21) holds, with $\tilde{\alpha}_h = \int_{|x|\le\tau_0} x^h dH(x)$. An application of Lemma 2.2 shows that the substitution of $\mathbf{T}$ by $\widetilde{\mathbf{T}}$ alters the ESD of the product by at most $\frac{1}{p}\#\{j : |\lambda_j(\mathbf{T})| \ge \tau_0\}$, which can be arbitrarily uniformly small by choosing τ_0 large.

We claim that Theorem 2.10 follows if we can prove that, with probability 1, $F^{\mathbf{S}\widetilde{\mathbf{T}}}$ converges to a non-degenerate distribution F_{τ_0} for each fixed τ_0. First, we can show the tightness of $\{F^{\mathbf{ST}}\}$ from $F^{\mathbf{T}} \to H$ and the inequality

$$\begin{aligned} F^{\mathbf{ST}}(M) - F^{\mathbf{ST}}(-M) &\ge F^{\mathbf{S}\widetilde{\mathbf{T}}}(M) - F^{\mathbf{S}\widetilde{\mathbf{T}}}(-M) - 2\|F^{\mathbf{ST}} - F^{\mathbf{S}\widetilde{\mathbf{T}}}\| \\ &\ge F^{\mathbf{S}\widetilde{\mathbf{T}}}(M) - F^{\mathbf{S}\widetilde{\mathbf{T}}}(-M) - 2(F^{\mathbf{T}}(-\tau_0) + 1 - F^{\mathbf{T}}(\tau_0)). \end{aligned}$$

Here, the second inequality follows by using Lemma 2.2. Second, we can show that any convergent subsequences of $\{F^{\mathbf{ST}}\}$ have the same limit by using the inequality

$$|F_1(x) - F_2(x)| \le \sum_{j=1}^{2} \left[|F_j(x) - F_{n_j}^{\mathbf{ST}}(x)| + \|F_{n_j}^{\mathbf{ST}} - F_{n_j}^{\mathbf{S}\widetilde{\mathbf{T}}}\|\right] + |F_{n_1}^{\mathbf{S}\widetilde{\mathbf{T}}}(x) - F_{n_2}^{\mathbf{S}\widetilde{\mathbf{T}}}(x)|,$$

where F_1 and F_2 denote the limits of two convergence subsequences $\{F_{n_1}^{\mathbf{ST}}\}$ and $\{F_{n_2}^{\mathbf{ST}}\}$ respectively. This completes the proof of the assertion.

Consequently, the proof of Theorem 2.10 reduces to showing that $\{F^{\mathbf{S}\widetilde{\mathbf{T}}}\}$ converge to a non-random limit. Again, using Lemma 2.2, we may assume that the entries of $\mathbf{X}$ are truncated at $\sqrt{n}\delta_n$ ($\delta_n \to 0$) and centralized. In the sequel, for convenience, we still use $\mathbf{X}$ and $\mathbf{T}$ to denote the truncated matrices.

After truncation and centralization, one can see that $|x_{jk}| \le \delta_n\sqrt{n}$ and

$$\mathrm{E}|x_{jk}|^2 \le \sigma^2 \quad \text{with} \quad \frac{1}{np}\sum_{jk}\mathrm{E}|x_{jk}|^2 \to \sigma^2. \tag{2.22}$$

To estimate the moments of the ESD, we have

$$\frac{1}{p}\mathrm{E}[(\mathbf{ST})^h|\mathbf{T}] = p^{-1}n^{-h}\sum_G \mathrm{E}((tx)_G|\mathbf{T}), \tag{2.23}$$

where

$$(tx)_G = x_{i_1j_1}\overline{x}_{i_2j_1}t_{i_2i_3}x_{i_3j_2}\overline{x}_{i_4j_2}\cdots x_{i_{2h-1}j_h}\overline{x}_{i_{2h}j_h}t_{i_{2h}i_1}.$$

The Q-graphs (named in Yin and Krishnaiah (1983)) are drawn as follows: as before, plot the vertices i's and j's on two parallel lines and draw h (down) edges from i_{2u-1} to j_u, h (up) edges from j_u to i_{2u} and h (horizontal) edges from i_{2u} to i_{2u+1} (with the convention $i_{2h+1} = i_1$). If there is a single vertical edge in G, then the corresponding term is zero. We split the sum of non-zero terms in (2.23) into subsums in accordance with isomorphic classes of graphs without single vertical edges. For a Q-graph G in a given isomorphic class, denote by s the number of non-coincident j-vertices and by r the number of disjoint connected blocks of horizontal edges. Glue the coincident vertical edges and denote the resulting graph by G^0. Let the $p\times p$-matrix $\mathbf{T}$ correspond to each horizontal edge of G^0 and let the $p\times n$-matrix $\mathbf{T}^{(x)}_{\mu,\nu} = (\mathrm{E}(x^{\mu}_{i,j}\overline{x}^{\nu}_{i,j}))$ correspond to each vertical edge of G^0 that consists of μ down edges and ν up edges of G. Note that $\mu+\nu\ge 2$ and $|\mathrm{E}(x^{\mu}_{i,j}\overline{x}^{\nu}_{i,j})| \le \sigma^2(\delta_n\sqrt{n})^{\mu+\nu-2}$. It is obvious that $\|\mathbf{T}\| \le \tau_0$, $\|\mathbf{T}^{(x)}_{\mu,\nu}\| \le \sqrt{np}\sigma^2(\delta_n\sqrt{n})^{\mu+\nu-2}$ and $\|\mathbf{T}^{(x)}_{\mu,\nu}\|_0 \le \max(n,p)\sigma^2(\delta_n\sqrt{n})^{\mu+\nu-2}$. Also, every horizontal edge of G^0 is non-cutting. Split the right hand side of (2.23) as J_1+J_2 where J_1 corresponds to the sum of those terms whose graphs G^0 contain at least one vertical edge with multiplicity greater than 2 and J_2 is the sum of all other terms. Applying Lemma 2.11, we get $J_1 = O(\delta_n^2) = o(1)$.

We further split J_2 as $J_{21}+J_{22}$, where J_{21} is the sum of all those terms whose G_0-graphs contain at least one non-cutting vertical edge and J_{22} is the sum of the rest. For graphs corresponding to the terms in J_{21}, we must have $s+r\le h$. When evaluating J_{21}, we fix the indices $j_1,\ldots,j_s$ and perform the summation for $i_1,\ldots,i_r$ first. Corresponding to the summation for fixed $j_1,\ldots,j_s$, we define a new graph $G(j_1,\ldots,j_s)$ as follows: If (i_g,j_h) is a vertical edge of G_0, consisting of μ-up and ν-down edges of G (note that $\mu+\nu=2$), then remove this edge and add to the vertex i_g a loop, to which there corresponds the $p\times p$ diagonal matrix $\mathbf{T}(j_h) = \mathrm{diag}(E(x^{\mu}_{1,j_h}\bar{x}^{\nu}_{1,j_h}),\ldots,E(x^{\mu}_{p,j_h}\bar{x}^{\nu}_{p,j_h}))$, see Figure 4. After all vertical edges of G^0 are removed, the r disjoint connected blocks of the resulting graph $G(j_1,\ldots,j_s)$ have no cutting edges. Note that the $\|\cdot\|$-norms of the diagonal matrices are not greater than σ^2. Applying Lemma 2.11 to each of connected blocks, we obtain

$$|J_{21}| \leq Cp^{-1}n^{-h} \sum_{s+r\leq h} \sum_{j_1,\ldots,j_s} p^r \sigma^{2h} \tau_0^h = O(1/n).$$

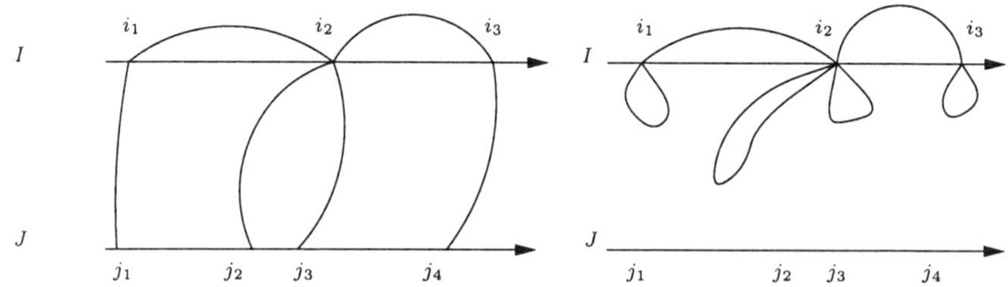

Figure 4. The left graph is the original one and the right one is the resulting graph.

Finally, consider J_{22}. Since all vertical edges are cutting edges, we have $s + r = h + 1$. There are exactly h non-coincident vertical edges, in which each down-edge (a, b) must coincide with one and only one up-edge (b, a). Thus, the contribution of the expectations of the x-variables is $\prod_{\ell=1}^{h} \mathrm{E}(|x^2_{i_{f_{ini}(e_\ell)}, j_{f_{end}(e_\ell)}}|)$. For a given vertical edge, if its corresponding matrix $\mathbf{T}^{(x)}$ is replaced by the $p \times n$ matrix of all entries σ^2, applying Lemma 2.11 again, this will cause a difference of $o(1)$ in J_{22}, since the norms ($\|\cdot\|$ or $\|\cdot\|_0$) of the difference matrix are only $o(n)$, by (2.22).

Now, denote by $\mu_1, \ldots, \mu_r$ the sizes (the numbers of edges) of the r disjoint blocks of horizontal edges. Then it is not difficult to show that for each class of isomorphic graphs, the sub-sum in J_{22} tends to $y^{r-1}\alpha_{\mu_1} \cdots \alpha_{\mu_r}(1 + o(1))$. Thus, to evaluate the right hand side of (2.23), one only needs to count the number of isomorphic classes.

Let i_m denote the number of disjoint blocks of horizontal subgraphs of size m. Then it can be shown that the number of isomorphic classes is $\frac{h!}{s! i_1! \cdots i_s!}$. For details, see Yin (1986). Hence,

$$\begin{aligned}
\frac{1}{p}\mathrm{E}[(\mathbf{ST})^h|\mathbf{T}] &= \sigma^{2h} \sum_{s=1}^{h} y^{h-s} \sum \frac{h!}{s!} \prod_{m=1}^{s} \left(\frac{[p^{-1}\mathrm{tr}(\mathbf{T}^m)]^{i_m}}{i_m!}\right) + o(1) \\
&= \sigma^{2h} \sum_{s=1}^{h} y^{h-s} \sum \frac{h!}{s!} \prod_{m=1}^{s} \frac{\alpha_m^{i_m}}{i_m!} + o(1), \qquad (2.24)
\end{aligned}$$

where the inner summation is taken with respect to all nonnegative integer solutions of $i_1 + \cdots + i_s = h + 1 - s$ and $i_1 + 2i_2 + \cdots + si_s = h$.

Similar to the proof of (2.11), to complete the proof of the theorem, one needs to show that

$$\mathrm{E}(|\frac{1}{p}(\mathbf{ST})^h - \mathrm{E}(\frac{1}{p}(\mathbf{ST})^h|\mathbf{T}]|^4|\mathbf{T}] = O(n^{-2}),$$

whose proof is similar to, and easier than, that of (2.24). The convergence of the ESD of **ST** and the non-randomness of the limiting distribution then follow by verifying Carleman's condition.

2.2. Limits of extreme eigenvalues

In multivariate analysis, many statistics involved with a random matrix can be written as functions of integrals with respect to the ESD of the random matrix. When applying the Helly-Bray theorem to find an approximate value of the statistics, one faces the difficulty of dealing with integrals with unbounded integrands. Thus, the study of strong limits of extreme eigenvalues is an important topic in spectral analysis of LDRM.

2.2.1. Limits of extreme eigenvalues of the Wigner matrix

The following theorem is a generalization of Bai and Yin (1988b) where the real case is considered. The complex case is treated here because the question often arises as to whether the result is true for the complex case.

Theorem 2.12. *Suppose that the diagonal elements of the Wigner matrix* $\mathbf{W}$ *are i.i.d. real random variables, the elements above the diagonal are i.i.d. complex random variables, and all these variables are independent. Then the largest eigenvalue of* $n^{-1/2}\mathbf{W}$ *tends to* $2\sigma > 0$ *with probability* 1 *if and only if the following four conditions are true.*

$$\begin{aligned}
&\text{(i) } \mathrm{E}((w_{11}^{+})^2) < \infty,\\
&\text{(ii) } \mathrm{E}(w_{12}) \textit{ is real and } \le 0,\\
&\text{(iii) } \mathrm{E}(|w_{12} - \mathrm{E}(w_{12}|^2) = \sigma^2,\\
&\text{(iv) } \mathrm{E}(|w_{12}^4|) < \infty,
\end{aligned} \tag{2.25}$$

where $x^+ = \max(x, 0)$.

The proof of the sufficiency part of Theorem 2.12 consists of the following steps. First, by Theorem 2.1, we have

$$\liminf_{n\to\infty} \lambda_{\max}(n^{-1/2}\mathbf{W}) \ge 2\sigma, \text{ a.s.} \tag{2.26}$$

Thus, the problem reduces to proving

$$\limsup_{n\to\infty} \lambda_{\max}(n^{-1/2}\mathbf{W}) \le 2\sigma, \text{ a.s.} \tag{2.27}$$

Let $\widetilde{\mathbf{W}}$ be the matrix obtained from $\mathbf{W}$ by replacing the diagonal elements with zero and centralizing the off diagonal elements. By Conditions (i) and (ii), we notice that $\limsup \frac{1}{\sqrt{n}} w_{kk}^+ = 0$, a.s. Then

$$\begin{aligned}
&\lambda_{\max}(n^{-1/2}\mathbf{W}) = n^{-1/2} \max_{||\mathbf{x}||=1} \Big(\sum_{j,k} x_j \bar{x}_k w_{jk} \Big) \\
&\le \max_{||\mathbf{x}||=1} \Big(\frac{1}{\sqrt{n}} \sum_{j \neq k} x_j \bar{x}_k (w_{jk} - \mathrm{E}(w_{jk})) + \frac{1}{\sqrt{n}} \max_k (w_{kk}^+ - \mathrm{E}(w_{12})) \Big) \\
&\le \lambda_{\max}(\widetilde{\mathbf{W}}) + o(1). \qquad (2.28)
\end{aligned}$$

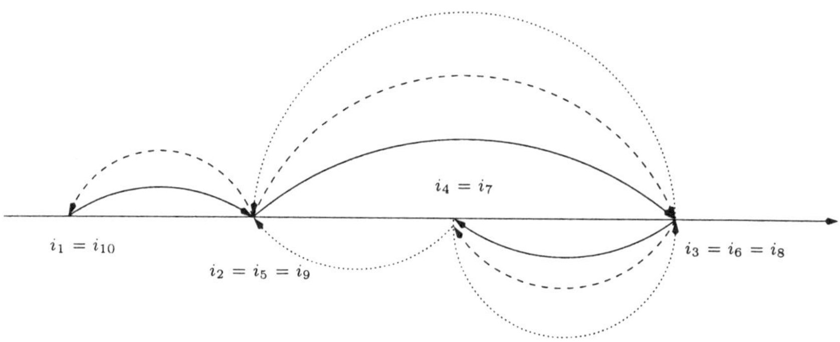

Figure 5. Four types of edges in a W-graph.

This further reduces the proof of (2.27) to showing that $\limsup_{n\to\infty} \lambda_{\max}(\widetilde{\mathbf{W}}) \le 2\sigma$, a.s.

For brevity of notation, we again use $\mathbf{W}$ for $\widetilde{\mathbf{W}}$, i.e., we assume that the diagonal elements and the means of off diagonal elements of $\mathbf{W}$ are zero. Then by condition (iv), we may select a sequence of constants $\delta_n \to 0$ such that

$$\mathrm{P}(\mathbf{W} \neq \widetilde{\mathbf{W}}, \text{ i.o.}) = 0,$$

where $\widetilde{\mathbf{W}}$ is redefined as $(w_{jk} I_{(|w_{jk}| \le \delta_n \sqrt{n})})$.

Note that $\mathrm{E}(w_{12}) = 0$ implies

$$\lambda_{\max}(n^{-1/2}\mathrm{E}(\widetilde{\mathbf{W}})) \le (1 + n^{-1/2}) |\mathrm{E}(w_{12} I_{(|w_{12}| \le \delta_n \sqrt{n})})| \to 0. \qquad (2.29)$$

Therefore, we only need to consider the upper limit of the largest eigenvalue of $\widetilde{\mathbf{W}} - \mathrm{E}(\widetilde{\mathbf{W}})$. For simplicity, we still use $\mathbf{W}$ to denote the truncated and recentralized matrix.

Select a sequence of even integers $h = h_n = 2s$ with the properties $h/\log n \to \infty$ and $h\delta_n^{1/4}/\log n \to 0$. We shall estimate

$$\mathrm{E}(\mathrm{tr}(\mathbf{W}^h)) = \sum_{i_1,\ldots,i_h} \mathrm{E}(w_{i_1 i_2} w_{i_2 i_3} \cdots w_{i_h i_1}) = \sum_G \mathrm{E}(w_G), \qquad (2.30)$$

where the graphs are constructed as in the proof of Theorem 2.1. Classify the edges into several types. An edge (i_a, i_{a+1}) is called an innovation or a Type 1 (T_1) edge if $i_{a+1} \notin \{i_1, \ldots, i_a\}$ and is called a Type 3 (T_3) edge if it coincides with an innovation which is single up to i_a. A T_3-edge (i_a, i_{a+1}) is called irregular if there is only one innovation which is single up to i_a and has a vertex coinciding with i_a in the chain $(i_1, \ldots, i_a)$. Otherwise, it is said to be regular. All other edges are called Type 4 (T_4) edges. A T_4-edge is also called a Type 2 (T_2) edge if it does not coincide with any edges prior to it. Examples of the four types of edges are given in Figure 5, in which the first three edges (in solid arcs) are innovations, the broken arcs are T_3 edges and the dotted arcs are T_4 edges. Among the T_3 edges, (i_5, i_6) is a regular T_3 edge (since the path may go to $i_6 = i_1$ instead of $i_6 = i_3$) and among the T_4 edges, (i_4, i_5) is a T_2 edge.

To estimate the right hand side of (2.30), we need the following lemmas whose proofs can be found in Bai and Yin (1988b) or Yin, Bai and Krishnaiah (1988).

Lemma 2.13. *Let t denote the number of non-coincident T_4-edges and u denote the number of innovations which are single up to i_a and have a vertex coinciding with i_a in the chain $(i_1, \ldots, i_a)$. Then $u \le t+1$.*

Lemma 2.14. *The number of regular T_3-edges is not greater than twice the number of T_4-edges.*

Now, we return to the proof of Theorem 2.12. Suppose that there are r innovations ($r \le s$) and t non-coincident T_4-edges in a graph. Then there are r T_3-edges, $2s-2r$ T_4-edges and $r+1$ non-coincident vertices. There are at most n^{r+1} ways to plot the non-coincident vertices and at most $\binom{2s}{r}$ ways to assign the innovations to the $2s$ edges. In a canonical graph (i.e., a graph which starts from the edge $(1,2)$ and the end-vertex of each innovation is one plus the end-vertex of the previous innovation), there is only one way to plot the innovations. There are at most $\binom{2s-r}{r}$ ways to select the T_3-edges from the remaining $2s-r$ edges and only one way to plot irregular T_3-edges. By Lemmas 2.13 and 2.14, there are at most $(t+1)^{4(s-r)}$ ways to plot the regular T_3-edges. After the T_1- and T_3-edges have been plotted, there are at most $\binom{2s-2r}{t}$ ways to select the t non-coincident T_4-edges and at most t^{2s-2r} ways to plot the $2s-2r$ T_4-edges into the t places. Finally, we note that the absolute value of each term is at most $\sigma^{2(r-t)}\mu^t(\sqrt{n}\delta_n)^{2(s-r)-t}$, where $\mu = \mathrm{E}(|w_{12}^3|)$. We obtain

$$\begin{aligned}
\mathrm{E}(\mathrm{tr}(\mathbf{W})^h) &\le \sum_{r=1}^{s}\sum_{t=0}^{2s-2r} n^{s+1}\binom{2s}{r}\binom{2s-r}{r}(t+1)^{8(s-r)}\sigma^{2r}(\mu/(\sigma\sqrt{n}\delta_n))^t\delta_n^{2(s-r)} \\
&\le 2sn^{s+1}[2\sigma + \delta_n(8s/\log(\sqrt{n}\delta_n\sigma/\mu))^4]^{2s} \\
&= 2sn^{s+1}[2\sigma + o(1)]^{2s}.
\end{aligned}$$

Then by the fact that $\lambda_{\max}^{2s}(n^{-1/2}\mathbf{W}) \le n^{-s}\mathrm{tr}(\mathbf{W}^{2s})$, for any $\varepsilon > 0$, we have

$$\mathrm{P}(\lambda_{\max}(n^{-1/2}\mathbf{W}) \ge 2\sigma + \varepsilon) \le 2sn[1 - \frac{\varepsilon - o(1)}{2\sigma + \varepsilon}]^{2s}.$$

The right hand side above is summable by the choice of s. Therefore, by the Borel Cantelli Lemma, $\limsup \lambda_{\max}(n^{-1/2}\mathbf{W}) \le 2\sigma$ almost surely. The sufficiency is proved.

Conversely, if $\limsup \lambda_{\max}(n^{-1/2}\mathbf{W}) \le 2a$ with $a > 0$, then, by $\lambda_{\max}(n^{-1/2}\mathbf{W}) \ge \max_k n^{-1/2}w_{kk}$, we have $\limsup \max_k n^{-1/2}w_{kk}^{+} \le 2a + \varepsilon$, which implies condition (i).

For $w_{jk} \ne 0$ and $|w_{kk}| \le n^{1/4}$, $|w_{jj}| \le n^{1/4}$, taking $x_k = w_{jk}/(\sqrt{2}|w_{jk}|)$ and $x_j = 1/\sqrt{2}$ in the first equality of (2.28), we have $\lambda_{\max}(n^{-1/2}\mathbf{W}) \ge n^{-1/2}|w_{jk}| - n^{-1/4}$. Thus, $\lambda_{\max}(n^{-1/2}\mathbf{W}) \ge n^{-1/2}\max_{\{|w_{kk}| \le n^{1/4}, |w_{jj}| \le n^{1/4}\}} |w_{jk}| - n^{-1/4}$. This implies Condition (iv), by noticing that $k_n = \#\{k \le n;\, |w_{kk}| > n^{1/4}\} = o_{a.s.}(n)$.

If $\mathrm{E}(\mathrm{Re}(w_{12})) > 0$, then take $\mathbf{x}$ with kth element $x_k = (n - k_n)^{-1/2}$ or 0 in accordance with $|w_{kk}| < n^{1/4}$ or not, respectively. Then applying (2.26) and noticing $k_n = o(n)$, one gets the following contradiction:

$$\begin{aligned}&\lambda_{\max}(n^{-1/2}\mathbf{W}) \ge n^{-1/2}\mathbf{x}^*\mathbf{W}\mathbf{x}\\ &\ge (n - k_n - 1)^{1/2}\mathrm{E}(\mathrm{Re}(w_{12})) - n^{-1/4} + \lambda_{\min}(n^{-1/2}[\widetilde{\mathbf{W}} - \mathrm{E}(\widetilde{\mathbf{W}})]) \to \infty,\end{aligned}$$

where $\widetilde{\mathbf{W}}$ is the matrix obtained from $\mathbf{W}$ with its diagonal elements replaced by zero. Here, we have used the fact that $\lambda_{\min}(n^{-1/2}[\widetilde{\mathbf{W}} - \mathrm{E}(\widetilde{\mathbf{W}})]) \to -2\sigma^2$, by the sufficiency part of the theorem. This proves that the real parts of the means of off-diagonal elements of $\mathbf{W}$ cannot be positive.

If $b = \mathrm{E}(\mathrm{Im}(w_{12})) \ne 0$, define $\mathbf{x}$ in such a way that $x_j = 0$ if $|w_{jj}| > n^{1/4}$, and the other $n - k_n$ elements are $(n - k_n)^{-1/2}(1, e^{i\pi \mathrm{sign}(b)(2\ell-1)/(n-k_n)}, \ldots, e^{i\pi \mathrm{sign}(b)(2\ell-1)(n-k_n-1)/(n-k_n)})$, respectively. Note that $\mathbf{x}$ is the eigenvector corresponding to the eigenvalue $\cot(\pi(2\ell - 1)/2(n - k_n))$ of the Hermitian matrix whose (j,k)th $(j < k)$ element is i if $|w_{jj}| \le n^{1/4}$ and $|w_{kk}| \le n^{1/4}$, or 0 otherwise. Therefore, we have, with $a = |\mathrm{E}(\mathrm{Re}(w_{12}))|$,

$$\begin{aligned}&\lambda_{\max}(n^{-1/2}\mathbf{W}) \ge n^{-1/2}\mathbf{x}^*\mathbf{W}\mathbf{x}\\ &\ge -\frac{|a|}{(n - k_n)\sqrt{n}\sin^2(\pi(2\ell - 1)/2(n - k_n))} + \frac{|b|}{\sqrt{n}\sin(\pi(2\ell - 1)/2(n - k_n))}\\ &\quad + \lambda_{\min}\Big(n^{-1/2}(\widetilde{\mathbf{W}} - \mathrm{E}(\widetilde{\mathbf{W}}))\Big) - n^{-1/4}\\ &:= I_1 + I_2 + I_3 - n^{-1/4}.\end{aligned}$$

Taking $\ell = [n^{1/3}]$ and noticing $k_n = o(n)$, we have

$$I_1 \sim -|a|n^{-1/6} \to 0, \quad I_2 \sim |b|n^{1/6} \to \infty \ \text{ and } \ I_3 \to -2\sigma^2.$$

This leads to the contradiction that $\lambda_{\max}(n^{-1/2}\mathbf{W}) \to \infty$, proving the necessity of Condition (ii).

Condition (iii) follows by applying the sufficiency part. The proof of Theorem 2.12 is now complete.

Remark 2.6. For the Wigner matrix, there is a symmetry between the largest and smallest eigenvalues. Thus, Theorem 2.12 actually proves that the necessary and sufficient conditions for both the largest and smallest eigenvalues to have finite limits almost surely are that the diagonal elements have finite second moments and the off-diagonal elements have zero mean and finite fourth moments.

Remark 2.7. In the proof of Theorem 2.12, if the entries of $\mathbf{W}$ depend on n but satisfy

$$\mathrm{E}(w_{jk}) = 0,\ \mathrm{E}(|w_{jk}^2|) \le \sigma^2,\ \mathrm{E}(|w_{jk}^\ell|) \le b(\delta_n n)^{\ell-3},\ (\ell \ge 3) \tag{2.31}$$

for some $b > 0$ and $\delta_n \downarrow 0$, then for fixed $\varepsilon > 0$ and $\ell > 0$, the following is true:

$$\mathrm{P}(\lambda_{\max}(n^{-1/2}\mathbf{W}) \ge 2\sigma + \varepsilon + x) = o(n^{-\ell}(2\sigma + \varepsilon + x)^{-2}), \tag{2.32}$$

uniformly for $x > 0$. This implies that the conclusion $\limsup \lambda_{\max}(n^{-1/2}\mathbf{W}) \le 2\sigma$ a.s. is still true.

2.2.2. Limits of extreme eigenvalues of sample covariance matrices

Geman (1980) proved that, as $p/n \to y$, the largest eigenvalue of a sample covariance matrix tends to $b(y)$ almost surely, assuming a certain growth condition on the moments of the underlying distribution, where $b(y) = \sigma^2(1+\sqrt{y})^2$ is defined in the statement of Theorem 2.5. Later, Yin, Bai and Krishnaiah (1988) and Bai, Silverstein and Yin (1988), respectively, proved that the necessary and sufficient condition for the largest eigenvalue of a sample covariance matrix to converge to a finite limit almost surely is that the underlying distribution has a zero mean and finite fourth moment, and that the limit must be $b(y)$. Silverstein (1989b) showed that the necessary and sufficient conditions for the weak convergence of the largest eigenvalue of a sample covariance matrix are $E(x_{11}) = 0$ and $n^2\mathrm{P}(|x_{11}| \ge \sqrt{n}) \to 0$. The most difficult problem in this direction is to establish the strong convergence of the smallest eigenvalue of a sample covariance matrix. Yin, Bai and Krishnaiah (1983) and Silverstein (1984) showed that when $y \in (0,1)$, there is a positive constant ε_0 such that the liminf of the smallest eigenvalue of $1/n$ times a Wishart matrix is larger than ε_0, a.s. In Silverstein (1985), this result is further improved to say that the smallest eigenvalue of a normalized Wishart matrix tends to $a(y) = \sigma^2(1-\sqrt{y})^2$ almost surely. Silverstein's approach strongly relies on the normality assumption and hence cannot be extended to the general case. The latest contribution is due to Bai and Yin (1993), in which a unified approach is presented, establishing the strong convergence of both the largest and smallest eigenvalues simultaneously under the existence of the fourth moment. Although only the real case is considered in Bai and Yin (1993), their results can easily be extended to the complex case.

Theorem 2.15. *In addition to the assumptions of Theorem* 2.5, *we assume that the entries of* $\mathbf{X}$ *have finite fourth moment. Then*

$$-2y\sigma^2 \le \liminf_{n\to\infty} \lambda_{\min}(\mathbf{S}-\sigma^2(1+y)\mathbf{I}) \le \limsup_{n\to\infty} \lambda_{\max}(\mathbf{S}-\sigma^2(1+y)\mathbf{I}) \le 2y\sigma^2, \quad \text{a.s.} \tag{2.33}$$

If we define the smallest eigenvalues as the $(p-n+1)$-st smallest eigenvalue of $\mathbf{S}$ when $p > n$, then from Theorem 2.15, one immediately gets the following Theorem.

Theorem 2.16. *Under the assumptions of Theorem* 2.15, *we have*

$$\lim_{n\to\infty} \lambda_{\min}(\mathbf{S}) = \sigma^2(1-\sqrt{y})^2, \quad a.s. \tag{2.34}$$

and

$$\lim_{n\to\infty} \lambda_{\max}(\mathbf{S}) = \sigma^2(1+\sqrt{y})^2, \quad \text{a.s.} \tag{2.35}$$

The proof of Theorem 2.15 relies on the following two lemmas.

Lemma 2.17. *Under the conditions of Theorem* 2.15, *we have*

$$\limsup_{n\to\infty} ||\mathbf{T}(\ell)|| \le (2\ell+1)(\ell+1)y^{(\ell-1)/2} \quad a.s.,$$

where $\mathbf{T}(\ell)$, $p\times p$, *has its* (a,b)*th entry* $n^{-\ell}(\sum' x_{av_1}\overline{x}_{u_1v_1}x_{u_1v_2}\overline{x}_{u_2v_2}\cdots x_{u_{\ell-1}v_\ell}\overline{x}_{bv_\ell})$ *and the summation* $\sum'$ *runs over* $v_1,\ldots,v_\ell = 1,\ldots,n$ *and* $u_1,\ldots,u_{\ell-1} = 1,\ldots,p$ *subject to the restriction*

$$a \ne u_1,\ u_1 \ne u_2,\ldots,u_{\ell-1}\ne b \quad and \quad v_1\ne v_2,\ v_2\ne v_3,\ldots,v_{\ell-1}\ne v_\ell.$$

Lemma 2.18. *Under the conditions of Theorem* 2.15, *we have*

$$(\mathbf{T}-y\mathbf{I})^h = \sum_{r=0}^{h}(-1)^{r+1}\mathbf{T}(r)\sum_{i=0}^{[(h-r)/2]} C_i(h,r)y^{h-r-i} + o(1), \tag{2.36}$$

where $\mathbf{T} = \mathbf{T}(1) = \mathbf{S}-\sigma^2(1+y)\mathbf{I}$ *and the constants* $|C_i(h,r)| \le 2^h$.

The proof of Lemma 2.17 is similar to that of Theorem 2.12, i.e., to consider the expectation of $\text{tr}(\mathbf{T}^{2s}(\ell))$. Construct the graphs as in the proof of Theorem 2.5. Using Lemmas 2.13 and 2.14 one gets an estimate

$$\text{E}(\text{tr}(\mathbf{T}^{2s}(\ell))) \le n^3[(2\ell+1)(\ell+1)y^{(\ell-1)/2}+o(1)]^{2s}.$$

From this, Lemma 2.17 can be proved; the details are omitted. The proof of Lemma 2.18 follows by induction.

2.3. Limiting behavior of eigenvectors

Relatively less work has been done on the limiting behavior of eigenvectors than eigenvalues in the spectral analysis of LDRM. Some work on eigenvectors of the Wigner matrix can be found in Girko, Kirsch and Kutzelnigg (1994), in which the first order properties are investigated. For eigenvectors of sample covariance matrices, some results can be found in Silverstein (1979, 1981, 1984b, 1989, 1990). Except for his first paper, the focus is on second order properties.

There is a good deal of evidence that the behavior of LDRM is asymptotically distribution-free, that is, it is asymptotically equivalent to the case where the basic entries are i.i.d. mean 0 normal, provided certain moment requirements are met. This phenomenon has been confirmed for distributions of eigenvalues. For the eigenvectors, the problem is how to formulate such a property. In the normal case, the matrix of orthonormal eigenvectors, which will be simply called the eigenmatrix, is Haar distributed. Since the dimension tends to infinity, it is difficult to compare the distribution of the eigenmatrix with the Haar measure. However, there are several different ways to characterize the similarity between these two distributions. The following approach is considered in the work referred to above.

Let $\mathbf{u}_n = (u_1, \ldots, u_p)'$ be a p-dimensional unit vector and $\mathbf{O}_n$ be the eigenmatrix of a covariance matrix. Define $\mathbf{y}_n = \mathbf{O}'_n \mathbf{u}_n = (y_1, \ldots, y_p)'$. If $\mathbf{O}_n$ is Haar distributed, then $\mathbf{y}$ is uniformly distributed on the unit sphere in a p-dimensional space. To this end, define a stochastic process $Y_n(t)$ as follows.

$$Y_n(t) = \sum_{i=1}^{[pt]} |y_i|^2.$$

Note that the process can also be viewed as a random measure of the uniformity of the distribution of $\mathbf{y}$. It is conceivable that $Y_n(F_n(t))$ converges to a common limiting stochastic process whatever the vector $\mathbf{u}_n$ is, where F_n is the ESD of the random matrix. This was proved in Girko, Kirsch and Kutzelnigg (1994) for the Wigner matrix and was the the main focus of Silverstein (1979) for large covariance matrices. This is implied by results in Silverstein's other work, in which second order properties are investigated. Here, we shall briefly introduce some of his results in this direction.

In the remainder of this subsection, we consider a real sample covariance matrix $\mathbf{S}$ with i.i.d. entries. Define

$$X_n(t) = \sqrt{p/2}(Y_n(t) - [pt]/p).$$

When $\mathbf{S}$ is a Wishart matrix, it is not difficult to show that $X_n(t)$ converges weakly to a Brownian bridge $W^0(t)$ in $D[0,1]$, the space of r.c.l.l. (right-continuous and left-limit) functions on $[0,1]$. In Silverstein (1989a), the following theorem is proved.

Theorem 2.19.

(i) *If*

$$\mathrm{E}(x_{11}) = 0, \ \mathrm{E}(|x_{11}^2|) = 1, \quad \mathrm{E}(|x_{11}^4|) = 3, \tag{2.37}$$

then for any integer k

$$\Big(\int_0^\infty x^r\, X_n(F^{\mathbf{S}}(dx)),\ r=1,\ldots,k\Big) \xrightarrow{\mathcal{D}} \Big(\int_0^\infty x^r\, W^0(F_y(dx)),\ r=1,\ldots,k\Big), \tag{2.38}$$

where F_y is the Marčenko-Pastur distribution with dimension-ratio y and parameter $\sigma^2 = 1$.

(ii) *If* $\int_0^\infty x\, X_n(F^{\mathbf{S}}(dx))$ *is to converge in distribution to a random variable for* $\mathbf{u}_n = (1,0,0,\ldots,0)'$ *and* $\mathbf{u}_n = p^{-1/2}(1,1,\ldots,1)'$, *then* $\mathrm{E}(|x_{11}^4|) < \infty$ *and* $\mathrm{E}(x_{11}) = 0$.

(iii) *If* $\mathrm{E}(|x_{11}^4|) < \infty$ *but* $\mathrm{E}(|x_{11} - \mathrm{E}(x_{11})|^4)/\mathrm{Var}(x_{11}) \neq 3$, *then there exist sequences* $\{\mathbf{u}_n\}$ *for which*

$$\Big(\int_0^\infty x\, X_n(F^{\mathbf{S}}(dx)),\ \int_0^\infty x^2\, X_n(F^{\mathbf{S}}(dx))\Big)$$

fails to converge in distribution.

Note that

$$\int_0^\infty x^k\, X_n(F^{\mathbf{S}}(dx)) = \sqrt{p/2}\Big(\mathbf{u}_n^*\mathbf{S}^k\mathbf{u}_n - p^{-1}\mathrm{tr}(\mathbf{S}^k)\Big).$$

The proof of (i) consists of the following three steps

1) $\sqrt{p}\mathrm{E}(\mathbf{u}_n^*\mathbf{S}^k\mathbf{u}_n - p^{-1}\mathrm{tr}(\mathbf{S}^k)) \xrightarrow{Pr} 0$;
2) $\sqrt{p}(p^{-1}\mathrm{tr}(\mathbf{S}^k) - \mathrm{E}(p^{-1}\mathrm{tr}(\mathbf{S}^k))) \xrightarrow{Pr} 0$;
3) $\sqrt{p/2}(\mathbf{u}_n^*\mathbf{S}^k\mathbf{u}_n - \mathrm{E}(\mathbf{u}_n^*\mathbf{S}^k\mathbf{u}_n)) \xrightarrow{\mathcal{D}} \int_0^\infty x^k\, W^0(F_y(dx))$.

The details are omitted. The proof of (ii) follows from standard limit theorems (see, e.g., Gnedenko and Kolmogorov (1954)). As for conclusion (iii), by elementary computation we have

$$\begin{aligned}&\mathrm{Cov}\Big(\int_0^\infty x\, X_n(F^{\mathbf{S}}(dx)),\ \int_0^\infty x^2\, X_n(F^{\mathbf{S}}(dx))\Big)\\ &= (2y+y^2)(1+o(1))\Big[1 + \frac{1}{2}(\mathrm{E}(|x_{11}^4|) - 3)\Big(\sum_{i=1}^n |\mathbf{u}_i|^4\Big)\Big].\end{aligned}$$

Then $\mathbf{u}_n$ can be chosen so that the right hand side of the above has no limit, unless $\mathrm{E}(|x_{11}^4|) = 3$.

Remark 2.8. The importance of the above theorem stems from the following. Assume $\mathrm{Var}\,(x_{11}) = 1$. If $\mathrm{E}(x_{11}) = 0$, $n^2\mathrm{P}(|x_{11}| \geq \sqrt{n}) \to 0$ (ensuring weak convergence of the largest eigenvalue of $\mathbf{S}$) and $X_n \xrightarrow{\mathcal{D}} W^0$, then it can be shown that (2.38) holds. Therefore, if weak convergence to a Brownian bridge is to hold for all choices of unit vectors $\mathbf{u}$, from (ii) and (iii) it must follow that $\mathrm{E}(|x_{11}^4|) = 3$. Thus it appears that similarity of the eigenmatrix to the Haar measure requires a certain amount of closeness of x_{11} to the standard normal distribution. At present, either of the two extremes, $X_n \xrightarrow{\mathcal{D}} W^0$ for all unit $\mathbf{u}$ and all x_{11} satisfying the above moment conditions, or $X_n \xrightarrow{\mathcal{D}} W^0$ only in the Wishart case, remains as a possibility. However, because of (i), verifying weak convergence to a Brownian bridge amounts to showing tightness of the sequence $\{X_n\}$ in $D[0,1]$.

The following theorem, found in Silverstein (1990), yields a partial solution to the problem, a case where tightness can be established.

Theorem 2.20. *Assume x_{11} is symmetrically distributed about 0 and $\mathrm{E}(x_{11}^4) < \infty$. Then $X_n \xrightarrow{\mathcal{D}} W^0$ holds for $\mathbf{u} = p^{-1/2}(\pm 1, \pm 1, \ldots)$.*

2.4. Miscellanea

Let $\mathbf{X}$ be an $n \times n$ matrix of i.i.d. complex random variables with mean zero and variance σ^2. In Bai and Yin (1986), large systems of linear equations and linear differential equations are considered. There, the norm of $(n^{-1/2}\mathbf{X})^k$ plays an important role for the stability of the solutions. The following theorem was proved.

Theorem 2.21. *If $\mathrm{E}(|x_{11}^4|) < \infty$, then*

$$\limsup_{n\to\infty} ||(n^{-1/2}\mathbf{X})^k|| \leq (1+k)\sigma^k, \quad \text{a.s., for all } k. \tag{2.39}$$

The proof of this theorem relies on, after truncation and centralization, the estimation of $\mathrm{E}([(n^{-1/2}\mathbf{X})^k(n^{-1/2}\mathbf{X}^*)^k]^\ell)$. The details are omitted. Here, we remark that when $k = 1$, the theorem reduces to a special case of Theorem 2.15 for $y = 1$. We also introduce an important consequence about the spectral radius of $n^{-1/2}\mathbf{X}$, which plays an important role in establishing the circular law (See Section 4). This was also independently proved by Geman (1986), under additional restrictions on the growth of moments of the underlying distribution.

Theorem 2.22. *If $\mathrm{E}(|x_{11}^4|) < \infty$, then*

$$\limsup_{n\to\infty} \max_{j\leq n} |\lambda_j(n^{-1/2}\mathbf{X})| \leq \sigma, \quad \text{a.s.} \tag{2.40}$$

Theorem 2.22 follows from the fact that for any k,

$$\limsup_{n\to\infty}\max_{j\le n}|\lambda_j(n^{-1/2}\mathbf{X})| = \limsup_{n\to\infty}\max_{j\le n}|\lambda_j[(n^{-1/2}\mathbf{X})^k]|^{1/k}$$
$$\le \limsup_{n\to\infty}||(n^{-1/2}\mathbf{X})^k||^{1/k} \le (1+k)^{1/k}\sigma \to \sigma,$$

by making $k\to\infty$.

Remark 2.9. Checking the proof of Theorem 2.21, one finds that, after truncation and centralization, the conditions for guaranteeing (2.39) are $|x_{jk}|\le \delta_n\sqrt{n}$, $\mathrm{E}(|x_{jk}^2|)\le\sigma^2$ and $\mathrm{E}(|x_{jk}^3|)\le b$, for some $b>0$. This is useful in extending the circular law to the case where the entries are not identically distributed.

3. Stieltjes Transform

Let G be a function of bounded variation defined on the real line. Then its Stieltjes transform is defined by

$$m(z) = \int_{-\infty}^{\infty}\frac{1}{x-z}G(dx), \tag{3.1}$$

where $z=u+iv$ with $v>0$. Throughout this section, z denotes $u+iv$ with $v>0$. Note that the integrand in (3.1) is bounded by $1/v$, the integral always exists, and

$$\frac{1}{\pi}\mathrm{Im}(m(z)) = \int_{-\infty}^{\infty}\Big[\frac{v}{\pi[(x-u)^2+v^2]}\Big]G(dx).$$

This is the convolution of G with a Cauchy density with a scale parameter v. If G is a distribution function, then the Stieltjes transform always has a positive imaginary part. Thus, one can easily verify that, for any continuity points $x_1 < x_2$ of G,

$$\lim_{v\downarrow 0}\int_{x_1}^{x_2}\frac{1}{\pi}\mathrm{Im}(m(z))du = G(x_2)-G(x_1). \tag{3.2}$$

Formula (3.2) obviously provides a continuity theorem between the family of distribution functions and the family of their Stieltjes transforms.

Also, if $\mathrm{Im}(m(z))$ is continuous at x_0+i0, then $G(x)$ is differentiable at $x=x_0$ and its derivative equals $\frac{1}{\pi}\mathrm{Im}(m(x_0+i0))$. This result was stated in Bai (1993a) and rigorously proved in Silverstein and Choi (1995). Formula (3.2) gives an easy way to find the density of a distribution function if its Stieltjes transform is known.

Now, let G be the ESD of a Hermitian matrix $\mathbf{W}$ of order p. Then it is easy to see that

$$m_G(z) = \frac{1}{p}\mathrm{tr}(\mathbf{W}-z\mathbf{I})^{-1}$$
$$= \frac{1}{p}\sum_{k=1}^{p}\frac{1}{w_{kk}-z-\boldsymbol{\alpha}_k^*(\mathbf{W}_k-z\mathbf{I}_{p-1})^{-1}\boldsymbol{\alpha}_k}, \tag{3.3}$$

where $\boldsymbol{\alpha}_k$ $((p-1)\times 1)$ is the kth column vector of $\mathbf{W}$ with the kth element removed and $\mathbf{W}_k$ is the matrix obtained from $\mathbf{W}$ with the kth row and column deleted. Formula (3.3) provides a powerful tool in the area of spectral analysis of LDRM.

As mentioned earlier, the mapping from distribution functions to their Stieltjes transforms is continuous. In Bai (1993a), this relation was more clearly characterized as a Berry-Esseen type inequality.

Theorem 3.1. *Let F be a distribution function and G be a function of bounded variation satisfying $\int |F(y)-G(y)|dy<\infty$. Then, for any $v>0$ and constants γ and a related to each other by the condition $\gamma=\frac{1}{\pi}\int_{|u|\le a}\frac{1}{u^2+1}du>1/2$,*

$$\|F-G\|\le\frac{1}{\pi(2\gamma-1)}\Big[\int|f(z)-g(z)|du+\frac{1}{v}\sup_x\int_{|y|\le 2va}|G(x+y)-G(x)|dy\Big],$$

where f and g are Stieltjes transforms of F and G respectively, and $z=u+iv$.

Sometimes, F and G have thin tails or even have bounded supports. In these cases, one may want to bound the difference between F and G in terms of an estimate of the difference of their Stieltjes transforms on a finite interval. We have the following theorem.

Theorem 3.2. *Under the conditions of Theorem 3.1, for any constants A and B restricted by $\kappa=\frac{4B}{\pi(A-B)(2\gamma-1)}\in(0,1)$, we have*

$$\|F-G\|\le\frac{1}{\pi(1-\kappa)(2\gamma-1)}\Big[\int_{-A}^{A}|f(z)-g(z)|du$$
$$+\frac{1}{v}\sup_x\int_{|y|\le 2va}|G(x+y)-G(x)|dy+2\pi v^{-1}\int_{|x|>B}|F(x)-G(x)|dx\Big].$$

Corollary 3.3. *In addition to the conditions of Theorem 3.1, assume further that, for some constant B, $F([-B,B])=1$ and $|G|((-\infty,-B))=|G|((B,\infty))=0$, where $|G|((b,c))$ denotes the total variation of G on the interval (b,c). Then for any A satisfying the constraint in Thereom 3.2, we have*

$$\|F-G\|\le\frac{1}{\pi(1-\kappa)(2\gamma-1)}\Big[\int_{-A}^{A}|f(z)-g(z)|du+\frac{1}{v}\sup_x\int_{|y|\le 2va}|G(x+y)-G(x)|dy\Big].$$

Remark 3.1. Corollary 3.3 is good enough for establishing the convergence rate of ESD's of LDRM since, in all known cases in the literature, the limiting distribution has a bounded support and the extreme eigenvalues have finite limits. It is more convenient than Theorem 3.1 since one does not need to estimate the integral of the difference of the Stieltjes transforms over the whole line.

3.1. Limiting spectral distributions

As an illustration, we use the Stieltjes transform (3.3) to derive the LSD's of Wigner and sample covariance matrices.

3.1.1. Wigner matrix

Now, as an illustration of how to use Formula (3.3) to find the LSD's, let us give a sketch of the proof of Theorem 2.1. Truncation and centralization are done first as in the proof of Theorem 2.1. That is, we may assume that $w_{kk} = 0$ and $|w_{jk}| \leq C$ for all $j \neq k$ and some constant C. Theorem 2.4 can be similarly proved but needs more tedious arguments.

Let $m_n(z)$ be the Stieltjes transform of the ESD of $n^{-1/2}\mathbf{W}$. By (3.3), and noticing $w_{kk} = 0$, we have

$$\begin{aligned} m_n(z) &= \frac{1}{n}\sum_{k=1}^{n} \frac{1}{-z - \frac{1}{n}\boldsymbol{\alpha}_k^*(n^{-1/2}\mathbf{W}_k - z\mathbf{I}_{n-1})^{-1}\boldsymbol{\alpha}_k} \\ &= \frac{1}{n}\sum_{k=1}^{n} \frac{1}{-z - \sigma^2 m_n(z) + \varepsilon_k} = -\frac{1}{z + \sigma^2 m_n(z)} + \delta_n, \end{aligned} \tag{3.4}$$

where

$$\varepsilon_k = \sigma^2 m_n(z) - \frac{1}{n}\boldsymbol{\alpha}_k^*(n^{-1/2}\mathbf{W}_k - z\mathbf{I}_{n-1})^{-1}\boldsymbol{\alpha}_k$$

and

$$\delta_n = \delta_n(z) = \frac{1}{n}\sum_{k=1}^{n} \frac{-\varepsilon_k}{(-z - \sigma^2 m_n(z) + \varepsilon_k)(-z - \sigma^2 m_n(z))}. \tag{3.5}$$

We first show that for any fixed $v_0 > 0$ and $B > 0$, with $z = u + iv$,

$$\sup_{|u| \leq B, v_0 \leq v \leq B} |\delta_n(z)| = o(1) \text{ a.s.} \tag{3.6}$$

By the uniform continuity of $m_n(z)$, the proof of (3.6) is equivalent to showing for each fixed z with $v > 0$,

$$|\delta_n(z)| = o(1) \text{ a.s.} \tag{3.7}$$

Note that

$$\begin{aligned} &|-z - \sigma^2 m_n(z) + \varepsilon_k| \geq \operatorname{Im}(-z - \frac{1}{n}\boldsymbol{\alpha}_k^*(n^{-1/2}\mathbf{W}_k - z\mathbf{I}_{n-1})^{-1}\boldsymbol{\alpha}_k)| \\ &= v(1 + \frac{1}{n}\boldsymbol{\alpha}_k^*((n^{-1/2}\mathbf{W}_k - u\mathbf{I}_{n-1})^2 + v^2\mathbf{I})^{-1}\boldsymbol{\alpha}_k) \geq v, \end{aligned}$$

and $|z + \sigma^2 m_n(z)| \geq v$. Then (3.7) follows if one can show

$$\max_k |\varepsilon_k(z)| = o(1) \text{ a.s.} \tag{3.8}$$

Let F_n and $F_{n(-k)}$ denote the ESD's of $n^{-1/2}\mathbf{W}$ and $n^{-1/2}\mathbf{W}_k$, respectively. Since $|nF_n(x) - (n-1)F_{n(-k)}(x)| \le 1$ by the interlacing theorem (see the proof of Lemma 2.2),

$$|m_n(z) - \frac{1}{n}\mathrm{tr}((n^{-1/2}\mathbf{W}_k - z\mathbf{I}_{n-1})^{-1})| = \Big|\frac{1}{n}\int \frac{nF_n(dx) - (n-1)F_{n(-k)}(dx)}{x-z}\Big|$$
$$\le \Big|\frac{1}{n}\int \frac{(nF_n(x) - (n-1)F_{n(-k)}(x))dx}{(x-z)^2}\Big| \le \pi/nv.$$

Based on this fact, in the proof of (3.8), we can replace ε_k's by $n^{-1}\boldsymbol{\alpha}_k^*(n^{-1/2}\mathbf{W}_k - z\mathbf{I}_{n-1})^{-1}\boldsymbol{\alpha}_k) - n^{-1}\mathrm{tr}((n^{-1/2}\mathbf{W}_k - z\mathbf{I}_{n-1})^{-1})$.

Since $\boldsymbol{\alpha}_k$ is independent of $\mathbf{W}_k$, it is not difficult to show that

$$\mathrm{E}\Big(\Big|n^{-1}\boldsymbol{\alpha}_k^*(n^{-1/2}\mathbf{W}_k - z\mathbf{I}_{n-1})^{-1}\boldsymbol{\alpha}_k - n^{-1}\mathrm{tr}(n^{-1/2}\mathbf{W}_k - z\mathbf{I}_{n-1})^{-1})\Big|^6\Big|\mathbf{W}_k\Big)$$
$$\le \frac{6!C^6}{n^6}\Big(\mathrm{tr}\Big[(n^{-1/2}\mathbf{W}_k - u\mathbf{I}_{n-1})^2 + v^2\mathbf{I}_{n-1}\Big]^{-1}\Big)\Big)^3 = O(n^{-3}).$$

This implies (3.8).

Solving equation (3.4) (in the variable m), one gets two solutions

$$m_n^{(1),(2)}(z) = -\frac{1}{2\sigma^2}[z + \delta_n\sigma^2 \mp \sqrt{(z - \delta_n\sigma^2)^2 - 4\sigma^2}],$$

where, for a complex number a, by convention $\sqrt{a}$ denotes the square root with positive imaginary part. We need to determine which solution is the Stieltjes transform of the spectrum of $n^{-1/2}\mathbf{W}$. By (3.4), we have

$$|\delta_n| \le |m_n| + 1/|z + \sigma^2 m_n| \le 2/v \to 0, \quad \text{as} \quad v \to \infty.$$

Thus, when z has a large imaginary part, $m_n = m_n^{(1)}(z)$. We claim this is true for all z with $v > 0$. Note that m_n and $m_n^{(1),(2)}$ are continuous in z on the upper half complex plane. We only need to show that $m_n^{(1)}$ and $m_n^{(2)}$ have no intersection. Suppose that they are equal at z_0 with $\mathrm{Im}(z_0) > 0$. Then we have $(z_0 - \sigma^2\delta_n)^2 - 4\sigma^2 = 0$ and

$$m_n(z_0) = -\frac{1}{2\sigma^2}(z_0 + \sigma^2\delta_n) = -z_0/\sigma^2 \pm 2/\sigma,$$

which contradicts with the fact that $m_n(z)$ has a positive imaginary part. Therefore, we have proved that

$$m_n(z) = -\frac{1}{2\sigma^2}[z + \delta_n\sigma^2 - \sqrt{(z - \delta_n\sigma^2)^2 - 4\sigma^2}].$$

Then from (3.6), it follows that with probability 1 for every fixed z with $v > 0$, $m_n(z) \to m(z) = -\frac{1}{2\sigma^2}[z - \sqrt{z^2 - 4\sigma^2}]$. Letting $v \downarrow 0$, we find the density of semicircular law as give in (2.2).

3.1.2. General sample covariance matrix

Note that a general form of sample covariance matrices can be considered as a special case of products of random matrices **ST** in Theorem 2.10. For generalization in another direction, as mentioned in Section 2.1.3, we present the following theorem.

Theorem 3.4. (Silverstein and Bai (1995)) *Suppose that for each* n*, the entries of* $\mathbf{X} = (\mathbf{x}_1, \ldots, \mathbf{x}_n)$*,* $p \times n$*, are i.i.d. complex random variables with* $\mathrm{E}(|x_{11} - \mathrm{E}(x_{11})|^2) = 1$*, and that* $\mathbf{T} = \mathbf{T}_n = \mathrm{diag}(\tau_1^n, \ldots, \tau_p^n)$*,* τ_i^n *real, and the ESD of* $\mathbf{T}$ *converges almost surely to a probability distribution function* H *as* $n \to \infty$*. Assume that* $\mathbf{B} = \mathbf{A} + \frac{1}{n}\mathbf{X}^*\mathbf{T}\mathbf{X}$*, where* $\mathbf{A} = \mathbf{A}_n$ *is Hermitian* $n \times n$ *satisfying* $F^{\mathbf{A}_n} \xrightarrow{v} F_a$ *almost surely, where* F_a *is a distribution function (possibly defective, i.e., of total variation less than* 1) *on the real line, and* $\xrightarrow{v}$ *means vague convergence, i.e., convergence without preservation of the total variation. Furthermore, assume that* $\mathbf{X}$*,* $\mathbf{T}$*, and* $\mathbf{A}$ *are independent. When* $p/n \to y > 0$ *as* $n \to \infty$*, we have almost surely* $F^{\mathbf{B}}$*, the ESD of* $\mathbf{B}$*, converges vaguely to a (non-random) d.f.* F*, whose Stieltjes transform* $m(z)$ *is given by*

$$m(z) = m_a\Big(z - y\int \frac{\tau dH(\tau)}{1 + \tau m(z)}\Big), \tag{3.9}$$

where z *is a complex number with a positive imaginary part and* m_a *is the Stieltjes transform of* F_a*.*

The set-up of Theorem 3.4 originated from nuclear physics, but is also encountered in multivariate statistics. In MANOVA, **A** can be considered as the between-covariance matrix, which may diverge in some directions under the alternative hypothesis. Examples of **B** can be found in the analysis of multivariate linear models and error-in-variables models, when the sample covariance matrix of the covariates is ill-conditioned. The role of **A** is to reduce the instability in the directions of the eigenvectors corresponding to small eigenvalues.

Remark 3.2. Note that Silverstein and Bai (1995) is more general than Yin (1986) in that it does not require the moment convergence of the ESD of **T** nor the positive definiteness of **T**. Also, it allows a perturbation matrix **A**. However, it is more restrictive than Yin (1986) in that it requires the matrix **T** to be diagonal. An extension of Yin's work in another direction is made in Silverstein (1995), who only assumes that **T** is positive definite and its ESD almost surely tends to a probability distribution, without requiring moment convergence. Weak convergence to (3.9) was established in Marčenko-Pastur (1967) under higher moment conditions than assumed in Theorem 3.4, but with mild dependence between the entries of **X**.

The assumption that the matrix $\mathbf{T}$ is diagonal in Theorem 3.4 is needed for the proof. It seems possible and is of interest to remove this restriction.

Now, we sketch a proof of Theorem 3.4 under more general conditions by using the Stieltjes transform. We replace the conditions for the x-variables with those given in Theorem 2.8. Remember that the entries of $\mathbf{X}$ and $\mathbf{T}$ depend on n. For brevity, we shall suppress the index n from these symbols and τ_i^n.

Denote by H_n and H the ESD of $\mathbf{T}_n$ and its LSD, and denote by m_{A_n} and m_A the Stieltjes transforms of the ESD of $\mathbf{A}_n$ and that of its LSD. Denote the Stieltjes transform of the ESD of $\mathbf{B}$ by $m_n(z)$.

Using the truncation and centralization techniques as in the proof of Theorem 2.10, without loss of generality, we may assume that the following additional conditions hold:

1. $|\tau_j| \le \tau_0$ for some positive constant τ_0,
2. $\mathrm{E}(x_{ij}) = 0$, $\mathrm{E}(|x_{ij}|^2) \le 1$ with $\frac{1}{pn}\sum_{ij}\mathrm{E}(|x_{ij}|^2) \to 1$ and $|x_{ij}| \le \delta_n\sqrt{n}$ for some sequence $\delta_n \to 0$.

If $F^{\mathbf{A}_n} \to c$, a.s. for some $c \in [0,1]$ (which is equivalent to almost all eigenvalues of $\mathbf{A}_n$ tending to infinity while the number of eigenvalues tending to negative infinity is about cn), then $F^{\mathbf{B}_n} \to c$ a.s., since the support of $\mathbf{XTX}^*$ remains bounded. Consequently, $m_n \to 0$ and $m_{A_n} \to 0$ a.s., and hence (3.9) is true. Thus, we only need to consider the case where the limit F^A of F^{A_n} has a positive mass over the real line. Then for any fixed z, there is a positive number η such that $\mathrm{Im}(m_n(z)) > \eta$.

Let $\mathbf{B}_{(i)} = \mathbf{B} - \tau_i \boldsymbol{\xi}_i \boldsymbol{\xi}_i^*$ and

$$\mu_n = \frac{1}{n}\sum_{i=1}^{n}\frac{\tau_i}{1+\tau_i m_n(z)} = \int \frac{\tau}{1+\tau m_n(z)} H_n(d\tau),$$

where $\boldsymbol{\xi}_i = n^{-1/2}\mathbf{x}_i$. Note that x has a non-positive imaginary part. Then by the identity

$$(\mathbf{A}_n - (z-\mu_n)\mathbf{I})^{-1} = (\mathbf{B} - z\mathbf{I})^{-1} + (\mathbf{A}_n - (z-\mu_n)\mathbf{I})^{-1}(\frac{1}{n}\mathbf{XTX}^* - \mu_n\mathbf{I})(\mathbf{B} - z\mathbf{I})^{-1},$$

we obtain

$$m_{A_n}(z-\mu_n) - m_n(z) = \frac{1}{n}\sum_{i=1}^{n}\frac{\tau_i d_i}{1+\tau_i m_n(z)}, \tag{3.10}$$

where

$$d_i = d_i^n(\mu_n) = \frac{(1+\tau_i m_n(z))\boldsymbol{\xi}_i^*(\mathbf{B}_{(i)} - z\mathbf{I})^{-1}(\mathbf{A}_n - (z-\mu_n)\mathbf{I})^{-1}\boldsymbol{\xi}_i}{1+\tau_i\boldsymbol{\xi}_i^*(\mathbf{B}_{(i)} - z\mathbf{I})^{-1}\boldsymbol{\xi}_i}$$
$$-\frac{1}{n}\mathrm{tr}[(\mathbf{B} - z\mathbf{I})^{-1}(\mathbf{A}_n - (z-\mu_n)\mathbf{I})^{-1}].$$

Note that for any fixed z, $\{m_n(z)\}$ is a bounded sequence. Thus, any subsequence of $\{m_n(z)\}$ contains a convergent subsequence. If m_n converges, then so does μ_n and hence $m_{A_n}(z-\mu_n)$. By (3.10), to prove (3.9), one only needs to show that equation (3.10) tends to (3.9) once $m_n(z)$ converges and that equation (3.9) has a unique solution. The proof of the latter is postponed to the next theorem. A proof of the former, i.e., the right hand side of (3.10) tends to zero, is presented here.

By (3.10) and the fact that $\mathrm{Im}(m_n(z)) > \eta$, we have $|1+\tau_i m_n(z)| \geq \min\{1/2, v\eta/2\tau_0\} > 0$. This implies that μ_n is uniformly bounded. Also, we know that μ_n has non-positive imaginary part from its definition. Therefore, to complete the proof of the convergence of (3.10), we can show the stronger conclusion that, with probability 1, the right hand side of (3.10) (with μ_n replaced by μ) tends to zero uniformly in μ over any compact set of the lower half complex plane. Due to the uniform continuity of both sides of (3.10) in u and μ, we only need to show (3.10) for any fixed z and non-random μ.

Note that the norms of $(A_n-(z-\mu)\mathbf{I})^{-1}$, $(\mathbf{B}-z\mathbf{I})^{-1}$ and $(\mathbf{B}_{(i)}-z\mathbf{I})^{-1}$ are bounded by $1/v$. Now we present an easier proof under the slightly stronger condition that $\delta_n^2 \log n \to 0$. (This holds if the random variables $|x_{jk}|^2\log(1+|x_{jk}|)$ are uniformly integrable or x_{jk} are identically distributed. For the second case, a second-step truncation is needed (see Silverstein and Bai (1995) for details)). Under this additional condition, it is sufficient to show that $\max_i\{|d_i|\} \to 0$, a.s. Using Lemma A.4 of Bai (1997), one can show that

$$\begin{aligned}&\mathrm{P}\Big(\Big|\boldsymbol{\xi}_i^*(\mathbf{B}_{(i)}-z\mathbf{I})^{-1}(A_n-(z-\mu)\mathbf{I})^{-1}\boldsymbol{\xi}_i-\frac{1}{n}\mathrm{tr}[(\mathbf{B}_{(i)}-z\mathbf{I})^{-1}(A_n-(z-\mu)\mathbf{I})^{-1}]\Big|\geq\varepsilon\Big)\\&\leq C\exp\{-b/\delta_n^2\}\end{aligned}$$

and

$$\mathrm{P}(|\boldsymbol{\xi}_i^*(\mathbf{B}_{(i)}-z\mathbf{I})^{-1}\boldsymbol{\xi}_i-\frac{1}{n}\mathrm{tr}(\mathbf{B}_{(i)}-z\mathbf{I})^{-1})|\geq\varepsilon)\leq C\exp\{-b/\delta_n^2\},\quad \text{for some } b>0.$$

These two inequalities show that for any fixed μ,

$$\max_{i\leq p}|\boldsymbol{\xi}_i^*(\mathbf{B}_{(i)}-z\mathbf{I})^{-1}(A_n-(z-\mu)\mathbf{I})^{-1}\boldsymbol{\xi}_i-n^{-1}\mathrm{tr}[(\mathbf{B}_{(i)}-z\mathbf{I})^{-1}(A_n-(z-\mu)\mathbf{I})^{-1}]|\to 0,\quad \text{a.s.}$$

and

$$\max_{i\leq p}|\boldsymbol{\xi}_i^*(\mathbf{B}_{(i)}-z\mathbf{I})^{-1}\boldsymbol{\xi}_i-\frac{1}{n}\mathrm{tr}(\mathbf{B}_{(i)}-z\mathbf{I})^{-1})|\to 0,\quad \text{a.s.}$$

Then the uniform continuity in μ implies that the above two limits hold uniformly when μ varies in any fixed compact subset of the lower half plane.

Note that the rank of $\mathbf{B}-\mathbf{B}_{(i)}$ is one. Also by Lemma 2.2, $||F^{\mathbf{B}}-F^{\mathbf{B}_{(i)}}||\leq 1/p$. Hence

$$|m_n(z)-m_{n(i)}(z)|\leq\frac{\pi}{pv},\tag{3.11}$$

where $m_{n(i)}(z)$ is the Stieltjes transform of the ESD of $\mathbf{B}_{(i)}$. Therefore,

$$\max_{i\le p}\left|\frac{1+\tau_i m_n(z)}{1+\tau_i\boldsymbol{\xi}_i^*(\mathbf{B}_{(i)}-z\mathbf{I})^{-1}\boldsymbol{\xi}_i}-1\right| = \max_{i\le p}\frac{o(1)}{1+\tau_i m_n(z)+o(1)} = o(1).$$

Here, the last step follows from the fact that $|1+\tau_i m_n(z)| \ge \min\{1/2,\ v\eta/2\tau_0\} > 0$. Finally, we get

$$\begin{aligned}\max_{i\le p}|d_i| &\le o(1)+\max_{i\le p}\left|\frac{1}{n}\mathrm{tr}\left([(\mathbf{B}-z\mathbf{I})^{-1}-(\mathbf{B}_{(i)}-z\mathbf{I})^{-1}](A_n-(z-\mu)\mathbf{I})^{-1}\right)\right|\\ &\le o(1)+\max_{i\le p}\frac{1}{n^2v^3}\sum_{k=1}^{n}|x_{ik}|^2 = o(1).\end{aligned}$$

For the general case without the additional condition, we need to rewrite the right hand side of (3.10) as

$$\begin{aligned}&\frac{1}{n}\sum_{i=1}^{n}\frac{\tau_i d_{i1}}{1+\tau_i m_n(z)}-\frac{1}{n}\sum_{i=1}^{n}\frac{\tau_i^2\eta_i d_{i2}}{(1+\tau_i m_n(z))^2}-\frac{1}{n}\sum_{i=1}^{n}\frac{\tau_i^2 d_{i1}d_{i2}}{(1+\tau_i m_n(z))^2}\\ &+\frac{1}{n}\sum_{i=1}^{n}\frac{\tau_i^3\eta_i d_{i2}^2}{(1+\tau_i m_n(z))^3}-\frac{1}{n}\sum_{i=1}^{n}\frac{\tau_i^4\eta_i d_{i2}^3}{(1+\tau_i m_n(z))^3(1+\tau_i\boldsymbol{\xi}_i^*(\mathbf{B}_{(i)}-z\mathbf{I})^{-1}\boldsymbol{\xi}_i)}\\ &+\frac{1}{n}\sum_{i=1}^{n}\frac{\tau_i^3 d_{i1}d_{i2}^2}{(1+\tau_i m_n(z))^2(1+\tau_i\boldsymbol{\xi}_i^*(\mathbf{B}_{(i)}-z\mathbf{I})^{-1}\boldsymbol{\xi}_i)}+o_{a.s.}(n^{-1+\varepsilon})\\ =&\sum_{k=1}^{6}\zeta_k+o_{a.s.}(n^{-1+\varepsilon}),\end{aligned}$$

where

$$d_{i1} = \boldsymbol{\xi}_i^*(\mathbf{B}_{(i)}-z\mathbf{I})^{-1}(\mathbf{A}_n-(z-\mu)\mathbf{I})^{-1}\boldsymbol{\xi}_i-\eta_i,$$

$$d_{i2} = \boldsymbol{\xi}_i^*(\mathbf{B}_{(i)}-z\mathbf{I})^{-1}\boldsymbol{\xi}_i-\frac{1}{n}\mathrm{tr}[(\mathbf{B}_{(i)}-z\mathbf{I})^{-1}]$$

and

$$\eta_i = \frac{1}{n}\mathrm{tr}[(\mathbf{B}_{(i)}-z\mathbf{I})^{-1}(\mathbf{A}_n-(z-\mu)\mathbf{I})^{-1}].$$

By elementary but tedious arguments, one can show that $\mathrm{E}(|\zeta_k|^4) = O(n^{-2})$, $k=1,2$, $\mathrm{E}(|\zeta_k|^2) = O(n^{-2})$, $k=3,4$, and $\mathrm{E}(|\zeta_k|) = O(n^{-3/2})$, $k=5,6$. Thus, the right hand side of (3.10) tends to zero almost surely. The proof of Theorem 3.4 is complete.

Theorem 3.5. *For any z with $\mathrm{Im}(z) > 0$, (3.9) has a unique solution $m(z)$ which has a positive imaginary part.*

The existence of a solution to equation (3.9) has already been proved in the proof of Theorem 3.4. To prove the uniqueness, rewrite equation (3.9) as

$$m = \int \frac{F^A(d\lambda)}{\lambda - z + xy}, \tag{3.12}$$

where

$$x = x(m) = \int \frac{\tau H(d\tau)}{1 + \tau m}.$$

Suppose that the equation has two roots $m_1 \neq m_2$. Let $x_j = x(m_j)$, $j = 1, 2$. Then by (3.12), we have

$$\begin{aligned} m_1 - m_2 &= \int \frac{y(x_2 - x_1)F^A(d\lambda)}{(\lambda - z + x_1 y)(\lambda - z + x_2 y)} \\ &= y(m_1 - m_2) \int \frac{\tau^2 H(d\tau)}{(1 + \tau m_1)(1 + \tau m_2)} \int \frac{F^A(d\lambda)}{(\lambda - z + x_1 y)(\lambda - z + x_2 y)}. \end{aligned}$$

Finally, a contradiction can be derived by Hölder's inequality, as follows,

$$\begin{aligned} 1 &= y \int \frac{\tau^2 H(d\tau)}{(1 + \tau m_1)(1 + \tau m_2)} \int \frac{F^A(d\lambda)}{(\lambda - z + x_1 y)(\lambda - z + x_2 y)} \\ &\le \Big(y \int \frac{\tau^2 H(d\tau)}{|1 + \tau m_1|^2} \int \frac{F^A(d\lambda)}{|\lambda - z + x_1 y|^2}\Big)^{1/2} \Big(y \int \frac{\tau^2 H(d\tau)}{|1 + \tau m_2|^2} \int \frac{F^A(d\lambda)}{|\lambda - z + x_2 y|^2}\Big)^{1/2} \\ &= \Big(y \int \frac{\tau^2 H(d\tau)}{|1 + \tau m_1|^2} \frac{\mathrm{Im}(m_1)}{v + \mathrm{Im}(x_1) y}\Big)^{1/2} \Big(y \int \frac{\tau^2 H(d\tau)}{|1 + \tau m_2|^2} \frac{\mathrm{Im}(m_2)}{v + \mathrm{Im}(x_2) y}\Big)^{1/2} < 1. \end{aligned}$$

Here, the last equality follows by comparing the imaginary part of equation (3.12) and the last inequality follows by observing that $\int \frac{F^A(d\lambda)}{|\lambda - z + x_j y|^2} = \frac{\mathrm{Im}(m_j)}{v - y \mathrm{Im}(x_j)}$ and $\mathrm{Im}(x_j) = \int \frac{\mathrm{Im}(m_j)\tau^2 H(d\tau)}{|1 + \tau m_j|^2}$. The proof of the theorem is complete.

3.2. Convergence rates of spectral distributions

The problem of convergence rates of ESD's of LDRM had been open for decades since no suitable tools were found. As seen in Section 2, most important works were done by employing the MCT. Carleman's criterion guarantees convergence but does not give any rate. A breakthrough was made in recent work of Bai (1993a,b) in which Theorem 3.1 - Corollary 3.3 were proved and some convergence rates were established. Although these rates are still far from expected, some solid rates have been established and, more importantly, we have found a way to establish them. Bai, Miao and Tsay (1996a,b, 1997) further investigated the convergence rates of the ESD of large dimensional Wigner matrices.

3.2.1. Wigner matrix

In this section, we first introduce a result in Bai (1993a). Consider the model of Theorem 2.4 and assume that the entries of $\mathbf{W}$ above or on the diagonal are independent and satisfy

$$\begin{aligned} &\text{(i) } \mathrm{E}(w_{jk}) = 0, \text{ for all } 1 \le i \le j \le n;\\ &\text{(ii) } \mathrm{E}(|w_{jk}^2|) = 1, \text{ for all } 1 \le i < j \le n;\\ &\text{(iii) } \mathrm{E}(|w_{jj}^2|) = \sigma^2, \text{ for all } 1 \le j \le n\\ &\text{(iv) } \sup_n \max_{1\le i\le j\le n} \mathrm{E}(|w_{jk}^4|) \le M < \infty. \end{aligned} \tag{3.13}$$

Theorem 3.6. *Under the conditions in* (3.13), *we have*

$$||\mathrm{E}F^{(1/\sqrt{n}\mathbf{W})} - F|| = O(n^{-1/4}), \tag{3.14}$$

where F is the semi-circular law with scale parameter 1.

Remark 3.3. The assertion (3.14) does not imply the complete convergence of $F^{(1/\sqrt{n}\mathbf{W})}$ to F. Here, we present a new result of Bai, Miao and Tsay (1997) in which a convergence rate in probability is established. Readers interested in the details of the proof of Theorem 3.6 are referred to Bai (1993a). Our purpose here is to illustrate how to use Theorem 3.1 - Corollary 3.3 to establish convergence rates of ESD's. Thus, we shall not pursue better rates through tedious arguments.

Theorem 3.7. *Under conditions* (i)-(iv) *in* (3.13), *we have*

$$||F^{(1/\sqrt{n}\mathbf{W})} - F|| = O_p(n^{-1/4}). \tag{3.15}$$

Truncate the diagonal entries of $\mathbf{W}$ at $n^{1/8}$ and off-diagonal elements at $n^{1/3}$. Let $F_n^{(t)}$ denote the ESD of the truncated matrix. Then by Lemma 2.2 and condition (iv), we have

$$\begin{aligned} \mathrm{E}||F_n - F_n^{(t)}|| &\le \frac{1}{n}\Big(\sum_{j\ne k} \mathrm{P}(|w_{jk}| \ge n^{1/3}) + \sum_k \mathrm{P}(|w_{kk}| \ge n^{1/8})\Big)\\ &\le \frac{Mn(n-1)n^{-4/3} + Mnn^{-1/2}}{n} \le 2Mn^{-1/3}. \end{aligned}$$

Centralize the off-diagonal elements of the truncated matrix, replace its diagonal elements by zero and denote the ESD of the resulting matrix by $F_n^{t,c}$. Then using Lemma 2.3, we obtain

$$\begin{aligned} L^3(F_n^{(t)} - F_n^{(t,c)}) &\le \frac{1}{n^2}\Big(\sum_{j\ne k} \mathrm{E}^2(|w_{jk}|I_{|w_{jk}|\ge n^{1/3}}) + \sum_k \mathrm{E}(|w_{kk}|^2 I_{|w_{kk}|\le n^{1/8}})\Big)\\ &\le \frac{M^2 n(n-1)n^{-2} + nn^{1/4}}{n^2} \le 2n^{-3/4} \end{aligned}$$

for all large n. Therefore, to prove Theorem 3.6, we may make the additional assumptions that the diagonal elements of $\mathbf{W}$ are zero and the off-diagonal elements are bounded by $n^{-1/3}$. Then the conditions in Remark 2.8 are satisfied. Therefore, we have

$$\begin{aligned} &\mathrm{E}\int_{|x|\geq 4}|F_n(x)-F(x)|dx \\ &\leq \int_4^\infty \Big(\mathrm{P}(\lambda_{\max}(n^{-1/2}\mathbf{W})\geq x)+\mathrm{P}(\lambda_{min}(n^{-1/2}\mathbf{W})\leq -x)\Big)dx = o(n^{-1}). \end{aligned}$$

Recalling Theorem 3.2, we have for any $v>0$,

$$\begin{aligned} &\mathrm{E}||F^{(1/\sqrt{n}\mathbf{W})}-F|| \\ &\leq C\Big(\int_{|u|<16}\mathrm{E}(|m_n(z)-m(z)|)du+\mathrm{E}\int_{|x|\geq 4}|F_n(x)-F(x)|dx+v\Big) \\ &= C\int_{|u|<16}\mathrm{E}(|m_n(z)-m(z)|)du+O(n^{-1/4}), \end{aligned}$$

if v is chosen to be $bn^{-1/4}$ for some $b>0$.

In Bai (1993a), it is proved that for the above chosen v,

$$\int|\mathrm{E}(m_n(z))-m(z)|du=O(v).$$

Thus, to prove (3.15), it is sufficient to prove

$$\int_{|u|<16}\mathrm{E}(|m_n(z)-\mathrm{E}(m_n(z))|)du=O(n^{-1/4}). \tag{3.16}$$

Define $\gamma_d=\mathrm{E}_d(m_n(z))-\mathrm{E}_{d-1}(m_n(z))$, $d=1,\ldots,n$, where E_d denotes the conditional expectation given the variables $\{w_{j,k}, 1\leq j\leq k\leq d\}$, with the convention that $\mathrm{E}_0=\mathrm{E}$. Note that $\{\gamma_1,\ldots,\gamma_n\}$ forms a martingale difference sequence and

$$m_n(z)-\mathrm{E}(m_n(z))=\frac{1}{n}\sum_{d=1}^n\gamma_d.$$

By noticing $|\gamma_k|\leq 2/v$ and the orthogonality of martingale differences, we get

$$\begin{aligned} \mathrm{E}|m_n(z)-\mathrm{E}(m_n(z))| &\leq \mathrm{E}^{1/2}|m_n(z)-\mathrm{E}(m_n(z))|^2 \\ &= \frac{1}{n}\left(\sum_{d=1}^n\mathrm{E}(|\gamma_d|^2)\right)^{1/2} \\ &\leq 2v^{-1}n^{-1/2}=O(n^{-1/4}). \end{aligned}$$

The proof of the theorem is complete.

In Bai, Miao and Tsay (1996a,b), the convergence rate of Wigner matrices is investigated further. The following results are established in the first of these works.

Theorem 3.8. *Suppose that the diagonal entries of* $\mathbf{W}$ *are i.i.d. with mean zero and finite sixth moment and that the elements above the diagonal are i.i.d. with mean zero, variance* 1 *and finite eighth moment. Then the following results are true:*

$$\|\mathrm{E}F_n - F\| = O(n^{-1/2})$$

and

$$\|F_n - F\| = O_p(n^{-2/5}).$$

If we further assume that the entries of $\mathbf{W}$ *have finite moments of all orders, then for any* $\varepsilon > 0$,

$$\|F_n - F\| = o_{a.s.}(n^{-2/5+\varepsilon}).$$

In Bai, Miao and Tsay (1996b), the convergence rate of the expected ESD of $\mathbf{W}$ is improved to $O(n^{-1/3})$ under the conditions of Theorem 3.6.

3.2.2. Sample covariance matrix

Assume the following conditions are true.

(i) $\mathrm{E}(x_{jk}) = 0$, $\mathrm{E}(|x_{jk}^2|) = 1$, for all j, k, n,

(ii) $\sup_n \sup_{j,k} \mathrm{E}|x_{jk}^4| I_{(|x_{jk}| \geq M)} \to 0$, as $M \to \infty$. (3.17)

In Bai (1993b), the following theorems are proved.

Theorem 3.9. *Under the assumptions in* (3.17), *for* $0 < \theta < \Theta < 1$ *or* $1 < \theta < \Theta < \infty$,

$$\sup_{y_p \in (\theta, \Theta)} \|\mathrm{E}F^{\mathbf{S}} - F_{y_p}\| = O(n^{-1/4}), \tag{3.18}$$

where $y_p = p/n$ *and* F_{y_p} *is defined in Theorem* 2.19.

Theorem 3.10. *Under the assumptions in* (3.17), *for any* $0 < \varepsilon < 1$,

$$\sup_{y_p \in (1-\varepsilon, 1+\varepsilon)} ||\mathrm{E}F^{\mathbf{S}} - F_{y_p}|| = O(n^{-5/48}). \tag{3.19}$$

By the same approach as in the proof of Theorem 3.8, Bai, Miao and Tsay (1996a) also generalized the results of Theorems 3.9 and 3.10 to the following theorem.

Theorem 3.11. *Under the assumptions in* (3.17), *the conclusions in Theorems* 3.9 *and* 3.10 *can be improved to*

$$\sup_{y_p \in (\theta, \Theta)} ||F^{\mathbf{S}} - F_{y_p}|| = O_p(n^{-1/4})$$

and

$$\sup_{y_p \in (1-\varepsilon, 1+\varepsilon)} ||F^{\mathbf{S}} - F_{y_p}|| = O_p(n^{-5/48}).$$

4. Circular Law - Non-Hermitian Matrices

In this section, we consider a kind of non-Hermitian matrix. Let $\mathbf{Q} = n^{-1/2}(x_{jk})$ be an $n \times n$ complex matrix with i.i.d. entries x_{jk} of mean zero and variance 1. The eigenvalues of $\mathbf{Q}$ are complex and thus the ESD of $\mathbf{Q}$, denoted by $F_n(x, y)$, is defined in the complex plane. Since the early 1950's, it has been conjectured that $F_n(x, y)$ tends to the uniform distribution over the unit disc in the complex plane, called the circular law. The major difficulty is that the major tools introduced in the previous two sections do not apply to non-Hermitian matrices.

Ginibre (1965) found the density of the eigenvalues of a matrix of i.i.d. complex $N(0, 1)$ entries to be

$$c \prod_{j \neq k} |\lambda_j - \lambda_k|^2 \exp\{-\frac{1}{2} \sum_{k=1}^{n} |\lambda_k|^2\}.$$

Based on this result, Mehta (1991) proved the circular law when the entries are i.i.d. complex normally distributed. Hwang (1986) reported that this result was also proved in an unpublished paper of Silverstein by the same approach.

Girko (1984a,b) presented a proof of the circular law under the condition that the entries have bounded densities on the complex plane and finite $(4 + \varepsilon)$th moments. Since they were published, many have tried to understand his mathematical arguments without success. The problem was considered open until Bai (1997) proved the following.

Theorem 4.1. *Suppose that the entries have finite* $(4 + \varepsilon)$*th moments, and that the joint distribution of the real and imaginary parts of the entries, or the conditional distribution of the real part given the imaginary part, has a uniformly bounded density. Then the circular law holds.*

Remark 4.1. The second part of Theorem 4.1 covers real random matrices. In this case, the joint distribution of the real and imaginary parts of the entries does not have a density in the complex plane. However, when the entries are real and have a bounded density, the real and imaginary parts are independent and hence the condition in the second part of Theorem 4.1 is satisfied. By considering the matrix $e^{i\theta}\mathbf{X}$, we can extend the density condition in the second part of Theorem 4.1 to: *the conditional density of* $\mathrm{Re}(x_{jk})\cos(\theta) - \mathrm{Im}(x_{jk})\sin(\theta)$ *given* $\mathrm{Re}(x_{jk})\sin(\theta) + \mathrm{Im}(x_{jk})\cos(\theta)$ *is bounded.*

Although Girko's arguments are hard to understand, or even deficient, he provided the following idea. Let $F_n(x, y)$ denote the ESD of $n^{-1/2}\mathbf{X}$, and $\nu_n(x, z)$

denote the ESD of the Hermitian matrix $\mathbf{H} = \mathbf{H}_n(z) = (n^{-1/2}\mathbf{X} - z\mathbf{I})(n^{-1/2}\mathbf{X} - z\mathbf{I})^*$ for given $z = s + it$.

Lemma 4.2. (Girko) *For $uv \neq 0$,*

$$\int\int e^{iux+iyv} F_n(dx, dy) = \frac{u^2+v^2}{4\pi iu} \int\int e^{isu+ivt} g_n(s,t) dtds, \tag{4.1}$$

and

$$\int\int e^{iux+iyv} F_{\text{cir}}(dx, dy) = \frac{u^2+v^2}{4\pi iu} \int\int e^{isu+ivt} g(s,t) dtds, \tag{4.2}$$

where

$$g_n(s,t) = \frac{\partial}{\partial s} \int_0^\infty \log x \nu_n(dx, z),$$

F_{cir} is the uniform distribution over the unit disc in the complex plane, and $g(s,t) = 2s$ or $2s/|z|^2$ in accordance with $|z| < 1$ or not.

Making use of the formula that for all $uv \neq 0$,

$$\frac{u^2+v^2}{2iu\pi} \int \left[\int \frac{s}{s^2+t^2} e^{ius+ivt} dt\right] ds = 1,$$

we obtain

$$\begin{aligned}
&\int\int e^{iux+ivy} F_n(dx, dy) \\
&= \frac{u^2+v^2}{2iu\pi} \int\int \frac{1}{n} \sum_{k=1}^n \frac{s}{s^2+t^2} e^{ius+ivt+iu\text{Re}(\lambda_k)+iv\text{Im}(\lambda_k)} dtds \\
&= \frac{u^2+v^2}{4iu\pi} \int\int \frac{1}{n} \sum_{k=1}^n \frac{2(s-\text{Re}(\lambda_k))}{(s-\text{Re}(\lambda_k))^2 + (t-\text{Im}(\lambda_k))^2} e^{ius+ivt} dtds \\
&= \frac{u^2+v^2}{4iu\pi} \int\int \frac{\partial}{\partial s} \frac{1}{n} \sum_{k=1}^n \log(|z-\lambda_k|^2) e^{ius+ivt} dtds \\
&= \frac{u^2+v^2}{4iu\pi} \int\int \left[\frac{\partial}{\partial s} \int_0^\infty \log x \nu_n(dx, z)\right] e^{ius+ivt} dtds.
\end{aligned} \tag{4.3}$$

Here, we have used the fact that $\prod_{k=1}^n |z-\lambda_k|^2 = \det(\mathbf{H})$. The proof of the first assertion of Lemma 4.2 is complete. The second assertion follows from the Green Formula.

Under the condition that the entries have finite $(4+\varepsilon)$th moments, it can be shown that, as mentioned in Subsection 2.2.2, the upper limit of the maximum absolute value of the eigenvalues of $n^{-1/2}\mathbf{X}$ is less than the maximum singular value, which tends to 2. Thus the distribution family $\{F_n(x,y)\}$ is tight. Hence going along some subsequence of integers, F_n and $\nu_n(x,z)$ tend to limits μ and ν respectively. It seems the circular law follows by making limit in (4.1) and getting (4.2) with μ and ν substituting F_{cir} and the ν defined by the circular law. However, there is no justifcation for passing the limit procedure $\nu_n \to \nu$ through the 3-fold integration since the outside integral range in (4.3) is the whole plane and the integrand of the inner integral is unbounded. To overcome the first difficulty, we need to reduce the integral range.

Let $T = \{z;\ |s| < A,\ |t| < A^2, |1 - |z|| > \varepsilon\}$.

Lemma 4.3. *For any $A > 0$ and $\varepsilon > 0$, with probability* 1,

$$\left| \int\int_{T^c} e^{isu+ivt} g_n(s,t) dtds \right| = O(A^{-1} + \varepsilon) \quad \textit{uniformly in } n.$$

The same is true if g_n is replaced by g, where g is defined in Lemma 4.2.

By the lemma and integration by parts, the problem is reduced to showing that

$$\int\int_T \left| \int_0^\infty \log x (\nu_n(dx, z) - \nu(dx, z)) \right| dtds \to 0 \quad a.s. \tag{4.4}$$

Since $z \in T$ and the norm of $\mathbf{Q}$ is bounded with probability 1, the support of $\nu_n(x, z)$ is bounded above by, say, M. Therefore, it is not a problem when dealing with the upper limit of the inner integral. However, since $\log x$ is not bounded at zero, (4.4) could not follow from $\nu_n \to \nu$. To overcome this difficulty, we estimate the convergence rate of $\nu_n - \nu$ and prove the following lemma.

Lemma 4.4. *Under the conditions of Theorem* 4.1, *we have*

$$\sup_{z \in T} ||\nu_n(\cdot, z) - \nu(\cdot, z)|| = o(n^{-\beta}), \quad a.s.,$$

where $\beta > 0$ depends on ε (in the moment condition) only.

Let $\varepsilon_n = e^{-n^\beta}$. Then by Lemma 4.4,

$$\sup_{z \in T} \left| \int_{\varepsilon_n}^\infty \log x (\nu_n(dx, z) - \nu(dx, z)) \right| \le n^\beta M \sup_{z \in T} ||\nu_n(\cdot, z) - \nu(\cdot, z)|| = o(1), \text{a.s.}$$

It remains to show that

$$\int\int_T \int_0^{\varepsilon_n} \log x \nu_n(dx, z) dtds \to 0 \quad \text{a.s.} \tag{4.5}$$

$$\int\int_T \int_0^{\varepsilon_n} \log x \nu(dx, z) dtds \to 0 \quad \text{a.s.}$$

The most difficult part is the proof of (4.5). For details, see Bai (1997).

5. Applications

In this section, we introduce some recent applications in multivariate statistical inference and signal processing. The examples discussed reveal that when the dimension of the data or parameters to be estimated is "very high", it causes non-negligible errors in many traditional multivariate statistical methods. Here, "very high" does not mean "incredibly" high, but "fairly" high. As simulation results for problems in the following sub-sections show (see cited papers), when the ratio of the degrees of freedom to dimension is less than 5, the non-exact test significantly beats the traditional T^2 in a two-sample problem (see Bai and Saranadasa (1996) for details); in the detection of the number of signals in a multivariate signal processing problem, when the number of sensors is greater than 10, the traditional MUSIC (MUltivariate SIgnal Classification) approach performs poorly, even when the sample size is as large as 1000. Such a phenomenon has been found in many different areas. In a normality test, say, the simplified W'-test beats Shapiro's W-test for most popular alternatives, although the latter is constructed by the Markov-Gaussian method, seemingly more reasonable than the usual least squares method. I was also told that when the number of regression coefficients in a multivariate regression problem is more than 6, the estimation becomes worse, and that when the number of parameters in a structured covariance matrix is more than 4, the estimates have serious errors. In applied time series analysis, models with orders greater than 6 (p in AR model, q in MA and $p+q$ in ARMA) are seldom considered. All these tell us that one has to be careful when dealing with high-dimensional data or a large number of parameters.

5.1. Non-exact test for the two-sample problem

Suppose that $\mathbf{x}_1, \ldots, \mathbf{x}_{n_1}$ and $\mathbf{y}_1, \ldots, \mathbf{y}_{n_2}$ are random samples from two populations with mean vectors $\boldsymbol{\mu}_1$ and $\boldsymbol{\mu}_2$, and a common covariance matrix Σ. Our problem is to test the hypothesis $H: \boldsymbol{\mu}_1 = \boldsymbol{\mu}_2$ against $K: \boldsymbol{\mu}_1 \neq \boldsymbol{\mu}_2$. The classical approach uses the Hotelling test (or T^2-test), with

$$T^2 = \frac{n_1 n_2}{n_1 + n_2}(\overline{\mathbf{x}} - \overline{\mathbf{y}})'\mathbf{A}^{-1}(\overline{\mathbf{x}} - \overline{\mathbf{y}}),$$

$\overline{\mathbf{x}} = \frac{1}{n_1}\sum_{i=1}^{n_1}\mathbf{x}_i$, $\overline{\mathbf{y}} = \frac{1}{n_2}\sum_{i=1}^{n_2}\mathbf{y}_i$ and

$$\mathbf{A} = \frac{1}{n_1 + n_2 - 2}\Big(\sum_{i=1}^{n_1}(\mathbf{x}_i - \overline{\mathbf{x}})(\mathbf{x}_i - \overline{\mathbf{x}})' + \sum_{i=1}^{n_2}(\mathbf{y}_i - \overline{\mathbf{y}})(\mathbf{y}_i - \overline{\mathbf{y}})'\Big).$$

The T^2 test has lots of good properties, but it is not well defined when the degrees of freedom $(n_1 + n_2 - 2)$ is less than the dimension (p) of the data.

As a remedy, Dempster (1959) proposed the so-called *non-exact test* (NET) by using the chi-square approximation technique. In recent research of Bai and Saranadasa (1996), it is found that Dempster's NET is also much more powerful than the T^2 test in many general situations when T^2 is well defined. One difficulty in computing Dempster's test statistic is the construction of a high dimensional orthogonal matrix and the other is the estimation of the degrees of freedom of the chi-square approximation. Bai and Saranadasa (1996) proposed a new test, the asymptotic normal test (ANT), in which the test statistic is based on $\|\overline{\mathbf{x}} - \overline{\mathbf{y}}\|^2$, normalized by consistent estimators of its mean and variance. It is known that ANT is asymptotically equivalent to NET, and simulations show that ANT is slightly more powerful than NET. It is easy to show that the type I errors for both NET and ANT tend to the prechosen level of the test. Simulation results show that NET and ANT gain a great amount of power with a slight loss of the exactness of the type I error. Note that *non-exact* does not mean that the error is larger.

Now, let us analyze why this happens. Under the normality assumption, if Σ were known, then the "most powerful test statistic" should be $(\overline{\mathbf{x}}-\overline{\mathbf{y}})'\Sigma^{-1}(\overline{\mathbf{x}}-\overline{\mathbf{y}})$. Since Σ is actually unknown, the matrix $\mathbf{A}$ plays the role of an estimator of Σ. Then there is the problem of how close $\mathbf{A}^{-1}$ is to Σ^{-1}. The matrix $\mathbf{A}^{-1}$ can be rewritten in the form $\Sigma^{-1/2}\mathbf{S}^{-1}\Sigma^{-1/2}$, where $\mathbf{S}$ is defined in Subsection 2.1.2, with $n = n_1 + n_2 - 2$. The approximation is good if $\mathbf{S}^{-1}$ is close to $\mathbf{I}$. Unfortunately, this is not really the case. For example, when $p/n = 0.25$, the ratio of the largest eigenvalue of $\mathbf{S}^{-1}$ to the smallest can be as large as 9. Even when p/n is as small as 0.01, the ratio can be as large as 1.493. This shows that it is practically impossible to get a "good" estimate of the inverse covariance matrix. In other words, if the ratio of the largest to the smallest eigenvalues of the population covariance matrix is not larger than $(\sqrt{n}+\sqrt{p})^2/(\sqrt{n}-\sqrt{p})^2$ (e.g. 9 for $p/n = 0.25$ and 1.493 for $p/n = 0.01$), NET or ANT give a better test than T^2.

A similar but simpler case is the one-sample problem. As in Bai and Saranadasa (1996), it can be shown that NET and ANT are better than the T^2 test. This phenomenon happens in many statistical inference problems, such as large contingency tables, MANOVA, discretized density estimation, linear models with large number of parameters and the Error in Variable Models. Once the dimension of the parameter is large, the performance of the classical estimators become poor and corrections may be needed.

5.2. Multivariate discrimination analysis

Suppose that $\mathbf{x}$ is a sample drawn from one of two populations with mean vectors $\boldsymbol{\mu}_1$ and $\boldsymbol{\mu}_2$ and a common covariance matrix Σ. Our problem is to classify the present sample $\mathbf{x}$ into one of the two populations. If $\boldsymbol{\mu}_1$ and $\boldsymbol{\mu}_2$ and Σ are known, then the best discriminant function is $d = (\mathbf{x}-\frac{1}{2}(\boldsymbol{\mu}_1+\boldsymbol{\mu}_2))'\Sigma^{-1}(\boldsymbol{\mu}_1-\boldsymbol{\mu}_2)$, i.e., assign $\mathbf{x}$ to Population 1 if $d > 0$.

When both the mean vectors and the covariance matrix are unknown, assume training samples $\mathbf{x}_1, \ldots, \mathbf{x}_{n_1}$ and $\mathbf{y}_1, \ldots, \mathbf{y}_{n_2}$ from the two populations are available. Then we can substitute the MLE $\overline{\mathbf{x}}$, $\overline{\mathbf{y}}$ and $\mathbf{A}$ of the mean vectors and covariance matrix into the discriminant function. Obviously, this is impossible if $n = n_1 + n_2 - 2 < p$. The problem is again whether this criterion has the smallest misclassification probability when p is large. If not, what discrimination criterion is better. Based on the same discussion in the last subsection, one may guess that the criterion $d = (\mathbf{x} - \frac{1}{2}(\overline{\mathbf{x}} + \overline{\mathbf{y}}))'(\overline{\mathbf{x}} - \overline{\mathbf{y}})$ should be better. Using the LSD of a large sample covariance matrix, this was theoretically proved in Saranadasa (1993). Simulation results presented in his paper strongly support the theoretical results, even for moderate n and p.

5.3. Detection of the number of signals

Consider the model

$$\mathbf{y}_j = \mathbf{A}\mathbf{s}_j + \mathbf{n}_j, \quad j = 1, \ldots, N,$$

where $\mathbf{y}_j$ is a $p \times 1$ complex vector of observations collected from p sensors, $\mathbf{s}_j$ a $q \times 1$ complex vector of unobservable signals emitted from q targets, $\mathbf{A}$ is an unknown $p \times q$ matrix whose columns are called the distance-direction vectors and $\mathbf{n}_j$ represents the noise generated by the sensors, usually assumed to be white. Usually, in detecting the number of signals (for non-coherent models), $\mathbf{A}$ is assumed to be of full rank and the number q is assumed to be less than p. In the estimation of DOA (Direction Of Arrivals), the (j,k)th element of $\mathbf{A}$ is assumed to be $r_k \exp(-2\pi d(j-1)\omega_0 \sin(\theta_k)/c)$, where r_k is the complex amplitude determined by the distance from the kth target to the jth sensor, d the spatial distance between adjacent sensors, ω_0 the central frequency, c the speed of light and θ_k the angle between the line of the sensors and the line from the jth sensor to the kth target, called the DOA. The most important problems are the detection of the number q of signals and the estimation of the DOA. In this section, we only consider the detection of the number of signals.

All techniques for solving the problem are based on the following:

$$\Sigma_{\mathbf{y}} = \mathbf{A}\mathbf{\Psi}\mathbf{A}^* + \sigma^2\mathbf{I},$$

where $\mathbf{\Psi}$ $(q \times q)$ is the covariance matrix of the signals. Denote the eigenvalues of $\Sigma_{\mathbf{y}}$ by $\lambda_1 \geq \cdots \geq \lambda_q > \lambda_{q+1} = \cdots = \lambda_p = \sigma^2$. This means that the multiplicity of the smallest eigenvalues σ^2 is $p - q$ and there is a gap between λ_q and λ_{q+1}.

Since the signals and noise have zero means, one can use $\widehat{\Sigma}_N = \frac{1}{N}\sum_{j=1}^N \mathbf{y}_j\mathbf{y}_j^*$ as an estimator of $\Sigma_{\mathbf{y}}$, and then compare a few of the smallest eigenvalues of $\widehat{\Sigma}_N$ to estimate the number of signals q. In the literature, AIC, BIC and GIC criteria are used to detect the number of signals. However, when p is large, the problem is then how big the gap between the qth and $(q+1)$-st largest eigenvalues of $\widehat{\Sigma}_N$ should be, so that q can be correctly detected by these criteria. Simulations in the literature usually take q to be 2 or 3 and p to be 4 or 5. Once $p = 10$ and $SNR = 0db$ (SNR (Signal to noise ratio) is defined as ten times the logarithm of the ratio of the variance of the signal (the kth component of $\mathbf{As}_1$) to the variance of the noise (the kth component of $\mathbf{n}_1$)), no criterion works well unless N is larger than 1000 (i.e. $y \sim 0.01$). Unreasonably, if we drop half of the data (i.e., reduce p to 5), the simulation results become good even for $n = 300$ or 400.

From the theory of LDRM, in the support of the LSD of $\widehat{\Sigma}_N$, there may be a gap at the $(1 - q/p)$th quantile or the gap may disappear, which depends on the original gap and the ratio $y = \lim p/N$ in a complicated manner. Some work was done in Silverstein and Combettes (1992). Their simulation results show that when the gap exists in the support of the LSD, the exact number q (not only the ratio q/p) can be exactly estimated for all large N. More precisely, suppose $p = p_N$ and $q = q_N$ tend to ∞ proportionally to N, then $\mathrm{P}(\hat{q}_N \neq q, i.o.) = 0$. Work on this is being done by Silverstein and myself (see Bai and Silverstein (1998)).

6. Unsolved Problems

6.1. Limiting spectral distributions

6.1.1. Existence of the LSD

Nothing is known about the existence of the LSD's of the following three matrices

$$\begin{pmatrix} x_1 & x_2 & \cdots & x_n \\ x_2 & x_1 & \ddots & x_{n-1} \\ \vdots & \ddots & \ddots & \vdots \\ x_n & x_{n-1} & \cdots & x_1 \end{pmatrix}, \quad \begin{pmatrix} x_1 & x_2 & \cdots & x_n \\ x_2 & x_3 & \cdots & x_{n+1} \\ \vdots & \vdots & \vdots & \vdots \\ x_n & x_{n+1} & \cdots & x_{2n-1} \end{pmatrix}$$

and

$$\begin{pmatrix} -\sum_{i=2}^{n} x_{1i} & x_{12} & \cdots & x_{1n} \\ x_{21} & -\sum_{i=1,\neq 2}^{n} x_{2i} & \cdots & x_{2n} \\ \vdots & \vdots & \vdots & \vdots \\ x_{n1} & x_{n2} & \cdots & -\sum_{i=1}^{n-1} x_{ni} \end{pmatrix},$$

where, in the first two matrices, x_j's are i.i.d. real random variables and in the third matrix, $x_{jk} = x_{kj}$, $j < k$, are i.i.d. real random variables. Consider the three matrices as limiting distributions of the form $\sqrt{n}(\mathbf{A}_n - \mathbf{A})$: the first is for the autocovariance matrix in time series analysis, the second is for the information matrix in a polynomial regression model and the third is for the derivative of a transition matrix in a Markov process.

6.1.2. Explicit forms of LSD

The only known explicit forms of densities of LSD's of LDRM are those of the semi-circular law, the circular law, the Marčenko-Pastur law and the Multivariate F-matrix. As shown in Theorem 3.4, there are a large class of random matrices whose LSD's exist but for which no explicit forms are known. It is of interest to find more explicit forms of the densities of the LSD's.

6.2. Limits of extreme eigenvalues

These are known for Wigner and sample covariance matrices. Nothing is known for multivariate F matrices.

As mentioned in Section 5.3, it is very interesting that there are no eigenvalues at all in the gap of the support (this is called the separation problem) of the LSD. More precisely, suppose that $q_N/N \to z$ and $p_N/N \to y$ with $0 < z < y < 1$, and $\lambda_{q_N} \to c$, $\lambda_{q_N+1} \to d$ with $d < c$. Under certain conditions, we conjecture that $\ell_{q_N} \to g_{z,y}(c)$ and $\ell_{q_N+1} \to g_{z,y}(d)$, where λ_{q_N}, λ_{q_N+1}, ℓ_{q_N} and ℓ_{q_N+1} are the q_Nth and (q_N+1)-st largest eigenvalues of Σ and $\mathbf{S}\Sigma$, respectively, and $g_{z,y}(c) > g_{z,y}(d)$ are the upper and lower bounds of the $(1-z)$-quantile of the LSD of $\mathbf{S}\Sigma$.

Remark 6.1. After this paper was written, the above mentioned problem has been partially solved in Bai and Silverstein (1998). For details, see Silverstein's discussion following this paper.

6.3. Convergence rates of spectral distributions

The only known results are introduced in Subsection 3.2. For Wigner and sample covariance matrices, some convergence rates of ESD's are given in Bai (1993a,b), Bai, Miao and Tsay (1996a,b, 1997) and the present paper. Of more interest is the rates of a.s. or in probability convergence. It is also of interest is to find the ideal convergence rates (the conjectured rates are of the order $O(1/n)$ or at least $O(1/\sqrt{n})$). Furthermore, nothing is known about other matrices.

6.4. Second order convergence

6.4.1. Second order convergence of spectral distributions

Of course, the convergence rates should be determined first. Suppose that the exact rate is found to be α_n. It is reasonable to conjecture that $\alpha_n^{-1}(F_n(x)-F(x))$ should tend to a limiting stochastic process. Based on this, it may be possible to find limiting distributions of statistics which are functionals of the ESD. Then statistical inference, such as testing of hypothesis and confidence intervals, can be performed.

6.4.2. Second order convergence of extreme eigenvalues

In Subsection 2.2, limits of extreme eigenvalues of some random matrices are presented. As mentioned in the last subsubsection, it is important to find the limiting distribution of $\alpha_n^{-1}(\ell_{\text{extr}} - \lambda_{\text{lim}})$, where ℓ_{extr} is the extreme eigenvalue and λ_{lim} is the limit of ℓ_{extr}. The normalizing constant α_n may be the same as, or different from, that for the corresponding ESD's. For example, for Wigner and sample covariance matrices with $y \neq 1$, the conjectured α_n is $\frac{1}{n}$, but for sample covariance matrices with $p = n$, the conjectured normalizing constant for the smallest eigenvalue of $\mathbf{S}$ is $1/n^2$. The smallest eigenvalue when $p = n$ is related to the condition number (the square-root of the ratio of the largest to the smallest eigenvalues of $\mathbf{S}$), important in numerical computation of linear equations. Reference is made to Edelman (1992).

6.4.3. Second order convergence of eigenvectors

Some results on the eigenvectors of large-dimensional sample covariance matrices were established in the literature and introduced in Subsection 2.3. A straightforward problem is to extend these results to other kinds of random matrices. Another problem is whether there are other ways to describe the similarity between the eigenmatrix and Haar measure.

6.5. Circular law

The conjectured condition for guaranteeing the circular law is finite second moment only, at least for the i.i.d. case. In addition to the difficulty of estimating (4.5), there are no similar results to Lemmas 2.2, 2.3, 2.6 and 2.7, so we cannot truncate the variable at $\sqrt{n}$ under the existence of the second moment of the underlying distributions.

Acknowledgement

The author would like to thank Professor J. W. Silverstein for indicating that the eigenvalues of the matrix with elements $i, -i$ and 0 above, below and on the diagonal are given by $\cot(\pi(2k-1)/2n)$, $k = 1, \ldots, n$. This knowledge plays a key role in dealing with the expected imaginary parts of the entries of a Wigner matrix in Theorems 2.1 and 2.12.

References

Arnold, L. (1967). On the asymptotic distribution of the eigenvalues of random matrices. *J. Math. Anal. Appl.* **20**, 262-268.

Arnold, L. (1971). On Wigner's semicircle law for the eigenvalues of random matrices. *Z. Wahrsch. Verw. Gebiete* **19**, 191-198.

Bai, Z. D. (1993a). Convergence rate of expected spectral distributions of large random matrices. Part I. Wigner Matrices. *Ann. Probab.* **21**, 625-648.

Bai, Z. D. (1993b). Convergence rate of expected spectral distributions of large random matrices. Part II. Sample Covariance Matrices. *Ann. Probab.* **21**, 649-672.

Bai, Z. D. (1997). Circular law. *Ann. Probab.* **25**, 494-529.

Bai, Z. D., Miao, B. Q. and Tsay, J. (1996a). Convergence rates of the spectral distributions of large Wigner matrices. Submitted.

Bai, Z. D., Miao, B. Q. and Tsay, J. (1996b). Remarks on the convergence rate of the spectral distributions of Wigner matrices. Submitted.

Bai, Z. D., Miao, B. Q. and Tsay, J. (1997). A note on the convergence rate of the spectral distributions of large random matrices. *Statist. Probab. Lett.* **34**, 95-102.

Bai, Z. D. and Saranadasa, H. (1996). Effect of high dimension: by an example of a two sample problem. *Statist. Sinica* **6**, 311-329.

Bai, Z. D. and Silverstein, J. W. (1998). No eigenvalues outside the support of the limiting spectral distribution of large dimensional sample covariance matrices. *Ann. Probab.* **26**, 316-345.

Bai, Z. D., Silverstein, Jack W. and Yin, Y. Q. (1988). A note on the largest eigenvalue of a large dimensional sample covariance matrix. *J. Multivariate Anal.* **26**, 166-168.

Bai, Z. D. and Yin, Y. Q. (1986). Limiting behavior of the norm of products of random matrices and two problems of Geman-Hwang. *Probab. Theory Related Fields* **73**, 555-569.

Bai, Z. D. and Yin, Y. Q. (1988a). Convergence to the semicircle law. *Ann. Probab.* **16**, 863-875.

Bai, Z. D. and Yin, Y. Q. (1988b). Necessary and sufficient conditions for the almost sure convergence of the largest eigenvalue of a Wigner matrix. *Ann. Probab.* **16**, 1729-1741.

Bai, Z. D. and Yin, Y. Q. (1993). Limit of the smallest eigenvalue of a large dimensional sample covariance matrix. *Ann. Probab.* **21**, 1275-1294.

Bai, Z. D., Yin, Y. Q. and Krishnaiah, P. R. (1986). On LSD' of product of two random matrices when the underlying distribution is isotropic. *J. Multivariate Anal.* **19**, 189-200.

Bai, Z. D., Yin, Y. Q. and Krishnaiah, P. R. (1987). On the limiting empirical distribution function of the eigenvalues of a multivariate F matrix. *Theory Probab. Appl.* **32**, 490-500.

Edelman, A. (1992). On the distribution of a scaled condition number. *Math. Comp.* **58**, 185-190.

Edelman, A. (1997). The circular law and the probability that a random matrix has k real eigenvalues. *J. Multivariate Anal.* **60**, 188-202.

Geman, S. (1980). A limit theorem for the norm of random matrices. *Ann. Probab.* **8**, 252-261.

Geman, S. (1986). The spectral radius of large random matrices. *Ann. Probab.* **14**, 1318-1328.

Ginibre, J (1965). Statistical ensembles of complex, quaterion and real matrices. *J. Math. Phys.* **6**, 440-449.

Girko, V. L. (1984a). Circle law. *Theory Probab. Appl.* **4**, 694-706.

Girko, V. L. (1984b). On the circle law. *Theory Probab. Math. Statist.* **28**, 15-23.

Girko, V. L. (1990). *Theory of Random Determinants.* Kluwer Academic Publishers, Dordrecht-Boston-London.

Girko, V., Kirsch, W. and Kutzelnigg, A. (1994). A necessary and sufficient condition for the semicircular law. *Random Oper. Stoch. Equ.* **2**, 195-202.

Gnedenko, B. V. and Kolmogorov, A. N. (1954). *Limit Distributions for Sums of Independent Random Variables.* Addison-Wesley, Reading.

Grenander, Ulf (1963). *Probabilities on Algebraic Structures.* John Wiley, New York-London.

Grenander, Ulf and Silverstein, Jack W. (1977). Spectral analysis of networks with random topologies. *SIAM J. Appl. Math.* **32**, 499-519.

Hoeffding, W. (1963). Probability inequalities for sums of bounded random variables. *J. Amer. Statist. Assoc.* **58**, 13-30.

Hwang, C. R. (1986). A brief survey on the spectral radius and the spectral distribution of large dimensional random matrices with i.i.d. entries. *Random Matrices and Their Applications*, Contemporary Mathematics **50**, 145-152, AMS, Providence.

Jonsson, D. (1982). Some limit theorems for the eigenvalues of a sample covariance matrix. *J. Multivariate Anal.* **12**, 1-38.

Loève, M. (1977). *Probability Theory.* 4th edition. Springer-Verlag, New York.

Marčenko, V. A. and Pastur, L. A. (1967). Distribution for some sets of random matrices. *Math. USSR-Sb.* **1**, 457-483.

Mehta, M. L. (1991). *Random Matrices.* Academic Press, New York.

Pastur, L. A. (1972). On the spectrum of random matrices. *Teoret. Mat. Fiz.* **10**, 102-112, (Teoret. Mat. Phys. **10**, 67-74).

Pastur, L. A. (1973). Spectra of random self-adjoint operators. *Uspekhi Mat. Nauk.* **28**, 4-63, (Russian Math. Surveys **28**, 1-67).

Prohorov, Ju. V. (1968). The extension of S. N. Bernstein's inequalities to a multi-dimensional case. (Russian) *Teor. Verojatnost. i Primenen.* **13**, 266-274.

Rao, C. R. (1976). *Linear Statistical Inference and Its Applications.* 2nd edition. John Wiley, New York.

Saranadasa, H. (1993). Asymptotic expansion of the misclassification probabilities of D- and A-criteria for discrimination from two high dimensional populations using the theory of large dimensional random matrices *J. Multivariate Anal.* **46**, 154-174.

Silverstein, J. W. (1979). On the randomness of eigenvectors generated from networks with random topologies. *SIAM J. Appl. Math.* **37**, 235-245.

Silverstein, J. W. (1981). Describing the behavior of eigenvectors of random matrices using sequences of measures on orthogonal groups. *SIAM J. Math. Anal.* **12**, 174-281.

Silverstein, J. W. (1984a). Comments on a result of Yin, Bai and Krishnaiah for large dimensional multivariate F matrices. *J. Multivariate Anal.* **15**, 408-409.

Silverstein, J. W. (1984b). Some limit theorems on the eigenvectors of large dimensional sample covariance matrices. *J. Multivariate Anal.* **15**, 295-324.

Silverstein, J. W. (1985a). The limiting eigenvalue distribution of a multivariate F matrix. *SIAM J. Appl. Math.* **16**, 641-646.

Silverstein, J. W. (1985b). The smallest eigenvalue of a large dimensional Wishart Matrix. *Ann. Probab.* **13**, 1364-1368.

Silverstein, J. W. (1989a). On the eigenvectors of large dimensional sample covariance matrices *J. Multivariate Anal.* **30**, 1-16.

Silverstein, J. W.(1989b). On the weak limit of the largest eigenvalue of a large dimensional sample covariance matrix *J. Multivariate Anal.* **30**, 307-311.

Silverstein, J. W. (1990). Weak convergence of random functions defined by the eigenvectors of sample covariance matrices. *Ann. Probab.* **18**, 1174-1194.

Silverstein, J. W. (1995). Strong convergence of the empirical distribution of eigenvalues of large dimensional random matrices. *J. Multivariate Anal.* **55**, 331-339.

Silverstein J. W. and Bai, Z. D. (1995). On the empirical distribution of eigenvalues of a class of large dimensional random matrices. *J. Multivariate Anal.* **54**, 175-192.

Silverstein, W. J. and Choi, S. I. (1995). Analysis of the limiting spectral distribution of large dimensional random matrices. *J. Multivariate Anal.* **54**, 295-309.

Silverstein, J. W. and Combettes, P. L. (1992). Signal detection via spectral theory of large dimensional random matrices. *IEEE ASSP* **40**, 2100-2104.

Wachter, K. W. (1978). The strong limits of random matrix spectra for sample matrices of independent elements. *Ann. Probab.* **6**, 1-18.
Wachter, K. W. (1980). The limiting empirical measure of multiple discriminant ratios. *Ann. Statist.* **8**, 937-957.
Wigner, E. P. (1955). Characteristic vectors bordered matrices with infinite dimensions. *Ann. Math.* **62**, 548-564.
Wigner, E. P. (1958). On the distributions of the roots of certain symmetric matrices. *Ann. Math.* **67**, 325-327.
Yin, Y. Q. (1986). LSD' for a class of random matrices. *J. Multivariate Anal.* **20**, 50-68.
Yin, Y. Q., Bai, Z. D. and Krishnaiah, P. R. (1983). Limiting behavior of the eigenvalues of a multivariate F matrix *J. Multivariate Anal.* **13**, 508-516.
Yin, Y. Q., Bai, Z. D. and Krishnaiah, P. R. (1988). On the limit of the largest eigenvalue of the large dimensional sample covariance matrix. *Probab. Theory Related Fields* **78**, 509-521.
Yin, Y. Q. and Krishnaiah, P. R. (1983). A limit theorem for the eigenvalues of product of two random matrices. *J. Multivariate Anal.* **13**, 489-507.
Yin, Y. Q. and Krishnaiah, P. R. (1985). Limit theorem for the eigenvalues of the sample covariance matrix when the underlying distribution is isotropic. *Theory Probab. Appl.* **30**, 861-867.

Department of Statistics and Applied Probability, National University of Singapore, Singapore 119260.
E-mail: matbaizd@leonis.nus.edu.sg

(Received January 1996; accepted March 1999)

COMMENT: SPECTRAL ANALYSIS OF RANDOM MATRICES USING THE REPLICA METHOD

G. J. Rodgers

Brunel University

Abstract: In this discussion paper, we give a brief review of the replica method applied to random matrices, and in particular to their spectral analysis. We illustrate the method by calculating the eigenvalue spectrum of the real random matrix ensemble describing the Hopfield model of autoassociative memory.

Key words and phrases: Random matrices, replica method, spectral analysis.

1. Introduction

In Bai (1999), the author reviews the theory of random matrices from the mathematical physics literature. In contrast to this rigorous analysis of spectral theory, there have been parallel, non-rigorous, developments in the theoretical physics literature. Here the replica method, and to a lesser extent super-symmetric methods, have been used to analyse the spectral properties of a variety of random matrices of interest to theoretical physicists. These matrices have applications in, for instance, random magnet theory, neural network theory and the conductor/insulator transition. In the present discussion we briefly review the work using the replica method. We then illustrate the use of this method by using it, for the first time, to obtain the spectral distribution of the sample covariance matrix. This problem is considered in Section 2.1.2 of Bai (1999) using a completely different approach.

The replica method was introduced by Edwards (1970) to study a polymer physics problem. It was first applied to a matrix model by Edwards and Jones (1976) who used it to obtain the Wigner semi-circular distribution for the spectrum of a random matrix with Gaussian distributed entries. Since then it was applied by Rodgers and Bray (1988) and Bray and Rodgers (1988) to obtain the spectral distribution of two different classes of sparse random matrices. Later, Sommers, Crisanti, Sompolinsky and Stein (1988) used an electrostatic method, which nevertheless relied on the replica method to demonstrate an assumption, to obtain the average eigenvalue distribution of random asymmetric matrices. Some of these approaches are analogous to the super-symmetric technique used on sparse random matrices by Rodgers and DeDominicis (1990) and Mirlin and Fyodorov (1991).

2. Illustration

We illustrate the replica method by using it to calculate the spectral distribution of the real version of the sample covariance matrix in Bai (1999, Section 2.1.2). The eigenvalue distribution of any $N \times N$ random matrix H_{jk} can be calculated by considering the generating function $Z(\mu)$ defined by

$$Z(\mu) = \int_{-\infty}^{\infty} \Big(\prod_j d\phi_j\Big) \exp\Big\{\frac{i}{2}\Big[\mu \sum_j \phi_j^2 - \sum_{jk} H_{jk}\phi_j\phi_k\Big]\Big\}, \tag{1}$$

where $\mu(x+i\varepsilon)$ implicitly contains a small positive imaginary part ε which ensures the convergence of the integrals. The integers j and k run from $1, \ldots, N$. The average normalised eigenvalue density is then given by

$$\rho(x) = \frac{2}{\pi N} \lim_{\varepsilon \to 0} \operatorname{Im} \frac{\partial}{\partial \mu} [\ln Z(\mu)]_{av}, \tag{2}$$

where $[]_{av}$ represents the average over the random variables in H_{jk}. We can connect this expression with Bai (1999) by observing that

$$\rho(x) = \frac{1}{\pi N} \lim_{\varepsilon \to 0} \operatorname{Im} \sum_{j=1}^{N} \frac{1}{\lambda_j - \mu} = \frac{1}{\pi} \lim_{\varepsilon \to 0} [\operatorname{Im} m_N(u)]_{av}, \tag{3}$$

where $m_N(\mu)$ is the Stieltjes defined in (3.1) of Bai (1999) and $\{\lambda_j, j = 1, N\}$ are the eignevalues of H_{jk}. The average in (2) is done using the replica method, which makes use of the indentity

$$[\ln Z]_{av} = \lim_{n\to 0} \frac{[Z^n]_{av} - 1}{n}. \tag{4}$$

In the right hand side of (4) the average is evaluated for integer n and then one must analytically continue to take the limit $n \to 0$. In random matrix problems this analytical continuation is straightforward, although in some physical problems, such as spin glasses, it can be more problematic. These problems occur in systems in which the phase space in the infinite system limit is partitioned so that the system is non-erogdic, see Mezard, Parisi and Virasoro (1988). This physical mechanism has no counterpart in studies of random matrices.

We will illustrate the replica method on the matrix

$$H_{jk} = \frac{1}{N} \sum_{v=1}^{p} \xi_j^v \xi_k^v, \tag{5}$$

where the real random variables $\{\xi_j^v\}$, $j = 1, \ldots, N$, $v = 1, \ldots, p$, are identically independently distributed with distribution $P(\xi_j^v)$, mean zero and variance σ^2. This matrix represents the patterns to be memorised in a neural network model of autoassociative memory, Hopfield (1982). It is also the real version of the sample covariance matrix studied in section 2.1.2 of Bai (1999). Here we have opted to study the real version because it is slightly simpler to analyse by the replica method and because the Hopfield model, which is the main application of this matrix, has real variables. To further connect with the theoretical physics literature, we have adopted the notation common within that field.

Introducing replica variables $\{\phi_{j\alpha}\}$, $j = 1, \ldots, N$ and $\alpha = 1, \ldots, n$, where n is an integer, allows us to write the average of the nth power of $Z(\mu)$ as

$$[Z^n]_{av} = \int_{-\infty}^{\infty} \Big[\prod_{j,\alpha} d\phi_{j\alpha}\Big]\Big[\prod_{j,v} P(\xi_j^v) d\xi_j^v\Big] \exp\{G\}, \tag{6}$$

where

$$G = \frac{i\mu}{2} \sum_{j,\alpha} \phi_{j\alpha}^2 - \frac{i}{2N} \sum_{\alpha,v} \Big(\sum_j \xi_j^v \phi_{j\alpha}\Big)^2. \tag{7}$$

We introduce the variables $\{x_{v\alpha}\}$, $v = 1, \ldots, p$ and $\alpha = 1, \ldots, n$ to linearise the second term in G using the Hubbard-Stratonovich transformation. This is just an integral generalisation of "completing the squares" such as

$$\exp\Big\{-\frac{i}{2N}\Big(\sum_j \xi_j^v \phi_{j\alpha}\Big)^2\Big\} = \sqrt{\frac{2}{\pi}} \int \exp\Big\{-\frac{1}{2} x_{v\alpha}^2 + \frac{1}{\sqrt{in}} x_{v\alpha} \sum_j \xi_j^v \phi_{j\alpha}\Big\} dx_{v\alpha}. \tag{8}$$

After repeatedly applying this transformation for all v and α we can intergrate over $\{\xi_j^v\}$ to obtain

$$[Z^n]_{av} = \int_{-\infty}^{\infty} \Big[\prod_{j,\alpha} d\phi_{j\alpha}\Big]\Big[\prod_{a,v} dx_{av}\Big] \exp\{\tilde{G}\}, \tag{9}$$

where

$$\tilde{G} = \frac{i\mu}{2}\sum_{j,\alpha}(\phi_{j\alpha})^2 - \frac{1}{2}\sum_{v,\alpha}(x_{va})^2 + \frac{1}{N}\sum_{j,v} f\Big(\sum_a x_{va}\phi_{j\alpha}\Big) \tag{10}$$

and $f(x) = -i\sigma^2x^2/2$. In order to illustrate the method we assume a general form for $f(x)$ for the time being so as to represent different types of randommess. We can expand (10) for a general $f(x)$ if $y_a = x_{v\alpha}\phi_{j\alpha}$ then

$$f\Big(\sum_a y_a\Big) = f(0) + \sum_{\alpha,r} b_r y_a^r + \sum_{\alpha<\beta,r,s} b_{rs} y_a^r y_\beta^s + \sum_{\alpha<\beta<y,r,s,t} b_{rst} y_\alpha^r y_\beta^s y_\gamma^t + \cdots \tag{11}$$

without loss of generality. (In our particular case of quadratic $f(x)$, the only non-zero terms are $b_2 = -i\sigma^2/2$ and $b_{11} = -i\sigma^2$.) This alows the third term in (10) to be rewritten as

$$\begin{aligned}\frac{1}{N}\sum_{j,v} f\Big(\sum_a x_{v\alpha}\phi_{j\alpha}\Big) = \frac{1}{N}\sum_{j,v} f(0) &+ \frac{1}{2N}\sum_{\alpha,r} b_r\Big[\Big(\sum_v x_{v\alpha}^r + \sum_j \phi_{j\alpha}^r\Big)^2 - \Big(\sum_v x_{va}^r\Big)^2 \\ &- \Big(\sum_j \phi_{j\alpha}^r\Big)^2\Big] + \frac{1}{2N}\sum_{\alpha<\beta,r,s} b_{rs}\Big[\Big(\sum_v x_{v\alpha}^r x_{v\beta}^s + \sum_j \phi_{j\alpha}^r\phi_{j\beta}^s\Big)^2 \\ &- \Big(\sum_v x_{v\alpha}^r x_{v\beta}^s\Big)^2 - \Big(\sum_j \phi_{j\alpha}^r\phi_{j\beta}^s\Big)^2\Big] + \cdots\end{aligned} \tag{12}$$

We now introduce conjugate variables to linearise these terms, again using the Hubbard-Stratonovich transformation. The variables and their conjugates are

$$\begin{array}{llllll} a_\alpha^{(r)} & \sum_v x_{v\alpha}^r + \sum_j \phi_{j\alpha}^r & b_\alpha^{(r)} & \sum_v x_{va}^r & c_a^{(r)} & \sum_j \phi_{j\alpha}^r \\ a_{\alpha\beta}^{(r,s)} & \sum_v x_{v\alpha}^r x_{v\beta}^s + \sum_j \phi_{j\alpha}^r\phi_{j\beta}^s, & b_{\alpha\beta}^{(r,s)} & \sum_v x_{va}^r x_{v\beta}^s \text{ and} & c_{\alpha\beta}^{(r)} & \sum_j \phi_{j\alpha}^r\phi_{j\beta}^s. \\ \vdots & \vdots & \vdots & \vdots & \vdots & \vdots \end{array} \tag{13}$$

Using these variables to linearise those in (12), then evaluating them by the method of steepest descents as p, $N \to \infty$, gives

$$\begin{array}{lll} a_\alpha^{(r)} = c\langle x_\alpha^r\rangle_2 + \langle\phi_\alpha^r\rangle_1 & b_a^{(r)} = ic\langle x_a^r\rangle_2 & c_a^{(r)} = i\langle\phi_\alpha^r\rangle_1 \\ a_{\alpha\beta}^{(r,s)} = c\langle x_\alpha^r x_\beta^s\rangle_2 + \langle\phi_a^r\phi_\beta^s\rangle_1, & b_{\alpha\beta}^{(r,s)} = ic\langle x_\alpha^r x_\beta^s\rangle_2 \text{ and} & c_{\alpha\beta}^{(r,s)} = i\langle\phi_\alpha^r\phi_\beta^s\rangle_1 \\ \vdots \quad \vdots \quad \vdots & \vdots \quad \vdots & \vdots \quad \vdots \end{array} \tag{14}$$

where $p/N \to c \in (0,\infty)$ in the limit $p \to \infty$ and $N \to \infty$. Using the notation $D\phi = \prod_\alpha d\phi_\alpha$, the brackets $\diamondsuit_1$ and $\diamondsuit_2$ are defined by

$$\langle h\{\phi_\alpha\}\rangle_1 = \frac{\int D\phi h\{\phi_\alpha\}\exp\left\{\frac{i\mu}{2}\sum_\alpha \phi_a^2 + g_1\{\phi_a\}\right\}}{\int D\phi \exp\left\{\frac{i\mu}{2}\sum_a \phi_\alpha^2 + g_1\{\phi_\alpha\}\right\}} \tag{15}$$

and

$$\langle h\{x_\alpha\}\rangle_2 = \frac{\int D_x h\{x_\alpha\}\exp\left\{-\frac{1}{2}\sum_\alpha x_a^2 + g_2\{x_\alpha\}\right\}}{\int Dx \exp\left\{-\frac{1}{2}\sum_a x_a^2 + g\{x_a\}\right\}}, \tag{16}$$

where

$$g_1\{\phi_a\} = cf(0) + c\langle f(\sum_a \phi_a x_a)\rangle_2 \tag{17}$$

and

$$g_2\{x_a\} = \langle f(\sum_a \phi_a x_a)\rangle_1. \tag{18}$$

We can rewrite our expression for the average normalised density of states as

$$\rho(x) = \lim_{n\to 0}\frac{1}{n\pi}\mathrm{Re}\langle\sum_\alpha(\phi_\alpha)^2\rangle_1. \tag{19}$$

Using the fact that $f(x) = -i\sigma^2x^2/2$ we can look for a self-consistent solution to (17) and (18) of the form $g_1\{\phi_\alpha\} = A\sum_\alpha(\phi_\alpha)^2$ and $g_2\{x_\alpha\} = B\sum_\alpha(x_\alpha)^2$. In this case $\rho(x)$ can be rewritten as $\rho(x) = \mathrm{Im}(A)/\pi\sigma^2$. Equations (17) and (18) can be solved self-consistently by performing the n-dimensional integrals as if n were a positive integer and then taking the limit $n \to 0$. This reveals expressions for A and B, and hence for $c > 1$,

$$\rho(x) = \begin{cases} \frac{1}{4\pi x\sigma^2}\sqrt{(b-x)(x-a)} & a \le x \le b \\ 0 & \text{otherwise} \end{cases} \tag{20}$$

with $a = 2\sigma^2(\sqrt{c}-1)^2$ and $b = 2\sigma^2(\sqrt{c}+1)^2$. This result is of the same form as Bai (1999, equation (2.12)), if we make the changes $c \to 1/y$ and $2c\sigma^2 \to \sigma^2$. These changes are caused by different definitions of the initial random matrices, and because we are treating the real version of the matrices whereas Bai (1999) considers the complex case.

3. Summary

We have shown how the replica method can be used to calculate the eigenvalue spectrum of real random matrices. It is also possible to use this method to analyse other problems discussed in Bai (1999). For instance, in Dhesi and Jones

(1990) there is an example of how to use a perturbative scheme with the replica method to find the corrections to the spectral distribution up to $O(1/N^2)$. In Weight (1998) the replica scheme is used to analyse the properties of products of random matrices. Thus the replica technique can be viewed as a useful addition to the analytical techniques presented in Bai (1999).

Department of Mathematics and Statistics, Brunel University, Uxbridge, Middlesex, UB8 3PH U.K.
E-mail: g.j.rodgers@brunel.ac.uk

COMMENT: COMPLEMENTS AND NEW DEVELOPMENTS

Jack W. Silverstein

North Carolina State University

My good friend and colleague has done a fine job in presenting the essential tools that have been used in understanding spectral behavior of various classes of large dimensional random matrices. The Stieltjes transform is by far the most important tool. As can be seen in the paper, some limit theorems are easier to prove using them and rates of convergence of the spectral distribution can be explored using Theorem 3.1. Moreover, as will be seen below, analysis of the Stieltjes transform of the limiting spectral distribution of matrices presented in Section 2.1.3 can explain much of the distribution's properties. Also, the conjecture raised in Section 6.2 has been proven using Stieltjes transforms.

However, this is not to say the moment method can be dispensed with. Indeed, there has been no alternative way of proving the behavior of the extreme eigenvalues. This paper shows further use of moments by proving Theorem 2.10 with no restriction on $\mathbf{T}$. An attempt to prove it in Silverstein (1995) without the assumption of positive definiteness was abandoned early on in the work. Another example will be seen below concerning the preliminary work done on the rate of convergence. Moments were used. In my opinion it would be nice to develop all random matrix spectral theory without relying on moment arguments. They reveal little of the underlying behavior, and the combinatorial arguments used are frequently horrendous. Unfortunately, it appears unlikely we can remove them from our toolbox.

The remaining comments are on the matrices appearing in Theorem 2.10 when $\mathbf{T}$ is non-negative definite. Their eigenvalues are the same as those of

$$B_p \equiv \frac{1}{n} T_p^{1/2} \mathbf{X}_p \mathbf{X}_p^* T_p^{1/2},$$

(note that at this stage it is necessary to change subscripts on the matrices) where $T_p^{1/2}$ is any Hermitian square root of $\mathbf{T}_p$, and differ from those of $\mathbf{B} = \mathbf{B}_p$ in Theorem 3.4 (with $\mathbf{A} = 0$) by $|p-n|$ zero eigenvalues. When the elements of $\mathbf{X}_p$ are standardized (mean zero and $\mathbf{E}(|X_{11}|^2) = 1$), B_p is (under the assumption of zero mean) the sample covariance matrix of n samples of the p-dimensional random vector $T_p^{1/2}\mathbf{X}_{\cdot 1}$, the population matrix being of course $\mathbf{T}_p$. This represents a broad class of random vectors which includes multivariate normal, resulting in Wishart matrices. Results on the spectral behavior of B_p are relevant in situations where p is high but sample size is not large enough to ensure sample and population eigenvalues are near each other, only large enough to be on the same order of magnitude as p. The following two sections provide additional information on what is known about the eigenvalues of B_p.

1. Understanding the Limiting Distribution Through Its Stieltjes Transform

For the following, let F denote the limiting spectral distribution of B_p with Stieltjes transform $m(z)$. Then it follows that F and $\mathbf{F}$, the limiting spectral distribution of $\mathbf{B}$, satisfy

$$\mathbf{F} = (1-y)I_{[0,\infty)} + yF$$

($I_{[0,\infty)}$ denoting the indicator function on $[0,\infty)$), while $m(z)$ and $\mathbf{m}(z)$, the Stieltjes transform of $\mathbf{F}$, satisfy

$$\mathbf{m}(z) = -\frac{(1-y)}{z} + ym(z).$$

From (3.9) we find that the inverse of $\mathbf{m} = \mathbf{m}(z)$ is known:

$$z = -\frac{1}{\mathbf{m}} + y\int \frac{\tau dH(\tau)}{1+\tau\mathbf{m}},$$

and from this it can be proven (see Silverstein and Choi (1995)):

1. On $\mathbb{R}^+$, F has a continuous derivative f given by $f(x) = (1/\pi)Im\, m(x) = (1/y\pi)\lim_{z\in\mathbb{C}^+\to x} Im\,\mathbf{m}(z)$ ($\mathbb{C}^+$ denoting the upper complex plane). The density $f(x)$ is analytic wherever it is positive, and for these x, $y\pi f(x)$ is the imaginary part of the unique $\mathbf{m} \in \mathbb{C}^+$ satisfying
$$x = -\frac{1}{\mathbf{m}} + y\int \frac{\tau dH(\tau)}{1+\tau\mathbf{m}}. \qquad (1)$$
2. Intervals outside the support of f are those on the vertical axis on the graph of (1), for $m \in \mathbb{R}$, corresponding to intervals where the graph is increasing (originally observed in Marčenko and Pastur (1967)). Thus, the graph of f can be obtained by first identifying intervals outside the support, and then applying Newton's method to (1) for values of x inside the support.

3. Let $a > 0$ be a boundary point in the support of f. If a is a relative extreme value of (1) (which is always the case whenever H is discrete), then near a and in the support of f, $f \sim \sqrt{|x-a|}$. More precisely, there exists a $C > 0$ such that

$$\lim_{x \in supp\, f \to a} \frac{f(x)}{C\sqrt{|x-a|}} = \lim_{x \in supp\, f \to a} \frac{\frac{d}{dx} f(x)}{\frac{d}{dx} C\sqrt{|x-a|}} = 1$$

4. y and F uniquely determine H.
5. $F \xrightarrow{\mathcal{D}} H$ as $y \to 0$, which complements the a.s. convergence of B_p to T_p for fixed p as $n \to \infty$.
6. If $0 < b_1 < b_2$ are boundary points of the support of H with $b_1-\epsilon$, $b_2+\epsilon$ outside the support of H for small $\epsilon > 0$, then for all y sufficiently small there exist corresponding boundary points $a_1(y)$, $a_2(y)$ of F such that $F\{[a_1(y), a_2(y)]\} = H\{[b_1, b_2]\}$ and $[a_1(y), a_2(y)] \to [b_1, b_2]$ as $y \to 0$.

Thus from the above properties relevant information on the spectrum of $\mathbf{T}_p$ for p large can be obtained from the eigenvalues of B_p with a sample size on the same order of magnitude as p. For the detection problem in Section 5.3 the properties tell us that for a large enough sample we should be able to estimate (at the very least) the proportion of targets in relation to the number of sensors. Finding the exact number of "signal" eigenvalues separated from the $p-q$ "noise" ones in our simulations, with the gap close to the gap we would expect from F, came as a delightful suprise (Silverstein and Combettes (1992)).

2. Separation of Eigenvalues

Verifying mathematically the observed phenomenon of exact separation of eigenvalues has been achieved by Zhidong Bai and myself. The proof is broken down into two steps. The first step is to prove that, almost surely, no eigenvalues lie in any interval that is outside the support of the limiting distribution for all p large (Bai and Silverstein (1998)). Define F^A to be the empirical distribution function of the eigenvalues of the matrix A, assumed to be Hermitian. Let $H_p = F^{\mathbf{T}_n}$, $y_p = p/n$, and F^{y_p,H_p} be the limiting spectral distribution of B_p with y, H replaced by y_p and H_p. We assume the entries of $\mathbf{X}_p$ have mean zero and finite fourth moment (which are necessary, considering the results in Section 2.2.2 on extreme eigenvalues) and the matrices $\mathbf{T}_p$ are bounded for all p in spectral norm. We have then

Theorem. (Theorem 1.1 of Bai and Silverstein (1998)) *For any interval* $[a, b]$ *with* $a > 0$ *which lies in an open interval outside the support of* $F(= F^{y,H})$ *and* F^{y_p,H_p} *for all large* p *we have*

$$\text{P(no eigenvalue of } B_p \text{ appears in } [a, b] \text{ for all large } p \text{)} = 1.$$

Note that the phrase "in an open interval" was inadvertently left out of the original paper.

The proof looks closely at properties of the Stieltjes transform of F^{B_p}, and uses moment bounds on both random quadratic forms (similar to Lemma A.4 of Bai (1997)) and martingale difference sequences.

The second step is to show the correct number of eigenvalues in each portion of the limiting support. This is achieved by appealing to the continuous dependence of the eigenvalues on their matrices. Let B_p^n denote the dependence of the matrix on n. Using the fact that the smallest and largest eigenvalues of $\frac{1}{n}\mathbf{X}_p\mathbf{X}_p^*$ are near $(1-\sqrt{p/n})^2$ and $(1+\sqrt{p/n})^2$ respectively, the eigenvalues of T_p and $B_p^M n$ are near each other for suitably large M. It is then a matter of showing eigenvalues do not cross over from one support region to another as the number of samples increases from n to Mn. This work is presently in preparation.

This work should be viewed as an extension of the results in Section 2.2.2 on the extreme eigenvalues of $S_p = (1/n)X_pX_p^*$. In particular, it handles the extreme eigenvalues of B_p (see the corollary to Theorem 1.1 in Bai and Silverstein (1998)). At the same time it should be noted that the proof of exact separation relies heavily on a.s. convergence of the extreme eigenvalues of S_p. As mentioned earlier, the moment method seems to be the only way in proving Theorem 2.15. On the other hand, the Stieltjes transform appears essential in proving exact separation, partly from what it reveals about the limiting distribution.

3. Results and Conjectures on the Rate of Convergence

I will finish up with my views on the rate of convergence issue concerning the spectral distribution of sample covariance matrices raised in Section 3.2.2. The natural question to ask is: what is the speed of convergence of $W_p \equiv F^{B_p} - F^{y_p,H_p}$ to 0? Here is some evidence the rate may be $1/p$ in the case $H_p = I_{[0,\infty)}$, that is, when $B_p = S_p = (1/n)XX^*$ (Section 2.1.2).

In Jonsson (1982) it is shown that the distribution of

$$\left\{ n \int x^r d(F^{S_p}(x) - \mathbf{E}(F^{S_p}(x))) \right\}_{r=1}^{\infty}$$

converges ($\mathbb{R}^\infty$) to that of a multivariate normal, suggesting an error rate of $1/p$. Continuing further, with the aid of moment analysis, the following has been observed.

Let $Y_p(x) = p\int_0^x [F^{S_p}(t) - (\mathbf{E}(F^{S_p}(t))] \, dt$. It appears that, as $p \to \infty$, $p(\mathbf{E}(F^{S_p}(x)) - F^{y_p,1_{[1,\infty)}}(x))$ converges to certain continuous function on $[0,(1+\sqrt{c})^2]$, and the covariance function $C_{Y_pY_p}(x_1,x_2) \equiv \mathbf{E}(Y_p(x_1)Y_p(x_2)) \to C_{YY}(x_1, x_2)$, continuous on $[0,(1+\sqrt{y})^2] \times [0,(1+\sqrt{y})^2]$. Both functions depend on y and $\mathbf{E}(X_{11}^4)$. Moreover, it can be verified that C_{YY} is the covariance function of a continuous mean zero Gaussian process on $[0,(1+\sqrt{c})^2]$. The uniqueness of any weakly convergent subsequence of $\{Y_p\}$ follows by the above result in Jonsson (1982) and the a.s. convergence of the largest eigenvalue of S_p (see Theorem 3.1 of Silverstein (1990)). Thus, if tightness can be proven, weak convergence of Y_p would follow, establishing the rate of convergence of 1/p for the partial sums of the eigenvalues of S_p. It should be noted that the conjecture on Y_p is substantiated by extensive simulations.

It seems that the integral making up Y_p is necessary because $\frac{\partial^2 C_{YY}}{\partial x_1 \partial x_2}(x_1,x_2)$, which would be the covariance function of $p(F^{S_p}(x)-[\mathbf{E}(F^{S_p}(x)])$ in the limit, turns out to be unbounded at $x_1=x_2$. As an illustration, when $\mathbf{E}(X_{11}^4)=3$ (as in the Gaussian case) $\frac{\partial^2 C_{YY}}{\partial x_1 \partial x_2}(x_1,x_2)=$

$$\frac{1}{2\pi^2}\ln\left[\frac{4y-((x_1-(1+y))(x_2-(1+y))+\sqrt{(4y-(x_1-(1+y))^2)(4y-(x_2-(1+y))^2)}}{4y-((x_1-(1+y))(x_2-(1+y))-\sqrt{(4y-(x_1-(1+y))^2)(4y-(x_2-(1+y))^2)}}\right]$$

for $(x_1,x_2)\in[(1-\sqrt{y})^2,(1+\sqrt{y})^2]\times[(1-\sqrt{y})^2,(1+\sqrt{y})^2]$, 0, otherwise. It therefore appears unlikely pW_p converges weakly.

Of course weak convergence of Y_p does not immediately imply $\alpha(p)W_p\to 0$ for $\alpha(p)=o(p)$. It only lends support to the conjecture that $1/p$ is the correct rate. Further work is definitely needed in this area.

Acknowledgement

This work is supported by NSF Grant DNS-9703591.

Department of Mathematics, North Carolina State University, Raleigh, NC, U.S.A.

REJOINDER

Z. D. Bai

Thanks to Professor Jack Silverstein and Dr. G. J. Rodgers for their additions to developments in the theory of spectral analysis of large dimensional random matrices not reported on in my review paper. I would like to make some remarks on the problems arising from their comments.

1. Spectrum Separation of Large Sample Covariance Matrices

Jack Silverstein reported a new result on spectrum separation of large sample covariance matrices obtained in Bai and Silverstein (1998), after my review paper was written. It is proved there that under very general conditions, for any closed interval outside the support of the limiting spectral distribution of a sequence of large dimensional sample covariance matrices, and with probability 1 for all large n, the sample covariance matrix has no eigenvalues falling in this interval. He also reported that a harder problem of exact spectrum separation is under our joint investigation. Now, I take this opportunity to report that this problem has been solved in Bai and Silverstein (1999). More specifically, the exact spectrum separation is established under the same conditions of Theorem 1.1 of Bai and Silverstein (1998).

1.1. Spectrum separation of large sample covariance matrices

Our setup and basic assumptions are the following.

(a) X_{ij}, $i, j = 1, 2, \ldots$ are independent and identically distributed (i.i.d.) complex random variables with mean 0, variance 1 and finite 4th moment;

(b) $n = n(p)$ with $y_n = p/n \to y > 0$ as $n \to \infty$;

(c) For each n, $\mathbf{T}_n$ is a $p \times p$ Hermitian nonnegative definite matrix satisfying $H_n \equiv F^{\mathbf{T}_n} \xrightarrow{\mathcal{D}} H$, a cumulative distribution function (c.d.f.);

(d) $\|\mathbf{T}_n\|$, the spectral norm of $\mathbf{T}_n$, is bounded in n;

(e) $\mathbf{S}_n = n^{-1}\mathbf{T}_n^{1/2}\mathbf{X}_n\mathbf{X}_n^*\mathbf{T}_n^{1/2}$, $\underline{\mathbf{S}}_n = n^{-1}\mathbf{X}_n^*\mathbf{T}_n\mathbf{X}_n$, where $\mathbf{X}_n = (X_{ij})$, $i = 1, \ldots, p$, $j = 1, \ldots, n$, and $\mathbf{T}_n^{1/2}$ is a Hermitian square root of T_n.

The matrix $\mathbf{S}_n$ is of major interest and the introduction of the matrix $\underline{\mathbf{S}}_n$ is for mathematical convenience. Note that

$$F^{\underline{\mathbf{S}}_n} = \left(1 - y_n\right) I_{[0,\infty)} + y_n F^{\mathbf{S}_n}$$

and

$$m_{F^{\underline{\mathbf{S}}_n}}(z) = -\frac{1 - y_n}{z} + y_n m_{F^{\mathbf{S}_n}}(z).$$

As previously mentioned, under conditions (a) - (e), the limiting spectral distribution (LSD) of $\mathbf{S}_n$ exists and the Stieltjes transform of the LSD of $\underline{\mathbf{S}}_n$ is the unique solution, with nonnegative imaginary part for z on the upper half plane, to the equation

$$z_{y,H}(m) = -\frac{1}{m} + y \int \frac{t}{1 + tm} dH(t).$$

The LSD of $\underline{\mathbf{S}}_n$ is denoted by $F^{y,H}$. Then, for each fixed n, F^{y_n,H_n} can be regarded as the LSD of a sequence of sample covariance matrices for which the LSD of the population covariance matrices is H_n and limit ratio of dimension to sample size is y_n. Its Stieltjes transform is then the unique solution with nonnegative imaginary part, for z on the upper half plane, to the equation

$$z_{y_n,H_n}(m) = \frac{1}{m}\left(-1 + y_n \int \left(\frac{tm}{1 + tm}\right) dH_n(t)\right) \tag{1}$$

$$z'_{y_n,H_n}(m) = \frac{1}{m^2}\left(1 - y_n \int \left(\frac{tm}{1 + tm}\right)^2 dH_n(t)\right). \tag{2}$$

It is easy to see that for any real $x \neq 0$, the function $m_{F^{y_n,H_n}}(x)$ and its derivative are well defined and continuous provided $-1/x$ is not a support point of H_n.

Under the further assumption that

(f) the interval $[a, b]$ with $a > 0$ lies in an open interval outside the support of F^{y_n,H_n} for all large n,

Bai and Silverstein (1998) proved that with probability one, for all large n, $\mathbf{S}_n$ has no eigenvalues falling in $[a, b]$.

To understand the meaning of exact separation, we give the following description.

1.2. Description of exact separation

From (1), it can be seen that F^{y_n,H_n} and its support tend to $F^{y,H}$ and the support of it, respectively. We use F^{y_n,H_n} to define the concept exact separation in the following.

Denote the eigenvalues of $\mathbf{T}_n$ by $0 = \lambda_1(\mathbf{T}_n) = \cdots = \lambda_h(\mathbf{T}_n) < \lambda_{h+1}(\mathbf{T}_n) \leq \cdots \leq \lambda_p(\mathbf{T}_n)$ ($h = 0$ if $\mathbf{T}_n$ has no zero eigenvalues). Applying Silverstein and Choi (1995), the following conclusions can be made. From (1) and (2), one can see that $mz_{y_n,H_n}(m) \to -1+y_n(1-H_n(0))$ as $m \to -\infty$, and $mz_{y_n,H_n}(m) > -1+y_n(1-H_n(0))$ for all $m < -M$ for some large M. Therefore, when m increases along the real axis from $-\infty$ to $-1/\lambda_{h+1}(\mathbf{T}_n)$, the function $z_{y_n,H_n}(m)$ increases from 0 to a maximum and then decreases to $-\infty$ if $-1 + y_n(1 - H_n(0)) \geq 0$; it decreases directly to $-\infty$ if $-1 + y_n(1 - H_n(0)) < 0$, where $H_n(0) = h/p$. In the latter case, we say that the maximum value of z_{y_n,H_n} in the interval $(-\infty, -\lambda_{h+1}(\mathbf{T}))$ is 0. Then, for $h < k < p$, when m increases from $-1/\lambda_k(\mathbf{T}_n)$ to $-1/\lambda_{k+1}(\mathbf{T}_n)$, the function z_{y_n,H_n} in (1) either decreases from ∞ to $-\infty$, or decreases from ∞ to a local minimum, then increases to a local maximum and finally decreases to $-\infty$. Once the latter case happens, the open interval of z_{y_n,H_n} values from the minimum to the maximum is outside the support of F^{y_n,H_n}. When m increases from $-1/\lambda_p(\mathbf{T}_n)$ to 0, the z value decreases from ∞ to a local minimum and then increases to ∞. This local minimum value determines the largest boundary of the support of F^{y_n,H_n}. Furthermore, when m increases from 0 to ∞, the function $z_{y_n,H_n}(m)$ increases from $-\infty$ to a local maximum and then decreases to 0 if $-1 + y_n(1 - H_n(0)) > 0$; it increases directly from $-\infty$ to 0 if $-1 + y_n(1 - H_n(0)) \leq 0$. In the latter case, we say that the local maximum value of z_{y_n,H_n} in the interval $(0, \infty)$ is 0. The maximum value of z_{y_n,H_n} in $(-\infty, -\lambda_{h+1}(\mathbf{T})) \cup (0, \infty)$ is the lower bound of the support of F^{y_n,H_n}.

Case 1. $y(1-H(0)) > 1$. For all large n, we can prove that the support of $F^{y,H}$ has a positive lower bound x_0 and $y_n(1 - H_n(0)) > 1$, $p > n$. In this case, we can prove that $\mathbf{S}_n$ has $p - n$ zero eigenvalues and the nth largest eigenvalues of $\mathbf{S}_n$ tend to x_0.

Case 2. $y(1-H(0)) \leq 1$ or $y(1-H(0)) > 1$ but $[a,b]$ is not in $[0,x_0]$. For large n, let $i_n \geq 0$ be the integer such that

$$\lambda_{i_n}^{T_n} > -1/m_{F^y,H}(b) \qquad \text{and} \qquad \lambda_{i_n+1}^{T_n} < -1/m_{F^y,H}(a).$$

It is seen that only when $m_{F^y,H}(b) < 0$, the exact separation occurs. In this case, we prove that

$$P(\lambda_{i_n}^{\mathbf{S}_n} > b \quad \text{and} \quad \lambda_{i_n+1}^{\mathbf{S}_n} < a \quad \text{for all large } n\ ) = 1.$$

This shows that with probability 1, when n is large, the number of eigenvalues of $\mathbf{S}_n$ which are greater than b is exactly the same as that of the eigenvalues of $\mathbf{T}_n$ which are greater than $-1/m_{F^y,H}(b)$, and contrarily, the number of eigenvalues of $\mathbf{S}_n$ which are smaller than a is exactly the same as that of the eigenvalues of $\mathbf{T}_n$ which are smaller than $-1/m_{F^y,H}(a)$.

1.3. Strategy of the proof of exact spectrum separation

Consider a number of sequences of sample covariance matrices of the form

$$\mathbf{S}_{n,k} = (n+kM)^{-1}\mathbf{T}_n^{1/2}\mathbf{X}_{n,k}\mathbf{X}_{n,k}^*\mathbf{T}_n^{1/2},$$

where $M = M_n$ is an integer such that $M/n \to c > 0$, for some small $c > 0$, and $\mathbf{X}_{n,k} = (X_{ij})$ with dimension $p \times (n+kM)$.

We need to prove the following.

(i) Define $y_k = y/(1+kc)$ and $a_k < b_k$ by

$$m_{F^y,H}(a) = m_{F^{y_k},H}(a_k) \quad \text{and} \quad m_{F^y,H}(b) = m_{F^{y_k},H}(b_k).$$

We show that when $c > 0$ is small enough,

$$P(\lambda_{\ell_n}(\mathbf{S}_{n,k}) < a_k \quad \text{and} \quad \lambda_{\ell_n+1}(\mathbf{S}_{n,k}) > b_k \quad \text{for all large } n) = 1,$$

and thus that

$$P(\lambda_{\ell_n}(\mathbf{S}_{n,k+1}) < a_{k+1} \quad \text{and} \quad \lambda_{\ell_n+1}(\mathbf{S}_{n,k+1}) > b_{k+1} \quad \text{for all large } n) = 1.$$

(ii) Let K be so large that $p/(n+KM) := y_{Kn} \to y_K = y/(1+Kc) < (-1/m_{F^y,H}(b) + 1/m_{F^y,H}(a)/(9\max_n \|\mathbf{T}_n\|)$. Then, by Corollary 7.3.8 of Horn and Johnson (1985), we have

$$\begin{aligned}
&\max_{i\leq p}|\lambda_i(\mathbf{S}_{n,K}) - \lambda_i(\mathbf{T}_n)| \leq \|T_n^{1/2}(\mathbf{S}_{n,K} - \mathbf{I})\mathbf{T}_n^{1/2}\| \\
&\leq \max_n \|\mathbf{T}_n\| \max(\lambda_p(\mathbf{S}_{n,K}) - 1, 1 - \lambda_1(\mathbf{S}_{n,K})) \\
&\to \max_n \|\mathbf{T}_n\|(2y_K + y_K^2) < (-1/m_{F^y,H}(b) + 1/m_{F^y,H}(a)/3. \qquad (3)
\end{aligned}$$

From (ii), it follows that with probability 1, for all large n, $\lambda_{i_n+1}(\mathbf{S}_{n,K}) > (b_K + a_K)/2$ and $\lambda_{i_n}(\mathbf{S}_{n,K}) < (b_K + a_K)/2$. Then by Bai and Silverstein (1998), $\lambda_{i_n+1}(\mathbf{S}_{n,K}) > b_K$ and $\lambda_{i_n}(\mathbf{S}_{n,K}) < a_K$. That is, the exact spectrum separation holds for the sequence $\{\mathbf{S}_{n,K}, n = 1, \ldots$ Applying (i), the exact spectrum separation remains true for any sequence $\{\mathbf{S}_{n,k}, n = 1, \ldots.\}$

2. On Replica Method

People working in the area of spectral analysis of large dimensional random matrices are aware that the theory was motivated by early findings, laws or conjectures, in theoretical physics, see the first paragraph of the introduction of my review paper (BaiP, hereafter). However, very few papers in pure probability or statistics refer to later developments in theoretical physics. Therefore, I greatly appreciate the relation of later developments in theoretical physics by G. J. Rodgers in his comments (RodC, hereafter), including the replica method and some valuable references.

From my point of view, the replica method starts at the same point as does the method of Stieltjes transform, analyzes with different approaches, and finds the same conclusions. At first, we note that the function $Z(\mu)$ defined in (1) of RodC is in fact $(2\pi i)^{N/2}\det^{1/2}(\mathbf{H} - \mu\mathbf{I})$. From this, one can derive that

$$\frac{2}{\pi N}\frac{\partial}{\partial \mu}\log Z(\mu) = \frac{1}{\pi N}\sum_{j=1}^{N}\frac{1}{\lambda_j - \mu} = \pi^{-1}m_n(\mu),$$

where $m_n(\cdot)$ is defined in (3.3) of BaiP. Note that $[Z^n(\mu)]_{av} = EZ^n(\mu)$. Consequently, the function in (2) of RodC is in fact

$$\rho(\mu) = \mathrm{Im}\frac{2}{\pi N}\frac{\partial}{\partial \mu}\log EZ^n(\mu).$$

For all large N, we should have $\rho(\mu) \sim \pi^{-1}\mathrm{Im}Em_n(\mu)$, which is asymptotically independent of n. This shows that the two methods start from the same point.

The method of Stieltjes transformation analyzes the resolvent of the random matrices by splitting

$$m_n(\mu) = \frac{1}{N}tr(\mathbf{H} - \mu\mathbf{I})^{-1}$$

into a sum of weakly dependent terms, while the replica method continues its analysis on the expected function $[Z^n(\mu)]_{av}$.

Now, we consider the Hubbard-Stratonovich transformation, in which a set of i.i.d standard normal variables $x_{\alpha j}$ are used to substitute for the variables

$$\sigma^{-1}\sum_{i=1}^{N}\xi_i^j\phi_{i\alpha}\left(\sum_{i=1}^{N}\phi_{i\alpha}^2\right)^{-1/2}.$$

The validity of this normal approximation is a key point in the replica method and might be the reason to call it "non-rigorous" in RodC.

For each fixed α and j, it is not difficult to show that as $N \to \infty$, the variable $\sigma^{-1} \sum_{i=1}^{N} \xi_i^j \phi_{i\alpha}$ is asymptotically normal for $\phi_{i,\alpha}$'s satisfying $\sum_{i=1}^{N} \phi_{i\alpha}^2 = 1$, except in a small portion on the unit sphere. However, I do not know how to show the asymptotic independence between $\sigma^{-1} \sum_{i=1}^{N} \xi_i^j \phi_{i\alpha}$ for different (j, α)'s. If this can be done, then many problems in the spectral analysis of large dimensional random matrices, say, the circular law under the only condition of the finite second moment, can be reduced to the normal case, under which the problems are well-known or easier to deal with. More specifically, the conjectures are the following.

Conjecture 1. Let $\mathbf{X}$ be an $n \times N$ matrix with i.i.d. entries of mean zero and variance 1, and let $\mathbf{H}$ be uniform distributed on the $p \times n$ $(p \leq n)$ matrix space of p orthonormal rows. Then as p, n, N proportionally tend to infinity, the $p \times N$ entries of $\mathbf{HX}$ are *asymptotically* **i.i.d.** normal.

Of course, there is a problem on how to define the terminology **asymptotically i.i.d.** since the number of variables goes to infinity. For use in spectral analysis of large dimensional random matrices, we restate Conjecture 1 as the following.

Conjecture 2. Let $\mathbf{X}$ be an $n \times N$ matrix with i.i.d. entries of mean zero and variance 1, and let $\mathbf{E}$ be uniform distributed on the $n \times n$ orthogonal matrix space. Then as n, N proportionally tend to infinity, the limiting behavior of all spectrum functionals of the matrix $\mathbf{HX}$ are the same as if all entries of $\mathbf{X}$ are i.i.d. normal.

More specifically, we have

Conjecture 3. Let $\mathbf{X}$ be an $n \times N$ matrix with i.i.d. entries of mean zero and variance 1. There exists an $n \times n$ orthogonal matrix $\mathbf{H}$ such that as n, N proportionally tend to infinity, the limiting behavior of all spectrum functionals of the matrix $\mathbf{HX}$ are the same as if all entries of $\mathbf{X}$ are i.i.d. normal.

This seems to be a very hard but interesting problem.

Additional References

Bai, Z. D. (1997). Circular Law. *Ann. Probab.* **25**, 494-529.

Bai, Z. D. (1999). Methodologies in spectral analysis of large dimensional random matrices, a review. *Statist. Sinica*, previous paper.

Bai, Z. D. and Silverstein, J. W. (1998). No eigenvalues outside the support of the limiting spectral distribution of large dimensional sample covariance matrices. *Ann. Probab.* **26**, 316-345.

Bai, Z. D. and Silverstein, J. W. (1999). Exact separation of eigenvalues of large dimensional sample covariance matrices. Accepted by *Ann. Probab.*

Bray, A. J. and Rodgers, G. J. (1988). Diffusion in a sparsely connected space: a model for glassy relaxation. *Phys. Rev. B* **38**, 11461-11470.

Dhesi, G. S. and Jones, R. C. (1990). Asymptotic corrections to the Wigner semicircular eigenvalue spectrum of a large real symmetric random matrix using the replica method. *J. Phys. A* **23**, 5577-5599.

Edwards, S. F. (1970). Statistical mechanics of polymerized materials. *Proc. 4th Int. Conf. on Amorphous Materials* (Edited by R. W. Douglas and B. Ellis), 279-300. Wiley, New York.

Edwards, S. F. and Jones, R. C. (1976). Eigenvalue spectrum of a large symmetric random matrix. *J. Phys. A* **9**, 1595-1603.

Hopfield, J. J. (1982). Neural networks and physical systems with emergent collective computational abilities. *Proc. Natl. Ac. Sci. USA* **79**, 2554-2558.

Jonsson, D. (1982). Some limit theorems for the eigenvalues of a sample covariance matrix. *J. Multivariate Anal.* **12**, 1-38.

Marchenko, V. A. and Pastur, L. A. (1967). Distribution of some sets of random matrices. *Math. USSR-Sb* **1**, 457-483.

Mezard, M., Parisi, G. and Virasoro, M. (1988). *Spin Glass Theory and Beyond.* World Scientific, Singapore.

Mirlin, A. D. and Fyodorov, Y. V. (1991). Universality of level correlation function of sparse random matrices. *J. Phys. A* **24**, 2273-2286.

Rodgers, G. J. and Bray, A. J. (1988). Density of states of a sparse random matrix *Phys. Rev. B* **37**, 3557-3562.

Rodgers, G. J. and De Dominicis, C. (1990). Density of states of sparse random matrices. *J. Phys. A* **23**, 1567-1573.

Silverstein, J. W. (1995). Strong convergence of the eimpirical distribution of eigenvalues of large dimensional random matrices *J. Multivariate Anal.* **5**, 331-339.

Silverstein, J. W. (1990). Weak convergence of random functions defined by the eigenvectors of sample covariance matrices. *Ann. Probab.* **18**, 1174-1194.

Silverstein, J. W. and Choi, S. I. (1995). Analysis of the limiting spectral distribution of large dimensional random matrices. *J. Multivariate Anal.* **54**, 295-309.

Silverstein, J. W. and Combettes, P. L. (1992). Signal detection via spectral theory of large dimensional random matrices. *IEEE Trans. Signal Processing* **40**, 2100-2105.

Sommers, H. J., Crisanti, A., Sompolinsky, H. and Stein, Y. (1988). Specturm of large random asymmetric matrices. *Phys. Rev. Lett.* **60**, 1895-1898.

Weight, M. (1998). A replica approach to products of random matrices. *J. Phys. A* **31**, 951-961.

The Annals of Statistics
1999, Vol. 27, No. 5, 1616–1637

ASYMPTOTIC DISTRIBUTIONS OF THE MAXIMAL DEPTH ESTIMATORS FOR REGRESSION AND MULTIVARIATE LOCATION

BY ZHI-DONG BAI[1] AND XUMING HE[2]

National University of Singapore and University of Illinois

We derive the asymptotic distribution of the maximal depth regression estimator recently proposed in Rousseeuw and Hubert. The estimator is obtained by maximizing a projection-based depth and the limiting distribution is characterized through a max–min operation of a continuous process. The same techniques can be used to obtain the limiting distribution of some other depth estimators including Tukey's deepest point based on half-space depth. Results for the special case of two-dimensional problems have been available, but the earlier arguments have relied on some special geometric properties in the low-dimensional space. This paper completes the extension to higher dimensions for both regression and multivariate location models.

1. Introduction. Multivariate ranking and depth have been of interest to statisticians for quite some time. The notion of depth plays an important role in data exploration, ranking, and robust estimation; see Liu, Parelius and Singh (1999) for some recent advances. The location depth of Tukey (1975) is the basis for a multivariate median; see Donoho and Gasko (1992). Recently, Rousseeuw and Hubert (1999) introduced a notion of depth in the linear regression setting. Both measures of depth are multivariate in nature and defined as the minimum of an appropriate univariate depth over all directions of projection. The maximal depth estimator is then obtained through a max–min operation which complicates the derivation of its asymptotic distribution. The present paper focuses on the asymptotics of maximal depth estimators.

First, we recall the definition of regression depth. Consider a regression model in the form of $y_i = \beta_0 + \mathbf{x}_i'\boldsymbol{\beta}_1 + e_i$ where $\mathbf{x}_i \in R^{p-1}$, $\boldsymbol{\beta}' = (\beta_0, \boldsymbol{\beta}_1') \in R^p$ and e_i are regression errors. A regression fit $\boldsymbol{\beta}$ is said to be a nonfit to the given data $\mathbf{Z}_n = \{(\mathbf{x}_i, y_i),\ i = 1, 2, \ldots, n\}$ if and only if there exists an affine hyperplane V in the design space such that no $\mathbf{x}_i$ belongs to V and such that the residuals $r_i > 0$ for all $\mathbf{x}_i$ in one of its open half-spaces and $r_i < 0$ for all $\mathbf{x}_i$ in the other open half-space. Then, the regression depth $rdepth(\boldsymbol{\beta}, \mathbf{Z}_n)$ is the smallest number of observations that need to be removed (of whose residuals need to change sign) to make $\boldsymbol{\beta}$ a nonfit. To put it into mathematical

Received August 1998; revised August 1999.

[1]Supported in part by National University of Singapore Grant RP397212.

[2]Supported in part by NSF Grant SBR 96-17278 and by the Wavelets Strategic Research Program (SWRP) funded by the National Science and Technology Board and the Ministry of Education of Singapore under Grant RP960 601/A.

AMS 1991 *subject classifications*. Primary 62G35, 62F12; secondary 62J05, 62H12.

Key words and phrases. Asymptotic distribution, consistency, estimator, median, multivariate location, regression depth, robustness.

formulation, let $\mathbf{w}_i' = (1, \mathbf{x}_i')$, $r_i(\boldsymbol{\beta}) = y_i - \mathbf{w}_i'\boldsymbol{\beta}$. Following Rousseeuw and Hubert (1999), we define

$$rdepth(\boldsymbol{\beta}, \mathbf{Z}_n) = \inf_{\substack{\|\mathbf{u}\|=1 \\ v\in R}} \min\left\{ \sum_{i=1}^{n} I(r_i(\boldsymbol{\beta})(\mathbf{u}'\mathbf{x}_i - v) > 0), \sum_{i=1}^{n} I(r_i(\boldsymbol{\beta})(\mathbf{u}'\mathbf{x}_i - v) < 0)\right\}. \tag{1.1}$$

The maximal depth estimate $\hat{\boldsymbol{\beta}}_n$ maximizes $rdepth(\boldsymbol{\beta}, \mathbf{Z}_n)$ over $\boldsymbol{\beta} \in R^p$. For convenience, we reformulate the objective function (1.1) as follows. Denote $S^p = \{\boldsymbol{\gamma} \in R^p,\ \|\boldsymbol{\gamma}\| = 1\}$ as the unit sphere in R^p. Then it is easy to show that

$$rdepth(\boldsymbol{\beta}, \mathbf{Z}_n) = \frac{n}{2} + \frac{1}{2} \inf_{\boldsymbol{\gamma}\in S^p} \sum_{i=1}^{n} \operatorname{sgn}(y_i - \mathbf{w}_i'\boldsymbol{\beta})\operatorname{sgn}(\mathbf{w}_i'\boldsymbol{\gamma}), \tag{1.2}$$

where $\operatorname{sgn}(x)$ is the sign of x. In the rest of the paper, we consider the problem of

$$\sup_{\boldsymbol{\beta}\in R^p} \inf_{\boldsymbol{\gamma}\in S^p} \sum_{i=1}^{n} \operatorname{sgn}(y_i - \mathbf{w}_i'\boldsymbol{\beta})\operatorname{sgn}(\mathbf{w}_i'\boldsymbol{\gamma}). \tag{1.3}$$

Note that the deepest point based on Tukey depth for multivariate data has a similar formulation. Given n observations $\mathbf{X}_n = (\mathbf{x}_1, \mathbf{x}_2, \dots, \mathbf{x}_n)$ in R^p, the deepest point $\hat{\boldsymbol{\theta}}_n$ solves

$$\begin{aligned} &\sup_{\boldsymbol{\theta}\in R^p} \inf_{\boldsymbol{\gamma}\in S^p} \sum_{i=1}^{n} I(\boldsymbol{\gamma}'(\mathbf{x}_i - \boldsymbol{\theta}) > 0) \\ &\quad = \frac{1}{n} + \frac{1}{2} \sup_{\boldsymbol{\theta}\in R^p} \inf_{\boldsymbol{\gamma}\in S^p} \sum_{i=1}^{n} \operatorname{sgn}(\boldsymbol{\gamma}'(\mathbf{x}_i - \boldsymbol{\theta})). \end{aligned} \tag{1.4}$$

Both (1.3) and (1.4) involve a max–min operation applied to a sum of data-dependent functions. Common techniques can be used to derive the asymptotic distributions of these estimators. In fact, the asymptotic distributions of both estimators have been derived for the case of $p = 2$ by He and Portnoy (1998) and Nolan (1999), respectively. The limiting distribution can be characterized by the random variable that solves $\max_{\boldsymbol{\beta}} \min_{\boldsymbol{\gamma}\in S^p}(W(\boldsymbol{\gamma}) + \boldsymbol{\mu}(\boldsymbol{\gamma})'\boldsymbol{\beta})$ for some Gaussian process W and smooth function $\boldsymbol{\mu}$. The difficulty in treating the higher-dimensional case lies mainly in proving uniqueness of the solution $\boldsymbol{\beta}$ to the above max–min problem. Both works cited above used arguments based on two-dimensional geometry and direct extensions to higher dimensions appear difficult. See Nolan (1999) for an explicit account of the difference between the two-dimensional and the higher-dimensional structures.

Limiting distributions as characterized by an arg-max or arg-min functional are not that uncommon in the statistics literature. A good recent reference is Kim and Pollard (1990). The problem we are concerned with here is complicated by the additional optimization over $\boldsymbol{\gamma} \in S^p$. This type of limiting distribution comes up naturally from the use of projections. We focus on the maximal depth regression and the deepest point (as a location estimate) in the present paper due to their importance as a natural generalization of *median* for regression and multivariate data. Both estimators enjoy some of the desirable properties that we expect from the median. For example, they are affine equivariant, have positive breakdown point (higher than that of an M-estimator), and are root-n consistent to their population counterparts. For confidence bands based on depth, see He (1999).

In Section 2, we show that the maximal depth regression estimate is consistent for the conditional median of y given $\mathbf{x}$ if it is linear. The conditional distribution of y given $\mathbf{x}$ may vary with $\mathbf{x}$. This property is shared with the least absolute deviation regression (LAD), commonly interpreted as the median regression; see Koenker and Bassett (1978). Because the breakdown robustness of the LAD is design-dependent [cf. He, Koenker and Portnoy (1990)], the maximal depth regression has the advantage of being robust against data contamination at the leverage points.

In Section 3, we derive the asymptotic distribution of the maximal depth estimate. In line with most other published results on the asymptotic distributions of regression estimators and to avoid being overshadowed by notational and technical complexity, we work with a more restrictive regression model with i.i.d. errors in this section. An almost sure LIL-type result for the estimator is also provided in this section. We then present the limiting distribution of the deepest point for multivariate data in Section 4, extending the work of Nolan (1999). The Appendix provides all the proofs needed in the paper. In particular, we provide a means to establish the uniqueness of solution to a max–min problem that arises from the projection-based depth in regression as well as multivariate location models. For computation of the regression and location depth, we refer to Rousseeuw and Struyf (1998).

2. Consistency of maximal depth regression. We assume that the conditional median of y given $\mathbf{x}$ is linear, that is, there exists $\boldsymbol{\beta}^* \in R^p$ such that

$$\text{Median}(y|\mathbf{x}) = \mathbf{w}'\boldsymbol{\beta}^*, \tag{2.1}$$

where $\mathbf{w}' = (1, \mathbf{x}')$. For a set of n design points $\mathbf{x}_1, \mathbf{x}_2, \ldots, \mathbf{x}_n$, independent observations of y_i are drawn from the conditional distributions of y given $\mathbf{x} = \mathbf{x}_i$.

If the conditional distribution of $y - \mathbf{w}'\boldsymbol{\beta}^*$ given $\mathbf{x}$ is the same for all $\mathbf{x}$, then the data can be modeled by the usual regression with i.i.d. errors. The above framework includes the case of random designs so that the data $(\mathbf{x}_i, y_i)$ come from the joint distribution of $(\mathbf{x}, y)$ as well as nonstochastic designs.

Since the maximal depth estimate $\hat{\boldsymbol{\beta}}_n$ is regression invariant, we assume without loss of generality that $\boldsymbol{\beta}^* = 0$ so that the conditional median of y is zero. To show that $\hat{\boldsymbol{\beta}}_n \to 0$, conditions on the design points and the error distributions are needed. For this purpose, let F_i be the conditional c.d.f. of y

given $\mathbf{x} = \mathbf{x}_i$. Also define for any $c > 0$,

$$Q_n(c) = \inf_{\mathbf{1} \in S^p} n^{-1} \sum_{i=1}^{n} I(|\mathbf{w}_i'\mathbf{1}| > c). \tag{2.2}$$

We now state our assumptions as follows. If the design points are random, then all the statements involving $\mathbf{w}_i$ are meant to be in the almost sure sense:

(D1) For some $b < \infty$, $\max_{1 \le i \le n} \|\mathbf{w}_i\| = O(n^b)$.
(D2) For any sequence $a_n \to 0$, $\lim_{n\to\infty} Q_n(a_n) = 1$.
(D3) For some $A < \infty$, $n^{-1}\sum_{i=1}^n \{1 - F_i(n^A) + F_i(-n^A)\} \to 0$ and $\max_{i \le n} \sup_x (F_i(x + n^{-A}) - F_i(x - n^{-A})) \to 0$ as $n \to \infty$.
(D4) For any $r > 0$, $\eta(r) = \inf_{i \ge 1} \min\{|1 - 2F_i(r)|, |1 - 2F_i(-r)|\} > 0$.

Condition (D2) is to avoid the degenerate case for the design points. This condition is satisfied if $\{\mathbf{x}_i\}$ is a random sample from a continuous multivariate distribution. Condition (D3) includes a weak requirement of the average tail thickness and a weak uniform continuity of all the conditional distribution functions, but (D4) requires that the error mass around the median is not too thin, which is satisfied if each F_i has a density with a common positive lower bound around the median. The following lemma is the basis for our consistency result.

LEMMA 2.1. *Under conditions* (D1)–(D3), *we have with probability* 1,

$$\sup_{\boldsymbol{\beta} \in R^p, \boldsymbol{\gamma} \in S^p} \left| n^{-1} \sum_{i=1}^{n} \{\operatorname{sgn}(y_i - \mathbf{w}_i'\boldsymbol{\beta})\operatorname{sgn}(\mathbf{w}_i'\boldsymbol{\gamma}) - E\operatorname{sgn}(y_i - \mathbf{w}_i'\boldsymbol{\beta})\operatorname{sgn}(\mathbf{w}_i'\boldsymbol{\gamma})\} \right| \to 0,$$

as $n \to \infty$.

Lemma 2.1 is a standard uniform approximation result except that the approximation is now over the whole space for $\boldsymbol{\beta}$. This is made possible by the fact that when $\|\boldsymbol{\beta}\|$ is large the function $\operatorname{sgn}(y_i - \mathbf{w}_i'\boldsymbol{\beta})$ does not change much. A proof of Lemma 2.1 for the possibly nonstochastic designs $\mathbf{w}_i$ is given in the Appendix.

By (D2) and (D4), for any given $c > 0$, there is a constant $r > 0$ such that $Q_n(r/c) > 1/2$ for sufficiently large n. Consequently, we have

$$\begin{aligned} &\inf_{\|\boldsymbol{\beta}\| \ge c} n^{-1} \sum_{i=1}^{n} |1 - 2F_i(\mathbf{w}_i'\boldsymbol{\beta})| \\ &\quad \ge \inf_{\mathbf{1} \in S^p} n^{-1} \sum_{i=1}^{n} \min\{|1 - 2F_i(r)|, |1 - 2F_i(-r)|\} I(|\mathbf{w}_i'\mathbf{1}| > r/c) \\ &\quad \ge \eta(r) Q_n(r/c) > \tfrac{1}{2}\eta(r) \end{aligned} \tag{2.3}$$

and with $\boldsymbol{\beta}^* = 0$ we have

$$n^{-1}\sum_{i=1}^{n} E\,\mathrm{sgn}(y_i - \mathbf{w}_i'\boldsymbol{\beta})\mathrm{sgn}(\mathbf{w}_i'\boldsymbol{\beta}) = -n^{-1}\sum_{i=1}^{n} |1 - 2F_i(\mathbf{w}_i'\boldsymbol{\beta})| < -\tfrac{1}{2}\eta(r)$$

for sufficiently large n. Thus, $\inf_{\boldsymbol{\gamma}\in S^p} n^{-1}\sum_{i=1}^{n} E\{\mathrm{sgn}(y_i - \mathbf{w}_i'\boldsymbol{\beta})\mathrm{sgn}(\mathbf{w}_i'\boldsymbol{\gamma})\} < 0$.

On the other hand, $E\{\mathrm{sgn}(y_i)\mathrm{sgn}(\mathbf{w}_i'\boldsymbol{\gamma})\} = 0$ for any $\boldsymbol{\gamma} \in S^p$, so

$$\inf_{\boldsymbol{\gamma}\in S^p}\{n^{-1}\sum_{i=1}^{n} E\,\mathrm{sgn}(y_i)\mathrm{sgn}(\mathbf{w}_i'\boldsymbol{\gamma})\} = 0.$$

Therefore, the maximal depth estimator has to be in the ball $\{\boldsymbol{\beta}: \|\boldsymbol{\beta}\| < c\}$. The consistency of $\hat{\boldsymbol{\beta}}_n$ follows from the fact that c can be arbitrarily small. We state the result formally as follows.

THEOREM 2.1. *Under conditions* (D1)–(D4), *the maximal depth regression estimate* $\hat{\boldsymbol{\beta}}_n \to \boldsymbol{\beta}^*$, *almost surely.*

Conditions (D1)–(D4) are sufficient but not necessary. It helps to note that the maximal depth regression estimator is consistent for the conditional median of y given $\mathbf{x}$ whenever the median is linear in $\mathbf{x}$. This is a property shared with L_1 regression but not other M-estimators. The limit of other M-estimators can only be identified with some additional information on the conditional distributions such as symmetry.

3. Limiting distribution of the maximal depth regression. In this section we derive the asymptotic distribution of the maximal depth estimator for the usual regression model

$$y_i = \beta_0 + \boldsymbol{\beta}_1'\mathbf{x}_i + e_i, \qquad i = 1, 2, \ldots, n,$$

where $\mathbf{x}_i$ is a random sample from a distribution in R^{p-1} with finite second moments, e_i's are independent of each other and of $\mathbf{x}_i$'s with a common distribution function F and density function f whose median is zero. We continue to use the same notation as in Section 2.

The following Lemma 3.1 is important for finding the limiting distribution of $\hat{\boldsymbol{\beta}}_n$. First, we itemize our assumptions for easy reference.

(C1) $E\|\mathbf{x}\|^2 \le B$ and $\sup_{\mathbf{1}\in S^p} P(|\mathbf{w}'\mathbf{1}| \le a\|\mathbf{w}\|) \le Ba^\delta$ for some $\delta \in (0, 2]$ and $B < \infty$.

(C2) $|F(x + r) - F(x)| \le B|r|^\delta$ for any x and r.

(C3) As $r \to 0$, $F(r) - F(0) = f(0)r + o(r)$ with $f(0) > 0$.

(C4) $E\{\mathrm{sgn}(\boldsymbol{\gamma}_1'\mathbf{w}\mathbf{w}'\boldsymbol{\gamma}_2)\}$ is continuous in $\boldsymbol{\gamma}_1, \boldsymbol{\gamma}_2 \in S^p$, and $E\{\mathbf{w}\,\mathrm{sgn}(\mathbf{w}'\boldsymbol{\gamma})\}$ is continuously differentiable in $\boldsymbol{\gamma} \in S^p$.

In typical cases, the constant $\delta = 1$ in (C1) and (C2).

REMARK 3.1. It is clear that conditions (D2) and (D4) are implied by (C1) and (C3). For independent and identically distributed errors whose distribution F has no positive mass at its median, condition (D3) is trivial. Condition

(D1) is true if $E\|\mathbf{x}\|^{1/b} < \infty$. Thus, the maximal depth estimator is consistent under conditions (C1)–(C3).

REMARK 3.2. If $\mathbf{x}_i$'s are not random or the e_i's may have different distributions F_i, the results of this section remain true if the above four conditions are replaced by:

(C1′) $n^{-1}\sum_{i=1}^n \mathbf{w}_i\mathbf{w}_i' \to \mathbf{A}$, a positive definite matrix, as $n \to \infty$, and $\sup_{\mathbf{1}\in S^p} n^{-1}\sum_{i=1}^n I(|\mathbf{w}_i'\mathbf{1}| \le a\|\mathbf{w}_i\|) \le Ba^\delta$ for some $\delta \in (0, 2]$ and $B < \infty$.
(C2′) For any x and r, $n^{-1}\sum_{i=1}^n |F_i(x+r) - F_i(x)| \le B|r|^\delta$.
(C3′) As $r \to 0$, $\max_{i\le n} |F_i(r) - F_i(0) - f_i(0)r| = o(r)$, as $r \to 0$, and $\bar{f} = \inf_n \bar{f}_n(0) = \inf_n n^{-1}\sum_{i=1}^n f_i(0) > 0$.
(C4′) The limit of $n^{-1}\sum_{i=1}^n \mathbf{w}_i \operatorname{sgn}(\mathbf{w}_i'\boldsymbol{\gamma})$ (as $n \to \infty$) exists and is continuously differentiable in $\boldsymbol{\gamma} \in S^p$, and the limit of $n^{-1}\sum_{i=1}^n \operatorname{sgn}(\boldsymbol{\gamma}_1'\mathbf{w}_i)\operatorname{sgn}(\mathbf{w}_i'\boldsymbol{\gamma}_2)$ exists uniformly and is continuous in $\boldsymbol{\gamma}_1, \boldsymbol{\gamma}_2 \in S^p$.

The proofs for our results in this section under conditions (C1′)–(C4′) are almost the same as those under (C1)–(C4) with averaging in place of expectations of $\mathbf{w}_i$. Let

$$S_n(\boldsymbol{\beta}, \boldsymbol{\gamma}) = \sum_{i=1}^n \{(\operatorname{sgn}(e_i - \mathbf{w}_i'\boldsymbol{\beta}) - \operatorname{sgn}(e_i))\operatorname{sgn}(\mathbf{w}_i'\boldsymbol{\gamma})\} \tag{3.1}$$
$$- E[(\operatorname{sgn}(e_i - \mathbf{w}_i'\boldsymbol{\beta}) - \operatorname{sgn}(e_i))\operatorname{sgn}(\mathbf{w}_i'\boldsymbol{\gamma})].$$

In this paper, we use $a_n \ll b_n \ll c_n$ to mean $a_n/b_n \to 0$ and $b_n/c_n \to 0$.

LEMMA 3.1. *If* (C1) *and* (C2) *hold, then for any constant* $\nu > 0$ *and any bounded sequence* $\Delta_n \gg n^{-1/(\delta+2\upsilon(1+\delta))}$, *we have*

$$\sup_{\|\boldsymbol{\beta}\|\le\Delta_n,\, \boldsymbol{\gamma}\in S^p} |S_n(\boldsymbol{\beta}, \boldsymbol{\gamma})| = O_p(n^{1/2}\Delta_n^{\delta/2-\nu}).$$

If we further assume $\Delta_n \to 0$ *slowly or regularly in the sense that there exist a constant* $\alpha > 0$ *and a function* $L(x)$ *such that* $\Delta_n = n^{-\alpha}L(n)$ *and* $L(bx)/L(x) \to 1$ *as* $x \to \infty$ *for any* $b > 0$, *then*

$$\limsup_{n\to\infty} \sup_{\|\boldsymbol{\beta}\|\le\Delta_n,\, \boldsymbol{\gamma}\in S^p} |S_n(\boldsymbol{\beta}, \boldsymbol{\gamma})|/(2n\Delta_n^\delta \log\log n)^{1/2} \le 1 \quad a.s.$$

In the Appendix, we actually prove a more general lemma in the form of an exponential inequality. This is often useful for asymptotic analyses in statistics. General results of this type may also be found in Pollard [(1984), page 144]. The following lemma allows for nonrandom designs as in He and Shao (1996), but is proved using a different chaining argument.

LEMMA 3.2. *Suppose that* $\Delta_n > 0$ *is a sequence of constants and* D *is a compact set in* R^p. *For each* $(\boldsymbol{\beta}, \boldsymbol{\gamma})$ *with* $\|\boldsymbol{\beta}\| \le \Delta_n$ *and* $\boldsymbol{\gamma} \in D$, $\{W_1(\boldsymbol{\beta}, \boldsymbol{\gamma}), W_2(\boldsymbol{\beta}, \boldsymbol{\gamma}), \ldots, W_n(\boldsymbol{\beta}, \boldsymbol{\gamma})\}$ *is a sequence of independent random variables of mean zero satisfying*:

(L1) *For some constants* $\delta > 0$ *and* $C_1 > 0$,

$$\frac{1}{n}\sum_{i=1}^{n} E|W_i(\boldsymbol{\beta}_1, \boldsymbol{\gamma}_1) - W_i(\boldsymbol{\beta}_2, \boldsymbol{\gamma}_2)|^2 \le C_1[\|\boldsymbol{\beta}_1 - \boldsymbol{\beta}_2\| + \|\boldsymbol{\gamma}_1 - \boldsymbol{\gamma}_2\|]^\delta.$$

(L2) *For some constant* $C_2 > 0$,

$$|W_i(\boldsymbol{\beta}_1, \boldsymbol{\gamma}_1) - W_i(\boldsymbol{\beta}_2, \boldsymbol{\gamma}_2)| \le C_2 \quad \textit{if } \|\boldsymbol{\beta}_j\| \le \Delta_n \textit{ and } \boldsymbol{\gamma}_j \in D, \ j = 1, 2.$$

(L3) *For some constant* C_3 *and for any* $d > 0$, $\|\boldsymbol{\beta}_1\| \le \Delta_n$ *and* $\boldsymbol{\gamma}_1 \in D$,

$$\frac{1}{n}\sum_{i=1}^{n} E \sup_{\|\boldsymbol{\beta}_1 - \boldsymbol{\beta}\| + \|\boldsymbol{\gamma}_1 - \boldsymbol{\gamma}\| \le d} |W_i(\boldsymbol{\beta}, \boldsymbol{\gamma}) - W_i(\boldsymbol{\beta}_1, \boldsymbol{\gamma}_1)|^2 \le C_3 d^\delta.$$

Then we have the following results:

(i) *If* $\Delta_n \to 0$ *and* $\Delta_n^\delta |\log \Delta_n| \ll \varepsilon_n \ll \sqrt{n}\Delta_n^{\delta(1+\nu)}$ *for some* $\nu \in (0, 1)$, *then, for any* $a > 2$, *there exists* $C_a < \infty$, *such that*

$$P\left(\sup_{\|\boldsymbol{\beta}\| \le \Delta_n, \boldsymbol{\gamma} \in D} \max_{m \le n} \left|\sum_{i=1}^{m} (W_i(\boldsymbol{\beta}, \boldsymbol{\gamma}) - W_i(0, \boldsymbol{\gamma}))\right| \ge \sqrt{n}\varepsilon_n\right) \tag{3.2}$$
$$\le C_a \exp\{-(aC_1\Delta_n^\delta)^{-1}\varepsilon_n^2\}.$$

(ii) *If* $\log(2 + \Delta_n) \ll \varepsilon_n^2 \ll n$, *then for any* $a > 2$, *there exists* $C_a < \infty$ *such that*

$$P\left(\sup_{\|\boldsymbol{\beta}\| \le \Delta_n, \boldsymbol{\gamma} \in D} \max_{m \le n} \left|\sum_{i=1}^{m} (W_i(\boldsymbol{\beta}, \boldsymbol{\gamma}) - W_i(0, \boldsymbol{\gamma}))\right| \ge \sqrt{n}\varepsilon_n\right) \tag{3.3}$$
$$\le C_a \exp\{-a^{-1}(\varepsilon_n/C_2)^2\}.$$

(iii) *If* $\varepsilon_n = c\sqrt{n}$ *for some constant* $c > 0$ *and* $|\log \Delta_n| = o(n)$, *then* (3.3) *continues to hold for some constant* $a \ge 12$ *even when* (L1) *and* (L3) *are replaced by one weaker condition* (L3′) *given below.*

(L3′) *There is a constant* $B > 0$ *such that*

$$n^{-1}\sum_{i=1}^{n} E \sup_{\|\boldsymbol{\beta} - \boldsymbol{\beta}_1\| + \|\boldsymbol{\gamma} - \boldsymbol{\gamma}_1\| \le n^{-B}} |W_i(\boldsymbol{\beta}, \boldsymbol{\gamma}) - W_i(\boldsymbol{\beta}_1, \boldsymbol{\gamma}_1)| \to 0. \tag{3.4}$$

Now back to the maximum depth regression. We first show that $\hat{\boldsymbol{\beta}}_n = O_p(n^{-1/2})$; that is, for any sequence $\zeta_n \to \infty$, we shall show that

$$P(\|\hat{\boldsymbol{\beta}}_n\| > \zeta_n/\sqrt{n}) \to 0. \tag{3.5}$$

We only need to consider the case with $\zeta_n/\sqrt{n} \to 0$ given the consistency of $\hat{\boldsymbol{\beta}}_n$. Note that for any $c > 0$,

$$\begin{aligned}&\sup_{\zeta_n n^{-1/2}\le\|\boldsymbol{\beta}\|\le 1}\ \min_{\boldsymbol{\gamma}\in S^p} n^{-1/2}\sum_{i=1}^n E[\operatorname{sgn}(e_i - \mathbf{w}_i'\boldsymbol{\beta}) - \operatorname{sgn}(e_i)]\operatorname{sgn}(\mathbf{w}_i'\boldsymbol{\gamma})\\ &\quad\le \sup_{\zeta_n n^{-1/2}\le\|\boldsymbol{\beta}\|\le 1} -2n^{-1/2}\sum_{i=1}^n E|F_i(\mathbf{w}_i'\boldsymbol{\beta}) - F_i(0)| \le -cf(0)\zeta_n n^{-1}N_n,\end{aligned}$$

where

$$N_n = \inf_{\mathbf{1}\in S^p}\sum_{i=1}^n P(|\mathbf{w}_i'\mathbf{1}| \ge c),$$

and we have used the fact that $|\mathbf{w}_i'\boldsymbol{\beta}|/\|\boldsymbol{\beta}\| \ge c$ implies, by condition (C3), $|F(\mathbf{w}_i'\boldsymbol{\beta}) - F(0)| \ge |F(c\zeta_n/\sqrt{n}) - F(0)| \ge \frac{1}{2}c\zeta_n f(0)/\sqrt{n}$. By condition (C1) and the fact that $\|\mathbf{w}_i\| \ge 1$, we have

$$\begin{aligned}n - N_n &= \sup_{\mathbf{1}\in S^p}\sum_{i=1}^n P(|\mathbf{w}_i'\mathbf{1}| < c)\\ &\le \sup_{\mathbf{1}\in S^p}\sum_{i=1}^n P(|\mathbf{w}_i'\mathbf{1}| < c\|\mathbf{w}_i\|) \le Bnc^{\delta}.\end{aligned}$$

Therefore, by choosing c small enough so that $Bc^{\delta} < 1/2$, we obtain

$$\sup_{\zeta_n n^{-1/2}\le\|\boldsymbol{\beta}\|\le 1}\ \min_{\boldsymbol{\gamma}\in S^p} n^{-1/2}\sum_{i=1}^n E[\operatorname{sgn}(e_i - \mathbf{w}_i'\boldsymbol{\beta}) - \operatorname{sgn}(e_i)]\operatorname{sgn}(\mathbf{w}_i'\boldsymbol{\gamma}) < -\eta(c)\zeta_n. \tag{3.6}$$

Lemma 3.1 then implies that

$$\begin{aligned}&\sup_{\zeta_n/\sqrt{n}\le\|\boldsymbol{\beta}\|\le 1}\ \min_{\boldsymbol{\gamma}\in S^p} n^{-1/2}\sum_{i=1}^n\{\operatorname{sgn}(e_i - \mathbf{w}_i'\boldsymbol{\beta}) - \operatorname{sgn}(e_i)\}\operatorname{sgn}(\mathbf{w}_i'\boldsymbol{\gamma})\\ &\quad< -\eta(c)\zeta_n + o_p(\zeta_n^{1/2}).\end{aligned}$$

This, together with Theorem 2.1, proves (3.5).

Now, define $\boldsymbol{\delta} = \sqrt{n}\boldsymbol{\beta}$ and $\hat{\boldsymbol{\delta}}_n = \sqrt{n}\hat{\boldsymbol{\beta}}_n = O_p(1)$. By condition (C1), we have $n^{-1/2}\max_{i\le n}|\hat{\boldsymbol{\delta}}_n'\mathbf{w}_i| = o_p(1)$. Then by condition (C3), we have, for $\|\boldsymbol{\delta}\| \le V$, any large constant,

$$\begin{aligned}&n^{-1/2}\sum_{i=1}^n E\{(\operatorname{sgn}(e_i - \mathbf{w}_i'\boldsymbol{\beta}) - \operatorname{sgn}(e_i))\operatorname{sgn}(\mathbf{w}_i'\boldsymbol{\gamma})\}\\ &\quad= -2n^{-1/2}\sum_{i=1}^n E\{(F(n^{-1/2}\mathbf{w}_i'\boldsymbol{\delta}) - F(0))\operatorname{sgn}(\mathbf{w}_i'\boldsymbol{\gamma})\} = \boldsymbol{\mu}'(\boldsymbol{\gamma})\boldsymbol{\delta} + o(1),\end{aligned}$$

where $\mu(\gamma) = -2f(0)E\{\mathbf{w}\,\mathrm{sgn}(\mathbf{w}'\gamma)\}$. Therefore, by Lemma 3.1, it holds uniformly for $\|\delta\| \le V$ and $\gamma \in S^p$,

$$\begin{aligned}(3.7)\qquad &\frac{1}{\sqrt{n}}\sum_{i=1}^{n}\mathrm{sgn}(e_i - \mathbf{w}_i'\beta_n)\mathrm{sgn}(\mathbf{w}_i'\gamma)\\ &= \frac{1}{\sqrt{n}}\sum_{i=1}^{n}\mathrm{sgn}(e_i)\mathrm{sgn}(\mathbf{w}_i\gamma) + \mu'(\gamma)\delta + o_p(1).\end{aligned}$$

Notice that $n^{-1/2}\sum_{i=1}^{n}\mathrm{sgn}(e_i)\mathrm{sgn}(\mathbf{w}_i\gamma)$ converges to a Gaussian process $W(\gamma)$ with mean 0 and covariance function $\Lambda(\gamma_1,\gamma_2) = E[\mathrm{sgn}(\mathbf{w}'\gamma_1)\mathrm{sgn}(\mathbf{w}'\gamma_2)]$.

Since $\Lambda(\gamma_1,\gamma_2)$ is continuous in γ_1 and γ_2, we may define $W(\gamma)$ so that almost all paths are continuous. Also, note that $\Lambda(\gamma_1,\gamma_2)$ satisfies the Hölder condition of order δ due to conditions (C1) and (C4). It follows from an application of Lemma 3.2 that the sequence of processes $\{n^{-1/2}\sum_{i=1}^{n}\mathrm{sgn}(e_i)\mathrm{sgn}(\mathbf{w}_i'\gamma)\}$ in $D(S^p)$-space is tight. Therefore, it converges weakly to $W(\gamma)$ with the Skorohod metric in $D(S^p)$-space. Similarly to Theorem 2.7 of Kim and Pollard (1990), it follows that the limiting distribution of $\hat{\delta}_n$ is characterized by the variable β that solves

$$(3.8)\qquad \max_{\beta}\min_{\gamma\in S^p}(W(\gamma) + \mu(\gamma)'\beta),$$

where

$$(3.9)\qquad \mu(\gamma) = -2f(0)E\{\mathrm{sgn}(\mathbf{w}'\gamma)\mathbf{w}\},$$

provided that the solution β to (3.8) is *unique*. Establishing this uniqueness property can be viewed as the most difficult part of the work we are undertaking in the present paper.

The following lemma, stated for each sample path, plays a fundamental role in the paper.

Suppose that $\mu(\gamma)$ is a continuously differentiable function defined on S^p. Extend $\mu(\gamma)$ to $R^p - \{0\}$ by $\mu(r\gamma) = \mu(\gamma)$ for any $r > 0$ and $\gamma \in S^p$. Let

$$D_\gamma = \left.\frac{\partial\mu'(\gamma + \mathbf{1})}{\partial\mathbf{1}}\right|_{\mathbf{1}=0},$$

which is a $p \times p$ matrix. Obviously, this matrix cannot be of full rank.

LEMMA 3.3. *Suppose that $W(\gamma)$ is continuous and $\mu(\gamma)$ is differentiable on S^p. Under the following conditions* (W1)–(W3), *the solution to* (3.8) *is unique.*

(W1) *For any $\mathbf{1} \in S^p$, the minimum of $\mathbf{1}'\mu(\gamma)$ is negative and achieved only at $\gamma = \mathbf{1}$.*
(W2) *There exists at most one direction $\pm\alpha \in S^p$ such that D_γ is well defined with* rank $p-1$ *and* $(\mathbf{D}_\gamma)\gamma = 0$ *for all γ not parallel to α.*
(W3) *There do not exist β and γ such that $W(\gamma) + \mu(\gamma)'\beta = W(-\gamma) + \mu(-\gamma)'\beta$.*

The same proof shows that Lemma 3.3 is true if $\mu(\gamma)$ is replaced by $-\mu(\gamma)$. It will be shown in the Appendix that $\boldsymbol{\mu}(\gamma) = -2f(0)E\{\mathrm{sgn}(\mathbf{w}'\gamma)\mathbf{w}\}$ satisfies (W1)–(W3) if conditions (C1)–(C4) hold. Our main purpose in the paper is to establish the following theorem.

THEOREM 3.1. *Under conditions* (C1)–(C4), $n^{1/2}(\hat{\boldsymbol{\beta}}_n - \boldsymbol{\beta})$ *converges in distribution to the random variable as the solution to*

$$\max_{\boldsymbol{\beta}\in R^p}\min_{\gamma\in S^p}\Big(W(\gamma) + \boldsymbol{\mu}(\gamma)'\boldsymbol{\beta}\Big),$$

where $\boldsymbol{\mu}(\gamma)$ *is given in* (3.9), $W(\gamma)$ *is the Gaussian process with mean* 0 *and covariance function* $\mathrm{cov}(W(\gamma_1), W(\gamma_2)) = E\{\mathrm{sgn}(\gamma_1'\mathbf{w}\mathbf{w}'\gamma_2)\}$.

In the case of $p = 2$, the limiting distribution of $n^{1/2}(\hat{\boldsymbol{\beta}}_n - \boldsymbol{\beta})$ simplifies to that derived in He and Portnoy (1998), even though the two forms look somewhat different. Except for the case of the usual median ($p = 1$) problem, the non-Gaussian limiting distributions given in Theorem 3.1 are typical for projection-based estimators but not convenient for inference. However, some properties of the limiting distributions may be understood; see He (1999) for more details. Tyler (1994) gives another example with the same type of limiting distributions.

Similar arguments to those used in Section 2 plus the second part of Lemma 3.1 allow us to get an almost sure bound on the estimator as follows.

THEOREM 3.2. *Under conditions* (C1) *and* (C2), *we have* $\hat{\boldsymbol{\beta}}_n - \boldsymbol{\beta} = O((\log\log n/n)^{1/2})$ *almost surely, provided that* $\inf_{\gamma\in S^p} E|\mathbf{w}'\gamma| > 0$. *If we further assume* (C3), *then*

$$\limsup_{n\to\infty}\sqrt{n/\log\log n}\,|\hat{\boldsymbol{\beta}}_n - \boldsymbol{\beta}| \le \Big(\sqrt{2}f(0)\inf_{\gamma\in S^p} E|\mathbf{w}'\gamma|\Big)^{-1}.$$

4. Asymptotics of the deepest point in R^P. The same techniques used in Section 3 apply to the asymptotic analysis of the deepest point for multivariate data. The result stated in this section completes the work of Nolan (1999). Let $\mathbf{X}_1, \ldots, \mathbf{X}_n$ be a random sample of p-dimensions. The deepest point $\mathbf{T}_n$ is defined as the solution to the max–min problem

$$\sup_{\mathbf{t}}\inf_{\mathbf{u}\in S^p}\sum_{i=1}^{n} I(\mathbf{u}'(\mathbf{X}_i - \mathbf{t}) > 0). \tag{4.1}$$

We assume that there exists $\boldsymbol{\theta}_0$ as the unique deepest point for the population such that $P(\mathbf{u}'(\mathbf{X} - \boldsymbol{\theta}_0) > 0) = 1/2$ for all $\mathbf{u} \in S^p$. Without loss of generality, assume $\boldsymbol{\theta}_0 = 0$. To get the asymptotic linearization results parallel to those in Section 3, let $P_{\mathbf{u}}$ be the one-dimensional marginal distribution of $\mathbf{u}'\mathbf{X}$, and $p_{\mathbf{u}}$ be its corresponding density function. Nolan (1999) showed that if

(N1) $P_{\mathbf{u}}$ has a unique median at 0, for all $\mathbf{u}$, and

(N2) $P_{\mathbf{u}}$ has a bounded positive density, $p_{\mathbf{u}}$, at 0, and $p_{\mathbf{u}}(x)$ is continuous in $\mathbf{u}$ and x at $x = 0$,

then $n^{1/2}\mathbf{T}_n$ converges to the random variable

$$\operatorname{argmax}_{\mathbf{t}} \min_{\mathbf{u}\in S^p}(Z(\mathbf{u}) - \mathbf{u}'\mathbf{t}p_{\mathbf{u}}(0)), \tag{4.2}$$

where $Z(\mathbf{u})$ is a Gaussian process on $\mathbf{u} \in S^p$ with mean zero and $\operatorname{Cov}[Z(\mathbf{u}), Z(\mathbf{v})] = P(\mathbf{u}'\mathbf{X} > 0, \mathbf{v}'\mathbf{X} > 0) - 1/4$, provided that the solution to (4.2) is unique. In the special case of $p = 2$, a proof is given in Nolan (1999) for the desired uniqueness based on some geometric properties in R^2. We now verify that the conditions of Lemma 3.3 hold so that the limiting distribution (4.2) is established for any dimension p. This is done under a mild assumption:

(N3) $\int \|\dot{f}(\mathbf{x})\| \|\mathbf{x}\| \, d\mathbf{x} < \infty$, where $\dot{f}$ is the gradient of f, the density function of $\mathbf{X}$.

THEOREM 4.1. *For any $p \geq 2$, under conditions* (N1)–(N3), $\sqrt{n}\mathbf{T}_n$ *tends in distribution to the random variable defined by the solution to* (4.2).

PROOF. We use Lemma 3.3 to prove the uniqueness of the solution to the max–min problem (4.2). Let $\boldsymbol{\mu}(\mathbf{u}) = -p_{\mathbf{u}}(0)\mathbf{u}$. We show that the derivative of $\boldsymbol{\mu}(\mathbf{u})$ is $\mathbf{D}_{\mathbf{u}} = -p_{\mathbf{u}}(0)(\mathbf{I} - \mathbf{u}\mathbf{u}') - (\mathbf{u}\mathbf{b}_{\mathbf{u}}')$. To get the directional derivative of $\boldsymbol{\mu}$ along any direction $\mathbf{1}$, we use the product rule. The derivative of $\mathbf{u}$ gives $p_{\mathbf{u}}(0)\mathbf{1}$ and the derivative of $p_{\mathbf{u}}(0)$ gives $-\mathbf{u}\mathbf{u}'\mathbf{1}p_{\mathbf{u}}(0) + (\mathbf{u}\mathbf{b}_{\mathbf{u}}')\mathbf{1}$, where $\mathbf{b}_{\mathbf{u}}$ will be calculated below.

Write $\mathbf{u}_t = (\mathbf{u} + t\mathbf{1})/\|\mathbf{u} + t\mathbf{1}\|$, and consider

$$P(\mathbf{u}_t'\mathbf{X} \leq a) = \int_{\mathbf{u}'\mathbf{x}+t\mathbf{1}'\mathbf{x}\leq a\|\mathbf{u}+t\mathbf{1}\|} f(\mathbf{x})\, d\mathbf{x}.$$

Let $\mathbf{B} = (\mathbf{u}, \mathbf{C})$ be an orthonormal matrix with the first column $\mathbf{u}$. Change the variable $\mathbf{x} = \mathbf{B}\mathbf{y}$ and partition $\mathbf{y}' = (v, \mathbf{z}')$ with $v \in R$. Then the above integral can be written as

$$\int \left[\int_{v \leq (a\|\mathbf{u}+t\mathbf{1}\| - t\mathbf{1}'\mathbf{C}\mathbf{z})/(1+t\mathbf{1}'\mathbf{u})} f(\mathbf{B}\mathbf{y})\, dv \right] d\mathbf{z}.$$

Taking derivative wrt a and evaluating it at $a = 0$ yields

$$p_{\mathbf{u}_t}(0) = \frac{1}{\mathbf{u}_t'\mathbf{u}} \int f\left(-\frac{t\mathbf{1}'\mathbf{C}\mathbf{z}\mathbf{u}}{(\mathbf{u}_t'\mathbf{u})\|\mathbf{u} + t\mathbf{1}\|} + \mathbf{C}\mathbf{z} \right) d\mathbf{z}.$$

The derivative of $1/(\mathbf{u}_t'\mathbf{u})$ wrt t at $t = 0$ is $-\mathbf{u}'\mathbf{1}$. Now taking the derivative of the inside under the integral wrt t at $t = 0$ we get

$$\mathbf{b}_{\mathbf{u}} = -\int [\mathbf{u}'\dot{f}(\mathbf{C}\mathbf{z})](\mathbf{C}\mathbf{z})\, d\mathbf{z}.$$

We have completed the proof of $\mathbf{D}_{\mathbf{u}} = -(\mathbf{I} - \mathbf{u}\mathbf{u}')p_{\mathbf{u}}(0) - \mathbf{u}\mathbf{b}_{\mathbf{u}}'$. The definition of $\mathbf{C}$ implies that $\mathbf{b}_{\mathbf{u}}'\mathbf{u} = 0$, and further that $\mathbf{D}_{\mathbf{u}}\mathbf{u} = 0$. Thus, $\{\mathbf{a}'\mathbf{D}_{\mathbf{u}} : \mathbf{a} \in R^p\} = \{\mathbf{a}'\mathbf{D}_{\mathbf{u}} : \mathbf{a}'\mathbf{u} = 0\} = \{p_{\mathbf{u}}(0)\mathbf{a}' : \mathbf{a}'\mathbf{u} = 0\}$, which means that the rank of $\mathbf{D}_u$ is $p-1$.

Here condition (W2) holds without having to exclude an exceptional direction α. The other conditions of Lemma 3.3 hold trivially. We then conclude that the asymptotic distribution for the deepest point estimator holds in any dimension and that the proof of Theorem 4.1 is complete. □

APPENDIX

PROOF OF LEMMA 2.1. We apply Lemma 3.2(iii) here. Under (D1)–(D3), we can verify condition (L3′) by taking $B = \max\{b+1, A\}$. It follows from

$$\begin{aligned}\frac{1}{n}\sum_{i=1}^{n} E \sup_{\|\beta-\beta_1\|+\|\gamma-\gamma_1\|\le n^{-B}} &\big|\operatorname{sgn}(y_i - \mathbf{w}_i'\beta)\operatorname{sgn}(\mathbf{w}_i'\gamma) \\ &\quad - \operatorname{sgn}(y_i - \mathbf{w}_i'\beta_1)\operatorname{sgn}(\mathbf{w}_i'\gamma_1)\big| \\ &\le \frac{2}{n}\sum_{i=1}^{n} E[F_i(\mathbf{w}_i'\beta_1 + n^{-A}) - F_i(\mathbf{w}_i'\beta_1 - n^{-A})] + I(|\mathbf{w}_i'\gamma_1| < n^{-1}) \\ &\le \frac{2}{n}\sum_{i=1}^{n} \sup_x [F_i(x + n^{-A}) - F_i(x - n^{-A})] + 2(1 - Q_n(1/n)) = o(1),\end{aligned}$$

where the first inequality uses a fact that $\mathbf{w}_i'\gamma$ and $\mathbf{w}_i'\gamma_1$ have the same sign when $|\mathbf{w}_1\gamma_1| \ge 1/n$ and $|\mathbf{w}_i'\gamma - \mathbf{w}_i'\gamma_1| \le n^{-2}$. Condition (L2) holds automatically.

Assume $\beta^* = 0$ without loss of generality and let

$$H_i(\beta, \gamma) = (\operatorname{sgn}(y_i - \mathbf{w}_i'\beta) - \operatorname{sgn}(y_i))\operatorname{sgn}(\mathbf{w}_i'\gamma). \tag{A.1}$$

By Lemma 3.2(iii) with $\varepsilon = \sqrt{n}$ and $b_n = n^{2A}$ for A given in (D3), we have

$$\sup_{\substack{\|\beta\|\le b_n \\ \gamma\in S^p}} \left|\sum_{i=1}^{n} \{H_i(\beta,\gamma) - EH_i(\beta,\gamma)\}\right| = o(n).$$

To complete the proof, it remains to show that

$$\sup_{\substack{\|\beta\|> b_n \\ \gamma\in S^p}} \left|\sum_{i=1}^{n} \{\operatorname{sgn}(y_i - \mathbf{w}_i'\beta)\operatorname{sgn}(\mathbf{w}_i'\gamma) - E\operatorname{sgn}(y_i - \mathbf{w}_i'\beta)\operatorname{sgn}(\mathbf{w}_i'\gamma)\}\right| = o(n),$$

which follows from

$$\sup_{\substack{\|\beta\|> b_n \\ \gamma\in S^p}} \left|\sum_{i=1}^{n} \{(\operatorname{sgn}(y_i - \mathbf{w}_i'\beta) - \operatorname{sgn}(-\mathbf{w}_i'\beta))\operatorname{sgn}(\mathbf{w}_i'\gamma)\}\right| = o(n).$$

To verify this, write $A_\beta = \{i \le n: |\mathbf{w}_i'\beta| > n^{-A}\|\beta\|\}$. By (D2), $\inf_\beta \operatorname{size}(A_\beta) = n - o(n)$. We further notice that for $i \in A_\beta$,

$$|(\operatorname{sgn}(y_i - \mathbf{w}_i'\beta) - \operatorname{sgn}(-\mathbf{w}_i'\beta))\operatorname{sgn}(\mathbf{w}_i'\gamma)| \le 2I(|y_i| > |\mathbf{w}_i'\beta|) \le 2I(|y_i| > n^A).$$

Therefore,

$$n^{-1} \sup_{\|\boldsymbol{\beta}\|>b_n,\, \boldsymbol{\gamma}\in S^p} \left| \sum_{i=1}^{n} \{(\mathrm{sgn}(y_i - \mathbf{w}_i'\boldsymbol{\beta}) - \mathrm{sgn}(-\mathbf{w}_i'\boldsymbol{\beta}))\mathrm{sgn}(\mathbf{w}_i'\boldsymbol{\gamma})\} \right|$$

$$\le n^{-1} \sup_{\|\boldsymbol{\beta}\|>b_n,\, \boldsymbol{\gamma}\in S^p} \left[\sum_{i=1}^{n} \{2I(|y_i| > n^A)\} + \mathrm{size}(A_{\boldsymbol{\beta}}^c) \right]$$

$$= 2n^{-1} \sum_{i=1}^{n} I(|y_i| > n^A) + o(1) = o_p(1),$$

where the last step is due to (D3). The proof is then complete. □

PROOF OF LEMMA 3.1. The proof of Lemma 3.1 is a direct application of Lemma 3.2 with $W_i(\boldsymbol{\beta}, \boldsymbol{\gamma}) = \mathrm{sgn}(e_i - \mathbf{w}_i'\boldsymbol{\beta})\mathrm{sgn}(\mathbf{w}_i'\boldsymbol{\gamma}) - E[\mathrm{sgn}(e_i - \mathbf{w}_i'\boldsymbol{\beta})\mathrm{sgn}(\mathbf{w}_i'\boldsymbol{\gamma})]$. Here we first verify that conditions (L1)–(L3) of Lemma 3.2 are satisfied. First, we notice that $|\mathrm{sgn}(\mathbf{w}_i'\boldsymbol{\gamma}_1) - \mathrm{sgn}(\mathbf{w}_i'\boldsymbol{\gamma}_2)| \ne 0$ ($= 2$ in fact) if and only if $\mathbf{w}_i'\boldsymbol{\gamma}_1$ and $\mathbf{w}_i'\boldsymbol{\gamma}_2$ have different signs. Consequently, $|\mathbf{w}_i'\boldsymbol{\gamma}_1| \le |\mathbf{w}_i'(\boldsymbol{\gamma}_1 - \boldsymbol{\gamma}_2)| \le \|\mathbf{w}_i\|\|\boldsymbol{\gamma}_1 - \boldsymbol{\gamma}_2\|$. This proves that $E|\mathrm{sgn}(\mathbf{w}_i'\boldsymbol{\gamma}_1) - \mathrm{sgn}(\mathbf{w}_i'\boldsymbol{\gamma}_2)| \le 2P(|\mathbf{w}_i'\boldsymbol{\gamma}_1| \le \|\mathbf{w}_i\|\|\boldsymbol{\gamma}_1 - \boldsymbol{\gamma}_2\|)$. Now, we can verify condition (L1) by

$$\frac{1}{n}\sum_{i=1}^{n} E\left|\mathrm{sgn}(e_i - \mathbf{w}_i'\boldsymbol{\beta}_1)\mathrm{sgn}(\mathbf{w}_i'\boldsymbol{\gamma}_1) - \mathrm{sgn}(e_i - \mathbf{w}_i'\boldsymbol{\beta}_2)\mathrm{sgn}(\mathbf{w}_i'\boldsymbol{\gamma}_2)\right|^2$$

$$\le \frac{4}{n}\sum_{i=1}^{n} \left[E\left|\mathrm{sgn}(e_i - \mathbf{w}_i'\boldsymbol{\beta}_1) - \mathrm{sgn}(e_i - \mathbf{w}_i'\boldsymbol{\beta}_2)\right| + E\left|\mathrm{sgn}(\mathbf{w}_i'\boldsymbol{\gamma}_1) - \mathrm{sgn}(\mathbf{w}_i'\boldsymbol{\gamma}_2)\right|\right]$$

$$\le \frac{8}{n}\sum_{i=1}^{n} \left[E\left|F_i(\mathbf{w}_i'\boldsymbol{\beta}_1) - F_i(\mathbf{w}_i'\boldsymbol{\beta}_2)\right| + P\left(\left|\mathbf{w}_i'\boldsymbol{\gamma}_1\right| \le \|\mathbf{w}_i\|\|\boldsymbol{\gamma}_1 - \boldsymbol{\gamma}_2\|\right)\right]$$

$$\le \frac{8}{n}\sum_{i=1}^{n} \left[BE\left|\mathbf{w}_i'(\boldsymbol{\beta}_1 - \boldsymbol{\beta}_2)\right|^{\delta} + \|\boldsymbol{\gamma}_1 - \boldsymbol{\gamma}_2\|^{\delta}\right]$$

$$\le 8(B(E\|\mathbf{w}\|^2)^{\delta/2} + 1)[\|\boldsymbol{\beta}_1 - \boldsymbol{\beta}_2\|^{\delta} + \|\boldsymbol{\gamma}_1 - \boldsymbol{\gamma}_2\|^{\delta}],$$

where the third inequality here uses (C2) for the first part and (C1) for the second part.

Condition (L2) is trivial, so it remains to verify condition (L3). For this purpose, we note that by conditions (C1) and (C2),

$$\frac{1}{n}\sum_{i=1}^{n} E \sup_{\|\boldsymbol{\beta}_1-\boldsymbol{\beta}\|+|\boldsymbol{\gamma}_1-\boldsymbol{\gamma}\|\le d} |\mathrm{sgn}(e_i - \mathbf{w}_i'\boldsymbol{\beta})\mathrm{sgn}(\mathbf{w}_i'\boldsymbol{\gamma}) - \mathrm{sgn}(e_i - \mathbf{w}_i'\boldsymbol{\beta}_1)\mathrm{sgn}(\mathbf{w}_i'\boldsymbol{\gamma}_1)|^2$$

$$\le \frac{4}{n}\sum_{i=1}^{n} E \sup_{\|\boldsymbol{\beta}_1-\boldsymbol{\beta}\|+\|\boldsymbol{\gamma}_1-\boldsymbol{\gamma}\|\le d} \left[\left|\mathrm{sgn}(e_i - \mathbf{w}_i'\boldsymbol{\beta}) - \mathrm{sgn}(e_i - \mathbf{w}_i'\boldsymbol{\beta}_1)\right| + \left|\mathrm{sgn}(\mathbf{w}_i'\boldsymbol{\gamma}) - \mathrm{sgn}(\mathbf{w}_i'\boldsymbol{\gamma}_1)\right|\right]$$

$$\leq \frac{8}{n}\sum_{i=1}^{n}[P(|e_i - \mathbf{w}_i'\boldsymbol{\beta}_1| \leq \|\mathbf{w}_i\|d) + P(|\mathbf{w}_i'\boldsymbol{\gamma}_1| \leq \|\mathbf{w}_i\|d)]$$

$$\leq \frac{8}{n}\sum_{i=1}^{n} E[B(\|\mathbf{w}_2\|d)^{\delta} + I(|\mathbf{w}_i'\boldsymbol{\gamma}_1| \leq \|\mathbf{w}_i\|d)] \leq 8[B(E\|\mathbf{w}\|^2)^{\delta/2} + 1]d^{\delta}.$$

The first conclusion of Lemma 3.1 follows from Lemma 3.2(i) or (ii) by taking $\varepsilon = \Delta_n^{\delta/2-v}$ in the cases of $\Delta_n \to 0$, but $\varepsilon = \zeta_n \to \infty$ and $\zeta_n \ll \sqrt{n}$ otherwise. For both cases, one can verify that $\Delta_n^{\delta}|\ln \Delta_n| \ll \varepsilon \ll \sqrt{n}\Delta_n^{\delta(1+v)}$ and $\log(2+\Delta_n) \ll \varepsilon \ll \sqrt{n}$.

Now we turn to the proof of the second conclusion. For any $t > 1$, choose ρ and a such that $1 < \rho < t^{2/(1+\delta(1+\alpha))}$ and $2 < a < 2t^2/\rho^{1+\delta(1+\alpha)}$, where α is the index of Δ_n given in the assumptions of Lemma 3.1. Also define $\overline{\Delta}(\ell) = \max\{\Delta_n,\ \rho^{\ell} \leq n < \rho^{\ell+1}\}$ and $\underline{\Delta}(\ell) = \min\{\Delta_n,\ \rho^{\ell} \leq n < \rho^{\ell+1}\}$. Note that, for all large ℓ, $\overline{\Delta}(\ell)/\underline{\Delta}(\ell) < \rho^{1+\alpha}$. Then Lemma 3.2(i) implies, for any large integer ℓ,

$$P\Bigl(\sup_{\|\boldsymbol{\beta}\|\leq\overline{\Delta}(\ell),\,\boldsymbol{\gamma}\in S^p}\ \max_{\rho^{\ell}\leq n<\rho^{\ell+1}} |S_n(\boldsymbol{\beta},\boldsymbol{\gamma})|/\sqrt{2n\Delta_n^{\delta}\log\log n} > t\Bigr)$$

$$\leq P\Bigl(\sup_{\|\boldsymbol{\beta}\|\leq\overline{\Delta}(\ell),\,\boldsymbol{\gamma}\in S^p}\ \max_{\rho^{\ell}\leq n<\rho^{\ell+1}} |S_n(\boldsymbol{\beta},\boldsymbol{\gamma})| > \sqrt{2t^2\rho^{\ell}\underline{\Delta}^{\delta}(\ell)[\log\ell + \log\rho]}\Bigr)$$

$$\leq M\Bigl[\exp\Bigl(-2(a\rho)^{-1}t^2[\log\ell + \log\rho](\underline{\Delta}/\overline{\Delta})^{\delta}\Bigr) + \exp(-\rho^{\ell(1/2-\nu)})\Bigr],$$

for some $M < \infty$ and $\nu < 1/2$. The above bound is summable in ℓ, so the desired result follows from the Borel–Cantelli lemma.

Before proceeding to the proof of Lemma 3.2, we quote the Lévy inequality from Loève [(1977), page 259].

LÉVY INEQUALITY. If $X_1, \dots, X_n$ are independent random variables and $S_k = \sum_{i=1}^{k} X_i$, then, for every ϵ,

$$P\Bigl\{\max_{k\leq n}|S_k - \mathrm{Median}(S_k - S_n)| \geq \epsilon\Bigr\} \leq 2P\{|S_n| \geq \epsilon\}.$$

PROOF OF LEMMA 3.2. The proof is based on chaining. It requires a sequence $M := M_n$ satisfying

$$\sqrt{n}M^{-3\delta}\varepsilon_n^{-1} \to \infty \quad \textit{and} \quad \log M = o(\varepsilon_n^2\Delta_n^{-\delta}). \tag{A.2}$$

For simplicity, we assume $\delta = 1$. If general, replace Δ_n by Δ_n^{δ} in the proof. We only give a detailed proof for the first case where $\Delta_n \to 0$. For the cases (ii) and (iii), see the remarks at the end of this proof. In the proof, we shall assume that $\Delta_n^{-1}\varepsilon^2$ has a positive lower bound, for, otherwise, the lemma becomes trivially true. In the case of $\Delta_n \to 0$, we have $\varepsilon_n = o(\sqrt{n}\Delta^{1+\nu})$, so we can simply choose $M = \Delta_n^{-\nu/3}$.

Under our conditions on Δ_n and $\varepsilon := \varepsilon_n$, there exists $\rho := \rho_n \to 0$ slowly enough such that $\min\{M\rho^4,\ M^w\rho\} \to \infty$ with $w = 3\nu/(21\nu+6)$ and for some $4 \le k_1 := k_{1n} \le k_2 := k_{2n}$,

$$\text{(A.3)} \qquad \frac{\Delta_n\sqrt{n}}{(M\rho)^{k_1-1}\varepsilon} \to \infty, \qquad \frac{\Delta_n\sqrt{n}}{(M\rho)^{k_1}\varepsilon} \to 0,$$

$$\text{(A.4)} \qquad \max\left\{\frac{\Delta_n\sqrt{n}}{(M\rho)^{k_2-1}\varepsilon}, \frac{\Delta_n^{1/2}\sqrt{n}}{(M^{1/2}\rho)^{k_2-1}\varepsilon}\right\} \to \infty$$

and

$$\text{(A.5)} \qquad \max\left\{\frac{\Delta_n\sqrt{n}}{(M\rho)^{k_2}\varepsilon}, \frac{\Delta_n^{1/2}\sqrt{n}}{(M^{1/2}\rho)^{k_2}\varepsilon}\right\} \to 0.$$

Note that the above choices of k_1 and k_2 are made for general cases. In our case of $\Delta_n \to 0$, the maxima in (A.4) and (A.5) are just equal to the second ratios there.

Our choice of ρ satisfies

$$M^{(k_2/4)-(3k_1/4)-(1/4)-(3/2\nu)} \ll M^{(k_2-1)/2-k_1}\rho^{k_2-1-k_1} = \frac{\Delta_n\sqrt{n}}{(M\rho)^{k_1}\varepsilon}\cdot\frac{(M^{1/2}\rho)^{k_2-1}\varepsilon}{\Delta_n^{1/2}\sqrt{n}} \to 0,$$

which implies

$$\text{(A.6)} \qquad k_2 \le 3k_1 + 1 + 6/\nu.$$

Without loss of generality, we assume that $W_i(0, \boldsymbol{\gamma}) \equiv 0$. This is equivalent to working with $W_i^*(\boldsymbol{\beta}, \boldsymbol{\gamma}) = W_i(\boldsymbol{\beta}, \boldsymbol{\gamma}) - W_i(0, \boldsymbol{\gamma})$ rather than $W_i(\boldsymbol{\beta}, \boldsymbol{\gamma})$.

We now use expanding collections of points denoted by $\{(\boldsymbol{\beta}_{j_1}, \boldsymbol{\gamma}_{\ell_1})\}$, $\{(\boldsymbol{\beta}_{j_1,j_2}, \boldsymbol{\gamma}_{\ell_1,\ell_2})\}, \dots, \{(\boldsymbol{\beta}_{j_1,\dots,j_{k_2}}, \boldsymbol{\gamma}_{\ell_1,\dots,\ell_{k_2}})\}$, with $j_t, \ell_t = 1, 2, \dots, J_t$ and $t = 1, 2, \dots k_2$, satisfying

$$\|\boldsymbol{\beta}_{j_1,\dots,j_{t-1}} - \boldsymbol{\beta}_{j_1,\dots,j_t}\| + \|\boldsymbol{\gamma}_{\ell_1,\dots,\ell_{t-1}} - \boldsymbol{\gamma}_{\ell_1,\dots,\ell_t}\| \le \Delta_n M^{-t+1}, \qquad 2 \le t \le k_2.$$

Also, for any $\|\boldsymbol{\beta}\| \le \Delta_n$, $\boldsymbol{\gamma} \in D$, there exist integers $j_1, \dots, j_{k_2}$ and $\ell_1, \dots, \ell_{k_2}$ such that

$$\|\boldsymbol{\beta} - \boldsymbol{\beta}_{j_1,\dots,j_{k_2}}\| + \|\boldsymbol{\gamma} - \boldsymbol{\gamma}_{\ell_1,\dots,\ell_{k_2}}\| \le \Delta_n M^{-k_2}.$$

Note that the tth set of points is constructed by adding J_t additional points around every point in the $(t-1)$th set. These expanding sets can be found with $J_1 \le KM^{2p}\Delta_n^{-p}$ and $J_t \le KM^{2p}$ $(t > 1)$ for some constant K. For brevity, we write $W_i(t) = W_i(\boldsymbol{\beta}_{j_1,\dots,j_t}, \boldsymbol{\gamma}_{\ell_1,\dots,\ell_t})$ and

$$U_i = U_{i,j_1,\dots,j_{k_2},\ell_1,\dots,\ell_{k_2}} = \sup\left|W_i(\boldsymbol{\beta}, \boldsymbol{\gamma}) - W_i(k_2)\right|$$

in the rest of the proof, where sup is taken over the set $\{(\boldsymbol{\beta}, \boldsymbol{\gamma}): \|\boldsymbol{\beta}-\boldsymbol{\beta}_{j_1,\dots,j_{k_2}}\|+\|\boldsymbol{\gamma}-\boldsymbol{\gamma}_{\ell_1,\dots,\ell_{k_2}}\| \le \Delta_n M^{-k_2}\}$. Then, we have

$$
\begin{aligned}
&P\Big(\sup_{\|\boldsymbol{\beta}\|\le\Delta_n,\,\boldsymbol{\gamma}\in D}\max_{m\le n}\Big|\sum_{i=1}^m W_i(\boldsymbol{\beta},\boldsymbol{\gamma})\Big| \ge \sqrt{n}\varepsilon\Big)\\
&\quad\le \sum_{1\le j,\,\ell\le J_1} P\Big(\max_{m\le n}\Big|\sum_{i=1}^m W_i(\boldsymbol{\beta}_j,\boldsymbol{\gamma}_\ell)\Big| \ge \varepsilon\sqrt{n}(1-\rho)\Big)\\
&\qquad+\sum_{t=2}^{k_2}\sum_{j_1,\ell_1,\dots,j_t,\ell_t} P\Big(\max_{m\le n}\Big|\sum_{i=1}^m (W_i(t-1)-W_i(t))\Big| \ge \rho^{t-1}\varepsilon\sqrt{n}(1-\rho)\Big)\\
&\qquad+\sum_{j_1,\ell_1,\dots,j_{k_2},\ell_{k_2}} P\Big(\Big|\sum_{i=1}^n U_i\Big| \ge \rho^{k_2}\varepsilon\sqrt{n}(1-\rho)\Big).
\end{aligned}
\tag{A.7}
$$

We shall show that in the right-hand side of (A.7) the first term dominates and gives the desired bound for Lemma 3.2.

For the case of $t = 1$, we have $|W_i(\boldsymbol{\beta}_j, \boldsymbol{\gamma}_\ell)| \le C_2$, and

$$E\big|W_i(\boldsymbol{\beta}_j,\boldsymbol{\gamma}_\ell)\big|^2 \le C_1\Delta_n. \tag{A.8}$$

Since $\varepsilon^2 \gg \Delta_n$, we have

$$\Big|\mathrm{median}\Big(\sum_{i=m}^n W_i(\boldsymbol{\beta}_j,\boldsymbol{\gamma}_\ell)\Big)\Big| \le C_1^{1/2}\sqrt{n}\Delta_n^{1/2} = o(\sqrt{n}\varepsilon).$$

Now by the Lévy inequality and Bernstein inequality, we obtain, for any $a > 2$, $\rho_1 > \rho > 0$ and for sufficiently large n,

$$
\begin{aligned}
&P\Big(\max_{m\le n}\Big|\sum_{i=1}^m W_i(\boldsymbol{\beta}_j,\boldsymbol{\gamma}_\ell)\Big| \ge \varepsilon\sqrt{n}(1-\rho)\Big)\\
&\quad\le 4\exp\Big(-\tfrac{1}{2}\varepsilon^2 n(1-\rho_1)^2[nC_1\Delta_n + C_2(1-\rho_1)\varepsilon\sqrt{n}]^{-1}\Big)\\
&\quad\le 4\exp\Big(-(aC_1\Delta_n)^{-1}\varepsilon^2\Big).
\end{aligned}
\tag{A.9}
$$

For any $1 < t \le k_2$, the same arguments show that for any $\rho_1 > \rho > 0$ and for sufficiently large n,

$$
\begin{aligned}
&P\Big(\max_{m\le n}\Big|\sum_{i=1}^m (W_i(t-1)-W_i(t))\Big| \ge \rho^{t-1}\varepsilon\sqrt{n}(1-\rho)\Big)\\
&\quad\le 2\exp\Big(-\frac{1}{4}n\rho^{2t-2}(1-\rho_1)^2\varepsilon^2[nC_1\Delta_n M^{-(t-1)}\\
&\qquad\qquad +C_2\rho^{t-1}(1-\rho_1)\varepsilon\sqrt{n}]^{-1}\Big)\\
&\quad\le \begin{cases} 4\exp\Big(-b\rho^{2t-2}M^{(t-1)}\Delta_n^{-1}\varepsilon^2\Big), & \text{if } 1 < t \le k_1,\\ 4\exp(-b\sqrt{n}\rho^{t-1}\varepsilon), & \text{if } k_1 < t \le k_2,\end{cases}
\end{aligned}
\tag{A.10}
$$

where $b > 0$ is a constant.

To bound the last term of (A.7), note that condition (L2) implies $U_i \le 2C_2$ and condition (L3) and (A.5) imply

$$\sum_{i=1}^{n} E(U_i^2) \le C_3 n\Delta_n M^{-k_2} = o(\rho^{k_2}\sqrt{n}\varepsilon).$$

which, together with (A.5), implies that

$$\sum_{i=1}^{n} E(U_i) = O(n\Delta_n^{1/2} M^{-k_2/2}) = o(\rho^{k_2}\sqrt{n}\varepsilon).$$

Then, for any constant $\rho_2 > \rho_1$ and for sufficiently large n,

$$\begin{aligned} P\bigg(\bigg|\sum_{i=1}^{n} U_i\bigg| &\ge \rho^{k_2}\varepsilon\sqrt{n}(1-\rho)\bigg) \\ &\le P\bigg(\bigg|\sum_{i=1}^{n}(U_i - E(U_i))\bigg| \ge \rho^{k_2}\varepsilon\sqrt{n}(1-\rho_1)\bigg) \\ &\le 2\exp(-\tfrac{1}{2}c\rho^{k_2}\varepsilon\sqrt{n}(1-\rho_2)) \le 2\exp(-b\rho^{k_2}\sqrt{n}\varepsilon), \end{aligned}$$

for some $b > 0$.

Therefore, we have a bound for (A.7) as

$$\begin{aligned} P\bigg(\sup_{\|\boldsymbol{\beta}\|\le\Delta_n,\, \boldsymbol{\gamma}\in D} \max_{m\le n} &\bigg|\sum_{i=1}^{m} W_i(\boldsymbol{\beta},\boldsymbol{\gamma})\bigg| \ge \sqrt{n}\varepsilon\bigg) \\ &\le 4J_1^2 \exp\Big(-(aC_1\Delta_n)^{-1}\varepsilon^2\Big) \\ &\quad +4\sum_{t=2}^{k_1}(J_1\cdots J_t)^2 \exp(-b\rho^{2t-2}M^{t-1}\Delta_n^{-1}\varepsilon^2) \qquad \text{(A.11)} \\ &\quad +4\sum_{t=k_1+1}^{k_2+1}(J_1\cdots J_t)^2 \exp(-b\sqrt{n}\rho^{t-1}\varepsilon), \end{aligned}$$

where we use the convention $J_{k_2+1} = 1$.

Our choices of J_t imply that the first term on the right-hand side of (A.11) is bounded by $\exp(-(a_1C_1\Delta_n)^{-1}\varepsilon^2)$ for any $a_1 > a > 2$.

Since $\rho^2 M \to \infty$, the second term on the right-hand side of (A.11) is of smaller order than $\exp(-(a_1C_1\Delta_n)^{-1}\varepsilon^2)$ as $n \to \infty$.

For the last term in (A.11), we use (A.6) and (A.3). For any $k_1 \le t \le k_2$, we have

$$\sqrt{n}\varepsilon\rho^t \gg \varepsilon^2\Delta_n^{-1}M^{k_1-1}\rho^{t+k_1-1} \gg \varepsilon^2\Delta_n^{-1}\Big(M^w\rho\Big)^{t+k_1-1},$$

where

$$w = \frac{3\nu}{21\nu+6} \le \min_{k_1\ge 4}\frac{k_1-1}{4k_1+5+6/\nu}.$$

Therefore, we bound the last sum of (A.11) by

$$4 \sum_{t=k_1+1}^{k_2+1} (J_1 \cdots J_t)^2 \exp(-b\sqrt{n}\rho^{t-1}\varepsilon)$$

$$\leq 4 \sum_{t=k_1}^{k_2} (J_1 \cdots J_{t+1})^2 \exp\Big(- b\varepsilon^2 \Delta_n^{-1} \Big(M^w \rho\Big)^{t+k_1-1}\Big)$$

$$\ll \exp\Big(- (a_1 C_1 \Delta_n)^{-1} \varepsilon^2\Big).$$

Putting things together, we have proved that (A.11) is bounded by $C \exp(-(a_1 C_1 \Delta_n)^{-1}\varepsilon^2)$ for any $a_1 > 2$, where C is a constant that may depend on a_1.

Finally, we add some remarks on the proofs for the other two cases. In case (ii), without loss of generality, we may assume that $\varepsilon \to \infty$, for otherwise the result becomes trivial if we choose a large constant C_a. As for case (iii), we only need one chain in the proof, that is, we only need to select $\{\boldsymbol{\beta}_j, \boldsymbol{\gamma}_\ell;\ j, \ell = 1, 2, \ldots, J, \}$ such that for any $\|\boldsymbol{\beta}\| \leq \Delta_n$ and $\boldsymbol{\gamma} \in D$, there are j and ℓ satisfying

$$\|\boldsymbol{\beta} - \boldsymbol{\beta}_j\| + \|\boldsymbol{\gamma} - \boldsymbol{\gamma}_\ell\| \leq n^{-B}.$$

By our assumptions, we can do so with $J \leq K\Delta^p n^{2pB}$ and thus $\log J = o(n)$. Also, we have

$$\sum_{i=1}^{n} E \sup_{\|\boldsymbol{\beta}-\boldsymbol{\beta}_j\|+\|\boldsymbol{\gamma}-\boldsymbol{\gamma}_\ell\| \leq n^{-B}} \big|W_i(\boldsymbol{\beta}, \boldsymbol{\gamma}) - W_i(\boldsymbol{\beta}_j, \boldsymbol{\gamma}_\ell)\big| = o(n).$$

The rest of the proof is similar to that for case (i). The proof of Lemma 3.2 ends here. □

We now prove Lemma 3.3. First, it is clear that the set of solutions to (3.8) is a nonempty convex set in R^p. Let $\boldsymbol{\beta}_0$ be one solution. Suppose that $\max_{\boldsymbol{\beta}} \min_{\boldsymbol{\gamma} \in S^p} (W(\boldsymbol{\gamma}) + \boldsymbol{\mu}(\boldsymbol{\gamma})'\boldsymbol{\beta}) = d_0$, and $G = \{\boldsymbol{\gamma} \in S^p \colon W(\boldsymbol{\gamma}) + \boldsymbol{\mu}(\boldsymbol{\gamma})'\boldsymbol{\beta}_0 = d_0\}$. By condition (W1), d_0 is finite almost surely. We now have the following lemma.

LEMMA A.1. *There does not exist* $\mathbf{1} \in S^p$ *such that* $\mathbf{1}'\boldsymbol{\mu}(\boldsymbol{\gamma}) > 0$ *for all* $\boldsymbol{\gamma} \in G$.

PROOF. Here $\boldsymbol{\mu}(\boldsymbol{\gamma})$ is continuous and G is a closed set. If the conclusion of Lemma A.1 is not true, then there is a vector $\mathbf{1}$ such that $\delta = \inf_{\boldsymbol{\gamma} \in G} \mathbf{1}'\boldsymbol{\mu}(\boldsymbol{\gamma}) > 0$ as G is obviously a compact set. Set $H = \{\boldsymbol{\gamma} \in S^p \colon \mathbf{1}'\boldsymbol{\mu}(\boldsymbol{\gamma}) > \delta/2\}$. Clearly, $H^c \cap S^p$ is a closed set and $H^c \cap G$ is empty.

Let $d_1 = \max_{\boldsymbol{\gamma} \in H^c \cap S^p} |\mathbf{1}'\boldsymbol{\mu}(\boldsymbol{\gamma})| \in (0, \infty)$ and $d_2 = \min_{\boldsymbol{\gamma} \in H^c \cap S^p} (W(\boldsymbol{\gamma}) + \boldsymbol{\mu}(\boldsymbol{\gamma})'\boldsymbol{\beta}_0) > d_0$. Consider the function

$$Q(\boldsymbol{\gamma}) = W(\boldsymbol{\gamma}) + \boldsymbol{\mu}(\boldsymbol{\gamma})'\boldsymbol{\beta}_0 + t\boldsymbol{\mu}(\boldsymbol{\gamma})'\mathbf{1},$$

with $t = (d_2 - d_0)/(2d_1)$. If $\boldsymbol{\gamma} \in H^c \cap S^p$, then $Q(\boldsymbol{\gamma}) \geq d_2 - td_1 = (d_0 + d_2)/2 > d_0$. If $\boldsymbol{\gamma} \in H$, then $Q(\boldsymbol{\gamma}) \geq d_0 + t\delta/2 > d_0$. These inequalities show that the solution should not be $\boldsymbol{\beta}_0$. The contradiction proves the lemma. □

Now we shall show that the solution to (3.8) is unique by establishing a set of linear equations that any solution to (3.8) must satisfy.

PROOF OF LEMMA 3.3. As in the proof of Lemma A.1, let $\boldsymbol{\beta}_0$ be a solution, and the minimum over $\boldsymbol{\gamma}$ in (3.8) is achieved at some $\boldsymbol{\gamma}^* \in S^p$ so that $W(\boldsymbol{\gamma}^*)+\boldsymbol{\mu}(\boldsymbol{\gamma}^*)'\boldsymbol{\beta}_0 = d_0$. By Lemma A.1, there are at least three different $\boldsymbol{\gamma}^* \in S^p$ in the set G. For otherwise, Lemma A.1 fails, because of (W1), by choosing $\mathbf{l} = -\boldsymbol{\gamma}^*$ if G contains only one vector or $\mathbf{l} = -(\boldsymbol{\gamma}_1^* + \boldsymbol{\gamma}_2^*)/\|\boldsymbol{\gamma}_1^* + \boldsymbol{\gamma}_2^*\|$ if G contains two vectors. This vector $\mathbf{l}$ is well defined since no pair of vectors in G can be in the opposite directions thanks to (W3). Also implied by (W3) is that we can always choose $\boldsymbol{\gamma}^* \in G$ such that it is not parallel to $\boldsymbol{\alpha}$.

At this $\boldsymbol{\gamma}^*$, consider the arc $\boldsymbol{\gamma} = (\boldsymbol{\gamma}^* + t\mathbf{l})/\|\boldsymbol{\gamma}^* + t\mathbf{l}\|$ as t varies for any direction $\mathbf{l}$ with $\mathbf{l}'\boldsymbol{\gamma}^* = 0$. Since $\boldsymbol{\gamma}^*$ is a minimizing point for $W(\boldsymbol{\gamma}) + \boldsymbol{\mu}(\boldsymbol{\gamma})'\boldsymbol{\beta}_0$ and the function is continuous, there must exist sequences $t_k \uparrow 0$ and $s_k \downarrow 0$ such that at least one sequence is strictly monotone and

$$W(\boldsymbol{\gamma}^* + t_k\mathbf{l}) + \boldsymbol{\mu}(\boldsymbol{\gamma}^* + t_k\mathbf{l})'\boldsymbol{\beta}_0 = W(\boldsymbol{\gamma}^* + s_k\mathbf{l}) + \boldsymbol{\mu}(\boldsymbol{\gamma}^* + s_k\mathbf{l})'\boldsymbol{\beta}_0.$$

Since $\boldsymbol{\mu}(\boldsymbol{\gamma})$ is differentiable, we know that along the sequence $k \to \infty$,

$$\lim_{k\to\infty} \frac{W(\boldsymbol{\gamma}^* + t_k\mathbf{l}) - W(\boldsymbol{\gamma}^* + s_k\mathbf{l})}{t_k - s_k} = \xi_l(\boldsymbol{\gamma}^*)$$

exists and is equal to $[\lim_{k\to\infty}(\boldsymbol{\mu}(\boldsymbol{\gamma}^* + t_k\mathbf{l}) - \boldsymbol{\mu}(\boldsymbol{\gamma}^* + s_k\mathbf{l}))/(t_k - s_k)]'\boldsymbol{\beta}_0$. That is,

$$\xi_l(\boldsymbol{\gamma}^*) = (\mathbf{l}'\mathbf{D}_{\boldsymbol{\gamma}^*})\boldsymbol{\beta}_0 \tag{A.12}$$

for any direction $\mathbf{l}$ orthogonal to $\boldsymbol{\gamma}^*$. By (W2) and the fact that $\boldsymbol{\gamma}^{*\prime}\mathbf{D}_{\boldsymbol{\gamma}^*} = 0$, $\{\mathbf{l}'\mathbf{D}_{\boldsymbol{\gamma}^*}\}$ spans the $(p-1)$-dimensional subspace orthogonal to $\boldsymbol{\gamma}^*$.

Lemma A.1 implies that there exists another $\tilde{\boldsymbol{\gamma}} \in G$ not parallel to $\boldsymbol{\gamma}^*$ such that $W(\tilde{\boldsymbol{\gamma}}) + \boldsymbol{\mu}(\tilde{\boldsymbol{\gamma}})'\boldsymbol{\beta}_0 = W(\boldsymbol{\gamma}^*) + \boldsymbol{\mu}(\boldsymbol{\gamma}^*)'\boldsymbol{\beta}_0$. This gives another equation,

$$W(\tilde{\boldsymbol{\gamma}}) - W(\boldsymbol{\gamma}^*) = (\boldsymbol{\mu}(\tilde{\boldsymbol{\gamma}}) - \boldsymbol{\mu}(\boldsymbol{\gamma}^*))'\boldsymbol{\beta}_0. \tag{A.13}$$

By (W1), $(\boldsymbol{\mu}(\tilde{\boldsymbol{\gamma}}) - \boldsymbol{\mu}(\boldsymbol{\gamma}^*))'\boldsymbol{\gamma}^* \neq 0$, so $\boldsymbol{\mu}(\tilde{\boldsymbol{\gamma}}) - \boldsymbol{\mu}(\boldsymbol{\gamma}^*)$ is not in the space of $\{\mathbf{l}: \mathbf{l}'\boldsymbol{\gamma}^* = 0\}$. This means that (A.12) and (A.13) put together include p linearly independent equations and they uniquely determine $\boldsymbol{\beta}_0$.

Conditions (W1) and (W3) are trivial for the above defined

$$\boldsymbol{\mu}(\boldsymbol{\gamma}) = -2f(0)E\{\text{sgn}(\mathbf{w}'\boldsymbol{\gamma})\mathbf{w}\}.$$

Thus, to complete the proof of Theorem 3.1, we only need to verify that condition (W2) is satisfied, which is shown in the following lemma.

LEMMA A.2. *Let* $\boldsymbol{\gamma}' = (\gamma_0, \boldsymbol{\gamma}_1') \in S^p$ *with* $\|\boldsymbol{\gamma}_1\| \neq 0$ *and let* $\mathbf{B}$ *be a* $(p-1) \times (p-1)$ *orthonormal matrix with* $\boldsymbol{\gamma}_1/\|\boldsymbol{\gamma}_1\|$ *as its first column. Assume that* $F_{\mathbf{B}}(y_0, \mathbf{y}_1)$, *the c.d.f. of* $\mathbf{B}'\mathbf{x} := (y_0, \mathbf{y}_1')'$, *is continuously differentiable with* *respect to y_0 with derivative $\dot{F}_{\mathbf{B}}(y_0, \mathbf{y}_1)$. Then, the derivative matrix of $\boldsymbol{\mu}(\boldsymbol{\gamma})$ is given by*

$$\mathbf{D}_{\gamma} = -4f(0)\|\boldsymbol{\gamma}_1\|^{-1}\int (1, \mathbf{y}'\mathbf{B}')'(1, \mathbf{y}'\mathbf{B}')\dot{F}_{\mathbf{B}}(-\gamma_0/\|\boldsymbol{\gamma}_1\|, d\mathbf{y}_1),$$

where $\mathbf{y}_1 \in R^{p-2}$ and $\mathbf{y} = (-\gamma_0/\|\boldsymbol{\gamma}_1\|, \mathbf{y}_1')'$. Consequently, the directional derivative of $\boldsymbol{\mu}(\boldsymbol{\gamma})$ along the direction $\mathbf{l} \in S^p$ is equal to $\mathbf{l}'\mathbf{D}_{\gamma}$.

It is seen from Lemma A.2 that $\mathbf{D}_{\gamma}$ is well defined if $\boldsymbol{\gamma}$ is not parallel to $\boldsymbol{\alpha} = (1, 0, \dots, 0)'$. Now, we verify that $\mathbf{D}_{\gamma}$ satisfies condition (W2) of Lemma 3.3. First, we note that $\mathbf{D}_{\gamma}\boldsymbol{\gamma} = 0$ holds for any $\boldsymbol{\gamma}$ not parallel to $\boldsymbol{\alpha}$ since $\boldsymbol{\gamma}_1'\mathbf{B} = (\|\boldsymbol{\gamma}_1\|, 0, \dots, 0)$ implies that $(1, \mathbf{y}'\mathbf{B}')\boldsymbol{\gamma} = 0$. Conversely, if $\mathbf{D}_{\gamma}\mathbf{l} = 0$ for some $\mathbf{l} \in S^p$, then $\mathbf{l}'\mathbf{D}_{\gamma}\mathbf{l} = 0$, which, together with the expression of $\mathbf{D}_{\gamma}$, implies that $(1, \mathbf{y}'\mathbf{B}')\mathbf{l} = 0$ for almost all $\mathbf{y} = (-\gamma_0/\|\boldsymbol{\gamma}_1\|, \mathbf{y}_1')$ with $\mathbf{y}_1 \in R^{p-2}$. Partitioning $\mathbf{l}' = (l_0, \mathbf{l}_1')$ and $\mathbf{B} = (\gamma_1/\|\gamma_1\| \vdots \mathbf{B}_1)$, we get

$$l_0 - (\mathbf{l}_1'\boldsymbol{\gamma}_1/\|\boldsymbol{\gamma}_1\|)\gamma_0 + \mathbf{y}_1'\mathbf{B}_1'\mathbf{l}_1 = 0.$$

Since $\mathbf{y}_1$ runs over $p-2$ linearly independent vectors in R^{p-2}, we obtain $l_0 = (\mathbf{l}_1'\boldsymbol{\gamma}_1/\|\boldsymbol{\gamma}_1\|)\gamma_0$ and $\mathbf{l}_1'\mathbf{B}_1 = 0$. Since $\mathbf{B}$ is orthogonal, we get $\mathbf{l}_1 = c\boldsymbol{\gamma}_1$ for some $c \in R$, and hence $l_0 = c\gamma_0$. Therefore, $\mathbf{l}$ must be parallel to $\boldsymbol{\gamma}$. Putting things together, we see that $\mathbf{D}_{\gamma}$ has rank $p-1$ and the set $\{\mathbf{l}'\mathbf{D}_{\gamma}: \mathbf{l}'\boldsymbol{\gamma} = 0\}$ forms a $(p-1)$-dimensional linear space orthogonal to $\boldsymbol{\gamma}$. This shows condition (W2) and completes the proof of Theorem 3.1. □

Now, let us prove Lemma A.2.

PROOF OF LEMMA A.2. For brevity, we suppress the constant factor $2f(0)$ from the definition of $\boldsymbol{\mu}$. We take any direction $\mathbf{l}$ such that $\mathbf{l}' = (l_0, \mathbf{l}_1') \in S^p$. Consider

$$\begin{aligned}\boldsymbol{\mu}(\boldsymbol{\gamma} + t\mathbf{l}) - \boldsymbol{\mu}(\boldsymbol{\gamma}) &= \Big(\int_{\mathbf{w}'\boldsymbol{\gamma}>0} - \int_{\mathbf{w}'(\boldsymbol{\gamma}+t\mathbf{l})>0}\Big)\mathbf{w}G(d\mathbf{x}) \\ &\quad -\Big(\int_{\mathbf{w}'\boldsymbol{\gamma}<0} - \int_{\mathbf{w}'(\boldsymbol{\gamma}+t\mathbf{l})<0}\Big)\mathbf{w}G(d\mathbf{x}) \\ &:= \Delta_1(t) - \Delta_2(t),\end{aligned}$$

where $G(\cdot)$ is the distribution of $\mathbf{x}$.

Note that $\mathbf{w}' = (1, \mathbf{x}')$. Use change of variables $\mathbf{x} = \mathbf{B}(y_0, \mathbf{y}_1')'$ with the orthogonal matrix $\mathbf{B}$ whose first column is $\boldsymbol{\gamma}_1/\|\boldsymbol{\gamma}_1\|$. Then $y_0 = \boldsymbol{\gamma}_1'\mathbf{x}/\|\boldsymbol{\gamma}_1\|$. Let $\mathbf{l}_1'\mathbf{B} = (a_0, \mathbf{a}_1') \in R \times R^{p-2}$. We have

$$\begin{aligned}\Delta_1(t) &:= \Big(\int_{\mathbf{w}'\boldsymbol{\gamma}>0} - \int_{\mathbf{w}'(\boldsymbol{\gamma}+t\mathbf{l})>0}\Big)\mathbf{w}G(d\mathbf{x}) \\ &= \Big(\int_{\gamma_0+\boldsymbol{\gamma}_1'\mathbf{x}>0} - \int_{\gamma_0+tl_0+\boldsymbol{\gamma}_1'\mathbf{x}+t\mathbf{l}_1'\mathbf{x}>0}\Big)(1, \mathbf{x}')'G(d\mathbf{x}) \\ &= -\int_{-(\gamma_0+tl_0+t\mathbf{a}_1'\mathbf{y}_1)/(\|\boldsymbol{\gamma}_1\|+ta_0)}^{-\gamma_0/\|\boldsymbol{\gamma}_1\|}(1, \mathbf{y}'\mathbf{B})F_{\mathbf{B}}(dy_0, d\mathbf{y}_1).\end{aligned}$$

It then follows that

$$\lim_{t\to 0}\frac{\Delta_1(t)}{t} = -\int \frac{-a_0\gamma_0 + l_0\|\boldsymbol{\gamma}_1\| + \mathbf{a}_1'\mathbf{y}_1\|\boldsymbol{\gamma}_1\|}{\|\boldsymbol{\gamma}_1\|^2}(1,\mathbf{y}'\mathbf{B})'\dot{F}_{\mathbf{B}}(-\gamma_0/\|\boldsymbol{\gamma}_1\|, d\mathbf{y}_1)$$

with $\mathbf{y}' = (-\gamma_0/\|\boldsymbol{\gamma}_1\|, \mathbf{y}_1')$. This is equivalent to

$$\lim_{t\to 0}\frac{\Delta_1(t)}{t} = -\|\boldsymbol{\gamma}_1\|^{-1}\Big[\int(1,\mathbf{y}'\mathbf{B})'(1,\mathbf{y}'\mathbf{B})\dot{F}_{\mathbf{B}}(-\gamma_0/\|\boldsymbol{\gamma}_1\|, d\mathbf{y}_1)\Big]\mathbf{1}.$$

Similarly, one can show that

$$\lim_{t\to 0}\frac{\Delta_2(t)}{t} = \|\boldsymbol{\gamma}_1\|^{-1}\Big[\int(1,\mathbf{y}'\mathbf{B})'(1,\mathbf{y}'\mathbf{B})\dot{F}_{\mathbf{B}}(-\gamma_0/\|\boldsymbol{\gamma}_1\|, d\mathbf{y}_1)\Big]\mathbf{1}.$$

The proof of Lemma A.2 is then complete. □

PROOF OF THEOREM 3.2. Since $\hat{\boldsymbol{\beta}}_n \to 0$, there exists a sequence of constants $\Delta_n \to 0$ such that $\limsup_{n\to\infty}\|\hat{\boldsymbol{\beta}}_n\|/\Delta_n \le 1$ almost surely. We only need to consider $\|\boldsymbol{\beta}\| \le \Delta_n$. Applications of Lemma 3.2 yield

$$\limsup \sup_{\gamma\in S^p}\left|\frac{1}{\sqrt{2n\log\log n}}\sum_{i=1}^n \operatorname{sgn}(e_i)\operatorname{sgn}(\mathbf{w}_i'\boldsymbol{\gamma})\right| \le 1 \quad \text{a.s.}$$

and

$$\begin{aligned}\frac{1}{\sqrt{n\log logn}}&\sum_{i=1}^n \operatorname{sgn}(e_i - \mathbf{w}_i'\boldsymbol{\beta})\operatorname{sgn}(\mathbf{w}_i'\boldsymbol{\gamma})\\ &= \sqrt{\frac{n}{\log\log n}}E\operatorname{sgn}(e - \mathbf{w}'\boldsymbol{\beta})\operatorname{sgn}(\mathbf{w}'\boldsymbol{\gamma})\\ &\quad + \frac{1}{\sqrt{n\log\log n}}\sum_{i=1}^n \operatorname{sgn}(e_i)\operatorname{sgn}(\mathbf{w}_i'\boldsymbol{\gamma}) + o(1) \quad \text{a.s.,}\end{aligned}$$

uniformly in $\boldsymbol{\gamma} \in S^p$. By (3.6), there exist $c > 0$ and $\eta(c) > 0$ such that

$$\inf_{\boldsymbol{\gamma}} E\operatorname{sgn}(e - \mathbf{w}'\boldsymbol{\beta})\operatorname{sgn}(\mathbf{w}'\boldsymbol{\gamma}) \le -\eta(c)\zeta_n$$

whenever $\|\boldsymbol{\beta}\| \ge \zeta_n/\sqrt{n}$. Thus, there exists $K < \infty$ such that $\|\boldsymbol{\beta}\| > \zeta_n = K\sqrt{\log\log n/n}$ implies

$$\inf_{\boldsymbol{\gamma}}\frac{1}{\sqrt{n\log\log n}}\sum_{i=1}^n \operatorname{sgn}(e_i - \mathbf{w}_i'\boldsymbol{\beta})\operatorname{sgn}(\mathbf{w}_i'\boldsymbol{\gamma}) < 0$$

for sufficiently large n. Similarly to the arguments in Section 2, we see that the estimate must satisfy $\|\hat{\boldsymbol{\beta}}_n\| \le K\sqrt{\log\log n/n}$ almost surely.

The second conclusion of Theorem 3.2 follows by noticing that, under (C3),

$$\inf_{\boldsymbol{\gamma}} E\operatorname{sgn}(e - \mathbf{w}'\boldsymbol{\beta})\operatorname{sgn}(\mathbf{w}'\boldsymbol{\gamma}) \le -2f(0)\|\boldsymbol{\beta}\|\inf_{\gamma\in S^p} E|\mathbf{w}'\boldsymbol{\gamma}|(1 + o(1)). \qquad \square$$

Acknowledgments. The authors thank two anonymous referees and one Associate Editor whose valuable comments and suggestions helped improve the paper.

REFERENCES

DONOHO, D. L. and GASKO, M. (1992). Breakdown properties of location estimates based on half space depth and projection outlyingness. *Ann. Statist.* **20** 1803–1827.

HE, X. (1999). Comment on "Regression depth," by P. J. Rousseeuw and M. Hubert. *J. Amer. Statist. Assoc.* **94** 403–404.

HE, X., JURECKOVA, J., KOENKER, R. and PORTNOY, S. (1990). Tail behavior of regression estimators and their breakdown points. *Econometrica* **58** 1195–1214.

HE, X. and PORTNOY, S. (1998). Asymptotics of the deepest line. In *Applied Statistical Science III: Nonparametric Statistics and Related Topics* (S.E. Ahmed, M. Ahsanullah, and B.K. Sinha, eds.) 71–81. Nova Science Publications, New York.

HE, X. and SHAO, Q. M. (1996). A general Bahadur representation of M-estimators and its application to linear regression with nonstochastic designs. *Ann. Statist.* **24** 2608–2630.

KIM, J. and POLLARD, D. (1990). Cube root asymptotics. *Ann. Statist.* **18** 191–219.

KOENKER, R. and BASSETT, G. (1978). Regression quantiles. *Econometrica* **46** 33–50.

LIU, R. Y., PARELIUS, J. M. and SINGH, K. (1999). Multivariate analysis by data depth: descriptive statistics, graphics and inference. *Ann. Statist.* **27**. 783–840.

LOÈVE, M. (1977). *Probability Theory*, 4th ed. Springer, New York.

NOLAN, D. (1999). On Min-Max Majority and Deepest Points. *Statist. Probab. Lett.* **43** 325–334.

POLLARD, D. (1984). *Convergence of Stochastic Processes*. Springer, New York.

ROUSSEEUW, P. J. and HUBERT, M. (1999). Regression depth (with discussions). *J. Amer. Statist. Assoc.* **94** 388–402.

ROUSSEEUW, P. J. and STRUYF, A. (1998). Computing location depth and regression depth in higher dimensions. *Statist. Comput.* **8** 193–203.

TUKEY, J. W. (1975). Mathematics and the picturing of data. In *Proceedings of the International Congress of Mathematicians, Vancouver* **2** 523–531.

TYLER, D. E. (1994). Finite sample breakdown points of projection based multivariate location and scatter statistics. *Ann. Statist.* **22** 1024–1044.

DEPARTMENT OF STATISTICS
AND APPLIED PROBABILITY
NATIONAL UNIVERSITY OF SINGAPORE
REPUBLIC OF SINGAPORE 119260

DEPARTMENT OF STATISTICS
UNIVERSITY OF ILLINOIS
725 S. WRIGHT STREET
CHAMPAIGN, ILLINOIS 61820
E-MAIL: he@stat.uiuc.edu

The Annals of Statistics
2002, Vol. 30, No. 1, 122–139

ASYMPTOTIC PROPERTIES OF ADAPTIVE DESIGNS FOR CLINICAL TRIALS WITH DELAYED RESPONSE

BY Z. D. BAI[1], FEIFANG HU[1] AND WILLIAM F. ROSENBERGER[2]

National University of Singapore, University of Virginia and University of Maryland

For adaptive clinical trials using a generalized Friedman's urn design, we derive the limiting distribution of the urn composition under staggered entry and delayed response. The stochastic delay mechanism is assumed to depend on both the treatment assigned and the patient's response. A very general setup is employed with K treatments and L responses. When $L = K = 2$, one example of a generalized Friedman's urn design is the randomized play-the-winner rule. An application of this rule occurred in a clinical trial of depression, which had staggered entry and delayed response. We show that maximum likelihood estimators from such a trial have the usual asymptotic properties.

1. Preliminaries.

1.1. *Introduction.* Adaptive designs for clinical trials use sequentially accruing outcome data to dynamically update the probability of assignment to one of two or more treatments. The idea is to skew these probabilities to favor the treatment that has been the most effective thus far in the trial, thus making the randomization strategy more attractive to physicians and their patients than standard equal allocation. A typical probability model for adaptive clinical trials is the generalized Friedman's urn model [cf. Athreya and Karlin (1968)]. Initially, a vector $\mathbf{Y}_1 = (Y_{11}, \dots, Y_{1K})$ of balls of type $1, \dots, K$ is placed in an "urn" (computer generated). Patients sequentially enter the trial. When a patient is ready to be randomized, a ball is drawn at random and replaced. If it was type i, the ith treatment is assigned. We then wait for a random variable ξ (whose probability distribution depends on i) to be observed. An additional d_{ij} balls are added to the urn of type $j = 1, \dots, K$, where $d_{ij}(\xi)$ is some function on the sample space of ξ. The algorithm is repeated through n stages.

Let $\mathbf{Y}_n = (Y_{n1}, \dots, Y_{nK})$ be the urn composition when the nth patient arrives to be randomized. Then the probability that the patient will be randomized to

Received February 2000; revised September 2001.

[1]Supported by a grant from the National University of Singapore.

[2]Supported by Grant R29-DK51017-05 from the National Institute of Diabetes and Digestive and Kidney Diseases.

AMS 2000 subject classification. 62G10.

Key words and phrases. Generalized Friedman's urn, martingales, randomization, randomized play-the-winner rule, staggered entry, treatment allocation, urn models.

treatment j is given by $Y_{nj}/|\mathbf{Y}_n|$, where $|\mathbf{Y}_n| = \sum_{i=1}^{K} Y_{ni}$. Let $\mathbf{D}(\xi) = ((d_{ij}(\xi), i, j = 1, \ldots, K))$. First-order asymptotics for the generalized Friedman's urn are determined by the generating matrix of the urn, given by $\mathbf{H} = E\{\mathbf{D}(\xi)\}$. Provided $\mathbf{H}$ is positive regular and $\Pr\{d_{ij} = 0\ \forall j\} = 0$ for all i,

$$\frac{Y_{nj}}{|\mathbf{Y}_n|} \to v_j \quad \text{almost surely,} \tag{1}$$

$j = 1, \ldots, K$, where $\mathbf{v} = (v_1, \ldots, v_K)$ is the normalized left eigenvector corresponding to the maximal eigenvalue of $\mathbf{H}$ [cf. Athreya and Karlin (1968)].

As a simple example, ξ might be the primary outcome of a clinical trial, such as death or cure. Assuming that $\mathbf{Y}_1$ is deterministic, let $d_{ij} = (K-1)\delta_{ij}$ if cure on treatment i, and $d_{ij} = (1 - \delta_{ij})$ if death on treatment i, where δ_{ij} is the Kronecker delta. Assuming that ξ is immediately observable after the patient is randomized, we have $|\mathbf{Y}_n| = |\mathbf{Y}_1| + (K-1)(n-1)$. When $K = 2$, this is the randomized play-the-winner rule of Wei and Durham (1978), which has been used occasionally in clinical trials [see, e.g., Bartlett, Roloff, Cornell, Andrews, Dillon and Zwischenberger (1985) and Tamura, Faries, Andersen and Heiligenstein (1994)]. Wei, Smythe, Lin and Park (1990) gave a simple probability model for the randomized play-the-winner rule, letting p_1 be the probability of success on treatment 1 and p_2 be the probability of success on treatment 2. Under this model, by (1),

$$\frac{Y_{n1}}{Y_{11} + Y_{12} + n - 1} \to \frac{1 - p_2}{2 - p_1 - p_2} \quad \text{almost surely}$$

[Rosenberger (1996), page 140] and hence the rule allocates according to the relative risk of failure on treatment 2 versus treatment 1. Wei (1979) first proposed using the generalized Friedman's urn to develop a broad class of allocation rules for clinical trials. The generalized Friedman's urn also has been used in other medical applications [see, e.g., Rosenberger (1996) and Rosenberger and Grill (1997)].

Typically, clinical trials do not result in immediate outcomes, and urn models are simply not appropriate for today's oft-performed long-term survival trials, where outcomes may not be ascertainable for many years. However, there are many trials where many or most outcomes are available during the recruitment period, even though individual patient outcomes may not be immediately available prior to the randomization of the next patient. Consequently, the urn can be updated when outcomes become available, and this does not involve any additional logistical complexities. Wei (1988) suggested such updating for the randomized play-the-winner rule and introduced an indicator function, δ_{jk}, $j < k$, that takes the value 1 if the response of patient j occurs before patient k is randomized and 0 otherwise. He did not explore its properties. Later, Bandyopadhyay and Biswas (1996) explored properties of a simple probability model, which assumes that $P(\delta_{jk} = 1)$ is a constant depending only on the lag $k - j$, for a modified version of the randomized play-the-winner rule. Delays in response can slow the adaptation process considerably, and simulation studies show that the expected allocation proportions generated by the design are more conservative than if outcomes are immediately ascertainable [see Rosenberger (1999)]. However, the adaptive nature of the design still accomplishes its purpose: more patients, on average, are assigned to the better treatment. In practice, time to response in clinical trials can depend on the treatment assigned and the response observed. Heretofore, what has not been known is how such delayed response with staggered entry affects the limiting distribution of the urn composition given by (1).

In this paper, we verify that stochastic staggered entry and delay mechanisms do not affect the limiting distribution of the urn for a wide class of designs defined by the generalized Friedman's urn. We then show that the maximum likelihood estimators have the usual asymptotic properties. This extends the work of Rosenberger, Flournoy and Durham (1997), who investigated properties of maximum likelihood estimators from a generalized Friedman's urn design with immediate response. In our proofs, we assume that patients arrive sequentially and their arrival process has independent increments, and that time to response has a distribution that depends on both the treatment assigned and the patient's response. We investigate a very general adaptive randomization scheme with K treatments and L outcomes, based on the generalized Friedman's urn model.

1.2. *Motivating example.* Tamura, Faries, Andersen and Heiligenstein (1994) described an adaptive placebo-controlled clinical trial of fluoxetine in depression. Patients were stratified by shortened or normal rapid eye movement latency (REML) and then were randomized according to the randomized play-the-winner rule. Separate urns were used in the REML strata. The outcome on which the adaptive randomization was based is a 50% reduction in the Hamilton Depression Scale ($HAMD_{17}$) score in two consecutive visits after at least three weeks of therapy. This outcome was obviously not ascertainable immediately, and hence the urn was updated based on the available data. Exploratory data analysis [Rosenberger and Hu (1999)] indicated that entry time was approximately uniformly distributed over a 270-day time interval. Time to response was similar among the two REML strata and two treatment groups, but differed according to patient outcomes. Time to response was approximately normal with mean 43 days and variance 122 days; in nonresponders, time to determination of nonresponse was approximately uniform on the interval (20 days, 75 days).

When there is immediate response, Wei, Smythe, Lin and Park (1990) showed that the usual maximum likelihood estimator of the treatment effect from a clinical trial employing the randomized play-the-winner rule has the usual asymptotic properties, namely, consistency and asymptotic normality. However, this result is not applicable to clinical trials with delayed response, such as the fluoxetine trial; hence, we have the motivation for this paper.

REMARK 1. The fluoxetine trial had a number of subtleties that necessitated nonstandard analyses; in particular, the outcome on which the adaptation was based was a surrogate outcome for the true primary outcome. We note that Tamura, Faries, Andersen and Heiligenstein (1994) simulated the joint distribution of these outcomes and the time delay in order to make inferences about the treatment effect, and they also performed a Bayesian analysis. They found that fluoxetine was modestly effective in the shortened REML stratum and not effective in the normal REML stratum. The results of our paper suggest a simple alternative analysis, but only for the surrogate outcome, based on the asymptotic distribution of the treatment effect. However, the accuracy of such a test may be appropriately questioned since there were approximately 40 patients in each stratum. With large numbers of patients, enumerating or simulating the exact distribution of the test statistic may be computationally intensive; in which case the asymptotic distribution given in this paper may be attractive.

1.3. *Organization of the paper.* In Section 2, we derive the limiting distribution of the urn composition under delayed response for a multi-armed trial. In Theorem 1, we prove that the urn composition, $\mathbf{Y}_n$, suitably normalized, tends to the same limit as in (1). In Theorem 2, we show that the urn composition tends to a multivariate normal distribution in law, and we give a form of the variance–covariance matrix. The main assumption of the theorems is that the delay cannot be very large relative to the patient entry stream. The important observation is made that the limiting distribution does not depend on the delay mechanism, but, in practice, the delay mechanism must be taken into account in estimating the variance–covariance matrix. In Section 3, we derive the full likelihood and show that the usual asymptotic inference results can be applied to data arising from a generalized Friedman's urn design when there is staggered entry and delayed response. We conclude our paper with some observations in Section 4. Proofs are relegated to the Appendix.

2. Asymptotic properties of the urn composition. We assume a multinomial response model with responses $\xi_n = l$ if patient n had response l, $l = 1, \dots, L$. Let J_n be the treatment indicator for the nth patient, that is, $J_n = j$ if patient n was randomized to treatment $j = 1, \dots, K$, and let $\mathbf{X}_n = (X_{n1}, \dots, X_{nK})$ satisfy $X_{nJ_n} = 1$ and all other elements 0. We assume that the entry time of the nth patient is t_n, where $\{t_n - t_{n-1}\}$ are i.i.d. for all n. The response time of the nth patient is denoted by $\tau_n(j, l)$, which has distribution g_{jl}, $j = 1, \dots, K$, $l = 1, \dots, L$, for all n, so that the distribution of the response times can depend on both the treatment assigned and the response observed. For the nth patient randomized to treatment j, we define an indicator function $M_{jl}(n, m)$ as follows: $M_{jl}(n, m) = 0$ if $\xi_n \neq l$ and $M_{jl}(n, m) = 1_{[\text{response time} \in (t_{n+m}, t_{n+m+1})]}$ if $\xi_n = l$. We assume for $n = 1, 2, \dots$ that, given j, $\{M_{j\xi_n}(n, m)\}$ are i.i.d. By definition, for every pair of n and j, there is only one pair (l, m) such that $M_{jl}(n, m) = 1$ and $M_{jl'}(n, m') = 0$ for all $(l, m) \neq (l', m')$. We can define $\mu_{jlm} = E\{M_{jl}(n, m)\}$ as the probability that a patient on treatment j with response l will respond after m

more patients are enrolled and before $m+1$ more patients are enrolled. Thus we have

$$\sum_{l,m} \mu_{jlm} = 1 \quad \text{for } j = 1, \dots, K.$$

For patient n, after observing $\xi_n = l$, $J_n = i$, we add $d_{ij}(\xi_n = l)$ balls of type j to the urn, where the total number of balls added at each stage is constant; that is, $\sum_{j=1}^{K} d_{ij}(\xi_n) = \beta$, where $\beta > 0$. Without loss of generality, we can assume $\beta = 1$; otherwise, we can consider the sequence $\{\mathbf{Y}_n/\beta\}$ instead. Note that the number of balls added to the urn does not have to be an integer, as in the models of Andersen, Faries and Tamura (1994). Let $\mathbf{D}(l) = ((d_{ij}(\xi_n = l),\ i, j = 1, \dots, K))$.

REMARK 2. It is possible to generalize our results to the case in which the total number of balls added at each stage is random, provided that the expected number of balls added is a positive constant.

For given n and m, if $M_{jl}(n,m) = 1$, then we add balls at the $(n+m)$th stage according to the rule $\mathbf{X}_n\mathbf{D}(l)$. $\mathbf{X}_n$ contains the randomness in J_n, and $\mathbf{D}(l)$ contains the randomness in ξ_n, conditioned on J_n. We can now write a recursive formula for the urn composition,

$$\mathbf{Y}_n = \mathbf{Y}_{n-1} + \mathbf{W}_n,$$

where $\mathbf{W}_n$ is the number of balls of each type added at the nth stage, given by

$$\mathbf{W}_n = \sum_{m=0}^{n-2} M_{J_{n-m-1},\xi_{n-m-1}}(n-m-1, m)\mathbf{X}_{n-m-1}\mathbf{D}(\xi_{n-m-1}).$$

Denote by $\mathcal{F}_n$ the sigma algebra generated by $\{\mathbf{Y}_1, \dots, \mathbf{Y}_n\}$ and let $E_n\{\cdot\} = E\{\cdot|\mathcal{F}_n\}$. We have

$$\begin{aligned} &E_{n-1}\{M_{J_{n-m-1},\xi_{n-m-1}}(n-m-1, m)\mathbf{X}_{n-m-1}\mathbf{D}(\xi_{n-m-1})\} \\ &\quad = \sum_{l=1}^{L} \frac{\mathbf{Y}_{n-m-1}}{|\mathbf{Y}_{n-m-1}|}\mu_{lm}\mathbf{D}(l), \end{aligned}$$

where μ_{lm} is a $K \times K$ diagonal matrix with the jth diagonal element μ_{jlm}. Then

$$E_{n-1}\{\mathbf{W}_n\} = \sum_{m=0}^{n-2} \frac{\mathbf{Y}_{n-m-1}}{|\mathbf{Y}_{n-m-1}|}\left(\sum_{l=1}^{L} \mu_{lm}\mathbf{D}(l)\right).$$

It turns out that it is easier to work with the recursive formula

$$\mathbf{Y}_n = \mathbf{Y}_{n-1} + E_{n-1}\{\mathbf{W}_n\} + (\mathbf{W}_n - E_{n-1}\{\mathbf{W}_n\}).$$

Setting $\mathbf{Q}_n = \mathbf{W}_n - E_{n-1}\{\mathbf{W}_n\}$, we obtain the recursive formula

$$\mathbf{Y}_n = \mathbf{Y}_{n-1} + \sum_{m=0}^{n-2} \frac{\mathbf{Y}_{n-m-1}}{|\mathbf{Y}_{n-m-1}|}\left(\sum_{l=1}^{L} \mu_{lm}\mathbf{D}(l)\right) + \mathbf{Q}_n. \tag{2}$$

We will use (2) as the pivotal recursion formula to prove asymptotic properties of $\mathbf{Y}_n$. But first we will require the following assumptions:

ASSUMPTION 1. For some $c \in (0, 1]$,

$$\sum_{i=m}^{\infty} \mu_{jli} = o(m^{-c}) \qquad \forall j, l. \tag{3}$$

REMARK 3. Assumption 1 implies that the probability that at least m additional patients will arrive prior to a patient's response is of order $o(m^{-c})$. Hence, in practical examples, the delay cannot be very large relative to the entry stream. In practice, it is convenient to verify this assumption by examining the time-to-response variable $\tau_n(j,l)$ and the entry times t_n. If (i) $E[\tau_n(j,l)]^{c_1} < \infty$ for each j, l and $c_1 > c$ and (ii) $E(t_i - t_{i-1}) > 0$ and $E(t_i - t_{i-1})^2 < \infty$, then Assumption 1 is satisfied. This is because

$$\mu_{jlm} = P\{\tau_n(j,l) \in (t_{n+m}, t_{n+m+1})\} = P\{\tau(j,l) \in (S_m, S_{m+1})\},$$

where $S_m = \sum_{i=1}^{m}(t_i - t_{i-1})$ $(t_0 = 0)$. Then

$$\sum_{i=m}^{\infty} \mu_{jli} = P\{\tau(j,l) \in (S_m, \infty)\}.$$

Since $S_m/m \to E(t_i - t_{i-1}) = E(t_1)$ almost surely as $m \to \infty$,

$$P\{\tau(j,l) \in (S_m, \infty)\} \le P\{\tau(j,l) \in (mE(t_1)/2, \infty)\} + P(S_m \le mE(t_1)/2).$$

By the Markov inequality, we have

$$\begin{aligned} P\{\tau(j,l) \in (mE(t_1)/2, \infty)\} &= P\{\tau(j,l) > mE(t_1)/2\} \\ &\le E[\tau(j,l)]^{c_1}/(mE(t_1)/2)^{c_1} \\ &= O(m^{-c_1}) = o(m^{-c}) \end{aligned}$$

and

$$P(S_m \le mE(t_1)/2) = P(S_m - ES_m \le -mE(t_1)/2) \le O(m^{-1}).$$

Consequently, Assumption 1 is not very stringent.

ASSUMPTION 2. Using the notation in Section 1, let $\mathbf{H} = E(\mathbf{D})$ and let $\mathbf{v}$ be the normalized left eigenvector of $\mathbf{H}$ corresponding to its maximal eigenvalue. Assume that $\mathbf{H}$ has the following Jordan decomposition:

$$\mathbf{T}^{-1}\mathbf{H}\mathbf{T} = \operatorname{diag}[1, \mathbf{\Psi}_1, \ldots, \mathbf{\Psi}_s],$$

where $\mathbf{\Psi}_t$ is a $\nu_t \times \nu_t$ matrix (defining ν_t to be the block size of the Jordan form), given by

$$\mathbf{\Psi}_t = \begin{bmatrix} \lambda_t & 1 & 0 & \cdots & 0 \\ 0 & \lambda_t & 1 & \cdots & 0 \\ 0 & 0 & \lambda_t & \cdots & 0 \\ \vdots & \vdots & \vdots & \cdots & \vdots \\ 0 & 0 & 0 & \cdots & \lambda_t \end{bmatrix}.$$

We may select the matrix $\mathbf{T}$ so that its first column is $\mathbf{1}'$ and the first row of $\mathbf{T}^{-1}$ is $\mathbf{v}$. Let $\lambda = \max\{\mathrm{Re}(\lambda_1), \ldots, \mathrm{Re}(\lambda_s)\}$ and $\nu = \max_j\{\nu_j$ such that $\mathrm{Re}(\lambda_j) = \lambda\}$, where Re( ) is the real part of the eigenvalue.

THEOREM 1. *Under Assumptions* 1 *and* 2, *if* $c > 0$ *and* $\lambda < 1$, *then* $\mathbf{Y}_n/|\mathbf{Y}_n| \to \mathbf{v}$ *almost surely.*

PROOF. See the Appendix.

We can extend Theorem 1 to apply not only to the urn composition, but also to the sample fractions assigned to each treatment. Let $\mathbf{N}_n = (N_{n1}, \ldots, N_{nK})$, where N_{nj} is the number of patients assigned to treatment j, $j = 1, \ldots, K$, after n stages.

COROLLARY 1. *Under the assumptions of Theorem* 1, $\mathbf{N}_n/n \to \mathbf{v}$ *almost surely.*

PROOF. See the Appendix.

We now give the central limit result.

THEOREM 2. *Under Assumptions* 1 *and* 2, *for* $c > 1/2$ *and* $\lambda < 1/2$, *we have* $n^{1/2}(\mathbf{Y}_n/|\mathbf{Y}_n| - \mathbf{v})$ *converges in law to* $N(0, \mathbf{\Sigma})$, *where the form of* $\mathbf{\Sigma}$ *is given in* (22).

PROOF. See the Appendix.

REMARK 4. If $\lambda = 1/2$, the asymptotic normality holds, but with a different norming, given by $n \log^{2\nu-1} n$. In this case, we can derive $\mathbf{\Sigma}$ using techniques similar to those in the proof of Theorem 2.

REMARK 5. Because $\boldsymbol{\Sigma}$ depends on $\sum_{m=0}^{\infty} \mu_{jlm} = \Pr\{\xi_n = l | J_n = j\}$ through (21), we see that $\boldsymbol{\Sigma}$ does not depend on the delay mechanism. But this is a limiting result. In practice, we need to estimate $\boldsymbol{\Sigma}$ using (19) and (22), and the estimate will involve the delayed-response mechanism, $M_{jl}(n, m)$. We can estimate $\boldsymbol{\Sigma}$ in practice using the following procedure:

(i) Estimate $\mathbf{H}$ by

$$\hat{\mathbf{H}} = \frac{\sum_{i=2}^{n} \sum_{m=0}^{i-2} M_{J_{i-m-1}, \xi_{i-m-1}}(i-m-1, m) \operatorname{diag}(\mathbf{X}_{i-m-1}) \mathbf{D}(\xi_{i-m-1})}{\sum_{i=2}^{n} \sum_{m=0}^{i-2} M_{J_{i-m-1}, \xi_{i-m-1}}(i-m-1, m)},$$

where $M_{J_{i-m-1}, \xi_{i-m-1}}(i-m-1, m)$, $\mathbf{X}_{i-m-1}$ and $\mathbf{D}(\xi_{i-m-1})$ are observed during the trial.

(ii) Estimate $\mathbf{B}_{ni}$ by

$$\hat{\mathbf{B}}_{ni} = \prod_{j=i+1}^{n} (\mathbf{I} + j^{-1} \hat{\mathbf{H}}).$$

(iii) Estimate $\boldsymbol{\Sigma}$ by

$$\hat{\boldsymbol{\Sigma}} = (\mathbf{I} - (\mathbf{Y}_n' / |\mathbf{Y}_n|) \mathbf{l}) \left[n^{-1} \sum_{i=1}^{n} \hat{\mathbf{B}}_{ni}' (\mathbf{W}_i - \bar{\mathbf{W}})' (\mathbf{W}_i - \bar{\mathbf{W}}) \hat{\mathbf{B}}_{ni} \right] (\mathbf{I} - \mathbf{l}' \mathbf{Y}_n / |\mathbf{Y}_n|),$$

where $\bar{\mathbf{W}} = n^{-1} \sum_{i=1}^{n} \mathbf{W}_i$. The $\mathbf{W}_i$ are the number of balls added to the urn at stage i, which are observed during the trial.

REMARK 6. For the sample fractions assigned to each treatment, we know that

$$n^{-1/2}(\mathbf{N}_n / n - \mathbf{v}) = n^{-1/2} \sum_{i=1}^{n} [\mathbf{X}_i - E(X_i | \mathcal{F}_i)] \tag{4}$$

$$+ n^{1/2} \sum_{i=1}^{n} (\mathbf{Y}_i / |\mathbf{Y}_i| - \mathbf{v}). \tag{5}$$

The asymptotic normality of the first term on the right-hand side of (4) follows from a multivariate version of the martingale central limit theorem. However, we still have not derived the asymptotic distribution of (5) and the correlation between the two terms, and we leave this as an additional research topic. Smythe (1996) proved the asymptotic joint normality of the sample fractions for the generalized Friedman's urn with immediate updating of the urn.

3. Likelihood results. Let $\mathbf{Y}^n = (\mathbf{Y}_1, \ldots, \mathbf{Y}_n)$ be the history of the urn composition, where $\mathbf{Y}_i$ is defined in Section 1. Let $\mathbf{J}^n = (J_1, \ldots, J_n)$ be the history of treatment assignments, $\boldsymbol{\xi}^n = (\xi_1, \ldots, \xi_n)$ be the history of patient responses,

$\boldsymbol{\tau}^n = (\tau_1, \ldots, \tau_n)$ be the history of response times and $\mathbf{t}^n = (t_1, \ldots, t_n)$ be the history of entry times. Then the full likelihood is given by

$$\begin{aligned}
\mathcal{L}_n &= \mathcal{L}(\boldsymbol{\tau}^n, \boldsymbol{\xi}^n, \mathbf{J}^n, \mathbf{Y}^n, \mathbf{t}^n) \\
&= \mathcal{L}(\tau_n | \boldsymbol{\tau}^{n-1}, \boldsymbol{\xi}^n, \mathbf{J}^n, \mathbf{Y}^n, \mathbf{t}^n) \mathcal{L}(\xi_n | \boldsymbol{\tau}^{n-1}, \boldsymbol{\xi}^{n-1}, \mathbf{J}^n, \mathbf{Y}^n, \mathbf{t}^n) \\
&\quad \times \mathcal{L}(J_n | \boldsymbol{\tau}^{n-1}, \boldsymbol{\xi}^{n-1}, \mathbf{J}^{n-1}, \mathbf{Y}^n, \mathbf{t}^n) \mathcal{L}(\mathbf{Y}_n | \boldsymbol{\tau}^{n-1}, \boldsymbol{\xi}^{n-1}, \mathbf{J}^{n-1}, \mathbf{Y}^{n-1}, \mathbf{t}^n) \\
&\quad \times \mathcal{L}(t_n | \boldsymbol{\tau}^{n-1}, \boldsymbol{\xi}^{n-1}, \mathbf{J}^{n-1}, \mathbf{Y}^{n-1}, \mathbf{t}^{n-1}) \mathcal{L}_{n-1} \\
&= \mathcal{L}(\tau_n | J_n, \xi_n) \mathcal{L}(\xi_n | J_n) \mathcal{L}(J_n | \mathbf{Y}_n) \mathcal{L}(t_n) \mathcal{L}_{n-1} \\
&= \prod_{i=1}^{n} \mathcal{L}(\tau_i | J_i, \xi_i) \mathcal{L}(\xi_i | J_i) \mathcal{L}(J_i | \mathbf{Y}_i) \mathcal{L}(t_i) \\
&\propto \prod_{i=1}^{n} \mathcal{L}(\xi_i | J_i).
\end{aligned}$$

Note that the allocation proportions are random and, together with treatment responses, form a sufficient statistic, unlike in the i.i.d. case with fixed allocation.

For the problem we have formulated, we have a product multinomial likelihood with $p_{jl} = \Pr\{\xi_n = l | J_n = j\}$ for all n, and $j = 1, \ldots, K$, $l = 1, \ldots, L-1$ and $p_{jL} = 1 - p_{j1} - \cdots - p_{j,L-1}$. Standard martingale techniques can be used to prove the consistency and asymptotic normality of the maximum likelihood estimators $\hat{p}_{jl}$ from this likelihood. Rosenberger, Flournoy and Durham (1997) gave a convenient set of sufficient conditions. In our case, only their condition (A3) is nontrival. Using their notation, let $L_i = \log(\mathcal{L}_i / \mathcal{L}_{i-1})$, where $\mathcal{L}_0 = 1$. Then condition (A3) requires

$$-n^{-1} \sum_{i=1}^{n} E_{i-1} \left\{ \frac{\partial^2 L_i}{\partial p_{jl} \partial p_{km}} \right\} \to \gamma_{jklm} \quad \text{almost surely,} \tag{6}$$

where γ_{jklm} is a constant, $j, k = 1, \ldots, K$, $l, m = 1, \ldots, L-1$. Using the multinomial likelihood, it is easy to show that the left-hand side of (6) is 0 when $j \neq k$ and, for $j = k$, is given by

$$\left(\frac{1}{p_{jl}} + \frac{1}{p_{jL}} \right) n^{-1} \sum_{i=1}^{n} \frac{Y_{ij}}{|\mathbf{Y}_i|}, \qquad l = m \tag{7}$$

and

$$\frac{1}{p_{jL}} n^{-1} \sum_{i=1}^{n} \frac{Y_{ij}}{|\mathbf{Y}_i|}, \qquad l \neq m. \tag{8}$$

From Theorem 1, (7) converges almost surely to $v_j / p_{jl} + v_j / p_{jL}$ and (8) converges almost surely to v_j / p_{jL}. Hence,

$$\boldsymbol{\Gamma} = ((\gamma_{jklm})) = \frac{v_j}{p_{jl}} \mathbf{I} + \frac{v_j}{p_{jL}} \mathbf{J},$$

where $\mathbf{I}$ is the identity matrix and $\mathbf{J} = \mathbf{1}\mathbf{1}'$. Then, by the theorem of Rosenberger, Flournoy and Durham (1997), page 71, we obtain the following result:

THEOREM 3. *For fixed $j = 1, \ldots, K$, the vector with components*

$$((n^{1/2}\{\hat{p}_{jl} - p_{jl}\}))_{l=1}^{L-1}$$

is asymptotically multivariate normal with mean vector $\mathbf{0}$ *and variance–covariance matrix*

$$\mathbf{\Gamma}^{-1} = \frac{p_{jl}}{v_j}\mathbf{I} - \frac{p_{jl}^2}{v_j(p_{jL} + (L-1)p_{jl})}\mathbf{J}.$$

Moreover, the K vectors are asymptotically independent.

Consequently, the usual asymptotic χ^2 tests can be used to investigate the treatment effect. For $K = L = 2$, we can use standard Z tests of the simple difference of proportions or the odds ratio.

4. Conclusions. Results on the asymptotic properties of the generalized Friedman's urn when there is a stochastic delay in updating the urn are interesting in their own right, from a probabilistic perspective. But the main contribution of this paper is in showing that randomized clinical trials using the generalized Friedman's urn for randomization can now use standard maximum likelihood estimation following the trial, under the standard clinical trial conditions of staggered entry and delayed response. We have also demonstrated, in Remark 3, that the assumptions on the entry stream and delay mechanism are typically not stringent.

We have not examined properties of estimators in this paper. For example, the joint distribution of sufficient statistics could be used to develop inferential tests, as an alternative to maximum likelihood. It would be interesting to develop several types of estimators and compare their efficiencies under different delay mechanisms, but we will leave that topic for future research.

Finally, asymptotic theory is becoming less important in this age of rapid algorithms for computing exact distributions. Hardwick and Stout (1998) performed seminal work in this area for adaptive designs and generally found samples as large as $n = 75$ to be amenable to exact computations, using parallel processing and networking algorithms. How one would implement such algorithms with a stochastic delay mechanism may be an interesting topic for further research. The third author has had some success with simulating the distribution of test statistics for adaptive designs with delayed response, using priority queues [see, e.g., Rosenberger and Seshaiyer (1997)]. However, the computational simplicity of an asymptotic normal test based on the maximum likelihood estimator, we presume, will always make it an attractive tool, and, in this paper, we have provided the necessary theory to justify its use.

APPENDIX

Because of the delayed response, the total number of balls in the urn at each stage will be a random variable, depending on which patients have already responded. To prove Theorem 1, we will need to take care of this complication, which we do in the following lemma:

LEMMA 1. (i) *For the total urn composition,* $\mathbf{Y}_n$, $n^{-1}|\mathbf{Y}_n| \to 1$ *in probability.* (ii) *If Assumption* 1 *is true,*

$$n^{-1}|\mathbf{Y}_n| = 1 + o_p(n^{-c'}) \quad \textit{for any } c' < c$$

and

$$n^{-1}|\mathbf{Y}_n| = 1 + o(n^{-c'}) \quad \textit{almost surely for any } c' < c/2,$$

where the constant c *is defined in Assumption* 1.

PROOF. Recall that we have assumed, without loss of generality, that the number of balls added to the urn at each stage is 1. Also, assume, without loss of generality, that $|\mathbf{Y}_1| = 1$. Then the number of balls at stage n will be n minus the balls not added due to a patient's nonresponse by stage n. We can write this mathematically as

$$|\mathbf{Y}_n| = n - \sum_{m=1}^{n-1} \sum_{i=n-m}^{\infty} M_{J_m,\xi_m}(m, i), \tag{9}$$

by noting that

$$\sum_{i=0}^{\infty} M_{J_m,\xi_m}(m, i) = 1.$$

Now, since

$$\sum_{i=m}^{\infty} \mu_{jli} \to 0\ [(= o(m^{-c}) \text{ under Assumption 1}] \qquad \text{as } m \to \infty,$$

we have

$$\begin{aligned} E\left(\sum_{m=1}^{n-1} \sum_{i=n-m}^{\infty} M_{J_m,\xi_m}(m, i)\right) &= \sum_{m=1}^{n-1} \sum_{i=n-m}^{\infty} E\{M_{J_m,\xi_m}(m, i)\} = \sum_{m=1}^{n-1} \sum_{i=n-m}^{\infty} \sum_{j=1}^{K} \sum_{l=1}^{L} \mu_{jli} \\ &= \begin{cases} o(n), & \text{without Assumption 1,} \\ o(n^{1-c}), & \text{under Assumption 1 and } 0 < c < 1, \\ o(\log n), & \text{under Assumption 1 and } c = 1. \end{cases} \end{aligned}$$

This proves conclusion (i) and the first part of conclusion (ii) of Lemma 1 by the Markov inequality.

Now, choose ρ such that $\rho(c-c') > 1$ and $\rho c' < 1$ ($c' < c/2$). Define $n_k = [k^\rho]$, where $[\cdot]$ is the greatest integer function. Then, for any $\varepsilon > 0$,

$$\begin{aligned} P\big(n_k^{-1+c'}\big||\mathbf{Y}_{n_k}| - n_k\big| \geq \varepsilon\big) &\leq \varepsilon^{-1} n_k^{-1+c'} E\big||\mathbf{Y}_{n_k}| - n_k\big| \\ &\leq \varepsilon^{-1} n_k^{-1+c'} n_k^{1-c)} \log n_k \\ &\leq Ck^{-\rho(c-c')} \log k \end{aligned}$$

for some constant C. Note that the right-hand side of the preceding inequality is summable. Thus, by the Borel–Cantelli lemma,

$$n_k^{-1+c'}(|\mathbf{Y}_{n_k}| - n_k) \to 0 \tag{10}$$

almost surely. To complete the proof of the lemma, we need to show that

$$\max_{n_{k-1} < n \leq n_k} |n^{-1+c'}(|\mathbf{Y}_n| - n)| \to 0$$

almost surely. It is easy to see that

$$|\mathbf{Y}_{n_{k-1}}| - n_{k-1} - (n_k - n_{k-1}) \leq |\mathbf{Y}_n| - n \leq |\mathbf{Y}_{n_k}| - n_k + (n_k - n_{k-1}).$$

From (10), we have

$$n^{-1+c'}(|\mathbf{Y}_{n_{k-1}}| - n_{k-1}) \to 0 \quad \text{and} \quad n^{-1+c'}(|\mathbf{Y}_{n_k}| - n_k) \to 0$$

almost surely. By the selection of ρ,

$$n^{-1+c'}(n_k - n_{k-1}) \leq n_{k-1}^{-1+c'} C n_k k^{-1} \leq C k^{\rho c' - 1} \to 0.$$

Therefore, we have proved the second part of (ii). $\square$

PROOF OF THEOREM 1. From (2), we have

$$\mathbf{Y}_n = \mathbf{Y}_{n-1} + \sum_{m=1}^{n-1} \frac{\mathbf{Y}_m}{m} \mathbf{G}(n-m-1) + \mathbf{Q}_n + \mathbf{R}_n, \tag{11}$$

where

$$\mathbf{G}(m) = \sum_{l=1}^{L} \mu_{lm} \mathbf{D}(l)$$

and

$$\mathbf{R}_n = \sum_{m=1}^{n-1} \frac{\mathbf{Y}_m(m - |\mathbf{Y}_m|)}{m|\mathbf{Y}_m|} \mathbf{G}(n-m-1).$$

Recalling the definition of $\mathbf{Q}_n$ from (2), we can derive the following using (11) [letting $\mathbf{G}(-1)$ be the identity matrix $\mathbf{I}$ for convenience]:

$$\begin{aligned}\mathbf{Y}_n &= \mathbf{Y}_1 + \sum_{i=2}^{n}\sum_{m=1}^{i-1}\frac{\mathbf{Y}_m}{m}\mathbf{G}(i-m-1) + \sum_{i=2}^{n}(\mathbf{Q}_i + \mathbf{R}_i) \\ &= \sum_{m=1}^{n-1}\sum_{i=m}^{n-1}\frac{\mathbf{Y}_m}{m}\mathbf{G}(i-m) + \sum_{i=1}^{n}(\mathbf{Q}_i + \mathbf{R}_i) + \mathbf{Y}_1 \qquad (12) \\ &= \sum_{m=1}^{n-1}\frac{\mathbf{Y}_m}{m}\sum_{i=0}^{n-m-1}\mathbf{G}(i) + \sum_{i=1}^{n}(\mathbf{Q}_i + \mathbf{R}_i) + \mathbf{Y}_1.\end{aligned}$$

Here $\mathbf{Q}_1 = \mathbf{R}_1 = 0$. By the definition of $\mathbf{G}(m)$, we have

$$\sum_{m=0}^{\infty}\mathbf{G}(m) = \sum_{m=0}^{\infty}\sum_{l=1}^{L}\mu_{lm}\mathbf{D}(l) = \mathbf{H},$$

recalling that $\mathbf{H} = E(\mathbf{D})$. Then, by (12), we have

$$\mathbf{Y}_n = \sum_{m=1}^{n-1}\frac{\mathbf{Y}_m\mathbf{H}}{m} + \sum_{i=1}^{n}\mathbf{Q}_i + \mathbf{Y}_1 + \mathbf{R}_n^{(1)}, \qquad (13)$$

where

$$\mathbf{R}_n^{(1)} = \sum_{i=1}^{n}\mathbf{R}_i - \sum_{m=1}^{n-1}\frac{\mathbf{Y}_m}{m}\sum_{i=n-m}^{\infty}\mathbf{G}(i).$$

Under condition (3) and by the result of Lemma 1, we further have

$$\mathbf{R}_n^{(1)} = \sum_{m=1}^{n-1}\frac{\mathbf{Y}_m(m-|\mathbf{Y}_m|)}{m|\mathbf{Y}_m|}\sum_{i=0}^{n-m-1}\mathbf{G}(i) - \sum_{m=1}^{n-1}\frac{\mathbf{Y}_m}{m}\sum_{i=n-m}^{\infty}\mathbf{G}(i) = o_p(n^{1-c'}),$$

and, by Lemma 1, we can strengthen this to almost sure convergence. From (13), we have

$$\mathbf{Y}_n = \mathbf{Y}_{n-1}(\mathbf{I} + (n-1)^{-1}\mathbf{H}) + \mathbf{Q}_n + \mathbf{R}_n^{(2)},$$

where

$$\mathbf{R}_n^{(2)} = \mathbf{R}_n^{(1)} - \mathbf{R}_{n-1}^{(1)} = o(n^{1-c'}).$$

Furthermore,

$$\mathbf{Y}_n = \sum_{i=2}^{n}\mathbf{Q}_i\mathbf{B}_{ni} + \mathbf{Y}_1\mathbf{B}_{n1} + \sum_{i=2}^{n}\mathbf{R}_i^{(2)}\mathbf{B}_{ni}, \qquad (14)$$

where

$$\mathbf{B}_{ni} = \prod_{j=i+1}^{n}(\mathbf{I} + (j-1)^{-1}\mathbf{H})$$

(with the convention that $\mathbf{B}_{nn}$ is the identity matrix $\mathbf{I}$).

Let $\mathbf{z}_n = n^{-1}\mathbf{Y}_n\mathbf{T}$, where $\mathbf{T}$ is defined in Assumption 2. We wish to show that $\mathbf{z}_n$ converges to $(1, 0, 0, \dots, 0)$ almost surely, which then implies that

$$n^{-1}\mathbf{Y}_n \to (1, 0, 0, \dots, 0)\mathbf{T}^{-1} = \mathbf{v} \quad \text{almost surely.} \tag{15}$$

We have already shown that the first element of $\mathbf{z}_n$ converges almost surely to 1. Hence, we can focus on $\mathbf{z}_n\mathbf{E}$, where $\mathbf{E}' = [\mathbf{0} : \mathbf{I}]_{K-1\times K}$. Then, from (14), we obtain

$$\mathbf{z}_n\mathbf{E} = n^{-1}\sum_{i=2}^{n}\mathbf{Q}_i\mathbf{B}_{ni}\mathbf{TE} + n^{-1}\mathbf{Y}_1\mathbf{B}_{n1}\mathbf{TE} + n^{-1}\sum_{i=2}^{n}\mathbf{R}_i^{(2)}\mathbf{B}_{ni}\mathbf{TE}. \tag{16}$$

Let $\tilde{\mathbf{B}}_{ni} = \mathbf{T}^{-1}\mathbf{B}_{ni}\mathbf{T}$. The second term of (16) can be written as $n^{-1}\mathbf{z}_1\tilde{\mathbf{B}}_{n1}\mathbf{E}$, which converges almost surely to 0, as $n^{-1}\mathbf{z}_1\tilde{\mathbf{B}}_{n1}$ converges almost surely to $(z_{11}, 0, \dots, 0)$, where z_{11} is the first element of $\mathbf{z}_1$. This is because $n^{-1}\tilde{\mathbf{B}}_{n1}$ converges to $(1, 0, \dots, 0)'(1, 0, \dots, 0)$ under the condition $\lambda < 1$.

The third term of (16) requires a careful analysis. We write

$$\begin{aligned} n^{-1}\sum_{i=2}^{n}\mathbf{R}_i^{(2)}\mathbf{B}_{ni}\mathbf{TE} &= n^{-1}\mathbf{R}_n^{(1)}\mathbf{B}_{nn}\mathbf{TE} + n^{-1}\sum_{i=2}^{n-1}\mathbf{R}_i^{(1)}(\mathbf{H}/i)\mathbf{B}_{n,i+1}\mathbf{TE} \\ &\quad - n^{-1}\mathbf{R}_1^{(1)}\mathbf{B}_{n2}\mathbf{TE}. \end{aligned} \tag{17}$$

The first term of (17) is $o(n^{-c'})$. We can write the second term of (17) as

$$n^{-1}\sum_{i=2}^{n-1}\mathbf{R}_i^{(1)}(\mathbf{H}/i)\mathbf{T}\tilde{\mathbf{B}}_{n,i+1}\mathbf{E}.$$

For the analysis of $\tilde{\mathbf{B}}_{n,i+1}$, recalling the definitions of λ and $\boldsymbol{\Psi}_t$ in Assumption 2, we see that, for $\varepsilon > 0$,

$$n^{-(\lambda+\varepsilon)}\prod_{j=i+1}^{n}(\mathbf{I} + j^{-1}\boldsymbol{\Psi}_t) = o(i^{-\lambda}).$$

Consequently, $\tilde{\mathbf{B}}_{ni}$ is of order $o(n^{\lambda+\varepsilon}/i^{\lambda})$, and the second term is of order $o(n^{-c'+\varepsilon})$ if $\lambda + c' < 1$ and $o(n^{\lambda+\varepsilon-1})$ if $\lambda + c' > 1$. Finally, the third term of (17) can be written as

$$-n^{-1}\mathbf{R}_1^{(1)}\mathbf{T}\tilde{\mathbf{B}}_{n2}\mathbf{E},$$

and this term is $o(n^{\lambda+\varepsilon-1})$.

To complete the analysis of (16), we inspect the first term. The variance of the jth ($j > 1$) term is given by

$$n^{-2}\sum_{i=1}^{n}\mathrm{Var}(\mathbf{Q}_i\mathbf{B}_{ni}\mathbf{Te}_j') = n^{-2}\sum_{i=1}^{n}\mathbf{e}_j'\mathbf{T}^*\mathbf{B}_{ni}'E(\mathbf{Q}_i'\mathbf{Q}_i)\mathbf{B}_{ni}\mathbf{Te}_j', \tag{18}$$

where $\mathbf{e}'_j$ is the jth column of $\mathbf{E}$ and $\mathbf{T}^*$ is the conjugate transpose of $\mathbf{T}$. The conditional expectation of $\mathbf{Q}'_n\mathbf{Q}_n$ is given by

$$\begin{aligned} E_{n-1}(\mathbf{Q}'_n\mathbf{Q}_n) &= E_{n-1}(\mathbf{W}'_n\mathbf{W}_n) - (E_{n-1}(\mathbf{W}_n))'(E_{n-1}(\mathbf{W}_n)) \\ &= \sum_{m=0}^{n-2}\sum_{l=1}^{L}\mathbf{D}(l)'\mathrm{Diag}\left(\frac{Y_{n-m-1,1}}{|\mathbf{Y}_{n-m-1}|}\mu_{1lm},\ldots,\frac{Y_{n-m-1,K}}{|\mathbf{Y}_{n-m-1}|}\mu_{Klm}\right)\mathbf{D}(l) \qquad (19) \\ &\quad - (E_{n-1}(\mathbf{W}_n))'(E_{n-1}(\mathbf{W}_n)), \end{aligned}$$

and hence $E(\mathbf{Q}'_i\mathbf{Q}_i)$ is bounded. [Equation (19) will be important in determining the variance–covariance structure of the limiting distribution, derived in the proof of Theorem 2.] Since the elements of $\tilde{\mathbf{B}}_{ni}\mathbf{e}'_j$ are controlled by $n^{\lambda+\varepsilon}/i^{\lambda}$, from the developments above, we can see that (18) is $o(n^{\lambda+2\varepsilon-1})$. Because ε can be made small, using the Chebyshev inequality, we conclude that

$$\Pr\left(\left|n^{-1}\sum_{i=1}^{n}\mathbf{Q}_i\mathbf{B}_{ni}\mathbf{T}\mathbf{e}'_j\right| \geq \delta\right) \leq Cn^{-b}$$

for some constant C and $b > 0$.

Now we define $n_k = [k^{\rho}]$, where ρ satisfies $b\rho > 1$. For the subsequence n_k, the first term of (16) converges almost surely to 0 by the Borel–Cantelli lemma. Hence, for $c' > 0$, and by choosing ε small, the terms of $\mathbf{z}_{n_k}\mathbf{E}$ converge almost surely to 0, and (15) holds on the subsequence n_k, implying that $\mathbf{Y}_{n_k}/n_k$ converges almost surely to $\mathbf{v}$. Applying the subsequence method (use the monotonicity of $\mathbf{Y}_n$), $\mathbf{Y}_n/n$ converges almost surely to $\mathbf{v}$. Then, under Assumption 1 and by Lemma 1, $\mathbf{Y}_n/|\mathbf{Y}_n|$ converges almost surely to $\mathbf{v}$. □

PROOF OF COROLLARY 1. We can write

$$\begin{aligned} \mathbf{N}_n = \mathbf{N}_{n-1} + \mathbf{X}_n &= \sum_{i=1}^{n}\mathbf{X}_i \\ &= \sum_{i=1}^{n}[\mathbf{X}_i - E(\mathbf{X}_i|\mathcal{F}_i)] + \sum_{i=1}^{n}E(\mathbf{X}_i|\mathcal{F}_i) \\ &= \sum_{i=1}^{n}[\mathbf{X}_i - E(\mathbf{X}_i|\mathcal{F}_i)] + \sum_{i=1}^{n}\mathbf{Y}_i/|\mathbf{Y}_i|. \end{aligned}$$

Then

$$\frac{\mathbf{N}_n}{n} = \frac{1}{n}\sum_{i=1}^{n}[\mathbf{X}_i - E(\mathbf{X}_i|\mathcal{F}_i)] + \frac{1}{n}\sum_{i=1}^{n}\mathbf{Y}_i/|\mathbf{Y}_i|.$$

From the martingale strong law [e.g., Theorem 2.18 of Hall and Heyde (1980), page 36], the first term converges to 0 almost surely. The second term

$$\frac{1}{n}\sum_{i=1}^{n}\mathbf{Y}_i/|\mathbf{Y}_i| \to \mathbf{v}$$

almost surely, which follows directly from Theorem 1. □

PROOF OF THEOREM 2. From Lemma 1, we have that

$$n^{1/2}(|\mathbf{Y}_n| - n) \to 0 \quad \text{in probability.}$$

Define $\mathbf{z}_n^{\#} = (n^{-1}\mathbf{Y}_n - \mathbf{v})\mathbf{T}$. We wish to examine the limit of $n^{1/2}\mathbf{z}_n^{\#}$. First, by Lemma 1, the first term converges almost surely to 0. We recall that $\mathbf{E}' = [\mathbf{0} : \mathbf{I}]_{K-1\times K}$ and, since $\mathbf{v}$ is the first row of $\mathbf{T}^{-1}$, we have $\mathbf{vT} = (1, 0, \ldots, 0)$, and $\mathbf{vTE} = 0$. Consequently, from (16),

$$\begin{aligned} n^{1/2}\mathbf{z}_n^{\#}\mathbf{E} &= n^{-1/2}\sum_{i=2}^{n}\mathbf{Q}_i\mathbf{B}_{ni}\mathbf{TE} + n^{-1/2}\mathbf{Y}_1\mathbf{B}_{n1}\mathbf{TE} \\ &\quad + n^{-1/2}\sum_{i=2}^{n}\mathbf{R}_i^{(2)}\mathbf{B}_{ni}\mathbf{TE}. \end{aligned} \tag{20}$$

The third term of (20), $n^{-1/2}\sum_{i=2}^{n}\mathbf{R}_i^{(2)}\mathbf{B}_{ni}\mathbf{TE}$, can be decomposed into three components, as in (17). From the proof of Theorem 1, it follows that the first term is of order $o_p(n^{-c_1+1/2})$, so it tends to 0 for $c > c_1 > 1/2$. The second term is $o_p(n^{-c_1+\varepsilon+1/2})$, and as ε is arbitrarily small, this term tends to 0. The third term is $o_p(n^{\lambda+\varepsilon-1/2})$, and as $\lambda < 1/2$, this also tends to 0.

From the proof of Theorem 1, $\tilde{\mathbf{B}}_{n1}\mathbf{E}$ is of order $o(n^{\lambda+\varepsilon})$, so that the second term of (20), $n^{-1/2}\mathbf{Y}_1\mathbf{B}_{n1}\mathbf{TE} = o(n^{\lambda+\varepsilon-1/2})$, also tends to 0.

Finally, we can use the martingale central limit theorem [e.g., Corollary 3.1 of Hall and Heyde (1980), page 58] to show that the first term of (20)

$$n^{-1/2}\sum_{i=1}^{n}\mathbf{Q}_i\mathbf{B}_{ni}\mathbf{TE} \to N(0, \boldsymbol{\Sigma}_1) \quad \text{in law.}$$

The form of $\boldsymbol{\Sigma}_1$ can be obtained by careful derivation, but it is quite messy. Using the same techniques as Bai and Hu (1999), we can derive an exact expression for $\boldsymbol{\Sigma}_1$. It is given by $\boldsymbol{\Sigma}_1 = ((\boldsymbol{\Sigma}_{gh},\ g, h, = 1, \ldots, s))$, where $\boldsymbol{\Sigma}_{gh}$ is a submatrix with (a, b) element

$$\sum_{a'=0}^{a-1}\sum_{b'=0}^{b-1}\frac{(a'+b')!}{a'!b'!(1-\lambda_g-\bar{\lambda}_h)^{a'+b'+1}}[\mathbf{T}_g^* E_\infty(\mathbf{Q}'\mathbf{Q})\mathbf{T}_h]_{a-a',b-b'}$$

($\mathbf{T}^*$ is the conjugate transpose of $\mathbf{T}$ and $\bar{\lambda}$ is the complex conjugate of λ), where $E_\infty(\mathbf{Q}'\mathbf{Q}) = \lim_{n\to\infty} E_{n-1}(\mathbf{Q}_n'\mathbf{Q}_n)$. From (19),

$$E_\infty(\mathbf{Q}'\mathbf{Q}) = \sum_{l=1}^{L} \mathbf{D}(l)'\mathrm{Diag}\left(v_1 \sum_{m=0}^{\infty} \mu_{1lm}, \ldots, v_K \sum_{m=0}^{\infty} \mu_{Klm}\right)\mathbf{D}(l) - \mathbf{v}'\mathbf{v}. \tag{21}$$

Finally, we obtain

$$\boldsymbol{\Sigma} = (\mathbf{T}^{-1})^* \begin{bmatrix} 0 & 0 \\ 0 & \boldsymbol{\Sigma}_1 \end{bmatrix} \mathbf{T}^{-1}. \tag{22}$$

□

Acknowledgments. Professor Rosenberger is also affiliated with the Department of Epidemiology and Preventive Medicine, University of Maryland School of Medicine. His research was done while visiting the Department of Statistics and Applied Probability, National University of Singapore. He thanks the Department for its hospitality and support. Professor Hu is also affiliated with Department of Statistics and Applied Probability, National University of Singapore. Special thanks go to the referees and Associate Editor for their constructive comments, which led to a much improved version of the paper.

REFERENCES

ANDERSEN, J., FARIES, D. and TAMURA, R. (1994). A randomized play-the-winner design for multi-arm clinical trials. *Comm. Statist. Theory Methods* **23** 309–323.

ATHREYA, K. B. and KARLIN, S. (1968). Embedding of urn schemes into continuous time Markov branching processes and related limit theorems. *Ann. Math. Statist.* **39** 1801–1817.

BAI, Z. D. and HU, F. (1999). Asymptotic theorems for urn models with nonhomogeneous generating matrices. *Stochastic Process. Appl.* **80** 87–101.

BANDYOPADHYAY, U. and BISWAS, A. (1996). Delayed response in randomized play-the-winner rule: a decision theoretic outlook. *Calcutta Statist. Assoc. Bull.* **46** 69–88.

BARTLETT, R. H., ROLOFF, D. W., CORNELL, R. G., ANDREWS, A. F., DILLON, P. W. and ZWISCHENBERGER, J. B. (1985). Extracorporeal circulation in neonatal respiratory failure: a prospective randomized study. *Pediatrics* **76** 479–487.

HALL, P. and HEYDE, C. C. (1980). *Martingale Limit Theory and Its Application.* Academic Press, San Diego.

HARDWICK, J. and STOUT, Q. (1998). Flexible algorithms for creating and analyzing adaptive sampling procedures. In *New Developments and Applications in Experimental Design* (N. Flournoy, W. F. Rosenberger, and W. K. Wong, eds.) 91–105. IMS, Hayward, CA.

ROSENBERGER, W. F. (1996). New directions in adaptive designs. *Statist. Sci.* **11** 137–149.

ROSENBERGER, W. F. (1999). Randomized play-the-winner clinical trials: review and recommendations. *Controlled Clinical Trials* **20** 328–342.

ROSENBERGER, W. F., FLOURNOY, N. and DURHAM, S. D. (1997). Asymptotic normality of maximum likelihood estimators from multiparameter response-driven designs. *J. Statist. Plann. Inference* **60** 69–76.

ROSENBERGER, W. F. and GRILL, S. G. (1997). A sequential design for psychophysical experiments: an application to estimating timing of sensory events. *Statistics in Medicine* **16** 2245–2260.

ROSENBERGER, W. F. and HU, F. (1999). Bootstrap methods for adaptive designs. *Statistics in Medicine* **18** 1757–1767.

ROSENBERGER, W. F. and SESHAIYER, P. (1997). Adaptive survival trials. *Journal of Biopharmaceutical Statistics* **7** 617–624.

SMYTHE, R. T. (1996). Central limit theorems for urn models. *Stochastic Process. Appl.* **65** 115–137.

TAMURA, R. N., FARIES, D. E., ANDERSEN, J. S. and HEILIGENSTEIN, J. H. (1994). A case study of an adaptive clinical trial in the treatment of out-patients with depressive disorder. *J. Amer. Statist. Assoc.* **89** 768–776.

WEI, L. J. (1979). The generalized Polya's urn design for sequential medical trials. *Ann. Statist.* **7** 291–296.

WEI, L. J. (1988). Exact two-sample permutation tests based on the randomized play-the-winner rule. *Biometrika* **75** 603–606.

WEI, L. J. and DURHAM, S. (1978). The randomized play-the-winner rule in medical trials. *J. Amer. Statist. Assoc.* **73** 840–843.

WEI, L. J., SMYTHE, R. T., LIN, D. Y. and PARK, T. S. (1990). Statistical inference with data-dependent treatment allocation rules. *J. Amer. Statist. Assoc.* **85** 156–162.

Z. D. BAI
DEPARTMENT OF STATISTICS
AND APPLIED PROBABILITY
NATIONAL UNIVERSITY OF SINGAPORE
SINGAPORE 119260

F. HU
DEPARTMENT OF STATISTICS
HALSEY HALL
UNIVERSITY OF VIRGINIA
CHARLOTTESVILLE, VIRGINIA 22904-4135
E-MAIL: fh6e@pitman.stat.virginia.edu

W. F. ROSENBERGER
DEPARTMENT OF MATHEMATICS AND STATISTICS
UNIVERSITY OF MARYLAND, BALTIMORE COUNTY
1000 HILLTOP CIRCLE
BALTIMORE, MARYLAND 21250

The Annals of Probability
2004, Vol. 32, No. 1A, 553–605

CLT FOR LINEAR SPECTRAL STATISTICS OF LARGE-DIMENSIONAL SAMPLE COVARIANCE MATRICES

BY Z. D. BAI[1] AND JACK W. SILVERSTEIN[2]

North East Normal University and National University of Singapore and North Carolina State University

Let $B_n = (1/N)T_n^{1/2}X_nX_n^*T_n^{1/2}$ where $X_n = (X_{ij})$ is $n \times N$ with i.i.d. complex standardized entries having finite fourth moment, and $T_n^{1/2}$ is a Hermitian square root of the nonnegative definite Hermitian matrix T_n. The limiting behavior, as $n \to \infty$ with n/N approaching a positive constant, of functionals of the eigenvalues of B_n, where each is given equal weight, is studied. Due to the limiting behavior of the empirical spectral distribution of B_n, it is known that these linear spectral statistics converges a.s. to a nonrandom quantity. This paper shows their rate of convergence to be $1/n$ by proving, after proper scaling, that they form a tight sequence. Moreover, if $\mathsf{E}X_{11}^2 = 0$ and $\mathsf{E}|X_{11}|^4 = 2$, or if X_{11} and T_n are real and $\mathsf{E}X_{11}^4 = 3$, they are shown to have Gaussian limits.

1. Introduction. Due to the rapid development of modern technology, statisticians are confronted with the task of analyzing data with ever increasing dimension. For example, stock market analysis can now include a large number of companies. The study of DNA can now incorporate a sizable number of its base pairs. Computers can easily perform computations with high-dimensional data. Indeed, within several milli-seconds, a mainframe can complete the spectral decomposition of a 1000×1000 symmetric matrix, a feat unachievable only 20 years ago. In the past, so-called *dimension reduction* schemes played the main role in dealing with high-dimensional data, but a large portion of information contained in the original data would inevitably get lost. For example, in variable selection of multivariate linear regression models, one will lose all information contained in the unselected variables; in principal component analysis, all information contained in the components deemed "nonprincipal" would be gone. Now when dimension reduction is performed it is usually not due to computational restrictions.

However, even though the technology exists to compute much of what is needed, there is a fundamental problem with the analytical tools used by statisticians.

Received June 2002; revised February 2003.

[1]Supported in part by NSFC Grant 201471000 and NUS Grant R-155-000-040-112.

[2]Supported in part by NSF Grant DMS-97-03591.

AMS 2000 subject classifications. Primary 15A52, 60F05; secondary 62H99.

Key words and phrases. Linear spectral statistics, random matrix, empirical distribution function of eigenvalues, Stieltjes transform.

Their use relies on their asymptotic behavior as the number of samples increase. It is to be expected that larger dimension will require larger samples in order to maintain the same level of behavior. But the required increase is typically orders of magnitude larger than the dimension, sample sizes that are simply unattainable in most situations. With a necessary limitation on the number of samples, many frequently used statistics in multivariate analysis perform in a completely different manner than they do on data of low dimension with no restriction on sample size. Some methods behave very poorly [see Bai and Saranadasa (1996)], and some are even not applicable [see Dempster (1958)]. Consider the following example.

Let X_{ij} be i.i.d. standard normal variables. Write

$$S_N = \left(\frac{1}{N}\sum_{k=1}^{N} X_{ik}X_{jk}\right)_{i,j=1}^{n},$$

which can be considered as a sample covariance matrix, N samples of an n-dimensional mean zero random vector with population matrix I. An important statistic in multivariate analysis is

$$L_N = \ln(\det S_N) = \sum_{j=1}^{n} \ln(\lambda_{N,j}),$$

where $\lambda_{N,j}$, $j = 1, \ldots, n$, are the eigenvalues of S_N. When n is fixed, $\lambda_{N,j} \to 1$ almost surely as $N \to \infty$ and thus $L_N \xrightarrow{\text{a.s.}} 0$.

Further, by taking a Taylor expansion on $\ln(1+x)$, one can show that

$$\sqrt{N/n}\, L_N \xrightarrow{D} N(0,2),$$

for any fixed n. This suggests the possibility that L_N is asymptotically normal, provided that $n = O(N)$. However, this is not the case. Let us see what happens when $n/N \to c \in (0,1)$ as $n \to \infty$. Using results on the limiting spectral distribution of $\{S_N\}$ [see Marčenko and Pastur (1967) and Bai (1999)], we have, with probability 1,

$$\begin{aligned} \frac{1}{n} L_N &\to \int_{a(c)}^{b(c)} \frac{\ln x}{2\pi x c}\sqrt{(b(c)-x)(x-a(c))}\,dx \\ &= \frac{c-1}{c}\ln(1-c) - 1 \equiv d(c) < 0, \end{aligned} \tag{1.1}$$

where $a(c) = (1-\sqrt{c})^2$, $b(c) = (1+\sqrt{c})^2$ (see Section 5 for a derivation of the integral). This shows that almost surely

$$\sqrt{N/n}\, L_N \sim d(c)\sqrt{Nn} \to -\infty.$$

Thus, any test which assumes asymptotic normality of $\sqrt{N/n}\, L_N$ will result in a serious error.

Besides demonstrating problems with relying on traditional methodology when sample size is restricted, the example introduces one of several results that can be used to handle data with large dimension n, proportional to N, the sample size. They are limit theorems, as n approaches infinity, on the eigenvalues of a class of random matrices of sample covariance type [Yin and Krishnaiah (1983), Yin (1986), Silverstein (1995) and Bai and Silverstein (1998, 1999)]. They take the form

$$B_n = \frac{1}{N} T_n^{1/2} X_n X_n^* T_n^{1/2},$$

where $X_n = (X_{ij}^n)$ is $n \times N$, $X_{ij}^n \in \mathbb{C}$ are i.i.d. with $\mathsf{E}|X_{11}^n - \mathsf{E}X_{11}^n|^2 = 1$, $T_n^{1/2}$ is $n \times n$ random Hermitian, with X_n and $T_n^{1/2}$ independent. When X_{11}^n is known to have mean zero and T_n is nonrandom, B_n can be viewed as a sample covariance matrix, which includes any Wishart matrix, formed from N samples of the random vector $T_n^{1/2} X_{\cdot 1}^n$ ($X_{\cdot 1}^n$ denoting the first column of X_n), which has population covariance matrix $T_n \equiv (T_n^{1/2})^2$. Besides sample covariance matrices, B_n, whose eigenvalues are the same as those of $(1/N) X_n X_n^* T_n$, models the spectral behavior of other matrices important to multivariate statistics, in particular multivariate F matrices, where X_{11}^n is $N(0, 1)$ and T_n is the inverse of another Wishart matrix.

The basic limit theorem on the eigenvalues of B_n concerns its empirical spectral distribution F^{B_n}, where for any matrix A with real eigenvalues, F^A denotes the empirical distribution function of the eigenvalues of A, that is, if A is $n \times n$ then

$$F^A(x) = \frac{1}{n}(\text{number of eigenvalues of } A \le x).$$

If:

1. for all n, i, j, X_{ij}^n are i.d.,
2. with probability 1, $F^{T_n} \xrightarrow{D} H$, a proper cumulative distribution function (c.d.f.) and
3. $n/N \to c > 0$ as $n \to \infty$,

then with probability 1 F^{B_n} converges in distribution to $F^{c,H}$, a nonrandom proper c.d.f.

The case when H distributes its mass at one positive number (called the Pastur–Marčenko law), as in the above example, is one of seven nontrivial cases where an explicit expression for $F^{c,H}$ is known (the multivariate F matrix case [Silverstein (1985)] and, as to be seen below, when H is discrete with at most three positive mass points with or without mass at zero). However, a good deal of information, including a way to compute $F^{c,H}$, can be extracted out of an equation satisfied by its *Stieltjes transform*, defined for any c.d.f. G to be

$$m_G(z) \equiv \int \frac{1}{\lambda - z} dG(\lambda), \qquad \Im z \neq 0.$$

We see that $m_G(\bar{z}) = \overline{m_G(z)}$. For each $z \in \mathbb{C}^+ \equiv \{z \in \mathbb{C} : \Im z > 0\}$, the Stieltjes transform $m(z) \equiv m_{F^{c,H}}(z)$ is the unique solution to

$$m = \int \frac{1}{\lambda(1 - c - czm) - z} \, dH(\lambda)$$

in the set $\{m \in \mathbb{C} : -\frac{1-c}{z} + cm \in \mathbb{C}^+\}$. The equation takes on a simpler form when $F^{c,H}$ is replaced by

$$\underline{F}^{c,H} \equiv (1 - c) I_{[0,\infty)} + c F^{c,H}$$

(I_A denoting the indicator function on the set A), which is the limiting empirical distribution function of $\underline{B}_n \equiv (1/N) X_n^* T_n X_n$ (the spectra of which differs from that of B_n by $|n - N|$ zeros). Its Stieltjes transform

$$\underline{m}(z) \equiv m_{\underline{F}^{c,H}}(z) = -\frac{1-c}{z} + cm(z)$$

has inverse

$$z = z(\underline{m}) = -\frac{1}{\underline{m}} + c \int \frac{t}{1 + t\underline{m}} \, dH(t). \tag{1.2}$$

Using (1.2) it is shown in Silverstein and Choi (1995) that, on $(0, \infty)$, $F^{c,H}$ has a continuous density, is analytic inside its support and is given by

$$\begin{aligned} f^{c,H}(x) &= c^{-1} \frac{d}{dx} \underline{F}^{c,H}(x) \\ &= (c\pi)^{-1} \Im \underline{m}(x) \equiv (c\pi)^{-1} \lim_{z \to x} \Im \underline{m}(z). \end{aligned} \tag{1.3}$$

Also, $F^{c,H}(0) = \max[1 - c^{-1}, H(0)]$. Moreover, considering (1.2) for $\underline{m}$ real, the range of values where it is increasing constitutes the complement of the support of $F^{c,H}$ on $(0, \infty)$ [Marčenko and Pastur (1967) and Silverstein and Choi (1995)]. From (1.2) and (1.3) $f^{c,H}(x)$ can be computed using Newton's method for each $x \in (0, \infty)$ inside its support [see Bai and Silverstein (1998) for an illustration of the density when $c = 0.1$ and H places mass 0.2, 0.4, and 0.4 at, resp., 1, 3 and 10].

Notice in (1.2) when H is discrete with at most three positive mass points the density has an explicit expression, since $\underline{m}(z)$ is the root of a polynomial of degree at most four.

Convergence in distribution of F^{B_n} of course reveals no information on the number of eigenvalues of B_n appearing on any interval $[a, b]$ outside the support of $F^{c,H}$, other than the number is almost surely $o(n)$. In Bai and Silverstein (1998) it is shown that, with probability 1, no eigenvalues of B_n appear in $[a, b]$ for all n large under the following additional assumptions:

1′. X_n is the first n rows and N columns of a doubly infinite array of i.i.d. random variables, with $\mathsf{E}X_{11} = 0$, $\mathsf{E}|X_{11}|^2 = 1$ and $\mathsf{E}|X_{11}|^4 < \infty$, and

2′. T_n is nonrandom, $\|T_n\|$, the spectral norm of T_n, is bounded in n, and
3′. $[a, b]$ lies in an open subset of $(0, \infty)$ which is outside the support of F^{c_n, H_n} for all n large, where $c_n \equiv n/N$ and $H_n \equiv F^{T_n}$.

The result extends what has been previously known on the extreme eigenvalues of $(1/N)X_nX_n^*$ $(T_n = I)$. Let $\lambda_{\max}^A$, $\lambda_{\min}^A$ denote, respectively, the largest and smallest eigenvalues of the Hermitian matrix A. Under condition 1′, Yin, Bai and Krishnaiah (1988) showed, as $n \to \infty$

$$\lambda_{\max}^{(1/N)X_nX_n^*} \xrightarrow{\text{a.s.}} (1 + \sqrt{c})^2,$$

while in Bai and Yin (1993) for $c \leq 1$

$$\lambda_{\min}^{(1/N)X_nX_n^*} \xrightarrow{\text{a.s.}} (1 - \sqrt{c})^2.$$

If $[a, b]$ separates the support of $F^{c,H}$ in $(0, \infty)$ into two nonempty sets, then associated with it is another interval J which separates the eigenvalues of T_n for all n large. The mass F^{c_n, H_n} places, say, to the right of b equals the proportion of eigenvalues of T_n lying to the right of J. In Bai and Silverstein (1999) it is proved that, with probability 1, the number of eigenvalues of B_n and T_n lying on the same side of their respective intervals is the same for all n large.

The above two results are intuitively plausible when viewing B_n as an approximation of T_n, especially when c_n is small (it can be shown that $F^{c,H} \xrightarrow{D} H$ as $c \to 0$). However, regardless of the size of c, when separation in the support of $F^{c,H}$ on $(0, \infty)$ associated with a gap in the spectrum of T_n occurs, there will be exact splitting of the eigenvalues of B_n.

These results can be used in applications where location of eigenvalues of the population covariance matrix is needed, as in the detection problem in array signal processing [see Silverstein and Combettes (1992)]. Here, each entry of the sampled vector is a reading off a sensor, due to an unknown number q of sources emitting signals in a noise-filled environment $(q < n)$. The problem is to determine q. The smallest eigenvalue of the population covariance matrix is positive with multiplicity $n - q$ (the so-called "noise" eigenvalues). The traditional approach has been to sample enough times so that the sample covariance matrix is close to the population matrix, relying on fixed dimension, large sample asymptotic analysis. However, it may be impossible to sample enough times if q is sizable. The above results show that in order to determine the number of sources, simply sample enough times so that the eigenvalues of B_n split into two discernable groups. The number on the right will, with high probability, equal q.

The results also enable us to understand the true behavior of statistics such as L_N in the above example when n and N are large but on the same order of magnitude; L_N is not close to zero, rather $n^{-1}L_N$ is close to the quantity $d(c)$ in (1.1), or perhaps more appropriately $d(c_n)$.

However, in order to fully utilize $n^{-1}L_N$, typically in hypothesis testing, it is important to establish the limiting distribution of $L_N - nd(c_n)$. We come to the main purpose of this paper, to study the limiting distribution of normalized spectral functionals like $L_N - nd(c)$, and as a by-product, the rate of convergence of statistics such as $n^{-1}L_N$, functionals of the eigenvalues of B_n where each is given equal weight. We will call them *linear spectral statistics*, quantities of the form

$$\frac{1}{n}\sum_{j=1}^{n} f(\lambda_j) \quad (\lambda_1, \ldots, \lambda_n \text{ denoting the eigenvalues of } B_n) = \int f(x)\,dF^{B_n}(x),$$

where f is a function on $[0, \infty)$.

We will show, under the assumption $\mathsf{E}|X_{11}|^4 < \infty$ and the analyticity of f, the rate $\int f(x)\,dF^{B_n}(x) - \int f(x)\,dF^{c_n,H_n}(x)$ approaches zero is essentially $1/n$. Define

$$G_n(x) = n\big[F^{B_n}(x) - F^{c_n,H_n}(x)\big].$$

The main result is stated in the following theorem.

THEOREM 1.1. *Assume*:

(a) *For each* n $X_{ij} = X_{ij}^n$, $i \le n$, $j \le N$ *are i.i.d., i.d. for all* n, i, j, $\mathsf{E}X_{11} = 0$, $\mathsf{E}|X_{11}|^2 = 1$, $\mathsf{E}|X_{11}|^4 < \infty$, $n/N \to c$, *and*

(b) T_n *is* $n \times n$ *nonrandom Hermitian nonnegative definite with spectral norm bounded in* n, *with* $F^{T_n} \xrightarrow{D} H$, *a proper c.d.f.*

Let $f_1, \ldots, f_k$ *be functions on* $\mathbb{R}$ *analytic on an open interval containing*

$$\left[\liminf_n \lambda_{\min}^{T_n} I_{(0,1)}(c)(1-\sqrt{c})^2, \limsup_n \lambda_{\max}^{T_n}(1+\sqrt{c})^2\right] \tag{1.4}$$

Then:

(i) *the random vector*

$$\left(\int f_1(x)\,dG_n(x), \ldots, \int f_k(x)\,dG_n(x)\right) \tag{1.5}$$

forms a tight sequence in n.

(ii) *If* X_{11} *and* T_n *are real and* $\mathsf{E}(X_{11}^4) = 3$, *then* (1.5) *converges weakly to a Gaussian vector* $(X_{f_1}, \ldots, X_{f_k})$, *with means*

$$\mathsf{E}X_f = -\frac{1}{2\pi i}\int f(z)\frac{c\int \underline{m}(z)^3 t^2(1+t\underline{m}(z))^{-3}\,dH(t)}{(1-c\int \underline{m}(z)^2 t^2 (1+t\underline{m}(z))^{-2}\,dH(t))^2}\,dz \tag{1.6}$$

and covariance function

$$\begin{aligned} &\operatorname{Cov}(X_f, X_g) \\ &\quad = -\frac{1}{2\pi^2}\iint \frac{f(z_1)g(z_2)}{(\underline{m}(z_1)-\underline{m}(z_2))^2}\frac{d}{dz_1}\underline{m}(z_1)\frac{d}{dz_2}\underline{m}(z_2)\,dz_1\,dz_2 \end{aligned} \tag{1.7}$$

$(f, g \in \{f_1, \dots, f_k\})$. *The contours in* (1.6) *and* (1.7) [*two in* (1.7), *which we may assume to be nonoverlapping*] *are closed and are taken in the positive direction in the complex plane, each enclosing the support of* $F^{c,H}$.

(iii) *If* X_{11} *is complex with* $\mathsf{E}(X_{11}^2) = 0$ *and* $\mathsf{E}(|X_{11}|^4) = 2$, *then* (ii) *also holds, except the means are zero and the covariance function is* $1/2$ *the function given in* (1.7).

This theorem can be viewed as an extension of results obtained in Jonsson (1982) where the entries of X_n are Gaussian and $T_n = I$ and is consistent with central limit theorem results on linear statistics of eigenvalues of other classes of random matrices [see, e.g., Johansson (1998), Sinai and Soshnikov (1998), Soshnikov (2000) and Diaconis and Evans (2001)]. As will be seen, the techniques and arguments used to prove the theorem, which rely heavily on properties of the Stieltjes transform of F^{B_n}, have nothing in common with any of the tools used in these other papers.

We begin the proof of Theorem 1.1 here with the replacement of the entries of X_n with truncated and centralized variables. For $m = 1, 2, \dots$ find n_m $(n_m > n_{m-1})$ satisfying

$$m^4 \int_{\{|X_{11}| \geq \sqrt{n}/m\}} |X_{11}|^4 < 2^{-m}$$

for all $n \geq n_m$. Define $\delta_n = 1/m$ for all $n \in [n_m, n_{m+1})$ $(= 1$ for $n < n_1)$. Then, as $n \to \infty$, $\delta_n \to 0$ and

$$\delta_n^{-4} \int_{\{|X_{11}| \geq \delta_n \sqrt{n}\}} |X_{11}|^4 \to 0. \tag{1.8}$$

Let now for each n δ_n be the larger of δ_n constructed above and the δ_n created in the proof of Lemma 2.2 of Yin, Bai and Krishnaiah (1988) with $r = 1/2$ and satisfying $\delta_n n^{1/4} \to \infty$. Let $\widehat{B}_n = (1/N) T_n^{1/2} \widehat{X}_n \widehat{X}_n^* T_n^{1/2}$ with $\widehat{X}_n$ $n \times N$ having (i, j)th entry $X_{ij} I_{\{|X_{ij}| < \delta_n \sqrt{n}\}}$. We have then

$$\mathsf{P}(B_n \neq \widehat{B}_n) \leq nNP(|X_{11}| \geq \delta_n \sqrt{n}) \leq K \delta_n^{-4} \int_{\{|X_{11}| \geq \delta_n \sqrt{n}\}} |X_{11}|^4 = o(1).$$

Define $\widetilde{B}_n = (1/N) T_n^{1/2} \widetilde{X}_n \widetilde{X}_n^* T_n^{1/2}$ with $\widetilde{X}_n$ $n \times N$ having (i, j)th entry $(\widehat{X}_{ij} - \mathsf{E}\widehat{X}_{ij})/\sigma_n$, where $\sigma_n^2 = \mathsf{E}|\widehat{X}_{ij} - \mathsf{E}\widehat{X}_{ij}|^2$. From Yin, Bai and Krishnaiah (1988) we know that both $\limsup_n \lambda_{\max}^{\widehat{B}_n}$ and $\limsup_n \lambda_{\max}^{\widetilde{B}_n}$ are almost surely bounded by $\limsup_n \|T_n\|(1 + \sqrt{c})^2$. We use $\widehat{G}_n(x)$ and $\widetilde{G}_n(x)$ to denote the analogues of $G_n(x)$ with the matrix B_n replaced by $\widehat{B}_n$ and $\widetilde{B}_n$, respectively. Let λ_i^A denote the ith smallest eigenvalue of Hermitian A. Using the same approach and bounds that are used in the proof of Lemma 2.7 of Bai (1999), we have, for each $j = 1, 2, \ldots, k$,

$$\left|\int f_j(x)\,d\widehat{G}_n(x) - \int f_j(x)\,d\widetilde{G}_n(x)\right|$$
$$\le K_j \sum_{k=1}^{n} |\lambda_k^{\widehat{B}_n} - \lambda_k^{\widetilde{B}_n}|$$
$$\le 2K_j (N^{-1} \operatorname{tr} T_n^{1/2}(\widehat{X}_n - \widetilde{X}_n)(\widehat{X}_n - \widetilde{X}_n)^* T_n^{1/2})^{1/2} (n(\lambda_{\max}^{\widehat{B}_n} + \lambda_{\max}^{\widetilde{B}_n}))^{1/2},$$

where K_j is a bound on $|f_j'(z)|$. From (1.8) we have

$$|\sigma_n^2 - 1| \le 2 \int_{\{|X_{11}| \ge 2\delta_n \sqrt{n}\}} |X_{11}|^2 \le 2\delta_n^{-2} n^{-1} \int_{\{|X_{11}| \ge \delta_n \sqrt{n}\}} |X_{11}|^4 = o(\delta_n^2 n^{-1}).$$

Moreover,

$$|\mathsf{E}\widehat{X}_{11}| = \left|\int_{\{|X_{11}| \ge \delta_n \sqrt{n}\}} X_{11}\right| = o(\delta_n n^{-3/2}).$$

These give us

$$(N^{-1} \operatorname{tr} T_n^{1/2}(\widehat{X}_n - \widetilde{X}_n)(\widehat{X}_n - \widetilde{X}_n)^* T_n^{1/2})^{1/2}$$
$$\le (N^{-1}(1 - 1/\sigma_n)^2 \operatorname{tr} \widehat{B}_n)^{1/2} + (N^{-1}\|T_n\|\sigma^{-2} \operatorname{tr} \mathsf{E}\widehat{X}_n \mathsf{E}\widehat{X}_n^*)^{1/2}$$
$$\le \left(\frac{(1-\sigma_n^2)^2}{\sigma^2(1+\sigma^2)^2} \frac{n}{N} \lambda_{\max}^{\widehat{B}_n}\right)^{1/2} + (n\|T_n\|)^{1/2} \sigma^{-1} |\mathsf{E}\widehat{X}_{11}|$$
$$= o(\delta_n n^{-1/2})(\lambda_{\max}^{\widehat{B}_n})^{1/2} + o(\delta_n n^{-1}).$$

From the above estimates, we obtain

$$\int f_j(x)\,dG_n(x) = \int f_j(x)\,d\widetilde{G}_n(x) + o_p(1)$$

[$o_p(1) \overset{\text{i.p.}}{\to} 0$ as $n \to \infty$.] Therefore, we only need to find the limiting distribution of $\{\int f_j(x)\,d\widetilde{G}_n(x),\ j = 1, \ldots, k\}$. Hence, in the sequel, we shall assume the underlying variables are truncated at $\delta_n \sqrt{n}$, centralized and renormalized. For simplicity, we shall suppress all sub- or superscripts on the variables and assume that $|X_{ij}| < \delta_n \sqrt{n}$, $\mathsf{E}X_{ij} = 0$, $\mathsf{E}|X_{ij}|^2 = 1$, $\mathsf{E}|X_{ij}|^4 < \infty$, and for assumption (ii) of Theorem 1.1 $\mathsf{E}|X_{11}|^4 = 3 + o(1)$, while for assumption (iii) $\mathsf{E}X_{11}^2 = o(1/n)$ and $\mathsf{E}|X_{11}|^4 = 2 + o(1)$.

Since the truncation steps are identical to those in Yin, Bai and Krishnaiah (1988) we have for any $\eta > (1 + \sqrt{c})^2$ the existence of $\{k_n\}$ for which

$$\frac{k_n}{\ln n} \to \infty \quad \text{and} \quad \mathsf{E}\|(1/N)X_n X_n^*\|^{k_n} \le \eta^{k_n}$$

for all n large. Therefore,

$$P(\|B_n\| \geq \eta) = o(n^{-\ell}), \tag{1.9a}$$

for any $\eta > \limsup \|T\|(1+\sqrt{c})^2$ and any positive ℓ. By modifying the proof in Bai and Yin (1993) on the smallest eigenvalue of $(1/N)X_nX_n^*$ it follows that when $\liminf_n \lambda_{\min}^T I_{(0,1)}(c)(1-\sqrt{c})^2 > 0$

$$P(\lambda_{\min}^{B_n} \leq \eta) = o(n^{-\ell}), \tag{1.9b}$$

whenever $0 < \eta < \liminf_n \lambda_{\min}^T I_{(0,1)}(c)(1-\sqrt{c})^2$. The modification is given in the Appendix.

After truncation and centralization, our proof of the main theorem relies on establishing limiting results on

$$M_n(z) = n[m_{F^{B_n}}(z) - m_{F^{c_n,H_n}}(z)] = N[m_{F^{\underline{B}_n}}(z) - m_{\underline{F}^{c_n,H_n}}(z)],$$

or more precisely, on $\widehat{M}_n(\cdot)$, a truncated version of $M_n(\cdot)$ when viewed as a random two-dimensional process defined on a contour $\mathcal{C}$ of the complex plane, described as follows. Let $v_0 > 0$ be arbitrary. Let x_r be any number greater than the right end point of interval (1.4). Let x_l be any negative number if the left end point of (1.4) is zero. Otherwise choose $x_l \in (0, \liminf_n \lambda_{\min}^{T_n} I_{(0,1)}(c)(1-\sqrt{c})^2)$. Let

$$\mathcal{C}_u = \{x + iv_0 : x \in [x_l, x_r]\}.$$

Then

$$\mathcal{C} \equiv \{x_l + iv : v \in [0, v_0]\} \cup \mathcal{C}_u \cup \{x_r + iv : v \in [0, v_0]\}.$$

We define now the subsets $\mathcal{C}_n$ of $\mathcal{C}$ on which $M_n(\cdot)$ agrees with $\widehat{M}_n(\cdot)$. Choose sequence $\{\varepsilon_n\}$ decreasing to zero satisfying for some $\alpha \in (0,1)$

$$\varepsilon_n \geq n^{-\alpha}. \tag{1.10}$$

Let

$$\mathcal{C}_l = \begin{cases} \{x_l + iv : v \in [n^{-1}\varepsilon_n, v_0]\}, & \text{if } x_l > 0, \\ \{x_l + iv : v \in [0, v_0]\}, & \text{if } x_l < 0, \end{cases}$$

and

$$\mathcal{C}_r = \{x_r + iv : v \in [n^{-1}\varepsilon_n, v_0]\}.$$

Then $\mathcal{C}_n = \mathcal{C}_l \cup \mathcal{C}_u \cup \mathcal{C}_r$. The process $\widehat{M}_n(\cdot)$ can now be defined. For $z = x + iv$ we have

$$\widehat{M}_n(z) = \begin{cases} M_n(z), & \text{for } z \in \mathcal{C}_n, \\ M_n(x_r + in^{-1}\varepsilon_n), & \text{for } x = x_r,\ v \in [0, n^{-1}\varepsilon_n], \\ & \text{and if } x_l > 0, \\ M_n(x_l + in^{-1}\varepsilon_n), & \text{for } x = x_l,\ v \in [0, n^{-1}\varepsilon_n]. \end{cases} \tag{1.11}$$

$\widehat{M}_n(\cdot)$ is viewed as a random element in the metric space $C(\mathcal{C}, \mathbb{R}^2)$ of continuous functions from $\mathcal{C}$ to $\mathbb{R}^2$. All of Chapter 2 of Billingsley (1968) applies to continuous functions from a set such as $\mathcal{C}$ (homeomorphic to $[0, 1]$) to finite-dimensional Euclidean space, with $|\cdot|$ interpreted as Euclidean distance.

Most of the paper will deal with proving the following lemma.

LEMMA 1.1. *Under conditions* (a) *and* (b) *of Theorem* 1.1 $\{\widehat{M}_n(\cdot)\}$ *forms a tight sequence on* $\mathcal{C}$. *Moreover, if assumptions in* (ii) *or* (iii) *of Theorem* 1.1 *on* X_{11} *hold, then* $\widehat{M}_n(\cdot)$ *converges weakly to a two-dimensional Gaussian process* $M(\cdot)$ *satisfying for* $z \in \mathcal{C}$ *under the assumptions in* (ii),

$$\mathsf{E}M(z) = \frac{c\int \underline{m}(z)^3 t^2 (1 + t\underline{m}(z))^{-3}\, dH(t)}{(1 - c\int \underline{m}(z)^2 t^2 (1 + t\underline{m}(z))^{-2}\, dH(t))^2} \tag{1.12}$$

and for $z_1, z_2 \in \mathcal{C} \cup \bar{\mathcal{C}}$, *with* $\bar{\mathcal{C}} \equiv \{\bar{z} : z \in \mathcal{C}\}$,

$$\begin{aligned} \operatorname{Cov}(M(z_1), M(z_2)) &\equiv \mathsf{E}\big[(M(z_1) - \mathsf{E}M(z_1))(M(z_2) - \mathsf{E}M(z_2))\big] \\ &= \frac{\underline{m}'(z_1)\underline{m}'(z_2)}{(\underline{m}(z_1) - \underline{m}(z_2))^2} - \frac{1}{(z_1 - z_2)^2}, \end{aligned} \tag{1.13}$$

while under the assumptions in (iii) $\mathsf{E}M(z) = 0$, *and the "covariance" function analogous to* (1.13) *is* $1/2$ *the right-hand side of* (1.13).

We show now how Theorem 1.1 follows from the above lemma. We use the identity

$$\int f(x)\, dG(x) = -\frac{1}{2\pi i} \int f(z) m_G(z)\, dz \tag{1.14}$$

valid for c.d.f. G and f analytic on an open set containing the support of G. The complex integral on the right-hand side is over any positively oriented contour enclosing the support of G and on which f is analytic. Choose v_0, x_r and x_l so that $f_1, \dots, f_k$ are all analytic on and inside the resulting $\mathcal{C} \cup \bar{\mathcal{C}}$.

Due to the a.s. convergence of the extreme eigenvalues of $(1/N)X_n X_n^*$ and the bounds

$$\lambda_{\max}^{AB} \le \lambda_{\max}^{A} \lambda_{\max}^{B} \qquad \lambda_{\min}^{AB} \ge \lambda_{\min}^{A} \lambda_{\min}^{B},$$

valid for $n \times n$ Hermitian nonnegative definite A and B, we have with probability 1

$$\liminf_{n\to\infty} \min(x_r - \lambda_{\max}^{B_n}, \lambda_{\min}^{B_n} - x_l) > 0.$$

It also follows that the support of F^{c_n, H_n} is contained in

$$[\lambda_{\min}^{T_n} I_{(0,1)}(c_n)(1 - \sqrt{c_n})^2, \lambda_{\max}^{T_n}(1 + \sqrt{c_n})^2].$$

Therefore for any $f \in \{f_1, \ldots, f_k\}$, with probability 1

$$\int f(x)\,dG_n(x) = -\frac{1}{2\pi i}\int f(z)M_n(z)\,dz$$

for all n large, where the complex integral is over $\mathcal{C} \cup \bar{\mathcal{C}}$. Moreover, with probability 1, for all n large,

$$\begin{aligned}&\left|\int f(z)(M_n(z) - \widehat{M}_n(z))\,dz\right|\\ &\quad \le 4K\varepsilon_n\Big(\big|\max(\lambda_{\max}^{T_n}(1+\sqrt{c_n})^2, \lambda_{\max}^{B_n}) - x_r\big|^{-1}\\ &\qquad\qquad + \big|\min(\lambda_{\min}^{T_n} I_{(0,1)}(c_n)(1-\sqrt{c_n})^2, \lambda_{\min}^{B_n}) - x_l\big|^{-1}\Big),\end{aligned}$$

which converges to zero as $n \to \infty$. Here K is a bound on f over $\mathcal{C}$.

Since

$$\widehat{M}_n(\cdot) \to \left(-\frac{1}{2\pi i}\int f_1(z)\widehat{M}_n(z)\,dz, \ldots, -\frac{1}{2\pi i}\int f_k(z)\widehat{M}_n(z)\,dz\right)$$

is a continuous mapping of $C(\mathcal{C}, \mathbb{R}^2)$ into $\mathbb{R}^k$, it follows that the above vector and, subsequently, (1.5) form tight sequences. Letting $M(\cdot)$ denote the limit of any weakly converging subsequence of $\{\widehat{M}_n(\cdot)\}$ we have the weak limit of (1.5) equal in distribution to

$$\left(-\frac{1}{2\pi i}\int f_1(z)M(z)\,dz, \ldots, -\frac{1}{2\pi i}\int f_k(z)\,M(z)\,dz\right).$$

The fact that this vector, under the assumptions in (ii) or (iii), is multivariate Gaussian follows from the fact that Riemann sums corresponding to these integrals are multivariate Gaussian and that weak limits of Gaussian vectors can only be Gaussian. The limiting expressions for the mean and covariance follow immediately.

Notice the assumptions in (ii) and (iii) require X_{11} to have the same first, second and fourth moments of either a real or complex Gaussian variable, the latter having real and imaginary parts i.i.d. $N(0, 1/2)$. We will use the terms "RG" and "CG" to refer to these conditions.

The reason why concrete results are at present only obtained for the assumptions in (ii) and (iii) is mainly due to the identity

$$\begin{aligned}&\mathsf{E}(X_{\cdot 1}^* A X_{\cdot 1} - \operatorname{tr} A)(X_{\cdot 1}^* B X_{\cdot 1} - \operatorname{tr} B)\\ &\quad = (\mathsf{E}|X_{11}|^4 - |\mathsf{E}X_{11}^2|^2 - 2)\sum_{i=1}^n a_{ii}b_{ii} + |\mathsf{E}X_{11}^2|^2 \operatorname{tr} AB^T + \operatorname{tr} AB\end{aligned} \tag{1.15}$$

valid for $n \times n$ $A = (a_{ij})$ and $B = (b_{ij})$, which is needed in several places in the proof of Lemma 1.1. The assumptions in (iii) leave only the last term on the right-hand side, whereas those in (ii) leave the last two, but in this case the matrix B will always be symmetric. This also accounts for the relation between the two covariance functions and the difficulty in obtaining explicit results more generally. As will be seen in the proof, whenever (1.15) is used, little is known about the limiting behavior of $\sum a_{ii} b_{ii}$.

Simple substitution reveals

$$\text{RHS of (1.7)} = -\frac{1}{2\pi^2} \iint \frac{f(z(m_1))g(z(m_2))}{(m_1 - m_2)^2} d(m_1)\, d(m_2). \tag{1.16}$$

However, the contours depend on the z_1, z_2 contours and cannot be arbitrarily chosen. It is also true that

$$\begin{aligned} (1.7) &= \frac{1}{\pi^2} \iint f'(x)g'(y) \ln\left|\frac{\underline{m}(x) - \overline{\underline{m}}(y)}{\underline{m}(x) - \underline{m}(y)}\right| dx\, dy \\ &= \frac{1}{2\pi^2} \iint f'(x)g'(y) \ln\left(1 + 4\frac{\underline{m}_i(x)\underline{m}_i(y)}{|\underline{m}(x) - \underline{m}(y)|^2}\right) dx\, dy \end{aligned} \tag{1.17}$$

and

$$\mathsf{E}X_f = \frac{1}{2\pi} \int f'(x) \arg\left(1 - c\int \frac{t^2 \underline{m}^2(x)}{(1 + t\underline{m}(x))^2} dH(t)\right) dx. \tag{1.18}$$

Here for $0 \neq x \in \mathbb{R}$

$$\underline{m}(x) = \lim_{z \to x} \underline{m}(z), \qquad z \in \mathbb{C}^+, \tag{1.19}$$

known to exist and to satisfy (1.2) [see Silverstein and Choi (1995)], and $\underline{m}_i(x) = \Im\, \underline{m}(x)$. The term

$$j(x) = \arg\left(1 - c\int \frac{t^2 \underline{m}^2(x)}{(1 + t\underline{m}(x))^2} dH(t)\right)$$

in (1.18) is well defined for almost every x and takes values in $(-\pi/2, \pi/2)$. Section 5 contains proofs of (1.17) and (1.18), along with showing

$$k(x, y) \equiv \ln\left(1 + 4\frac{\underline{m}_i(x)\underline{m}_i(y)}{|\underline{m}(x) - \underline{m}(y)|^2}\right) \tag{1.20}$$

to be Lebesgue integrable on $\mathbb{R}^2$. It is interesting to note that the support of $k(x, y)$ matches the support of $f^{c,H}$ on $\mathbb{R} - \{0\}$: $k(x, y) = 0 \Leftrightarrow \min(f^{c,H}(x), f^{c,H}(y)) = 0$. We also have $f^{c,H}(x) = 0 \Rightarrow j(x) = 0$.

Section 5 also contains derivations of the relevant quantities associated with the example given at the beginning of this section. The linear spectral statistic $(1/n)L_N$ has a.s. limit $d(c)$ as stated in (1.1). The quantity $L_N - nd(n/N)$ converges weakly to a Gaussian random variable $X_{\ln}$ with

$$\mathsf{E}X_{\ln} = \tfrac{1}{2}\ln(1 - c) \tag{1.21}$$

and

$$\operatorname{Var} X_{\ln} = -2\ln(1-c). \tag{1.22}$$

Results on both $L_N - \mathsf{E}L_N$ and $n[\int x^r \, dF^{S_N}(x) - \mathsf{E}\int x^r \, dF^{S_N}(x)]$ for positive integer r are derived in Jonsson (1982). Included in Section 5 are derivations of the following expressions for means and covariances, in this case ($H = I_{[1,\infty)}$). We have

$$\mathsf{E}X_{x^r} = \tfrac{1}{4}((1-\sqrt{c})^{2r} + (1+\sqrt{c})^{2r}) - \tfrac{1}{2}\sum_{j=0}^{r}\binom{r}{j}^2 c^j \tag{1.23}$$

and

$$\begin{aligned}\operatorname{Cov}(X_{x^{r_1}}, X_{x^{r_2}}) = 2c^{r_1+r_2}\sum_{k_1=0}^{r_1-1}\sum_{k_2=0}^{r_2}\binom{r_1}{k_1}\binom{r_2}{k_2}\left(\frac{1-c}{c}\right)^{k_1+k_2} \\ \times \sum_{\ell=1}^{r_1-k_1}\ell\binom{2r_1-1-(k_1+\ell)}{r_1-1} \\ \times \binom{2r_2-1-k_2+\ell}{r_2-1}.\end{aligned} \tag{1.24}$$

It is noteworthy to mention here a consequence of (1.17), namely that if the assumptions in (ii) or (iii) of Theorem 1.1 were to hold, then G_n, considered as a random element in $D[0,\infty)$ (the space of functions on $[0,\infty)$ that are right-continuous with left-hand side limits, together with the Skorohod metric) cannot form a tight sequence in $D[0,\infty)$. Indeed, under the assumptions of either one, if $G(x)$ were a weak limit of a subsequence, then, because of Theorem 1.1, it is straightforward to conclude that for any x_0 in the interior of the support of F and positive ε,

$$\int_{x_0}^{x_0+\varepsilon} G(x)\,dx$$

would be Gaussian, and therefore so would

$$G(x_0) = \lim_{\varepsilon\to 0}\frac{1}{\varepsilon}\int_{x_0}^{x_0+\varepsilon} G(x)\,dx.$$

However, the variance would necessarily be

$$\lim_{\varepsilon\to 0}\frac{1}{2\pi^2}\frac{1}{\varepsilon^2}\int_{x_0}^{x_0+\varepsilon}\int_{x_0}^{x_0+\varepsilon} k(x,y)\,dx\,dy = \infty.$$

Still, under the assumptions in (ii) or (iii), a limit may exist for $\{G_n\}$ when G_n is viewed as a linear functional

$$f \longrightarrow \int f(x)\,dG_n(x),$$

that is, a limit expressed in terms of a measure in a space of generalized functions. The characterization of the limiting measure of course depends on the space, which in turn relies on the set of test functions, which for now is restricted to functions analytic on the support of F. Work in this area is currently being pursued.

We emphasize here the importance of studying $G_n(x)$ which essentially balances $F^{B_n}(x)$ with F^{c_n,H_n}, and not $F^{c,H}$ or $\mathsf{E}F^{B_n}(x)$. $F^{c,H}$ cannot be used simply because the convergence of $c_n \to c$ and that of $H_n \to H$ can be arbitrarily slow. It should be viewed as a mathematical convenience because the result is expressed as a limit theorem. From the point of view of statistical inference, the choice of F^{c_n,H_n} over $\mathsf{E}F^{B_n}(x)$ is made simply because much is known of the former, while little is analytically known about the latter.

The proof of Lemma 1.1 is divided into three sections. Sections 2 and 3 handle the limiting behavior of the centralized M_n, while Section 4 analyzes the nonrandom part. In each of the three sections the reader will be referred to work done in Bai and Silverstein (1998).

2. Convergence of finite-dimensional distributions. Write for $z \in \mathcal{C}_n$, $M_n(z) = M_n^1(z) + M_n^2(z)$ where

$$M_n^1(z) = n[m_{F^{B_n}}(z) - \mathsf{E}m_{F^{B_n}}(z)]$$

and

$$M_n^2(z) = n[m_{\mathsf{E}F^{B_n}}(z) - m_{F^{c_n,H_n}}(z)].$$

In this section we will show for any positive integer r, the sum

$$\sum_{i=1}^{r} \alpha_i M_n^1(z_i) \qquad (\Im z_i \neq 0)$$

whenever it is real, is tight, and, under the assumptions in (ii) or (iii) of Theorem 1.1, will converge in distribution to a Gaussian random variable. Formula (1.13) will also be derived. We begin with a list of results.

LEMMA 2.1 [Burkholder (1973)]. *Let $\{X_k\}$ be a complex martingale difference sequence with respect to the increasing σ-field $\{\mathcal{F}_k\}$. Then, for $p > 1$,*

$$\mathsf{E}\Big|\sum X_k\Big|^p \le K_p \mathsf{E}\Big(\sum |X_k|^2\Big)^{p/2}.$$

(Note: The reference considers only real variables. Extending to complex variables is straightforward.)

LEMMA 2.2 [Lemma 2.7 in Bai and Silverstein (1998)]. *For $X = (X_1, \ldots, X_n)^T$ i.i.d. standardized (complex) entries, C $n \times n$ matrix (complex) we have, for any $p \ge 2$,*

$$\mathsf{E}|X^*CX - \operatorname{tr} C|^p \le K_p((\mathsf{E}|X_1|^4 \operatorname{tr} CC^*)^{p/2} + \mathsf{E}|X_1|^{2p} \operatorname{tr}(CC^*)^{p/2}).$$

LEMMA 2.3. *Let $f_1, f_2, \dots$ be analytic in D, a connected open set of $\mathbb{C}$, satisfying $|f_n(z)| \le M$ for every n and z in D, and $f_n(z)$ converges, as $n \to \infty$ for each z in a subset of D having a limit point in D. Then there exists a function f, analytic in D for which $f_n(z) \to f(z)$ and $f_n'(z) \to f'(z)$ for all $z \in D$. Moreover, on any set bounded by a contour interior to D the convergence is uniform and $\{f_n'(z)\}$ is uniformly bounded by $2M/\varepsilon$, where ε is the distance between the contour and the boundary of D.*

PROOF. The conclusions on $\{f_n\}$ are from Vitali's convergence theorem [see Titchmarsh (1939), page 168]. Those on $\{f_n'\}$ follow from the dominated convergence theorem (d.c.t.) and the identity

$$f_n'(z) = \frac{1}{2\pi i}\int_{\mathcal{C}} \frac{f_n(w)}{(w-z)^2}\,dw. \qquad \square$$

LEMMA 2.4 [Theorem 35.12 of Billingsley (1995)]. *Suppose for each n $Y_{n1}, Y_{n2}, \dots, Y_{nr_n}$ is a real martingale difference sequence with respect to the increasing σ-field $\{\mathcal{F}_{nj}\}$ having second moments. If as $n \to \infty$,*

$$\text{(i)} \qquad \sum_{j=1}^{r_n} \mathsf{E}(Y_{nj}^2 \mid \mathcal{F}_{n,j-1}) \xrightarrow{i.p.} \sigma^2,$$

where σ^2 is a positive constant, and for each $\varepsilon > 0$,

$$\text{(ii)} \qquad \sum_{j=1}^{r_n} \mathsf{E}(Y_{nj}^2 I_{(|Y_{nj}|\ge\varepsilon)}) \to 0$$

then

$$\sum_{j=1}^{r_n} Y_{nj} \xrightarrow{D} N(0,\sigma^2).$$

Recalling the truncation and centralization steps, we get from Lemma 2.2

$$\begin{aligned}(2.1)\qquad \mathsf{E}|X_{\cdot 1}^* C X_{\cdot 1} - \operatorname{tr} C|^p &\le K_p \|C\|^p [n^{p/2} + \delta_n^{(2p-4)} n^{(p-1)}] \\ &\le \underline{K}_p \|C\|^p \delta_n^{(2p-4)} n^{(p-1)}, \qquad p \ge 2.\end{aligned}$$

Let $v = \Im z$. For the following analysis we will assume $v > 0$. To facilitate notation, we will let $T = T_n$. Because of assumption (2′) we may assume $\|T\| \le 1$ for all n. Constants appearing in inequalities will be denoted by K and may take on different values from one expression to the next. Let $r_j = (1/\sqrt{N}) T^{1/2} X_{\cdot j}$, $D(z) = B_n - zI$, $D_j(z) = D(z) - r_j r_j^*$,

$$\varepsilon_j(z) = r_j^* D_j^{-1}(z) r_j - \frac{1}{N} \operatorname{tr} T D_j^{-1}(z),$$

$$\delta_j(z) = r_j^* D_j^{-2}(z) r_j - \frac{1}{N} \operatorname{tr} T D_j^{-2}(z) = \frac{d}{dz}\varepsilon_j(z)$$

and

$$\beta_j(z) = \frac{1}{1 + r_j^* D_j^{-1}(z) r_j},$$

$$\bar{\beta}_j(z) = \frac{1}{1 + N^{-1} \operatorname{tr} T_n D_j^{-1}(z)},$$

$$b_n(z) = \frac{1}{1 + N^{-1} \mathsf{E} \operatorname{tr} T_n D_1^{-1}(z)}.$$

All of the three latter quantities are bounded in absolute value by $|z|/v$ [see (3.4) of Bai and Silverstein (1998)]. We have

$$D^{-1}(z) - D_j^{-1}(z) = -D_j^{-1}(z) r_j r_j^* D_j^{-1}(z) \beta_j(z)$$

and from Lemma 2.10 of Bai and Silverstein (1998) for any $n \times n$ A

$$\left| \operatorname{tr}(D^{-1}(z) - D_j^{-1}(z)) A \right| \le \frac{\|A\|}{\Im z}. \tag{2.2}$$

For nonrandom $n \times n$ $A_k, k = 1, \dots, p$ and $B_l, l = 1, \dots, q$, we shall establish the following inequality:

$$\begin{aligned} &\left| \mathsf{E}\left(\prod_{k=1}^{p} r_1^* A_k r_1 \prod_{l=1}^{q} (r_1^* B_l r_1 - N^{-1} \operatorname{tr} T B_l) \right) \right| \\ &\qquad \le K N^{-(1 \wedge q)} \delta_n^{(2q-4) \vee 0} \prod_{k=1}^{p} \|A_k\| \prod_{l=1}^{q} \|B_l\|, \qquad p \ge 0,\ q \ge 0. \end{aligned} \tag{2.3}$$

When $p = 0,\ q = 1$, the left-hand side is 0. When $p = 0,\ q > 1$, (2.3) is a consequence of (2.1) and Hölder's inequality. If $p \ge 1$, then by induction on p we have

$$\begin{aligned} &\left| \mathsf{E}\left(\prod_{k=1}^{p} r_1^* A_k r_1 \prod_{l=1}^{q} (r_1^* B_l r_1 - N^{-1} \operatorname{tr} T B_l) \right) \right| \\ &\quad \le \left| \mathsf{E}\left(\prod_{k=1}^{p-1} r_1^* A_k r_1 (r_1^* A_p r_1 - N^{-1} \operatorname{tr} T A_p) \prod_{l=1}^{q} (r_1^* B_l r_1 - N^{-1} \operatorname{tr} T B_l) \right) \right| \\ &\qquad + n N^{-1} \|A_p\| \left| \mathsf{E}\left(\prod_{k=1}^{p-1} r_1^* A_k r_1 \prod_{l=1}^{q} (r_1^* B_l r_1 - N^{-1} \operatorname{tr} T B_l) \right) \right| \\ &\quad \le K N^{-1} \delta_n^{(2q-4) \vee 0} \prod_{k=1}^{p} \|A_k\| \prod_{l=1}^{q} \|B_l\|. \end{aligned}$$

We have proved the case where $q > 0$. When $q = 0$, (2.3) is a trivial consequence of (2.1).

Let $\mathsf{E}_0(\cdot)$ denote expectation and $\mathsf{E}_j(\cdot)$ denote conditional expectation with respect to the σ-field generated by $r_1, \ldots, r_j$.

We have

$$\begin{aligned}
n\big[m_{F^{B_n}}(z) - \mathsf{E}m_{F^{B_n}}(z)\big] &= \operatorname{tr}[D^{-1}(z) - \mathsf{E}D^{-1}(z)] \\
&= \sum_{j=1}^{N} \operatorname{tr}\mathsf{E}_j D^{-1}(z) - \operatorname{tr}\mathsf{E}_{j-1} D^{-1}(z) \\
&= \sum_{j=1}^{N} \operatorname{tr}\mathsf{E}_j [D^{-1}(z) - D_j^{-1}(z)] - \operatorname{tr}\mathsf{E}_{j-1}[D^{-1}(z) - D_j^{-1}(z)] \\
&= -\sum_{j=1}^{N} (\mathsf{E}_j - \mathsf{E}_{j-1})\beta_j(z) r_j^* D_j^{-2}(z) r_j.
\end{aligned}$$

Write $\beta_j(z) = \bar{\beta}_j(z) - \beta_j(z)\bar{\beta}_j(z)\varepsilon_j(z) = \bar{\beta}_j(z) - \bar{\beta}_j^2(z)\varepsilon_j(z) + \bar{\beta}_j^2(z) \times \beta_j(z)\varepsilon_j^2(z)$. We have then

$$\begin{aligned}
&(\mathsf{E}_j - \mathsf{E}_{j-1})\beta_j(z) r_j^* D_j^{-2}(z) r_j \\
&\quad= (\mathsf{E}_j - \mathsf{E}_{j-1})\Big(\bar{\beta}_j(z)\delta_j(z) - \bar{\beta}_j^2(z)\varepsilon_j(z)\delta_j(z) \\
&\qquad\qquad - \bar{\beta}_j^2(z)\varepsilon_j(z)\frac{1}{N}\operatorname{tr} T D_j^{-2}(z) + \bar{\beta}_j^2(z)\beta_j(z)\varepsilon_j^2(z) r_j^* D_j^{-2}(z) r_j\Big) \\
&\quad= \mathsf{E}_j\Big(\bar{\beta}_j(z)\delta_j(z) - \bar{\beta}_j^2(z)\varepsilon_j(z)\frac{1}{N}\operatorname{tr} T D_j^{-2}(z)\Big) \\
&\qquad - (\mathsf{E}_j - \mathsf{E}_{j-1})\bar{\beta}_j^2(z)\big(\varepsilon_j(z)\delta_j(z) - \beta_j(z) r_j D_j^{-2}(z) r_j \varepsilon_j^2(z)\big).
\end{aligned}$$

Using (2.3), we have

$$\begin{aligned}
\mathsf{E}\left|\sum_{j=1}^{N} (\mathsf{E}_j - \mathsf{E}_{j-1})\bar{\beta}_j^2(z)\varepsilon_j(z)\delta_j(z)\right|^2 &= \sum_{j=1}^{N} \mathsf{E}|(\mathsf{E}_j - \mathsf{E}_{j-1})\bar{\beta}_j^2(z)\varepsilon_j(z)\delta_j(z)|^2 \\
&\le 4\sum_{j=1}^{N} \mathsf{E}|\bar{\beta}_j^2(z)\varepsilon_j(z)\delta_j(z)|^2 = o(1).
\end{aligned}$$

Therefore, $\sum_{j=1}^{N}(\mathsf{E}_j - \mathsf{E}_{j-1})\bar{\beta}_j^2(z)\varepsilon_j(z)\delta_j(z)$ converges to zero in probability.

By the same argument, we have

$$\sum_{j=1}^{N} (\mathsf{E}_j - \mathsf{E}_{j-1})\beta_j(z) r_j D_j^{-2}(z) r_j \varepsilon_j^2(z) \stackrel{\text{i.p.}}{\to} 0.$$

Therefore we need only consider the sum

$$\sum_{i=1}^{r}\alpha_i\sum_{j=1}^{N}Y_j(z_i)=\sum_{j=1}^{N}\sum_{i=1}^{r}\alpha_i Y_j(z_i),$$

where

$$Y_j(z)=-\mathsf{E}_j\left(\bar\beta_j(z)\delta_j(z)-\bar\beta_j^2(z)\varepsilon_j(z)\frac{1}{N}\operatorname{tr}TD_j^{-2}(z)\right)=-\mathsf{E}_j\frac{d}{dz}\bar\beta_j(z)\varepsilon_j(z).$$

Again, by using (2.3), we obtain

$$\mathsf{E}|Y_j(z)|^4\le K\left(\frac{|z|^4}{v^4}\mathsf{E}|\delta_j(z)|^4+\frac{|z|^8}{v^{16}}\left(\frac{n}{N}\right)^4\mathsf{E}|\varepsilon_j(z)|^4\right)=o(N^{-1}),$$

which implies for any $\varepsilon>0$

$$\sum_{j=1}^{N}\mathsf{E}\left(\left|\sum_{i=1}^{r}\alpha_i Y_j(z_i)\right|^2 I_{(|\sum_{i=1}^{r}\alpha_i Y_j(z_i)|\ge\varepsilon)}\right)\le\frac{1}{\varepsilon^2}\sum_{j=1}^{N}\mathsf{E}\left|\sum_{i=1}^{r}\alpha_i Y_j(z_i)\right|^4\to 0$$

as $n\to\infty$. Therefore condition (ii) of Lemma 2.4 is satisfied and it is enough to prove, under the assumptions in (ii) or (iii) of Theorem 1.1, for z_1, z_2 with nonzero imaginary parts

$$\sum_{j=1}^{N}\mathsf{E}_{j-1}[Y_j(z_1)Y_j(z_2)] \tag{2.4}$$

converges in probability to a constant (and to determine the constant).

We show here for future use the tightness of the sequence $\{\sum_{i=1}^{r}\alpha_i M_n^1(z_i)\}$. From (2.3) we easily get $\mathsf{E}|Y_j(z)|^2=O(N^{-1})$, so that

$$\begin{aligned}\mathsf{E}\left|\sum_{i=1}^{r}\alpha_i\sum_{j=1}^{N}Y_j(z_i)\right|^2&=\sum_{j=1}^{N}\mathsf{E}\left|\sum_{i=1}^{r}\alpha_i Y_j(z_i)\right|^2\\&\le r\sum_{j=1}^{N}\sum_{i=1}^{r}|\alpha_i|^2\mathsf{E}|Y_j(z_i)|^2\le K.\end{aligned} \tag{2.5}$$

Consider the sum

$$\sum_{j=1}^{N}\mathsf{E}_{j-1}[\mathsf{E}_j(\bar\beta_j(z_1)\varepsilon_j(z_1))\mathsf{E}_j(\bar\beta_j(z_2)\varepsilon_j(z_2))]. \tag{2.6}$$

In the jth term (viewed as an expectation with respect to $r_{j+1},\dots,r_N$) we apply the d.c.t. to the difference quotient defined by $\bar\beta_j(z)\varepsilon_j(z)$ to get

$$\frac{\partial^2}{\partial z_2\,\partial z_1}(2.6)=(2.4).$$

Let v_0 be a lower bound on $|\Im z_i|$. For each j let $A^i_j = (1/N)T^{1/2}\mathsf{E}_j D_j^{-1}(z_i) \times T^{1/2}$, $i = 1, 2$. Then $\operatorname{tr} A^i_j A^{i\,*}_j \le n(v_0 N)^{-2}$. Using (2.1) we see, therefore, that (2.6) is bounded, with a bound depending only on $|z_i|$ and v_0.

We can then appeal to Lemma 2.3. Suppose (2.6) converges in probability to a nonrandom limit for each $z_k, z_l \in \{z_i\} \subset \underline{D} \equiv \{z : v_0 < |\Im z| < K\}$ ($K > v_0$ arbitrary), a sequence having two limit points, one on each side of the real axis. Then by a diagonalization argument, for any subsequence of the natural numbers, there is a further subsequence such that, with probability one, (2.6) converges for each pair z_k, z_l. Write (2.6) as $f_n(z_1, z_2)$. We concentrate on this subsequence and on one realization for which convergence holds. For each $z_l \in \{z_i\}$ we apply Lemma 2.3 on each of $\{z : v_0/2 < \Im z < K\}$ and $\{z : -K < \Im z < -v_0/2\}$ to get convergence of $f_n(z, z_l)$ to a function $f(z, z_l)$, analytic for $z \in \underline{D}$ satisfying $\frac{\partial}{\partial z} f_n(z, z_l) \to \frac{\partial}{\partial z} f(z, z_l)$. From Lemma 2.3 we see that $\frac{\partial}{\partial z} f_n(z, w)$ is bounded in w and n for all $w \in \underline{D}$. Applying again Lemma 2.3 on the remaining variable we see that $f_n(z, w) \to f(z, w)$, analytic for $w \in \underline{D}$ and $\frac{\partial^2}{\partial u\, \partial z} f_n(z, w) \to \frac{\partial^2}{\partial w\, \partial z} f_n(z, w)$. Since $f(z, w)$ does not depend on the realization nor on the subsequence, we have convergence in probability of (2.6) to f and (2.4) to the mixed partials of f. Therefore we need only show (2.6) converges in probability and to determine its limit.

From the derivation above (4.3) of Bai and Silverstein (1998) we get

$$\mathsf{E}|\bar{\beta}_j(z_i) - b_n(z_i)|^2 \le K \frac{|z_i|^4}{v_0^6} N^{-1}.$$

This implies

$$\begin{aligned}\mathsf{E}\Big|&\mathsf{E}_{j-1}[\mathsf{E}_j(\bar{\beta}_j(z_1)\varepsilon_j(z_1))\mathsf{E}_j(\bar{\beta}_j(z_2)\varepsilon_j(z_2))]\\ &- \mathsf{E}_{j-1}[\mathsf{E}_j(b_n(z_1)\varepsilon_j(z_1))\mathsf{E}_j(b_n(z_2)\varepsilon_j(z_2))]\Big| = O(N^{-3/2})\end{aligned}$$

from which we get

$$\begin{aligned}&\sum_{j=1}^{N} \mathsf{E}_{j-1}[\mathsf{E}_j(\bar{\beta}_j(z_1)\varepsilon_j(z_1))\mathsf{E}_j(\hat{\beta}_j(z_2)\varepsilon_j(z_2))]\\ &\qquad - b_n(z_1)b_n(z_2)\sum_{j=1}^{N} \mathsf{E}_{j-1}[\mathsf{E}_j(\varepsilon_j(z_1))\mathsf{E}_j(\varepsilon_j(z_2))] \xrightarrow{\text{i.p.}} 0.\end{aligned}$$

Thus the goal is to show

$$b_n(z_1)b_n(z_2)\sum_{j=1}^{N} \mathsf{E}_{j-1}[\mathsf{E}_j(\varepsilon_j(z_1))\mathsf{E}_j(\varepsilon_j(z_2))] \tag{2.7}$$

converges in probability, and to determine its limit. The latter's second mixed partial derivative will yield the limit of (2.4).

We now assume the CG case, namely $\mathsf{E}X_{11}^2 = o(1/n)$ and $\mathsf{E}|X_{11}|^4 = 2 + o(1)$, so that, using (1.15), (2.7) becomes

$$b_n(z_1)b_n(z_2)\frac{1}{N^2}\sum_{j=1}^{N}(\operatorname{tr} T^{1/2}\mathsf{E}_j(D_j^{-1}(z_1))T\mathsf{E}_j(D_j^{-1}(z_2))T^{1/2} + o(1)A_n),$$

where

$$|A_n| \le K\Big(\operatorname{tr} T\mathsf{E}_j(D_j^{-1}(z_1))T\mathsf{E}_j(\bar{D}_j^{-1}(z_1)) \\ \times \operatorname{tr} T\mathsf{E}_j(D_j^{-1}(z_2))T\mathsf{E}_j(\bar{D}_j^{-1}(z_2))\Big)^{1/2} = O(N).$$

Thus we study

$$b_n(z_1)b_n(z_2)\frac{1}{N^2}\sum_{j=1}^{N}\operatorname{tr} T^{1/2}\mathsf{E}_j(D_j^{-1}(z_1))T\mathsf{E}_j(D_j^{-1}(z_2))T^{1/2}. \tag{2.8}$$

The RG case [T_n, X_{11} real, $\mathsf{E}|X_{11}|^4 = 3 + o(1)$] will be double that of the limit of (2.8).

Let $D_{ij}(z) = D(z) - r_ir_i^* - r_jr_j^*$,

$$\beta_{ij}(z) = \frac{1}{1 + r_i^*D_{ij}^{-1}(z)r_i} \quad \text{and} \quad b_1(z) = \frac{1}{1 + N^{-1}\mathsf{E}\operatorname{tr} TD_{12}^{-1}(z)}.$$

We write

$$D_j(z_1) + z_1I - \frac{N-1}{N}b_1(z_1)T = \sum_{i\neq j}^{N} r_ir_i^* - \frac{N-1}{N}b_1(z_1)T.$$

Multiplying by $(z_1I - \frac{N-1}{N}b_1(z_1)T)^{-1}$ on the left-hand side, $D_j^{-1}(z_1)$ on the right-hand side and using

$$r_i^*D_j^{-1}(z_1) = \beta_{ij}(z_1)r_i^*D_{ij}^{-1}(z_1)$$

we get

$$\begin{aligned} D_j^{-1}(z_1) &= -\Big(z_1I - \frac{N-1}{N}b_1(z_1)T\Big)^{-1} \\ &\quad + \sum_{i\neq j}\beta_{ij}(z_1)\Big(z_1I - \frac{N-1}{N}b_1(z_1)T\Big)^{-1} r_ir_i^*D_{ij}^{-1}(z_1) \\ &\quad - \frac{N-1}{N}b_1(z_1)\Big(z_1I - \frac{N-1}{N}b_1(z_1)T\Big)^{-1} TD_j^{-1}(z_1) \\ &= -\Big(z_1I - \frac{N-1}{N}b_1(z_1)T\Big)^{-1} + b_1(z_1)A(z_1) + B(z_1) + C(z_1), \end{aligned} \tag{2.9}$$

where

$$A(z_1) = \sum_{i \neq j} \left(z_1 I - \frac{N-1}{N} b_1(z_1) T \right)^{-1} (r_i r_i^* - N^{-1} T) D_{ij}^{-1}(z_1),$$

$$B(z_1) = \sum_{i \neq j} (\beta_{ij}(z_1) - b_1(z_1)) \left(z_1 I - \frac{N-1}{N} b_1(z_1) T \right)^{-1} r_i r_i^* D_{ij}^{-1}(z_1)$$

and

$$C(z_1) = N^{-1} b_1(z_1) \left(z_1 I - \frac{N-1}{N} b_1(z_1) T \right)^{-1} T \\ \times \sum_{i \neq j} (D_{ij}^{-1}(z_1) - D_j^{-1}(z_1)).$$

It is easy to verify for any real t,

$$\left| 1 - \frac{t}{z(1 + N^{-1} \mathsf{E} \operatorname{tr} T D_{12}^{-1}(z))} \right|^{-1} \leq \frac{|z(1 + N^{-1} \mathsf{E} \operatorname{tr} T D_{12}^{-1}(z))|}{\Im z(1 + N^{-1} \mathsf{E} \operatorname{tr} T D_{12}^{-1}(z))} \\ \leq \frac{|z|(1 + n/(N v_0))}{v_0}.$$

Thus

$$\left\| \left(z_1 I - \frac{N-1}{N} b_1(z_1) T \right)^{-1} \right\| \leq \frac{1 + n/(N v_0)}{v_0}. \tag{2.10}$$

Let M be $n \times n$ and let $|||M|||$ denote a nonrandom bound on the spectral norm of M for all parameters governing M and under all realizations of M. From (4.3) of Bai and Silverstein (1998), (2.3) and (2.10) we get

$$\begin{aligned} &\mathsf{E}|\operatorname{tr} B(z_1) M| \\ &\quad \leq N \mathsf{E}^{1/2}(|\beta_{12}(z_1) - b_1(z_1)|^2) \\ &\qquad \times \mathsf{E}^{1/2}\left(\left| r_i^* D_{ij}^{-1}(z_1) M \left(z_1 I - \frac{N-1}{N} b_1(z_1) T \right)^{-1} r_i \right|^2 \right) \\ &\quad \leq K |||M||| \frac{|z_1|^2 (1 + n/(N v_0))}{v_0^5} N^{1/2}. \end{aligned} \tag{2.11}$$

From (2.2) we have

$$|\operatorname{tr} C(z_1) M| \leq |||M||| \frac{|z_1|(1 + n/(N v_0))}{v_0^3}. \tag{2.12}$$

From (2.3) and (2.10) we get, for M nonrandom,

$$\begin{aligned}
&\mathsf{E}|\operatorname{tr} A(z_1)M| \\
&\le K\mathsf{E}^{1/2}\operatorname{tr} T^{1/2}D_{ij}^{-1}(z_1)M\left(z_1 I - \frac{N-1}{N}b_1(z_1)T\right)^{-1} \\
&\quad\times T\left(\bar z_1 I - \frac{N-1}{N}b_1(\bar z_1)T\right)^{-1} M^* D_{ij}^{-1}(\bar z_1)T^{1/2} \qquad (2.13)\\
&\le K\|M\|\frac{(1+n/(Nv_0))}{v_0^2}N^{1/2}.
\end{aligned}$$

We write [using the identity above (2.2)]

$$\operatorname{tr}\mathsf{E}_j(A(z_1))TD_j^{-1}(z_2)T = A_1(z_1,z_2) + A_2(z_1,z_2) + A_3(z_1,z_2), \qquad (2.14)$$

where

$$\begin{aligned}
A_1(z_1,z_2) &= -\operatorname{tr}\sum_{i<j}\left(z_1 I - \frac{N-1}{N}b_1(z_1)T\right)^{-1} r_i r_i^* \mathsf{E}_j(D_{ij}^{-1}(z_1)) \\
&\qquad\times T\beta_{ij}(z_2)D_{ij}^{-1}(z_2)r_i r_i^* D_{ij}^{-1}(z_2)T \\
&= -\sum_{i<j}\beta_{ij}(z_2)r_i^*\mathsf{E}_j(D_{ij}^{-1}(z_1))TD_{ij}^{-1}(z_2)r_i r_i^* \\
&\qquad\times D_{ij}^{-1}(z_2)T\left(z_1 I - \frac{N-1}{N}b_1(z_1)T\right)^{-1} r_i, \\
A_2(z_1,z_2) &= -\operatorname{tr}\sum_{i<j}\left(z_1 I - \frac{N-1}{N}b_1(z_1)T\right)^{-1} N^{-1}T \\
&\qquad\times \mathsf{E}_j(D_{ij}^{-1}(z_1))T(D_j^{-1}(z_2) - D_{ij}^{-1}(z_2))T
\end{aligned}$$

and

$$\begin{aligned}
A_3(z_1,z_2) &= \operatorname{tr}\sum_{i<j}\left(z_1 I - \frac{N-1}{N}b_1(z_1)T\right)^{-1}(r_i r_i^* - N^{-1}T) \\
&\qquad\times \mathsf{E}_j(D_{ij}^{-1}(z_1))TD_{ij}^{-1}(z_2)T.
\end{aligned}$$

We get from (2.2) and (2.10)

$$|A_2(z_1,z_2)| \le \frac{(1+n/(Nv_0))}{v_0^2} \qquad (2.15)$$

and similarly to (2.11) we have

$$\mathsf{E}|A_3(z_1,z_2)| \le \frac{(1+n/(Nv_0))}{v_0^3} N^{1/2}.$$

Using (2.1), (2.3) and (4.3) of Bai and Silverstein (1998) we have for $i < j$

$$\begin{aligned}
\mathsf{E}\bigg|&\beta_{ij}(z_2) r_i^* \mathsf{E}_j(D_{ij}^{-1}(z_1)) T D_{ij}^{-1}(z_2) r_i r_i^* \\
&\times D_{ij}^{-1}(z_2) T \left(z_1 I - \frac{N-1}{N} b_1(z_1) T\right)^{-1} r_i \\
&- b_1(z_2) N^{-2} \operatorname{tr}(\mathsf{E}_j(D_{ij}^{-1}(z_1)) T D_{ij}^{-1}(z_2) T) \\
&\times \operatorname{tr}\left(D_{ij}^{-1}(z_2) T \left(z_1 I - \frac{N-1}{N} b_1(z_1) T\right)^{-1} T\right)\bigg| \\
&\le K N^{-1/2}
\end{aligned}$$

(K now depending as well on the z_i and on n/N). Using (2.2) we have

$$\begin{aligned}
&\bigg|\operatorname{tr}(\mathsf{E}_j(D_{ij}^{-1}(z_1)) T D_{ij}^{-1}(z_2) T) \operatorname{tr}\left(D_{ij}^{-1}(z_2) T \left(z_1 I - \frac{N-1}{N} b_1(z_1) T\right)^{-1} T\right) \\
&- \operatorname{tr}(\mathsf{E}_j(D_j^{-1}(z_1)) T D_j^{-1}(z_2) T) \operatorname{tr}\left(D_j^{-1}(z_2) T \left(z_1 I - \frac{N-1}{N} b_1(z_1) T\right)^{-1} T\right)\bigg| \\
&\le KN.
\end{aligned}$$

It follows that

$$\begin{aligned}
\mathsf{E}\bigg|&A_1(z_1,z_2) + \frac{j-1}{N^2} b_1(z_2) \operatorname{tr}(\mathsf{E}_j(D_j^{-1}(z_1)) T D_j^{-1}(z_2) T) \\
&\times \operatorname{tr}\left(D_j^{-1}(z_2) T \left(z_1 I - \frac{N-1}{N} b_1(z_1) T\right)^{-1} T\right)\bigg| \qquad (2.16)\\
&\le K N^{1/2}.
\end{aligned}$$

Therefore, from (2.9)–(2.16) we can write

$$\begin{aligned}
&\operatorname{tr}(\mathsf{E}_j(D_j^{-1}(z_1)) T D_j^{-1}(z_2) T) \\
&\quad\times \left[1 + \frac{j-1}{N^2} b_1(z_1) b_1(z_2) \operatorname{tr}\left(D_j^{-1}(z_2) T \left(z_1 I - \frac{N-1}{N} b_1(z_1) T\right)^{-1} T\right)\right] \\
&\quad= -\operatorname{tr}\left(\left(z_1 I - \frac{N-1}{N} b_1(z_1) T\right)^{-1} T D_j^{-1}(z_2) T\right) + A_4(z_1,z_2),
\end{aligned}$$

where

$$\mathsf{E}|A_4(z_1,z_2)| \le K N^{1/2}.$$

Using the expression for $D_j^{-1}(z_2)$ in (2.9) and (2.11)–(2.13) we find that

$$\begin{aligned}
&\operatorname{tr}(\mathsf{E}_j(D_j^{-1}(z_1))TD_j^{-1}(z_2)T)\\
&\quad\times\left[1-\frac{j-1}{N^2}b_1(z_1)b_1(z_2)\right.\\
&\qquad\left.\times\operatorname{tr}\left(\left(z_2I-\frac{N-1}{N}b_1(z_2)T\right)^{-1}T\left(z_1I-\frac{N-1}{N}b_1(z_1)T\right)^{-1}T\right)\right]\\
&=\operatorname{tr}\left(\left(z_2I-\frac{N-1}{N}b_1(z_2)T\right)^{-1}T\left(z_1I-\frac{N-1}{N}b_1(z_1)T\right)^{-1}T\right)\\
&\quad+A_5(z_1,z_2),
\end{aligned}$$

where

$$\mathsf{E}|A_5(z_1,z_2)|\le KN^{1/2}.$$

From (2.2) we have

$$|b_1(z)-b_n(z)|\le KN^{-1}.$$

From (4.3) of Bai and Silverstein (1998) we have

$$|b_n(z)-\mathsf{E}\beta_1(z)|\le KN^{-1/2}.$$

From the formula

$$\underline{m}_n=-\frac{1}{zN}\sum_{j=1}^{N}\beta_j(z)$$

[(2.2) of Silverstein (1995)] we get $\mathsf{E}\beta_1(z)=-z\mathsf{E}\underline{m}_n(z)$. Section 5 of Bai and Silverstein (1998) proves that

$$|\mathsf{E}\underline{m}_n(z)-\underline{m}_n^0(z)|\le KN^{-1}.$$

Therefore we have

$$|b_1(z)+z\underline{m}_n^0(z)|\le KN^{-1/2}, \tag{2.17}$$

so that we can write

$$\begin{aligned}
&\operatorname{tr}(\mathsf{E}_j(D_j^{-1}(z_1))TD_j^{-1}(z_2)T)\\
&\quad\times\left[1-\frac{j-1}{N^2}\underline{m}_n^0(z_1)\underline{m}_n^0(z_2)\right.\\
&\qquad\left.\times\operatorname{tr}((I+\underline{m}_n^0(z_2)T)^{-1}T(I+\underline{m}_n^0(z_1)T)^{-1}T)\right]\\
&=\frac{1}{z_1z_2}\operatorname{tr}((I+\underline{m}_n^0(z_2)T)^{-1}T(I+\underline{m}_n^0(z_1)T)^{-1}T)+A_6(z_1,z_2),
\end{aligned} \tag{2.18}$$

where

$$\mathsf{E}|A_6(z_1,z_2)| \le KN^{1/2}.$$

Rewrite (2.18) as

$$\begin{aligned}
&\operatorname{tr}\big(\mathsf{E}_j(D_j^{-1}(z_1))TD_j^{-1}(z_2)T\big)\\
&\quad\times\left[1-\frac{j-1}{N}c_n\underline{m}_n^0(z_1)\underline{m}_n^0(z_2)\int\frac{t^2\,dH_n(t)}{(1+t\underline{m}_n^0(z_1))(1+t\underline{m}_n^0(z_2))}\right]\\
&\quad=\frac{Nc_n}{z_1z_2}\int\frac{t^2\,dH_n(t)}{(1+t\underline{m}_n^0(z_1))(1+t\underline{m}_n^0(z_2))}+A_6(z_1,z_2).
\end{aligned}$$

Using (3.9) and (3.16) in Bai and Silverstein (1998) we find

$$\begin{aligned}
(2.19)\quad &\left|c_n\underline{m}_n^0(z_1)\underline{m}_n^0(z_2)\int\frac{t^2\,dH_n(t)}{(1+t\underline{m}_n^0(z_1))(1+t\underline{m}_n^0(z_2))}\right|\\
&=\left|c_n\left[\int\frac{t^2\,dH_n(t)}{(1+t\underline{m}_n^0(z_1))(1+t\underline{m}_n^0(z_2))}\right]\right.\\
&\qquad\left.\times\left[\left(-z_1+c_n\int\frac{t\,dH_n(t)}{1+t\underline{m}_n^0(z_1)}\right)\left(-z_2+c_n\int\frac{t\,dH_n(t)}{1+t\underline{m}_n^0(z_2)}\right)\right]^{-1}\right|\\
&\le\left(c_n\int\frac{t^2\,dH_n(t)}{|1+t\underline{m}_n^0(z_1)|^2}\left|-z_1+c_n\int\frac{t\,dH_n(t)}{1+t\underline{m}_n^0(z_1)}\right|^{-2}\right)^{1/2}\\
&\qquad\times\left(c_n\int\frac{t^2\,dH_n(t)}{|1+t\underline{m}_n^0(z_2)|^2}\left|-z_2+c_n\int\frac{t\,dH_n(t)}{1+t\underline{m}_n^0(z_2)}\right|^{-2}\right)^{1/2}\\
&=\left(\left(\Im\underline{m}_n^0(z_1)c_n\int\frac{t^2\,dH_n(t)}{|1+t\underline{m}_n^0(z_1)|^2}\right)\right.\\
&\qquad\left.\times\left(\Im z_1+\Im\underline{m}_n^0(z_1)c_n\int\frac{t^2\,dH_n(t)}{|1+t\underline{m}_n^0(z_1)|^2}\right)^{-1}\right)^{1/2}\\
&\qquad\times\left(\left(\Im\underline{m}_n^0(z_2)c_n\int\frac{t^2\,dH_n(t)}{|1+t\underline{m}_n^0(z_2)|^2}\right)\right.\\
&\qquad\left.\times\left(\Im z_2+\Im\underline{m}_n^0(z_2)c_n\int\frac{t^2\,dH_n(t)}{|1+t\underline{m}_n^0(z_2)|^2}\right)^{-1}\right)^{1/2}<1
\end{aligned}$$

since

$$\Im z\left(\Im\underline{m}_n^0(z)c_n\int\frac{t^2\,dH_n(t)}{|1+t\underline{m}_n^0(z)|^2}\right)^{-1}$$

is bounded away from 0. Therefore using (2.17) and letting $a_n(z_1,z_2)$ denote the expression inside the absolute value sign in (2.19) we find that (2.8) can be written as

$$a_n(z_1, z_2)\frac{1}{N}\sum_{j=1}^{N}\frac{1}{1-((j-1)/N)a_n(z_1, z_2)} + A_7(z_1, z_2),$$

where

$$\mathsf{E}|A_7(z_1, z_2)| \le KN^{-1/2}.$$

We see then that

$$(2.8) \xrightarrow{\text{i.p.}} a(z_1, z_2)\int_0^1 \frac{1}{1-ta(z_1, z_2)}\,dt = \int_0^{a(z_1,z_2)} \frac{1}{1-z}\,dz$$

where

$$\begin{aligned} a(z_1, z_2) &= c\underline{m}(z_1)\underline{m}(z_2)\int \frac{t^2\,dH(t)}{(1+t\underline{m}(z_1))(1+t\underline{m}(z_2))} \\ &= \frac{\underline{m}(z_1)\underline{m}(z_2)}{\underline{m}(z_2)-\underline{m}(z_1)}\left(c\int \frac{t\,dH(t)}{1+t\underline{m}(z_1)} - c\int \frac{t\,dH(t)}{1+t\underline{m}(z_2)}\right) \\ &= 1 + \frac{\underline{m}(z_1)\underline{m}(z_2)(z_1-z_2)}{\underline{m}(z_2)-\underline{m}(z_1)}. \end{aligned}$$

Thus the i.p. limit of (2.4) under the CG case is

$$\begin{aligned} &\frac{\partial^2}{\partial z_2\,\partial z_1}\int_0^{a(z_1,z_2)} \frac{1}{1-z}\,dz \\ &= \frac{\partial}{\partial z_2}\left(\frac{\partial a(z_1, z_2)/\partial z_1}{1-a(z_1, z_2)}\right) \\ &= \frac{\partial}{\partial z_2}\left[\frac{(\underline{m}(z_2)-\underline{m}(z_1))(\underline{m}'(z_1)\underline{m}(z_2)(z_1-z_2)+\underline{m}(z_1)\underline{m}(z_2))}{(\underline{m}(z_2)-\underline{m}(z_1))^2}\right. \\ &\qquad\qquad \left. + \frac{\underline{m}(z_1)\underline{m}(z_2)(z_1-z_2)\underline{m}'(z_1)}{(\underline{m}(z_2)-\underline{m}(z_1))^2}\right] \\ &\qquad \times \frac{\underline{m}(z_2)-\underline{m}(z_1)}{\underline{m}(z_1)\underline{m}(z_2)(z_2-z_1)} \\ &= -\frac{\partial}{\partial z_2}\left(\frac{\underline{m}'(z_1)}{\underline{m}(z_1)} + \frac{1}{z_1-z_2} + \frac{\underline{m}'(z_1)}{\underline{m}(z_2)-\underline{m}(z_1)}\right) \\ &= \frac{\underline{m}'(z_1)\underline{m}'(z_2)}{(\underline{m}(z_2)-\underline{m}(z_1))^2} - \frac{1}{(z_1-z_2)^2} \end{aligned}$$

which is (1.13).

3. Tightness of $M_n^1(z)$. We proceed to prove tightness of the sequence of random functions $\widehat{M}_n^1(z)$ for $z \in \mathcal{C}$ defined by (1.11). We will use Theorem 12.3 [Billingsley (1968), page 96]. It is easy to verify from the proof of the Arzela–Ascoli theorem [Billingsley (1968), page 221] that condition (i) of Theorem 12.3 can be replaced with the assumption of tightness at any point in [0, 1]. From (2.5) we see that this condition is satisfied. We will verify condition (ii) of Theorem 12.3 by proving the moment condition (12.51) of Billingsley (1968). We will show

$$\sup_{n;z_1,z_2\in\mathcal{C}_n} \frac{\mathsf{E}|M_n^1(z_1) - M_n^1(z_2)|^2}{|z_1 - z_2|^2}$$

is finite.

We claim that moments of $\|D^{-1}(z)\|$, $\|D_j^{-1}(z)\|$ and $\|D_{ij}^{-1}(z)\|$ are bounded in n and $z \in \mathcal{C}_n$. This is clearly true for $z \in \mathcal{C}_u$ and for $z \in \mathcal{C}_l$ if $x_l < 0$. For $z \in \mathcal{C}_r$ or, if $x_l > 0$, $z \in \mathcal{C}_l$, we use (1.9) and (1.10) on, for example $B_{(1)} = B_n - r_1 r_1^*$, to get

$$\begin{aligned}\mathsf{E}\|D_j^{-1}(z)\|^p &\le K_1 + v^{-p}\mathsf{P}(\|B_{(1)}\| \ge \eta_r \text{ or } \lambda_{\min}^{B_{(1)}} \le \eta_l)\\ &\le K_1 + K_2 n^p \varepsilon^{-p} n^{-\ell} \le K\end{aligned}$$

for suitably large ℓ. Here, η_r is any fixed number between $\limsup_n \|T\|(1+\sqrt{c})^2$ and x_r, and, if $x_l > 0$, η_l is any fixed number between x_l and $\liminf_n \lambda_{\min}^T (1 - \sqrt{c})^2$ (take $\eta_l < 0$ if $x_l < 0$). Therefore for any positive p,

$$\max(\mathsf{E}\|D^{-1}(z)\|^p, \mathsf{E}\|D_j^{-1}(z)\|^p, \mathsf{E}\|D_{ij}^{-1}(z)\|^p) \le K_p. \tag{3.1}$$

We can use the above argument to extend (2.3). Using (1.8) and (2.3) we get

$$\begin{aligned}&\left|\mathsf{E}\left(a(v)\prod_{l=1}^{q}(r_1^* B_l(v) r_1 - N^{-1}\operatorname{tr} T B_l(v))\right)\right|\\ &\quad\le K N^{-(1\wedge q)} \delta_n^{(2q-4)\vee 0}, \qquad q \ge 0,\end{aligned} \tag{3.2}$$

where now the matrices $B_l(v)$ are independent of r_1 and

$$\max(|a(v)|, \|B_l(v)\|) \le K(1 + n^s I(\|B_n\| \ge \eta_r \text{ or } \lambda_{\min}^{\widetilde{B}} \le \eta_l))$$

for some positive s, with $\widetilde{B}$ being B_n or B_n with one or two of the r_j's removed.

We would like to inform the reader that in applications of (3.2), $a(v)$ is a product of factors of the form $\beta_1(z)$ or $r_1^* A(z) r_1$ and A is a product of one or several $D_1^{-1}(z)D_1^{-1}(z_j)$, $j = 1, 2$ or similarly defined D^{-1} matrices. The matrices B_l also have this form. For example, we have $|r_1^* D_1^{-1}(z_1) D_1^{-1}(z_2) r_1| \le |r_1|^2 \|D_1^{-1}(z_1) D_1^{-1}(z_2)\| \le K^2 \eta_r + |z| n^{3+2\alpha} I(\|B_n\| \ge \eta_r \text{ or } \lambda_{\min}^{B_{(1)}} \le \eta_l)$, where K can be taken to be $\max((x_r - \eta_r)^{-1}, (\eta_l - x_l)^{-1}, v_0^{-1})$, and where we have used (1.10) and the fact that $|r_1|^2 \le \eta_r$ if $\|B_n\| < \eta_r$ and $|r_1|^2 \le n$ otherwise. We have $\|B_l\|$ obviously satisfying this condition. We also have $\beta_1(z)$ satisfying this condition since from (3.3) (see below) $|\beta_1(z)| = |1 - r_1^* D^{-1} r_1| \le 1 + K\eta_r + |z| n^{2+\alpha} I(\|B_n\| \ge \eta_r \text{ or } \lambda_{\min}^{B_n} \le \eta_l)$. In the sequel, we shall freely use (3.2) without verifying these conditions, even similarly defined β_j functions and A, B matrices.

We have

$$D^{-1}(z) - D_j^{-1}(z) = -\frac{D_j^{-1}(z) r_j r_j^* D_j^{-1}(z)}{1 + r_j^* D_j^{-1}(z) r_j} \tag{3.3}$$
$$= -\beta_j(z) D_j^{-1}(z) r_j r_j^* D_j^{-1}(z).$$

Let

$$\gamma_j(z) = r_j^* D_j^{-1}(z) r_j - N^{-1} \mathsf{E}(\operatorname{tr}(D_j^{-1}(z) T)).$$

We first derive bounds on the moments of $\gamma_j(z)$ and $\varepsilon_j(z)$. Using (3.2) we have

$$\mathsf{E}|\varepsilon_j(z)|^p \le K_p N^{-1} \delta_n^{2p-4} \qquad p \text{ even.} \tag{3.4}$$

It should be noted that constants obtained do not depend on $z \in \mathcal{C}_n$.

Using Lemma 2.1, (3.2), and Hölder's inequality, we have, for all even p,

$$\mathsf{E}|\gamma_j(z) - \varepsilon_j(z)|^p = \mathsf{E}|\gamma_1(z) - \varepsilon_1(z)|^p$$
$$= \mathsf{E}\left|\frac{1}{N}\sum_{j=2}^N \mathsf{E}_j \operatorname{tr} T D_1(z)^{-1} - \mathsf{E}_{j-1} \operatorname{tr} T D_1^{-1}(z)\right|^p$$
$$= \mathsf{E}\left|\frac{1}{N}\sum_{j=2}^N \mathsf{E}_j \operatorname{tr} T(D_1^{-1}(z) - D_{1j}^{-1}(z)) - \mathsf{E}_{j-1} \operatorname{tr} T(D_1^{-1}(z) - D_{1j}^{-1}(z))\right|^p$$
$$= \frac{1}{N^p}\mathsf{E}\left|\sum_{j=2}^N (\mathsf{E}_j - \mathsf{E}_{j-1}) \beta_{1j}(z) r_j^* D_{1j}^{-1}(z) T D_{1j}^{-1}(z) r_j\right|^p$$
$$\le \frac{K_p}{N^p}\mathsf{E}\left(\sum_{j=2}^N |(\mathsf{E}_j - \mathsf{E}_{j-1}) \beta_{1j}(z) r_j^* D_{1j}^{-1}(z) T D_{1j}^{-1}(z) r_j|^2\right)^{p/2}$$
$$\le \frac{K_p}{N^{1+p/2}} \sum_{j=2}^N \mathsf{E}|(\mathsf{E}_j - \mathsf{E}_{j-1}) \beta_{1j}(z) r_j^* D_{1j}^{-1}(z) T D_{1j}^{-1}(z) r_j|^p$$
$$\le \frac{K_p}{N^{p/2}} \mathsf{E}|\beta_{12}(z) r_2^* D_{12}^{-1}(z) T D_{12}^{-1}(z) r_2|^p$$
$$\le \frac{K_p}{N^{p/2}}.$$

Therefore

$$\mathsf{E}|\gamma_j|^p \le K_p N^{-1}\delta_n^{2p-4}, \qquad p \ge 2. \tag{3.5}$$

We next prove that $b_n(z)$ is bounded for all n. From (3.2) we find, for any $p \ge 1$,

$$\mathsf{E}|\beta_1(z)|^p \le K_p. \tag{3.6}$$

Since $b_n = \beta_1(z) + \beta_1(z)b_n(z)\gamma_1(z)$ we get from (3.5), (3.6)

$$|b_n(z)| = |\mathsf{E}\beta_1(z) + \mathsf{E}\beta_1(z)b_n(z)\gamma_1(z)| \le K_1 + K_2|b_n(z)|N^{-1/2}.$$

Thus for all n large,

$$|b_n(z)| \le \frac{K_1}{1 - K_2 N^{-1/2}}$$

and subsequently $b_n(z)$ is bounded for all n.

From (3.3) we have

$$\begin{aligned}
&D^{-1}(z_1)D^{-1}(z_2) - D_j^{-1}(z_1)D_j^{-1}(z_2)\\
&\quad = \big(D^{-1}(z_1) - D_j^{-1}(z_1)\big)\big(D^{-1}(z_2) - D_j^{-1}(z_2)\big)\\
&\qquad + \big(D^{-1}(z_1) - D_j^{-1}(z_1)\big)D_j^{-1}(z_2)\\
&\qquad + D_j^{-1}(z_1)\big(D^{-1}(z_2) - D_j^{-1}(z_2)\big)\\
&\quad = \beta_j(z_1)\beta_j(z_2)D_j^{-1}(z_1)r_jr_j^*D_j^{-1}(z_1)D_j^{-1}(z_2)r_jr_j^*D_j^{-1}(z_2)\\
&\qquad - \beta_j(z_1)D_j^{-1}(z_1)r_jr_j^*D_j^{-1}(z_1)D_j^{-1}(z_2)\\
&\qquad - \beta_j(z_2)D_j^{-1}(z_1)D_j^{-1}(z_2)r_jr_j^*D_j^{-1}(z_2).
\end{aligned}$$

Therefore

$$\begin{aligned}
&\operatorname{tr}\big(D^{-1}(z_1)D^{-1}(z_2) - D_j^{-1}(z_1)D_j^{-1}(z_2)\big)\\
&\quad = \beta_j(z_1)\beta_j(z_2)\big(r_j^*D_j^{-1}(z_1)D_j^{-1}(z_2)r_j\big)^2 \qquad (3.7)\\
&\qquad - \beta_j(z_1)r_j^*D_j^{-2}(z_1)D_j^{-1}(z_2)r_j - \beta_j(z_2)r_j^*D_j^{-2}(z_2)D_j^{-1}(z_1)r_j.
\end{aligned}$$

We write

$$\begin{aligned}
m_n(z_1) - m_n(z_2) &= \frac{1}{n}\operatorname{tr}\big(D^{-1}(z_1) - D^{-1}(z_2)\big)\\
&= \frac{1}{n}(z_1 - z_2)\operatorname{tr} D^{-1}(z_1)D^{-1}(z_2).
\end{aligned}$$

Therefore, from (3.7) we have

$$
\begin{aligned}
&n\frac{m_n(z_1)-m_n(z_2)-\mathsf{E}(m_n(z_1)-m_n(z_2))}{z_1-z_2}\\
&\quad=\sum_{j=1}^{N}(\mathsf{E}_j-\mathsf{E}_{j-1})\operatorname{tr}D^{-1}(z_1)D^{-1}(z_2)\\
&\quad=\sum_{j=1}^{N}(\mathsf{E}_j-\mathsf{E}_{j-1})\beta_j(z_1)\beta_j(z_2)(r_j^*D_j^{-1}(z_1)D_j^{-1}(z_2)r_j)^2\\
&\qquad-\sum_{j=1}^{N}(\mathsf{E}_j-\mathsf{E}_{j-1})\beta_j(z_1)r_j^*D_j^{-2}(z_1)D_j^{-1}(z_2)r_j\\
&\qquad-\sum_{j=1}^{N}(\mathsf{E}_j-\mathsf{E}_{j-1})\beta_j(z_2)r_j^*D_j^{-2}(z_2)D_j^{-1}(z_1)r_j.
\end{aligned}
\tag{3.8}
$$

Our goal is to show that the absolute second moment of (3.8) is bounded. We begin with the second sum in (3.8). We have

$$
\begin{aligned}
&\sum_{j=1}^{N}(\mathsf{E}_j-\mathsf{E}_{j-1})\beta_j(z_1)r_j^*D_j^{-2}(z_1)D_j^{-1}(z_2)r_j\\
&\quad=\sum_{j=1}^{N}(\mathsf{E}_j-\mathsf{E}_{j-1})\big(b_n(z_1)r_j^*D_j^{-2}(z_1)D_j^{-1}(z_2)r_j\\
&\qquad\qquad-\beta_j(z_1)b_n(z_1)r_j^*D_j^{-2}(z_1)D_j^{-1}(z_2)r_j\gamma_j(z_1)\big)\\
&\quad=b_n(z_1)\sum_{j=1}^{N}\mathsf{E}_j\big(r_j^*D_j^{-2}(z_1)D_j^{-1}(z_2)r_j\\
&\qquad\qquad-N^{-1}\operatorname{tr}T^{1/2}D_j^{-2}(z_1)D_j^{-1}(z_2)T^{1/2}\big)\\
&\qquad-b_n(z_1)\sum_{j=1}^{N}(\mathsf{E}_j-\mathsf{E}_{j-1})\beta_j(z_1)r_j^*D_j^{-2}(z_1)D_j^{-1}(z_2)r_j\gamma_j(z_1)\\
&\quad\equiv b_n(z_1)W_1-b_n(z_1)W_2.
\end{aligned}
$$

Using (3.2) we have

$$
\begin{aligned}
\mathsf{E}|W_1|^2&=\sum_{j=1}^{N}\mathsf{E}\big|\mathsf{E}_j\big(r_j^*D_j^{-2}(z_1)D_j^{-1}(z_2)r_j-N^{-1}\operatorname{tr}T^{1/2}D_j^{-2}(z_1)D_j^{-1}(z_2)T^{1/2}\big)\big|^2\\
&\le K.
\end{aligned}
$$

Using (3.5), and the bounds for $\beta_1(z_1)$ and $r_1^* D_1^{-2}(z_1) D_1^{-1}(z_2) r_1$ given in the remark to (3.2), we have

$$\begin{aligned}
\mathsf{E}|W_2|^2 &= \sum_{j=1}^{N} \mathsf{E}|(\mathsf{E}_j - \mathsf{E}_{j-1})\beta_j(z_1) r_j^* D_j^{-2}(z_1) D_j^{-1}(z_2) r_j \gamma_j(z_1)|^2 \\
&\le KN[\mathsf{E}|\gamma_1(z_1)|^2 + v^{-10} n^2 P(\|B_n\| > \eta_r \text{ or } \lambda_{\min}^{B_{(1)}} < \eta_l)] \\
&\le K.
\end{aligned}$$

This argument of course handles the third sum in (3.8).

For the first sum in (3.8) we have

$$\begin{aligned}
&\sum_{j=1}^{N} (\mathsf{E}_j - \mathsf{E}_{j-1})\beta_j(z_1)\beta_j(z_2)(r_j^* D_j^{-1}(z_1) D_j^{-1}(z_2) r_j)^2 \\
&\quad = b_n(z_1) b_n(z_2) \sum_{j=1}^{N} (\mathsf{E}_j - \mathsf{E}_{j-1})[(r_j^* D_j^{-1}(z_1) D_j^{-1}(z_2) r_j)^2 \\
&\qquad\qquad - (N^{-1} \operatorname{tr} T^{1/2} D_j^{-1}(z_1) D_j^{-1}(z_2) T^{1/2})^2] \\
&\qquad - b_n(z_2) \sum_{j=1}^{N} (\mathsf{E}_j - \mathsf{E}_{j-1})\beta_j(z_1)\beta_j(z_2)(r_j^* D_j^{-1}(z_1) D_j^{-1}(z_2) r_j)^2 \gamma_j(z_2) \\
&\qquad - b_n(z_1) b_n(z_2) \sum_{j=1}^{N} (\mathsf{E}_j - \mathsf{E}_{j-1})\beta_j(z_1)(r_j^* D_j^{-1}(z_1) D_j^{-1}(z_2) r_j)^2 \gamma_j(z_1) \\
&\quad = b_n(z_1) b_n(z_2) Y_1 - b_n(z_2) Y_2 - b_n(z_1) b_n(z_2) Y_3.
\end{aligned}$$

Both Y_2 and Y_3 are handled the same way as W_2 above. Using (3.2) we have

$$\begin{aligned}
\mathsf{E}|Y_1|^2 &\le N\mathsf{E}|(r_1^* D_1^{-1}(z_1) D_1^{-1}(z_2) r_1)^2 \\
&\qquad - (N^{-1} \operatorname{tr} T^{1/2} D_1^{-1}(z_1) D_1^{-1}(z_2) T^{1/2})^2|^2 \\
&\le N\Big(2\mathsf{E}|(r_1^* D_1^{-1}(z_1) D_1^{-1}(z_2) r_1 \\
&\qquad\qquad - N^{-1} \operatorname{tr} T^{1/2} D_1^{-1}(z_1) D_1^{-1}(z_2) T^{1/2})|^4 \\
&\qquad + 4(nN^{-1})^2 \mathsf{E}|(r_1^* D_1^{-1}(z_1) D_1^{-1}(z_2) r_1 \\
&\qquad\qquad - N^{-1} \operatorname{tr} T^{1/2} D_1^{-1}(z_1) D_1^{-1}(z_2) T^{1/2}) \\
&\qquad\qquad\qquad \times \|D_1^{-1}(z_1) D_1^{-1}(z_2)\||^2\Big) \\
&\le K.
\end{aligned}$$

Therefore, condition (ii) of Theorem 12.3 in Billingsley (1968) is satisfied, and we conclude that $\{\widehat{M}_n^1(z)\}$ is tight.

4. Convergence of $M_n^2(z)$. The proof of Lemma 1.1 is complete with the verification of $\{M_n^2(z)\}$ for $z \in \mathcal{C}_n$ to be bounded and form an equicontinuous family, and convergence to (1.12) under the assumptions in (ii) of Theorem 1.1 and to zero under those in (iii).

In order to simplify the exposition, we let $\mathcal{C}_1 = \mathcal{C}_u$ or $\mathcal{C}_u \cup \mathcal{C}_l$ if $x_l < 0$, and $\mathcal{C}_2 = \mathcal{C}_2(n) = \mathcal{C}_r$ or $\mathcal{C}_r \cup \mathcal{C}_l$ if $x_l > 0$. We begin with proving

$$\sup_{z\in\mathcal{C}_n} |\mathsf{E}\underline{m}_n(z) - \underline{m}(z)| \to 0 \qquad \text{as } n \to \infty. \tag{4.1}$$

Since $F^{\underline{B}_n} \xrightarrow{D} F^{c,H}$ almost surely, we get from d.c.t. $\mathsf{E}F^{\underline{B}_n} \xrightarrow{D} F^{c,H}$. It is easy to verify that $\mathsf{E}F^{\underline{B}_n}$ is a proper c.d.f. Since, as z ranges in $\mathcal{C}_1$, the functions $(\lambda - z)^{-1}$ in $\lambda \in [0, \infty)$ form a bounded, equicontinuous family, it follows [see, e.g., Billingsley (1968), Problem 8, page 17] that

$$\sup_{z\in\mathcal{C}_1} |\mathsf{E}\underline{m}_n(z) - \underline{m}(z)| \to 0.$$

For $z \in \mathcal{C}_2$ we write (η_l, η_r defined as in the previous section)

$$\begin{aligned}\mathsf{E}\underline{m}_n(z) - \underline{m}(z) &= \int \frac{1}{\lambda - z} I_{[\eta_l,\eta_r]}(\lambda)\, d\big(\mathsf{E}F^{\underline{B}_n}(\lambda) - F^{c,H}(\lambda)\big)\\ &\quad + \mathsf{E}\int \frac{1}{\lambda - z} I_{[\eta_l,\eta_r]^c}(\lambda)\, dF^{\underline{B}_n}(\lambda).\end{aligned}$$

As above, the first term converges uniformly to zero. For the second term we use (1.9) with $\ell \geq 2$ to get

$$\begin{aligned}\sup_{z\in\mathcal{C}_2}& \left|\mathsf{E}\int \frac{1}{\lambda - z} I_{[\eta_l,\eta_r]^c}(\lambda)\, dF^{\underline{B}_n}(\lambda)\right|\\ &\leq (\varepsilon_n/n)^{-1}\mathsf{P}\big(\|B_n\| \geq \eta_r \text{ or } \lambda_{\min}^{B_n} \leq \eta_l\big)\\ &\leq Kn\varepsilon^{-1}n^{-\ell} \to 0.\end{aligned}$$

Thus (4.1) holds.

From the fact that $F^{c_n,H_n} \xrightarrow{D} F^{c,H}$ [see Bai and Silverstein (1998), below (3.10)] along with the fact that $\mathcal{C}$ lies outside the support of $F^{c,H}$, it is straightforward to verify that

$$\sup_{z\in\mathcal{C}} |\underline{m}_n^0(z) - \underline{m}(z)| \to 0 \qquad \text{as } n \to \infty. \tag{4.2}$$

We now show that

$$\sup_{n\, z\in\mathcal{C}_n} \|(\mathsf{E}\underline{m}_n(z)T + I)^{-1}\| < \infty. \tag{4.3}$$

From Lemma 2.11 of Bai and Silverstein (1998), $\|(\mathsf{E}\underline{m}_n(z)T + I)^{-1}\|$ is bounded by $\max(2, 4v_0^{-1})$ on $\mathcal{C}_u$. Let $x = x_l$ or x_r. Since x is outside the support of $\underline{F}^{c,H}$ it follows from Theorem 4.1 of Silverstein and Choi (1995) that for any t in the support of H $\underline{m}(x)t + 1 \neq 0$. Choose any t_0 in the support of H. Since $\underline{m}(z)$ is continuous on $\mathcal{C}^0 \equiv \{x + iv : v \in [0, v_0]\}$, there exist positive constants δ_1 and μ_0 such that

$$\inf_{z\in\mathcal{C}^0} |\underline{m}(z)t_0 + 1| > \delta_1 \quad \text{and} \quad \sup_{z\in\mathcal{C}^0} |\underline{m}(z)| < \mu_0.$$

Using $H_n \xrightarrow{D} H$ and (4.1), for all large n, there exists an eigenvalue λ^T of T such that $|\lambda^T - t_0| < \delta_1/4\mu_0$ and $\sup_{z\in\mathcal{C}_l\cup\mathcal{C}_r} |\mathsf{E}\underline{m}_n(z) - \underline{m}(z)| < \delta_1/4$. Therefore, we have

$$\inf_{z\in\mathcal{C}_l\cup\mathcal{C}_r} |\mathsf{E}\underline{m}_n(z)\lambda^T + 1| > \delta_1/2,$$

which completes the proof of (4.3).

Next we show the existence of $\xi \in (0, 1)$ such that for all n large

$$\sup_{z\in\mathcal{C}_n} \left| c_n \mathsf{E}\underline{m}_n(z)^2 \int \frac{t^2}{(1 + t\mathsf{E}\underline{m}_n(z))^2} \, dH_n(t) \right| < \xi. \tag{4.4}$$

From the identity (1.1) of Bai and Silverstein (1998),

$$\underline{m}(z) = \left(-z + c \int \frac{t}{1 + t\underline{m}(z)} \, dH(t)\right)^{-1}$$

valid for $z = x + iv$ outside the support of $F^{c,H}$; we find

$$\begin{aligned} \Im\, \underline{m}(z) &= \left(v + \Im\, \underline{m}(z) c \int \frac{t^2}{|1 + t\underline{m}(z)|^2} \, dH(t)\right) \\ &\quad \times \left| -z + c \int \frac{t}{1 + t\underline{m}(z)} \, dH(t) \right|^{-2}. \end{aligned}$$

Therefore

$$\begin{aligned} &\left| c\underline{m}(z)^2 \int \frac{t^2}{(1 + t\underline{m}(z))^2} \, dH(t) \right| \\ &\quad \le \left(c \int \frac{t^2}{|1 + t\underline{m}(z)|^2} \, dH(t)\right) \left| -z + c \int \frac{t}{1 + t\underline{m}(z)} \, dH(t) \right|^{-2} \\ &\quad = \left(\Im\, \underline{m}(z) c \int \frac{t^2}{|1 + t\underline{m}(z)|^2} \, dH(t)\right) \\ &\qquad \times \left[v + \Im\, \underline{m}(z) c \int \frac{t^2}{|1 + t\underline{m}(z)|^2} \, dH(t)\right]^{-1} \end{aligned} \tag{4.5}$$

$$= \left(c\int \frac{1}{|x-z|^2}\,dF^{c,H}(x)\int \frac{t^2}{|1+t\underline{m}(z)|^2}\,dH(t)\right)$$

$$\times\left[1+c\int \frac{1}{|x-z|^2}\,dF^{c,H}(x)\int \frac{t^2}{|1+t\underline{m}(z)|^2}\,dH(t)\right]^{-1}<1,$$

for all $z \in \mathcal{C}$. By continuity, we have the existence of $\xi_1 < 1$ such that

$$\sup_{z\in\mathcal{C}}\left|c\underline{m}(z)^2\int \frac{t^2}{(1+t\underline{m}(z))^2}\,dH(t)\right| < \xi_1. \tag{4.6}$$

Therefore, using (4.1), (4.4) follows.

We proceed with some improved bounds on quantities appearing earlier.

Let M be nonrandom $n \times n$. Then, using (3.2) and the argument used to derive the bound on $\mathsf{E}|W_2|$, we find

$$\begin{aligned}
&\mathsf{E}|\operatorname{tr} D^{-1}M - \mathsf{E}\operatorname{tr} D^{-1}M|^2\\
&\quad= \mathsf{E}\left|\sum_{j=1}^N \mathsf{E}_j \operatorname{tr} D^{-1}M - \mathsf{E}_{j-1}\operatorname{tr} D^{-1}M\right|^2\\
&\quad= \mathsf{E}\left|\sum_{j=1}^N (\mathsf{E}_j - \mathsf{E}_{j-1})\operatorname{tr}(D^{-1} - D_j^{-1})M\right|^2\\
&\quad= \sum_{j=1}^N \mathsf{E}|(\mathsf{E}_j - \mathsf{E}_{j-1})\beta_j r_j^* D_j^{-1} M D_j^{-1} r_j|^2 \qquad (4.7)\\
&\quad\le 2\sum_{j=1}^N \mathsf{E}|[\beta_j(r_j^* D_j^{-1} M D_j^{-1} r_j) - N^{-1}\operatorname{tr}(TD_j^{-1}MD_j^{-1})|^2\\
&\qquad\qquad + \mathsf{E}|\beta_j - \bar\beta_j|^2 |N^{-1}\operatorname{tr}(TD_j^{-1}MD_j^{-1})|^2]\\
&\quad\le K\|M\|^2.
\end{aligned}$$

The same argument holds for D_1^{-1} so we also have

$$\mathsf{E}|\operatorname{tr} D_1^{-1}M - \mathsf{E}\operatorname{tr} D_1^{-1}M|^2 \le K\|M\|^2. \tag{4.8}$$

Our next task is to investigate the limiting behavior of

$$\begin{aligned}
&N\left(c_n\int \frac{dH_n(t)}{1+t\mathsf{E}\underline{m}_n} + zc_n\mathsf{E}m_n\right)\\
&\quad= N\mathsf{E}\beta_1\left[r_1^* D_1^{-1}(\mathsf{E}\underline{m}_n T + I)^{-1} r_1 - \frac{1}{N}\mathsf{E}\operatorname{tr}(\mathsf{E}\underline{m}_n T + I)^{-1} T D^{-1}\right]
\end{aligned}$$

for $z \in \mathcal{C}_n$ [see (5.2) in Bai and Silverstein (1998)]. Throughout the following, all bounds, including $O(\cdot)$ and $o(\cdot)$ expressions, and convergence statements hold uniformly for $z \in \mathcal{C}_n$.

We have

$$
\begin{aligned}
&\mathsf{E}\operatorname{tr}(\mathsf{E}\underline{m}_n T + I)^{-1} T D_1^{-1} - \mathsf{E}\operatorname{tr}(\mathsf{E}\underline{m}_n T + I)^{-1} T D^{-1} \\
&\quad = \mathsf{E}\beta_1 \operatorname{tr}(\mathsf{E}\underline{m}_n T + I)^{-1} T D_1^{-1} r_1 r_1^* D_1^{-1} \qquad (4.9) \\
&\quad = b_n \mathsf{E}(1 - \beta_1\gamma_1) r_1^* D_1^{-1} (\mathsf{E}\underline{m}_n T + I)^{-1} T D_1^{-1} r_1.
\end{aligned}
$$

From (3.2), (3.5) and (4.3) we get

$$
|\mathsf{E}\beta_1\gamma_1 r_1^* D_1^{-1}(\mathsf{E}\underline{m}_n T + I)^{-1} T D_1^{-1} r_1| \le K N^{-1}.
$$

Therefore

$$
|(4.9) - N^{-1} b_n \mathsf{E}\operatorname{tr} D_1^{-1}(\mathsf{E}\underline{m}_n T + I)^{-1} T D_1^{-1} T| \le K N^{-1}.
$$

Since $\beta_1 = b_n - b_n^2\gamma_1 + \beta_1 b_n^2 \gamma_1^2$ we have

$$
\begin{aligned}
N&\mathsf{E}\beta_1 r_1^* D_1^{-1}(\mathsf{E}\underline{m}_n T + I)^{-1} r_1 - \mathsf{E}\beta_1 \mathsf{E}\operatorname{tr}(\mathsf{E}\underline{m}_n T + I)^{-1} T D_1^{-1} \\
&= -b_n^2 N \mathsf{E}\gamma_1 r_1^* D_1^{-1}(\mathsf{E}\underline{m}_n T + I)^{-1} r_1 \\
&\quad + b_n^2 (N\mathsf{E}\beta_1\gamma_1^2 r_1^* D_1^{-1}(\mathsf{E}\underline{m}_n T + I)^{-1} r_1 \\
&\qquad\qquad - (\mathsf{E}\beta_1\gamma_1^2)\mathsf{E}\operatorname{tr}(\mathsf{E}\underline{m}_n T + I)^{-1} T D_1^{-1}) \\
&= -b_n^2 N \mathsf{E}\gamma_1 r_1^* D_1^{-1}(\mathsf{E}\underline{m}_n T + I)^{-1} r_1 \\
&\quad + b_n^2 (\mathsf{E}[N\beta_1\gamma_1^2 r_1^* D_1^{-1}(\mathsf{E}\underline{m}_n T + I)^{-1} r_1 \\
&\qquad\qquad - \beta_1\gamma_1^2 \operatorname{tr} D_1^{-1}(\mathsf{E}\underline{m}_n T + I)^{-1} T]) \\
&\quad + b_n^2 \operatorname{Cov}(\beta_1\gamma_1^2, \operatorname{tr} D_1^{-1}(\mathsf{E}\underline{m}_n T + I)^{-1} T)
\end{aligned}
$$

[$\operatorname{Cov}(X, Y) = \mathsf{E}XY - \mathsf{E}X\mathsf{E}Y$]. Using (3.2), (3.5) and (4.3), we have

$$
|\mathsf{E}[N\beta_1\gamma_1^2 r_1^* D_1^{-1}(\mathsf{E}\underline{m}_n T + I)^{-1} r_1 - \beta_1\gamma_1^2 \operatorname{tr} D_1^{-1}(\mathsf{E}\underline{m}_n T + I)^{-1} T]| \le K\delta_n^2.
$$

Using (3.5), (3.6), (4.3) and (4.8) we have

$$
\begin{aligned}
&|\operatorname{Cov}(\beta_1\gamma_1^2, \operatorname{tr} D_1^{-1}(\mathsf{E}\underline{m}_n T + I)^{-1} T)| \\
&\quad \le (\mathsf{E}|\beta_1|^4)^{1/4} (\mathsf{E}|\gamma_1|^8)^{1/4} \\
&\qquad \times (\mathsf{E}|\operatorname{tr} D_1^{-1}(\mathsf{E}\underline{m}_n T + I)^{-1} T - \mathsf{E}\operatorname{tr} D_1^{-1}(\mathsf{E}\underline{m}_n T + I)^{-1} T|^2)^{1/2} \\
&\quad \le K\delta_n^3 N^{-1/4}.
\end{aligned}
$$

Since $\beta_1 = b_n - b_n\beta_1\gamma_1$, we get from (3.5) and (3.6) $\mathsf{E}\beta_1 = b_n + O(N^{-1/2})$.

Write

$$\begin{aligned}
&\mathsf{E}N\gamma_1 r_1^* D_1^{-1}(\mathsf{E}\underline{m}_n T + I)^{-1} r_1 \\
&\quad = N\mathsf{E}[(r_1^* D_1^{-1} r_1 - N^{-1}\operatorname{tr} D_1^{-1} T) \\
&\qquad\qquad \times (r_1^* D_1^{-1}(\mathsf{E}\underline{m}_n T + I)^{-1} r_1 - N^{-1}\operatorname{tr} D_1^{-1}(\mathsf{E}\underline{m}_n T + I)^{-1} T)] \\
&\qquad + N^{-1}\operatorname{Cov}(\operatorname{tr} D_1^{-1} T, \operatorname{tr} D_1^{-1}(\mathsf{E}\underline{m}_n T + I)^{-1} T).
\end{aligned}$$

From (4.8) we see the second term above is $O(N^{-1})$. Therefore, we arrive at

$$\begin{aligned}
&N\left(c_n \int \frac{dH_n(t)}{1 + t\mathsf{E}\underline{m}_n} + zc_n \mathsf{E}m_n\right) \\
&\qquad = b_n^2 N^{-1}\mathsf{E}\operatorname{tr} D_1^{-1}(\mathsf{E}\underline{m}_n T + I)^{-1} T D_1^{-1} T \\
&\qquad\quad - b_n^2 N\mathsf{E}[(r_1^* D_1^{-1} r_1 - N^{-1}\operatorname{tr} D_1^{-1} T) \\
&\qquad\qquad\qquad \times (r_1^* D_1^{-1}(\mathsf{E}\underline{m}_n T + I)^{-1} r_1 \\
&\qquad\qquad\qquad\quad - N^{-1}\operatorname{tr} D_1^{-1}(\mathsf{E}\underline{m}_n T + I)^{-1} T)] \\
&\qquad\quad + o(1).
\end{aligned} \tag{4.10}$$

Using (1.15) on (4.10) and arguing the same way (1.15) is used in Section 2 [below (2.7)], we see that under the assumptions in (iii) of Theorem 1.1, the CG case

$$N\left(c_n \int \frac{dH_n(t)}{1 + t\mathsf{E}\underline{m}_n} + zc_n \mathsf{E}m_n\right) \to 0 \qquad \text{as } n \to \infty, \tag{4.11}$$

while under the assumptions in (ii) of Theorem 1.1, the RG case

$$\begin{aligned}
&N\left(c_n \int \frac{dH_n(t)}{1 + t\mathsf{E}\underline{m}_n} + zc_n \mathsf{E}m_n\right) \\
&\qquad = -b_n^2 N^{-1}\mathsf{E}\operatorname{tr} D_1^{-1}(\mathsf{E}\underline{m}_n T + I)^{-1} T D_1^{-1} T + o(1).
\end{aligned}$$

Let $A_n(z) = c_n \int \frac{dH_n(t)}{1 + t\mathsf{E}\underline{m}_n(z)} + zc_n \mathsf{E}m_n(z)$. Using the identity

$$\mathsf{E}\underline{m}_n(z) = -\frac{(1 - c_n)}{z} + c_n \mathsf{E}m_n$$

we have

$$\begin{aligned}
A_n(z) &= c_n \int \frac{dH_n(t)}{1 + t\mathsf{E}\underline{m}_n(z)} - c_n + z\mathsf{E}\underline{m}_n(z) + 1 \\
&= -\mathsf{E}\underline{m}_n(z)\left(-z - \frac{1}{\mathsf{E}\underline{m}_n(z)} + c_n \int \frac{t\,dH_n(t)}{1 + t\mathsf{E}\underline{m}_n(z)}\right).
\end{aligned}$$

It follows that

$$\mathsf{E}\underline{m}_n(z) = \left[-z + c_n \int \frac{t\,dH_n(t)}{1 + t\mathsf{E}\underline{m}_n(z)} + A_n/\mathsf{E}\underline{m}_n(z)\right]^{-1}.$$

From this, together with the analogous identity (4.4) we get

$$\begin{aligned} &\mathsf{E}\underline{m}_n(z) - \underline{m}_n^0(z) \\ &\quad = -\underline{m}_n^0 A_n \left[1 - c_n \mathsf{E}\underline{m}_n \underline{m}_n^0 \int \frac{t^2\,dH_n(t)}{(1 + t\mathsf{E}\underline{m}_n)(1 + t\underline{m}_n^0)}\right]^{-1}. \end{aligned} \tag{4.12}$$

We see from (4.4) and the corresponding bound involving $\underline{m}_n^0(z)$, that the denominator of (4.12) is bounded away from zero.

Therefore from (4.11), in the CG case

$$\sup_{z \in \mathcal{C}_n} M_n^2(z) \to 0 \qquad \text{as } n \to \infty.$$

We now find the limit of $N^{-1}\mathsf{E}\operatorname{tr} D_1^{-1}(\mathsf{E}\underline{m}_n T + I)^{-1} T D_1^{-1} T$. Applications of (3.1)–(3.3), (3.6) and (4.3) show that both

$$\mathsf{E}\operatorname{tr} D_1^{-1}(\mathsf{E}\underline{m}_n T + I)^{-1} T D_1^{-1} T - \mathsf{E}\operatorname{tr} D^{-1}(\mathsf{E}\underline{m}_n T + I)^{-1} T D_1^{-1} T$$

and

$$\mathsf{E}\operatorname{tr} D^{-1}(\mathsf{E}\underline{m}_n T + I)^{-1} T D_1^{-1} T - \mathsf{E}\operatorname{tr} D^{-1}(\mathsf{E}\underline{m}_n T + I)^{-1} T D^{-1} T$$

are bounded. Therefore it is sufficient to consider

$$N^{-1}\mathsf{E}\operatorname{tr} D^{-1}(\mathsf{E}\underline{m}_n T + I)^{-1} T D^{-1} T.$$

Write

$$D(z) + zI - b_n(z)T = \sum_{j=1}^{N} r_j r_j^* - b_n(z)T.$$

It is straightforward to verify that $zI - b_n(z)T$ is nonsingular. Taking inverses we get

$$\begin{aligned} D^{-1}(z) &= -(zI - b_n(z)T)^{-1} \\ &\qquad + \sum_{j=1}^{N} \beta_j(z)(zI - b_n(z)T)^{-1} r_j r_j^* D_j(z) \\ &\qquad - b_n(z)(zI - b_n(z)T)^{-1} T D^{-1}(z) \\ &= -(zI - b_n(z)T)^{-1} + b_n(z)A(z) + B(z) + C(z), \end{aligned} \tag{4.13}$$

where

$$A(z) = \sum_{j=1}^{N} (zI - b_n(z)T)^{-1}(r_j r_j^* - N^{-1}T)D_j^{-1}(z),$$

$$B(z) = \sum_{j=1}^{N} (\beta_j(z) - b_n(z))(zI - b_n(z)T)^{-1} r_j r_j^* D_j^{-1}(z)$$

and

$$C(z) = N^{-1} b_n(z)(zI - b_n(z)T)^{-1} T \sum_{j=1}^{N} (D_j^{-1}(z) - D^{-1}(z))$$

$$= N^{-1} b_n(z)(zI - b_n(z)T)^{-1} T \sum_{j=1}^{N} \beta_j(z) D_j^{-1}(z) r_j r_j^* D_j^{-1}(z).$$

Since $\mathsf{E}\beta_1 = -z\mathsf{E}\underline{m}_n$ and $\mathsf{E}\beta_1 = b_n + O(N^{-1})$ we have $b_n \to -z\underline{m}$. From (4.3) it follows that $\|(zI - b_n(z)T)^{-1}\|$ is bounded.

We have by (3.5) and (3.6)

$$\mathsf{E}|\beta_1 - b_n|^2 = |b_n|^2 \mathsf{E}|\beta_1 \gamma_1|^2 \le KN^{-1}. \tag{4.14}$$

Let M be $n \times n$. From (3.1), (3.2), (3.6) and (4.14) we get

$$\begin{aligned} |N^{-1}\mathsf{E}\operatorname{tr} B(z)M| &\le K(\mathsf{E}|\beta_1 - b_n|^2)^{1/2}(\mathsf{E}|r_1^* r_1 \|D_1^{-1}M\||^2)^{1/2} \\ &\le KN^{-1/2}(\mathsf{E}\|M\|^4)^{1/4} \end{aligned} \tag{4.15}$$

and

$$\begin{aligned} |N^{-1}\mathsf{E}\operatorname{tr} C(z)M| &\le KN^{-1}\mathsf{E}|\beta_1|r_1^* r_1 \|D_1^{-1}\|^2 \|M\| \\ &\le KN^{-1}(\mathsf{E}\|M\|^2)^{1/2}. \end{aligned} \tag{4.16}$$

For the following M, $n \times n$, is nonrandom, bounded in norm. Write

$$\operatorname{tr} A(z)TD^{-1}M = A_1(z) + A_2(z) + A_3(z), \tag{4.17}$$

where

$$A_1(z) = \operatorname{tr} \sum_{j=1}^{N} (zI - b_n T)^{-1} r_j r_j^* D_j^{-1} T(D^{-1} - D_j^{-1})M,$$

$$A_2(z) = \operatorname{tr} \sum_{j=1}^{N} (zI - b_n T)^{-1}(r_j r_j^* D_j^{-1} T D_j^{-1} - N^{-1} T D_j^{-1} T D_j^{-1})M$$

and

$$A_3(z) = \operatorname{tr} \sum_{j=1}^{N} (zI - b_n T)^{-1} N^{-1} T D_j^{-1} T(D_j^{-1} - D^{-1})M.$$

We have $\mathsf{E}A_2(z) = 0$ and similarly to (4.16) we have

$$|\mathsf{E}N^{-1}A_3(z)| \le KN^{-1}. \tag{4.18}$$

Using (3.2) and (4.14) we get

$$\begin{aligned}\mathsf{E}N^{-1}A_1(z) &= -\mathsf{E}\beta_1 r_1^* D_1^{-1} T D_1^{-1} r_1 r_1^* D_1^{-1} M(zI - b_n T)^{-1} r_1 \\ &= -b_n \mathsf{E}(N^{-1}\operatorname{tr} D_1^{-1} T D_1^{-1} T)(N^{-1}\operatorname{tr} D_1^{-1} M(zI - b_n T)^{-1} T) + o(1) \\ &= -b_n \mathsf{E}(N^{-1}\operatorname{tr} D^{-1} T D^{-1} T)(N^{-1}\operatorname{tr} D^{-1} M(zI - b_n T)^{-1} T) + o(1).\end{aligned}$$

Using (3.1) and (4.7) we find

$$\begin{aligned}&|\operatorname{Cov}(N^{-1}\operatorname{tr} D^{-1}TD^{-1}T, N^{-1}\operatorname{tr} D^{-1}M(zI - b_nT)^{-1}T)| \\ &\quad \le (\mathsf{E}|N^{-1}\operatorname{tr} D^{-1}TD^{-1}T|^2)^{1/2} N^{-1} \\ &\qquad \times (\mathsf{E}|\operatorname{tr} D^{-1}M(zI - b_nT)^{-1}T - \mathsf{E}D^{-1}M(zI - b_nT)^{-1}T|^2)^{1/2} \\ &\quad \le KN^{-1}.\end{aligned}$$

Therefore

$$\begin{aligned}&\mathsf{E}N^{-1}A_1(z) \\ &\quad = -b_n(\mathsf{E}N^{-1}\operatorname{tr} D^{-1}TD^{-1}T) \\ &\qquad \times (\mathsf{E}N^{-1}\operatorname{tr} D^{-1}M(zI - b_nT)^{-1}T) + o(1).\end{aligned} \tag{4.19}$$

From (4.13), (4.15) and (4.16) we get

$$\begin{aligned}&\mathsf{E}N^{-1}\operatorname{tr} D^{-1}T(zI - b_nT)^{-1}T \\ &\quad = N^{-1}\operatorname{tr}(-(zI - b_nT)^{-1} + \mathsf{E}B(z) + \mathsf{E}C(z))T(zI - b_nT)^{-1}T \\ &\quad = -\frac{c_n}{z^2}\int \frac{t^2\, dH_n(t)}{(1 + t\mathsf{E}\underline{m}_n)^2} + o(1).\end{aligned} \tag{4.20}$$

Similarly,

$$\begin{aligned}&\mathsf{E}N^{-1}\operatorname{tr} D^{-1}(\mathsf{E}\underline{m}_n T + I)^{-1}T(zI - b_nT)^{-1}T \\ &\quad = -\frac{c_n}{z^2}\int \frac{t^2\, dH_n(t)}{(1 + t\mathsf{E}\underline{m}_n)^3} + o(1).\end{aligned} \tag{4.21}$$

Using (4.13) and (4.15)–(4.20) we get

$$\begin{aligned}&\mathsf{E}N^{-1}\operatorname{tr} D^{-1}TD^{-1}T \\ &\quad = -\mathsf{E}N^{-1}D^{-1}T(zI - b_nT)^{-1}T \\ &\qquad - b_n^2(\mathsf{E}N^{-1}\operatorname{tr} D^{-1}TD^{-1}T)(\mathsf{E}N^{-1}\operatorname{tr} D^{-1}T(zI - b_nT)^{-1}T) + o(1) \\ &\quad = \frac{c_n}{z^2}\int \frac{t^2\, dH_n(t)}{(1 + t\mathsf{E}\underline{m}_n)^2}(1 + z^2\mathsf{E}\underline{m}_n^2\mathsf{E}N^{-1}\operatorname{tr} D^{-1}TD^{-1}T) + o(1).\end{aligned}$$

Therefore

$$
\begin{aligned}
&\mathsf{E}N^{-1}\operatorname{tr} D^{-1}TD^{-1}T \\
(4.22)\quad &=\left[\frac{c_n}{z^2}\int\frac{t^2\,dH_n(t)}{(1+t\mathsf{E}\underline{m}_n)^2}\right]\left[1-c_n\int\frac{\mathsf{E}\underline{m}_n^2 t^2\,dH_n(t)}{(1+t\mathsf{E}\underline{m}_n)^2}\right]^{-1}+o(1).
\end{aligned}
$$

Finally we have from (4.13)–(4.19), (4.21) and (4.22)

$$
\begin{aligned}
N^{-1}&\mathsf{E}\operatorname{tr} D^{-1}(\mathsf{E}\underline{m}_nT+I)^{-1}TD^{-1}T \\
&= -\mathsf{E}N^{-1}D^{-1}(\mathsf{E}\underline{m}_nT+I)^{-1}T(zI-b_nT)^{-1}T \\
&\quad - b_n^2(\mathsf{E}N^{-1}\operatorname{tr} D^{-1}TD^{-1}T) \\
&\quad\quad \times (\mathsf{E}N^{-1}\operatorname{tr} D^{-1}(\mathsf{E}\underline{m}_nT+I)^{-1}T(zI-b_nT)^{-1}T)+o(1) \\
&= \frac{c_n}{z^2}\int\frac{t^2\,dH_n(t)}{(1+t\mathsf{E}\underline{m}_n)^3} \\
&\quad \times\left(1+z^2\mathsf{E}\underline{m}_n^2\left[\frac{c_n}{z^2}\int\frac{t^2\,dH_n(t)}{(1+t\mathsf{E}\underline{m}_n)^2}\right]\left[1-c_n\int\frac{\mathsf{E}\underline{m}_n^2t^2\,dH_n(t)}{(1+t\mathsf{E}\underline{m}_n)^2}\right]^{-1}\right)+o(1) \\
&= \left[\frac{c_n}{z^2}\int\frac{t^2\,dH_n(t)}{(1+t\mathsf{E}\underline{m}_n)^3}\right]\left[1-c_n\int\frac{\mathsf{E}\underline{m}_n^2t^2\,dH_n(t)}{(1+t\mathsf{E}\underline{m}_n)^2}\right]^{-1}+o(1).
\end{aligned}
$$

Therefore, from (4.12) we conclude that in the RG case

$$
\sup_{z\in\mathcal{C}_n} M_n^2(z)\to c\int\frac{\underline{m}(z)^3t^2\,dH(t)}{(1+t\underline{m}(z))^3}\left(1-c\int\frac{\underline{m}(z)^2t^2\,dH(t)}{(1+t\underline{m}(z))^2}\right)^{-2}\qquad \text{as } n\to\infty,
$$

which is (1.12).

Finally, for general standardized X_{11}, we see that in light of the above work, in order to show $\{M_n^2(z)\}$ for $z\in\mathcal{C}_n$ is bounded and equicontinuous, it is sufficient to prove $\{f_n'(z)\}$, where

$$
\begin{aligned}
f_n(z)\equiv N\mathsf{E}[&(r_1^*D_1^{-1}r_1-N^{-1}\operatorname{tr} D_1^{-1}T) \\
&\times (r_1^*D_1^{-1}(\mathsf{E}\underline{m}_nT+I)^{-1}r_1-N^{-1}\operatorname{tr} D_1^{-1}(\mathsf{E}\underline{m}_nT+I)^{-1}T)]
\end{aligned}
$$

is bounded. Using (2.3) we find

$$
\begin{aligned}
|f'(z)|\le KN^{-1}\Big(&(\mathsf{E}(\operatorname{tr} D_1^{-2}T\overline{D}_1^{-2}T) \\
&\times \mathsf{E}(\operatorname{tr} D_1^{-1}(\mathsf{E}\underline{m}_nT+I)^{-1}T(\mathsf{E}\overline{\underline{m}}_nT+I)^{-1}\overline{D}_1^{-1}T))^{1/2} \\
&+(\mathsf{E}(\operatorname{tr} D_1^{-1}T\overline{D}_1^{-1}T) \\
&\times \mathsf{E}(\operatorname{tr} D_1^{-2}(\mathsf{E}\underline{m}_nT+I)^{-1}T(\mathsf{E}\overline{\underline{m}}_nT+I)^{-1}\overline{D}_1^{-2}T))^{1/2}
\end{aligned}
$$

$$+ |\mathsf{E}\underline{m}'_n| \big(\mathsf{E}(\operatorname{tr} D_1^{-1} T \overline{D}_1^{-1} T)$$
$$\times \mathsf{E}\big(\operatorname{tr} D_1^{-1} (\mathsf{E}\underline{m}_n T + I)^{-2} T^3$$
$$\times (\mathsf{E}\overline{\underline{m}}_n T + I)^{-2} \overline{D}_1^{-1} T)\big)^{1/2}\big).$$

Using the same argument resulting in (3.1) it is a simple matter to conclude that $\mathsf{E}\underline{m}'_n(z)$ is bounded for $z \in \mathcal{C}_n$. All the remaining expected values are $O(N)$ due to (3.1) and (4.3), and we are done.

5. Some derivations and calculations. This section contains proofs of formulas stated in Section 1. We begin with deriving some properties of $\underline{m}(z)$. We claim that for any bounded subset S of $\mathbb{C}$,

$$\inf_{z \in S} |\underline{m}(z)| > 0. \tag{5.1}$$

Suppose not. Then there exists a sequence $\{z_n\} \subset \mathbb{C}^+$ which converges to a number for which $\underline{m}(z_n) \to 0$. From (1.2) we must have

$$c \int \frac{t\underline{m}(z_n)}{1 + t\underline{m}(z_n)} dH(t) \to 1.$$

However, because H has bounded support, the limit of the left-hand side of the above is obviously 0. The contradiction proves our assertion.

Next, we find a lower bound on the size of the difference quotient $(\underline{m}(z_1) - \underline{m}(z_2))/(z_1 - z_2)$ for distinct $z_1 = x + iv_1$, $z_2 = y + iv_2$, $v_1, v_2 \neq 0$. From (1.2) we get

$$z_1 - z_2 = \frac{\underline{m}(z_1) - \underline{m}(z_2)}{\underline{m}(z_1)\underline{m}(z_2)} \left(1 - c \int \frac{\underline{m}(z_1)\underline{m}(z_2) t^2 \, dH(t)}{(1 + t\underline{m}(z_1))(1 + t\underline{m}(z_2))}\right).$$

Therefore, from (2.19) we can write

$$\frac{\underline{m}(z_1) - \underline{m}(z_2)}{z_1 - z_2}$$
$$= [\underline{m}(z_1)\underline{m}(z_2)] \left[1 - c \int \frac{\underline{m}(z_1)\underline{m}(z_2) t^2 \, dH(t)}{(1 + t\underline{m}(z_1))(1 + t\underline{m}(z_2))}\right]^{-1}$$

and conclude that

$$\left|\frac{\underline{m}(z_1) - \underline{m}(z_2)}{z_1 - z_2}\right| \geq \frac{1}{2} |\underline{m}(z_1)\underline{m}(z_2)|. \tag{5.2}$$

We proceed to show (1.17). Choose $f, g \in \{f_1, \dots, f_k\}$. Let S_F denote the support of $F^{c,H}$, and let $a \neq 0$, b be such that S_F is a subset of (a, b), on whose closure f and g are analytic. Assume the z_1 contour encloses the z_2 contour. Using integration by parts twice, first with respect to z_2 and then with respect to z_1, we get

$$\begin{aligned}
\text{RHS of (1.7)} &= \frac{1}{2\pi^2}\iint \frac{f(z_1)g'(z_2)}{(\underline{m}(z_1)-\underline{m}(z_2))}\frac{d}{dz_1}\underline{m}(z_1)\,dz_2\,dz_1\\
&= -\frac{1}{2\pi^2}\iint f'(z_1)g'(z_2)\log(\underline{m}(z_1)-\underline{m}(z_2))\,dz_1\,dz_2\\
&\qquad\qquad\text{(where log is any branch of the logarithm)}\\
&= -\frac{1}{2\pi^2}\iint f'(z_1)g'(z_2)[\ln|\underline{m}(z_1)-\underline{m}(z_2)|\\
&\qquad\qquad + i\arg(\underline{m}(z_1)-\underline{m}(z_2))]\,dz_1\,dz_2.
\end{aligned}$$

We choose the contours to be rectangles with sides parallel to the axes. The inside rectangle intersects the real axis at a and b, and the horizontal sides are a distance $v<1$ away from the real axis. The outside rectangle intersects the real axis at $a-\varepsilon, b+\varepsilon$ (points where f and g remain analytic), with height twice that of the inside rectangle. We let $v\to 0$.

We need only consider the logarithm term and show its convergence, since the real part of the arg term disappears (f and g are real valued on $\mathbb{R}$) in the limit, and the sum (1.7) is real. Therefore the arg term also approaches zero.

We split up the log integral into 16 double integrals, each one involving a side from each of the two rectangles. We argue that any portion of the integral involving a vertical side can be neglected. This follows from (5.1), (5.2) and the fact that z_1 and z_2 remain a positive distance apart, so that $|\underline{m}(z_1)-\underline{m}(z_2)|$ is bounded away from zero. Moreover, at least one of $|\underline{m}(z_1)|$, $|\underline{m}(z_2)|$ is bounded, while the other is bounded by $1/v$, so the integral is bounded by $Kv\ln v^{-1}\to 0$.

Therefore we arrive at

$$\begin{aligned}
-\frac{1}{2\pi^2}\int_a^b\int_{a-\varepsilon}^{b+\varepsilon} &[(f'(x+i2v)g'(y+iv)+\bar{f}'(x+i2v)\bar{g}'(y+iv))\\
&\times\ln|\underline{m}(x+i2v)-\underline{m}(y+iv)|\\
&-(f'(x+i2v)\bar{g}'(y+iv)\\
&\quad+\bar{f}'(x+i2v)g'(y+iv))\\
&\times\ln|\underline{m}(x+i2v)-\overline{\underline{m}}(y+iv)|]\,dx\,dy.
\end{aligned}\tag{5.3}$$

Using subscripts to denote real and imaginary parts, we find

$$(5.3) = -\frac{1}{\pi^2}\int_a^b\int_{a-\varepsilon}^{b+\varepsilon}\big[\big(f_r'(x+i2v)g_r'(y+iv) - f_i'(x+i2v)g_i'(y+iv)\big) \times \ln|\underline{m}(x+i2v)-\underline{m}(y+iv)| - \big(f_r'(x+i2v)g_r'(y+iv) + f_i'(x+i2v)g_i'(y+iv)\big) \times \ln|\underline{m}(x+i2v)-\overline{\underline{m}}(y+iv)|\big]\,dx\,dy$$

$$(5.4)\quad = \frac{1}{\pi^2}\int_a^b\int_{a-\varepsilon}^{b+\varepsilon} f_r'(x+i2v)g_r'(y+iv)\ln\left|\frac{\underline{m}(x+i2v)-\overline{\underline{m}}(y+iv)}{\underline{m}(x+i2v)-\underline{m}(y+iv)}\right|dx\,dy$$

$$+\frac{1}{\pi^2}\int_a^b\int_{a-\varepsilon}^{b+\varepsilon} f_i'(x+i2v)g_i'(y+iv)$$

$$(5.5)\quad \times \ln\big|\big(\underline{m}(x+i2v)-\underline{m}(y+iv)\big) \times \big(\underline{m}(x+i2v)-\overline{\underline{m}}(y+iv)\big)\big|\,dx\,dy.$$

We have for any real-valued h, analytic on the bounded interval $[\alpha,\beta]$ for all v sufficently small

$$(5.6)\quad \sup_{x\in[\alpha,\beta]}|h_i(x+iv)| \le K|v|,$$

where K is a bound on $|h'(z)|$ for z in a neighborhood of $[\alpha,\beta]$. Using this and (5.1), (5.2) we see that (5.5) is bounded in absolute value by $Kv^2\ln v^{-1}\to 0$.

For (5.4) we write

$$(5.7)\quad \ln\left|\frac{\underline{m}(x+i2v)-\overline{\underline{m}}(y+iv)}{\underline{m}(x+i2v)-\underline{m}(y+iv)}\right| = \frac{1}{2}\ln\left(1+\frac{4\underline{m}_i(x+i2v)\underline{m}_i(y+iv)}{|\underline{m}(x+i2v)-\underline{m}(y+iv)|^2}\right).$$

From (5.2) we get

$$\text{RHS of (5.7)} \le \frac{1}{2}\ln\left(1+\frac{16\underline{m}_i(x+i2v)\underline{m}_i(y+iv)}{(x-y)^2|\underline{m}(x+i2v)\underline{m}(y+iv)|^2}\right).$$

From (5.1) we have

$$\sup_{\substack{x,y\in[a-\varepsilon,b+\varepsilon]\\ v\in(0,1)}}\frac{\underline{m}_i(x+i2v)\underline{m}_i(y+iv)}{|\underline{m}(x+i2v)\underline{m}(y+iv)|^2}<\infty.$$

Therefore, there exists a $K > 0$ for which the right-hand side of (5.7) is bounded by

$$\frac{1}{2}\ln\left(1+\frac{K}{(x-y)^2}\right) \tag{5.8}$$

for $x, y \in [a-\varepsilon, b+\varepsilon]$. It is straightforward to show that (5.8) is Lebesgue integrable on bounded subsets of $\mathbb{R}^2$. Therefore, from (1.19) and the dominated convergence theorem we conclude that (1.20) is Lebesgue integrable and that (1.17) holds.

We now verify (1.18). From (1.2) we have

$$\frac{d}{dz}\underline{m}(z) = \underline{m}^2(z)\left[1 - c\int \frac{t^2\underline{m}^2(z)}{(1+t\underline{m}(z))^2}\,dH(t)\right]^{-1}.$$

In Silverstein and Choi (1995) it is argued that the only place where $\underline{m}'(z)$ can possibly become unbounded are near the origin and the boundary, ∂S_F, of S_F. It is a simple matter to verify

$$\begin{aligned}\mathsf{E}X_f &= \frac{1}{4\pi i}\int f(z)\frac{d}{dz}\log\left(1 - c\int \frac{t^2\underline{m}^2(z)}{(1+t\underline{m}(z))^2}\,dH(t)\right)dz\\ &= -\frac{1}{4\pi i}\int f'(z)\log\left(1 - c\int \frac{t^2\underline{m}^2(z)}{(1+t\underline{m}(z))^2}\,dH(t)\right)dz,\end{aligned}$$

where, because of (2.19), the arg term for log can be taken from $(-\pi/2, \pi/2)$. We choose a contour as above. From (3.17) of Bai and Silverstein (1998) there exists a $K > 0$ such that for all small v,

$$\inf_{x\in\mathbb{R}}\left|1 - c\int \frac{t^2\underline{m}^2(x+iv)}{(1+t\underline{m}(x+iv))^2}\,dH(t)\right| \ge Kv^2. \tag{5.9}$$

Therefore, we see the integrals on the two vertical sides are bounded by $Kv\ln v^{-1} \to 0$. The integral on the two horizontal sides is equal to

$$\begin{aligned}&\frac{1}{2\pi}\int_a^b f_i'(x+iv)\ln\left|1 - c\int \frac{t^2\underline{m}^2(x+iv)}{(1+t\underline{m}(x+iv))^2}\,dH(t)\right|dx\\ &\quad + \frac{1}{2\pi}\int_a^b f_r'(x+iv)\arg\left(1 - c\int \frac{t^2\underline{m}^2(x+iv)}{(1+t\underline{m}(x+iv))^2}\,dH(t)\right)dx.\end{aligned} \tag{5.10}$$

Using (2.19), (5.6) and (5.9) we see the first term in (5.10) is bounded in absolute value by $Kv\ln v^{-1} \to 0$. Since the integrand in the second term converges for all $x \notin \{0\} \cup \partial S_F$ (a countable set) we get, therefore, (1.18) from the dominated convergence theorem.

We now derive $d(c)$ ($c \in (0,1)$) in (1.1), (1.21) and the variance in (1.22). The first two rely on Poisson's integral formula

$$u(z) = \frac{1}{2\pi}\int_0^{2\pi} u(e^{i\theta})\frac{1-r^2}{1+r^2-2r\cos(\theta-\phi)}\,d\theta, \tag{5.11}$$

where u is harmonic on the unit disk in $\mathbb{C}$, and $z = re^{i\phi}$ with $r \in [0, 1)$. Making the substitution $x = 1 + c - 2\sqrt{c}\cos\theta$ we get

$$d(c) = \frac{1}{\pi}\int_0^{2\pi} \frac{\sin^2\theta}{1+c-2\sqrt{c}\cos\theta} \ln(1+c-2\sqrt{c}\cos\theta)\,d\theta$$

$$= \frac{1}{2\pi}\int_0^{2\pi} \frac{2\sin^2\theta}{1+c-2\sqrt{c}\cos\theta} \ln|1-\sqrt{c}e^{i\theta}|^2\,d\theta.$$

It is straightforward to verify that

$$f(z) \equiv -(z - z^{-1})^2(\log(1-\sqrt{c}z) + \sqrt{c}z) - \sqrt{c}(z - z^3)$$

is analytic on the unit disk, and that

$$\Re f(e^{i\theta}) = 2\sin^2\theta \ln|1-\sqrt{c}e^{i\theta}|^2.$$

Therefore from (5.11) we have

$$d(c) = \frac{f(\sqrt{c})}{1-c} = \frac{c-1}{c}\ln(1-c) - 1.$$

For (1.21) we use (1.18). From (1.2), with $H(t) = I_{[1,\infty)}(t)$ we have for $z \in \mathbb{C}^+$

$$z = -\frac{1}{\underline{m}(z)} + \frac{c}{1+\underline{m}(z)}. \tag{5.12}$$

Solving for $\underline{m}(z)$ we find

$$\underline{m}(z) = \frac{-(z+1-c) + \sqrt{(z+1-c)^2 - 4z}}{2z}$$

$$= \frac{-(z+1-c) + \sqrt{(z-1-c)^2 - 4c}}{2z},$$

the square roots defined to yield positive imaginary parts for $z \in \mathbb{C}^+$. As $z \to x \in [a(y), b(y)]$ [limits defined below (1.1)] we get

$$\underline{m}(x) = \frac{-(x+1-c) + \sqrt{4c - (x-1-c)^2}\,i}{2x}$$

$$= \frac{-(x+1-c) + \sqrt{(x-a(c))(b(c)-x)}\,i}{2x}.$$

The identity (5.12) still holds with z replaced by x and from it we get

$$\frac{\underline{m}(x)}{1+\underline{m}(x)} = \frac{1+x\underline{m}(x)}{c},$$

so that

$$\begin{aligned} 1 - c\frac{\underline{m}^2(x)}{(1+\underline{m}(x))^2} &= 1 - \frac{1}{c}\left(\frac{-(x-1-c)+\sqrt{4c-(x-1-c)^2}\,i}{2}\right)^2 \\ &= \frac{\sqrt{4c-(x-1-c)^2}}{2c}\left(\sqrt{4c-(x-1-c)^2}+(x-1-c)i\right). \end{aligned}$$

Therefore, from (1.18)

$$\begin{aligned} \mathsf{E}X_f &= \frac{1}{2\pi}\int_{a(c)}^{b(c)} f'(x)\tan^{-1}\left(\frac{x-1-c}{\sqrt{4c-(x-1-c)^2}}\right)dx \\ &= \frac{f(a(c))+f(b(c))}{4} - \frac{1}{2\pi}\int_{a(c)}^{b(c)}\frac{f(x)}{\sqrt{4c-(x-1-c)^2}}\,dx. \end{aligned} \tag{5.13}$$

To compute the last integral when $f(x) = \ln x$ we make the same substitution as before, arriving at

$$\frac{1}{4\pi}\int_0^{2\pi} \ln|1-\sqrt{c}e^{i\theta}|^2\,d\theta.$$

We apply (5.11) where now $u(z) = \ln|1-\sqrt{c}z|^2$, which is harmonic, and $r = 0$. Therefore, the integral must be zero, and we conclude

$$\mathsf{E}X_{\ln} = \frac{\ln(a(c)b(c))}{4} = \frac{1}{c}\ln(1-c).$$

To derive (1.22) we use (1.16). Since the z_1, z_2 contours cannot enclose the origin (because of the logarithm), neither can the resulting m_1, m_2 contours. Indeed, either from the graph of $x(\underline{m})$ or from $\underline{m}(x)$ we see that $x > b(c) \Leftrightarrow \underline{m}(x) \in (-(1+\sqrt{c})^{-1}, 0)$ and $x \in (0, a(y)) \Leftrightarrow \underline{m}(x) < (\sqrt{c}-1)^{-1}$. For our analysis it is sufficient to know that the m_1, m_2 contours, nonintersecting and both taken in the positive direction, enclose $(c-1)^{-1}$ and -1, but not 0. Assume the m_2 contour encloses the m_1 contour. For fixed m_2, using (5.12) we have

$$\begin{aligned} &\int \frac{\log(z(m_1))}{(m_1-m_2)^2}\,dm_1 \\ &\quad = \int \frac{1/m_1^2 - c/(1+m_1)^2}{-1/m_1 + c/(1+m_1)}\frac{1}{(m_1-m_2)}\,dm_1 \\ &\quad = \int \frac{(1+m_1)^2 - cm_1^2}{cm_1(m_1-m_2)}\left(\frac{-1}{m_1+1} + \frac{1}{m_1 - 1/(c-1)}\right)dm_1 \\ &\quad = 2\pi i\left(\frac{1}{m_2+1} - \frac{1}{m_2-1/(c-1)}\right). \end{aligned}$$

Therefore

$$\operatorname{Var} X_{\ln} = \frac{1}{\pi i}\int\left(\frac{1}{m+1} - \frac{1}{m-1/(c-1)}\right)\log(z(m))\,dm$$
$$= \frac{1}{\pi i}\int\left[\frac{1}{m+1} - \frac{1}{m-1/(c-1)}\right]\log\left(\frac{m-1/(c-1)}{m+1}\right)dm$$
$$- \frac{1}{\pi i}\int\left[\frac{1}{m+1} - \frac{1}{m-1/(c-1)}\right]\log(m)\,dm.$$

The first integral is zero since the integrand has antiderivative

$$-\frac{1}{2}\left[\log\left(\frac{m-1/(c-1)}{m+1}\right)\right]^2,$$

which is single valued along the contour. Therefore we conclude that

$$\operatorname{Var} X_{\ln} = -2[\log(-1) - \log((c-1)^{-1})] = -2\ln(1-c).$$

Finally, we compute expressions for (1.23) and (1.24). Using (5.13) we have

$$\mathsf{E}X_{x^r} = \frac{(a(c))^r + (b(c))^r}{4} - \frac{1}{2\pi}\int_{a(c)}^{b(c)} \frac{x^r}{\sqrt{4c-(x-1-c)^2}}\,dx$$
$$= \frac{(a(c))^r + (b(c))^r}{4} - \frac{1}{4\pi}\int_0^{2\pi} \left|1-\sqrt{c}e^{i\theta}\right|^{2r}d\theta$$
$$= \frac{(a(c))^r + (b(c))^r}{4} - \frac{1}{4\pi}\int_0^{2\pi} \sum_{j,k=0}^{r}\binom{r}{j}\binom{r}{k}(-\sqrt{c})^{j+k}e^{i(j-k)\theta}\,d\theta$$
$$= \frac{1}{4}\left((1-\sqrt{c})^{2r} + (1+\sqrt{c})^{2r}\right) - \frac{1}{2}\sum_{j=0}^{r}\binom{r}{j}^2 c^j,$$

which is (1.23).

For (1.24) we use (1.16) and rely on observations made in deriving (1.22). For $c \in (0,1)$ the contours can again be made enclosing -1 and not the origin. However, because of the fact that (1.7) derives from (1.14) and the support of $F^{c,I_{[1,\infty)}}$ on $\mathbb{R}^+$ is $[a(c), b(c)]$, we may also take the contours taken in the same way when $c > 1$. The case $c = 1$ simply follows from the continuous dependence of (1.16) on c.

Keeping m_2 fixed, we have on a contour within 1 of -1

$$\int \frac{(-1/m_1 + c/(1+m_1))^{r_1}}{(m_1-m_2)^2}\,dm_1$$
$$= c^{r_1}\int\left(\frac{1}{m_1+1} + \frac{1-c}{c}\right)^{r_1}(1-(m_1+1))^{-r_1}$$
$$\times (m_2+1)^{-2}\left(1 - \frac{m_1+1}{m_2+1}\right)^{-2}dm_1$$

$$= c^{r_1} \int \sum_{k_1=0}^{r_1} \binom{r_1}{k_1} \left(\frac{1-c}{c}\right)^{k_1} (1+m_1)^{k-r_1}$$

$$\times \sum_{j=0}^{\infty} \binom{r_1+j-1}{j} (m_1+1)^j (m_2+1)^{-2} \sum_{\ell=1}^{\infty} \ell \left(\frac{m_1+1}{m_2+1}\right)^{\ell-1} dm_1$$

$$= 2\pi i c^{r_1} \sum_{k_1=0}^{r_1-1} \sum_{\ell=1}^{r_1-k_1} \binom{r_1}{k_1} \left(\frac{1-c}{c}\right)^{k_1} \binom{2r_1-1-(k_1+\ell)}{r_1-1} \ell (m_2+1)^{-\ell-1}.$$

Therefore,

$\mathrm{Cov}(X_{x^{r_1}}, X_{x^{r_2}})$

$$= -\frac{i}{\pi} c^{r_1+r_2} \sum_{k_1=0}^{r_1-1} \sum_{\ell=1}^{r_1-k_1} \binom{r_1}{k_1} \left(\frac{1-c}{c}\right)^{k_1} \binom{2r_1-1-(k_1+\ell)}{r_1-1} \ell$$

$$\times \int (m_2+1)^{-\ell-1} \sum_{k_2=0}^{r_2} \binom{r_2}{k_2} \left(\frac{1-c}{c}\right)^{k_2} (m_2+1)^{k_2-r_2}$$

$$\times \sum_{j=0}^{\infty} \binom{r_2+j-1}{j} (m_2+1)^j \, dm_2$$

$$= 2c^{r_1+r_2} \sum_{k_1=0}^{r_1-1} \sum_{k_2=0}^{r_2} \binom{r_1}{k_1} \binom{r_2}{k_2} \left(\frac{1-c}{c}\right)^{k_1+k_2}$$

$$\times \sum_{\ell=1}^{r_1-k_1} \ell \binom{2r_1-1-(k_1+\ell)}{r_1-1} \binom{2r_2-1-k_2+\ell}{r_2-1},$$

which is (1.24), and we are done.

APPENDIX

We verify (1.9b) by modifying the proof in Bai and Yin (1993) [hereafter referred to as BY (1993)]. To avoid confusion we maintain as much as possible the original notation used in BY (1993).

THEOREM. *For* $Z_{ij} \in \mathbb{C}$, $i = 1, \ldots, p$, $j = 1, \ldots, n$ *i.i.d.* $\mathsf{E}Z_{11} = 0$, $\mathsf{E}|Z_{11}|^2 = 1$, *and* $\mathsf{E}|Z_{11}|^4 < \infty$; *let* $S_n = (1/n)XX^*$ *where* $X = (X_{ij})$ *is* $p \times n$ *with*

$$X_{ij} = X_{ij}(n) = Z_{ij} I_{\{|Z|_{ij} \le \delta_n \sqrt{n}\}} - \mathsf{E}Z_{ij} I_{\{|Z|_{ij} \le \delta_n \sqrt{n}\}},$$

where $\delta_n \to 0$ *more slowly than that constructed in the proof of Lemma* 2.2 *of Yin, Bai and Krishnaiah* (1988) *and satisfying* $\delta_n n^{1/3} \to \infty$. *Assume* $p/n \to y \in (0, 1)$

as $n \to \infty$. *Then for any* $\eta < (1-\sqrt{y})^2$ *and any* $\ell > 0$

$$\mathsf{P}(\lambda_{\min}(S_n) < \eta) = o(n^{-\ell}).$$

PROOF. We follow along the proof of Theorem 1 in BY (1993). The conclusions of Lemmas 1 and 3–8 need to be improved from "almost sure" statements to ones reflecting tail probabilities. We shall denote the augmented lemmas with primes (′) after the number. We remark here that the proof in BY (1993) assumes entries of Z_{11} to be real, but all the arguments can be easily modified to allow complex variables.

For Lemma 1 it has been shown that for the Hermitian matrices $T(l)$ defined in (2.2), and integers m_n satisfying $m_n/\ln n \to \infty$, $m_n\delta_n^{1/3}/\ln n \to 0$ and $m_n/(\delta_n\sqrt{n}) \to 0$

$$\mathsf{E}\,\mathrm{tr}\, T^{2m_n}(l) \le n^2((2l+1)(l+1))^{2m_n}(p/n)^{m_n(l-1)}(1+o(1))^{4m_n l}.$$

[(2.13) of BY (1993)]. Therefore, writing $m_n = k_n \ln n$, for any $\varepsilon > 0$ there exists an $a \in (0,1)$ such that for all n large,

$$\text{(A.1)}\quad \mathsf{P}(\mathrm{tr}\, T(l) > (2l+1)(l+1)y^{(l-1)/2} + \varepsilon) \le n^2 a^{m_n} = n^{2+k_n \log a} = o(n^{-\ell})$$

for any positive ℓ. We call (A.1) Lemma 1′.

We next replace Lemma 2 of BY (1993) with the following:

LEMMA 2′. *Let for every* n $X_1, X_2, \dots, X_n$ *be i.i.d. with* $X_1 = X_1(n) \sim X_{11}(n)$. *Then for any* $\varepsilon > 0$ *and* $\ell > 0$,

$$\mathsf{P}\left(\left|n^{-1}\sum_{i=1}^{n}|X_i|^2 - 1\right| > \varepsilon\right) = o(n^{-\ell})$$

and for any $f > 1$,

$$\mathsf{P}\left(n^{-f}\sum_{i=1}^{n}|X_i|^{2f} > \varepsilon\right) = o(n^{-\ell}).$$

PROOF. Since as $n \to \infty$ $\mathsf{E}|X_1|^2 \to 1$,

$$n^{-f}\sum_{i=1}^{n}\mathsf{E}|X_i|^{2f} \le 2^{2f}\mathsf{E}|Z_{11}|^{2f}n^{1-f} \to 0 \qquad \text{for } f \in (1,2]$$

and

$$n^{-f}\sum_{i=1}^{n}\mathsf{E}|X_i|^{2f} \le 2^{2f}\mathsf{E}|Z_{11}|^4 n^{1-f+(2f-4)/2} = Kn^{-1} \to 0 \qquad \text{for } f > 2,$$

it is sufficient to show for $f \geq 1$,

$$\mathsf{P}\left(n^{-f}\left|\sum_{i=1}^{n}(|X_i|^{2f} - \mathsf{E}|X_i|^{2f})\right| > \varepsilon\right) = o(n^{-\ell}). \tag{A.2}$$

For any positive integer m we have this probability bounded by

$$\begin{aligned}
&n^{-2mf}\varepsilon^{-2m}\mathsf{E}\left[\sum_{i=1}^{n}(|X_i|^{2f} - \mathsf{E}|X_i|^{2f})\right]^{2m} \\
&= n^{-2mf}\varepsilon^{-2m} \sum_{\substack{i_1\geq 0,\ldots,i_n\geq 0\\ i_1+\cdots+i_n=2m}} \binom{2m}{i_1\cdots i_n}\prod_{t=1}^{n}\mathsf{E}(|X_t|^{2f} - \mathsf{E}|X_t|^{2f})^{i_t} \\
&= n^{-2mf}\varepsilon^{-2m}\sum_{k=1}^{m}\binom{n}{k} \sum_{\substack{i_1\geq 2,\ldots,i_k\geq 2\\ i_1+\cdots+i_k=2m}} \binom{2m}{i_1\cdots i_k}\prod_{t=1}^{k}\mathsf{E}(|X_1|^{2f} - \mathsf{E}|X_1|^{2f})^{i_t} \\
&\leq 2^{2m}n^{-2mf}\varepsilon^{-2m}\sum_{k=1}^{m}n^k \sum_{\substack{i_1\geq 2,\ldots,i_k\geq 2\\ i_1+\cdots+i_k=2m}} \binom{2m}{i_1\cdots i_k}\prod_{t=1}^{k}\mathsf{E}|X_1|^{2fi_t} \\
&\leq 2^{2m}n^{-2mf}\varepsilon^{-2m}\sum_{k=1}^{m}n^k \sum_{\substack{i_1\geq 2,\ldots,i_k\geq 2\\ i_1+\cdots+i_k=2m}} \binom{2m}{i_1\cdots i_k}\prod_{t=1}^{k}(2\delta_n\sqrt{n})^{2fi_t-4}\mathsf{E}|Z_{11}|^4 \\
&\leq 2^{2m}n^{-2mf}\varepsilon^{-2m}\sum_{k=1}^{m}k^{2m}(2\delta_n\sqrt{n})^{4fm-4k}n^k(\mathsf{E}|Z_{11}|^4)^k \\
&= 2^{2m}\varepsilon^{-2m}\sum_{k=1}^{m}(2\delta_n)^{4fm}(\mathsf{E}|Z_{11}|^4)^k(4\delta_n^2 n)^{-k}k^{2m} \\
&\leq \text{(for all } n \text{ large)}\quad m\left(\frac{(2\delta_n)^{2f}4m}{\varepsilon\ln(4\delta_n^2 n/\mathsf{E}|Z_{11}|^4)}\right)^{2m},
\end{aligned}$$

where we have used the inequality $a^{-x}x^b \leq (b/\ln a)^b$, valid for all $a > 1$, $b > 0$, $x \geq 1$. Choose $m_n = k_n \ln n$ with $k_n \to \infty$ and $\delta_n^{2f}k_n \to 0$. Since $\delta_n n^{1/3} \geq 1$ for n large we get for these $n \ln(\delta_n^2 n) \geq (1/3)\ln n$. Using this and the fact that $\lim_{x\to\infty} x^{1/x} = 1$, we have the existence of $a \in (0, 1)$ for which

$$m\left(\frac{(2\delta_n)^{2f}4m}{\varepsilon\ln(4\delta_n^2 n/\mathsf{E}|Z_{11}|^4)}\right)^{2m} \leq a^{2k_n\ln n} = n^{2k_n\ln a}$$

for all n large. Therefore (A.2) holds.

Redefining the matrix $X^{(f)}$ in BY (1993) to be $[|X_{uv}|^f]$, Lemma 3′ states for any positive integer f

$$\mathsf{P}(\lambda_{\max}\{n^{-f}X^{(f)}X^{(f)*}\} > 7+\varepsilon) = o(n^{-\ell}) \qquad \text{for any positive } \varepsilon \text{ and } \ell.$$

Its proof relies on Lemmas 1′, 2′ (for $f = 1, 2$) and on the bounds used in the proof of Lemma 3 in BY (1993). In particular we have the Geršgorin bound

$$\begin{aligned}
&\lambda_{\max}\{n^{-f}X^{(f)}X^{(f)*}\} \\
&\quad \le \max_i n^{-f}\sum_{j=1}^{n}|X_{ij}|^{2f} + \max_i n^{-f}\sum_{k\neq i}\sum_{j=1}^{n}|X_{ij}|^f|X_{kj}|^f \qquad \text{(A.3)}\\
&\quad \le \max_i n^{-f}\sum_{j=1}^{n}|X_{ij}|^{2f} + \left(\max_i n^{-1}\sum_{j=1}^{n}|X_{ij}|^f\right)\left(\max_j n^{-1}\sum_{k=1}^{p}|X_{kj}|^f\right).
\end{aligned}$$

We show the steps involved for $f = 2$. With $\varepsilon_1 > 0$ satisfying $(p/n+\varepsilon_1)(1+\varepsilon_1) < 7+\varepsilon$ for all n we have from Lemma 2′ and (A.3)

$$\begin{aligned}
&\mathsf{P}(\lambda_{\max}\{n^{-2}X^{(2)}X^{(2)*}\} > 7+\varepsilon) \\
&\quad \le p\mathsf{P}\left(n^{-2}\sum_{j=1}^{n}|X_{1j}|^4 > \varepsilon_1\right) \\
&\qquad + p\mathsf{P}\left(n^{-1}\sum_{j=1}^{n}|X_{1j}|^2 - 1 > \varepsilon_1\right) + n\mathsf{P}\left(p^{-1}\sum_{k=1}^{p}|X_{k1}|^2 - 1 > \varepsilon_1\right) \\
&\quad = o(n^{-\ell}).
\end{aligned}$$

The same argument can be used to prove Lemma 4′, which states for integer $f > 2$

$$\mathsf{P}(\|n^{-f/2}X^{(f)}\| > \varepsilon) = o(n^{-\ell}) \qquad \text{for any positive } \varepsilon \text{ and } \ell.$$

The proofs of Lemmas 4′–8′ are handled using the arguments in BY (1993) and those used above: each quantity L_n in BY (1993) that is $o(1)$ a.s. can be shown to satisfy $\mathsf{P}(|L_n| > \varepsilon) = o(n^{-\ell})$.

From Lemmas 1′ and 8′ there exists a positive C such that for every integer $k > 0$ and positive ε and ℓ,

$$\mathsf{P}(\|T - yI\|^k > Ck^4 2^k y^{k/2} + \varepsilon) = o(n^{-\ell}). \qquad \text{(A.4)}$$

For given $\varepsilon > 0$ let integer $k > 0$ be such that

$$|2\sqrt{y}(1-(Ck^4)^{1/k})| < \varepsilon/2.$$

Then

$$2\sqrt{y}+\varepsilon > 2\sqrt{y}(Ck^4)^{1/k} + \varepsilon/2 \ge (Ck^4 2^k y^{k/2} + (\varepsilon/2)^k)^{1/k}.$$

Therefore from (A.4) we get, for any $\ell > 0$,

$$\mathsf{P}(\|T - yI\| > 2\sqrt{y} + \varepsilon) = o(n^{-\ell}). \tag{A.5}$$

From Lemma 2′ and (A.5) we get for positive ε and ℓ

$$\begin{aligned}
&\mathsf{P}(\|S_n - (1+y)I\| > 2\sqrt{y} + \varepsilon) \\
&\quad \le \mathsf{P}(\|S_n - I - T\| > \varepsilon/2) + o(n^{-\ell}) \\
&\quad = \mathsf{P}\left(\max_{i \le p}\left|n^{-1}\sum_{j=1}^{n}|X_{ij}|^2 - 1\right| > \varepsilon/2\right) + o(n^{-\ell}) = o(n^{-\ell}).
\end{aligned}$$

Finally, for any positive $\eta < (1-\sqrt{y})^2$ and $\ell > 0$

$$\begin{aligned}
\mathsf{P}(\lambda_{\min}(S_n) < \eta) &= \mathsf{P}(\lambda_{\min}(S_n - (1+y)I) < \eta - (1-\sqrt{y})^2 - 2\sqrt{y}) \\
&\le \mathsf{P}((\|S_n - (1+y)I\|) > 2\sqrt{y} + (1-\sqrt{y})^2 - \eta) = o(n^{-\ell})
\end{aligned}$$

and we are done. □

Acknowledgments. Part of this work was done while J. W. Silverstein visited the Department of Statistics and Applied Probability at National University of Singapore. He thanks the members of the department for their hospitality.

REFERENCES

BAI, Z. D. (1999). Methodologies in spectral analysis of large dimensional random matrices, A review. *Statist. Sinica* **9** 611–677.

BAI, Z. D. and SARANADASA, H. (1996). Effect of high dimension comparison of significance tests for a high dimensional two sample problem. *Statist. Sinica* **6** 311–329.

BAI, Z. D. and SILVERSTEIN, J. W. (1998). No eigenvalues outside the support of the limiting spectral distribution of large dimensional random matrices. *Ann. Probab.* **26** 316–345.

BAI, Z. D. and SILVERSTEIN, J. W. (1999). Exact separation of eigenvalues of large dimensional sample covariance matrices. *Ann. Probab.* **27** 1536–1555.

BAI, Z. D. and YIN, Y. Q. (1993). Limit of the smallest eigenvalue of a large dimensional sample covariance matrix. *Ann. Probab.* **21** 1275–1294.

BILLINGSLEY, P. (1968). *Convergence of Probability Measures.* Wiley, New York.

BILLINGSLEY, P. (1995). *Probability and Measure*, 3rd ed. Wiley, New York.

BURKHOLDER, D. L. (1973). Distribution function inequalities for martingales. *Ann. Probab.* **1** 19–42.

DEMPSTER, A. P. (1958). A high dimensional two sample significance test. *Ann. Math. Statist.* **29** 995–1010.

DIACONIS, P. and EVANS, S. N. (2001). Linear functionals of eigenvalues of random matrices. *Trans. Amer. Math. Soc.* **353** 2615–2633.

JOHANSSON, K. (1998). On fluctuations of eigenvalues of random Hermitian matrices. *Duke Math. J.* **91** 151–204.

JONSSON, D. (1982). Some limit theorems for the eigenvalues of a sample covariance matrix *J. Multivariate Anal.* **12** 1–38.

MARČENKO, V. A. and PASTUR, L. A. (1967). Distribution of eigenvalues for some sets of random matrices. *Math. USSR-Sb.* **1** 457–483.

SILVERSTEIN, J. W. (1985). The limiting eigenvalue distribution of a multivariate F matrix. *SIAM J. Math. Anal.* **16** 641–646.

SILVERSTEIN, J. W. (1995). Strong convergence of the empirical distribution of eigenvalues of large dimensional random matrices. *J. Multivariate Anal.* **55** 331–339.

SILVERSTEIN, J. W. and CHOI, S. I. (1995). Analysis of the limiting spectral distribution of large dimensional random matrices. *J. Multivariate Anal.* **54** 295–309.

SILVERSTEIN, J. W. and COMBETTES, P. L. (1992). Signal detection via spectral theory of large dimensional random matrices. *IEEE Trans. Signal Process.* **40** 2100–2105.

SINAI, YA. and SOSHNIKOV, A. (1998). Central limit theorem for traces of large symmetric matrices with independent matrix elements. *Bol. Soc. Brasil Mat. (N.S.)* **29** 1–24.

SOSHNIKOV, A. (2000). The central limit theorem for local linear statistics in classical compact groups and related combinatorial identities. *Ann. Probab.* **28** 1353–1370.

TITCHMARSH, E. C. (1939). *The Theory of Functions*, 2nd ed. Oxford Univ. Press.

YIN, Y. Q. (1986). Limiting spectral distribution for a class of random matrices. *J. Multivariate Anal.* **20** 50–68.

YIN, Y. Q., BAI, Z. D. and KRISHNAIAH, P. R. (1988). On limit of the largest eigenvalue of the large dimensional sample covariance matrix. *Probab. Theory Related Fields* **78** 509–521.

YIN, Y. Q. and KRISHNAIAH, P. R. (1983). A limit theorem for the eigenvalues of product of two random matrices. *J. Multivariate Anal.* **13** 489–507.

DEPARTMENT OF MATHEMATICS
NORTH EAST NORMAL UNIVERSITY
CHANGCHUN
CHINA 130024
E-MAIL: stabaizd@leonis.nus.edu.sg

DEPARTMENT OF MATHEMATICS
BOX 8205
NORTH CAROLINA STATE UNIVERSITY
RALEIGH, NORTH CAROLINA 27695-8205
USA
E-MAIL: jack@math.ncsu.edu

The Annals of Applied Probability
2005, Vol. 15, No. 1B, 914–940
DOI 10.1214/105051604000000774

ASYMPTOTICS IN RANDOMIZED URN MODELS

BY ZHI-DONG BAI[1] AND FEIFANG HU[2]

Northeast Normal University and National University of Singapore, and University of Virginia

This paper studies a very general urn model stimulated by designs in clinical trials, where the number of balls of different types added to the urn at trial n depends on a random outcome directed by the composition at trials $1, 2, \dots, n-1$. Patient treatments are allocated according to types of balls. We establish the strong consistency and asymptotic normality for both the urn composition and the patient allocation under general assumptions on random generating matrices which determine how balls are added to the urn. Also we obtain explicit forms of the asymptotic variance–covariance matrices of both the urn composition and the patient allocation. The conditions on the nonhomogeneity of generating matrices are mild and widely satisfied in applications. Several applications are also discussed.

1. Introduction. In designing a clinical trial, the limiting behavior of the patient allocation to several treatments during the process is of primary consideration. Suppose patients arrive sequentially from a population. Adaptive designs in clinical trials are inclining to assign more patients to better treatments, while seeking to maintain randomness as a basis for statistical inference. Thus the cumulative information of the responses of treatments on previous patients will be used to adjust treatment assignment to coming patients. For this purpose, various urn models [Johnson and Kotz (1977)] have been proposed and used extensively in adaptive designs [for more references, see Zelen (1969), Wei (1979), Flournoy and Rosenberger (1995) and Rosenberger (1996)].

One large family of randomized adaptive designs is based on the generalized Friedman's urn (GFU) model [Athreya and Karlin (1967, 1968), also called the generalized Pólya urn (GPU) in the literature]. The model can be described as follows. Consider an urn containing balls of K types, respectively, representing K "treatments" in a clinical trial. These treatments are to be assigned sequentially in n stages. At the beginning, the urn contains $\mathbf{Y}_0 = (Y_{01}, \dots, Y_{0K})$ balls, where Y_{0k} denotes the number of balls of type k, $k = 1, \dots, K$. At stage i, $i = 1, \dots, n$,

Received June 2003; revised March 2004.

[1]Supported by NSFC Grant 201471000 and NUS Grant R-155-000-030-112.
[2]Supported by NSF Grant DMS-0204232 and NUS Grant R-155-000-030-112.
AMS 2000 subject classifications. Primary 62E20, 62L05; secondary 62F12.
Key words and phrases. Asymptotic normality, extended Pólya's urn models, generalized Friedman's urn model, martingale, nonhomogeneous generating matrix, response-adaptive designs, strong consistency.

a ball is randomly drawn from the urn and then replaced. If the ball is of type q, then the treatment q is assigned to the ith patient, $q = 1, \dots, K$, $i = 1, \dots, n$. We then wait until we observe a random variable $\xi(i)$, which may include the response and/or other covariates of patient i. After that, an additional $D_{qk}(i)$ balls of type k, $k = 1, \dots, K$, are added to the urn, where $D_{qk}(i)$ is some function of $\xi(i)$. This procedure is repeated throughout the n stages. After n splits and generations, the urn composition is denoted by the row vector $\mathbf{Y}_n = (Y_{n1}, \dots, Y_{nK})$, where Y_{nk} represents the number of balls of type k in the urn after the nth split. This relation can be written as the following recursive formula:

$$\mathbf{Y}_n = \mathbf{Y}_{n-1} + \mathbf{X}_n \mathbf{D}_n,$$

where $\mathbf{X}_n$ is the result of the nth draw, distributed according to the urn composition at the previous stage; that is, if the nth draw is a type-k ball, then the kth component of $\mathbf{X}_n$ is 1 and other components are 0. Furthermore, write $\mathbf{N}_n = (N_{n1}, \dots, N_{nK})$, where N_{nk} is the number of times a type-k ball was drawn in the first n stages, or equivalently, the number of patients who receive the treatment k in the first n patients.

For notation, let $\mathbf{D}_i = \langle\langle D_{qk}(i), q, k = 1, \dots, K\rangle\rangle$ and let $\mathcal{F}_i$ be the sequence of increasing σ-fields generated by $\{\mathbf{Y}_j\}_{j=0}^{i}$, $\{\mathbf{X}_j\}_{j=1}^{i}$ and $\{\mathbf{D}_j\}_{j=1}^{i}$. Define $\mathbf{H}_i = \langle\langle E(D_{qk}(i)|\mathcal{F}_{i-1}), q, k = 1, \dots, K\rangle\rangle$, $i = 1, \dots, n$. The matrices $\mathbf{D}_i$ are called *addition rules* and $\mathbf{H}_i$ *generating matrices*. In practice, the addition rule $\mathbf{D}_i$ often depends only on the treatment on the ith patient and its outcome. In these cases, the addition rules $\mathbf{D}_i$ are i.i.d. (independent and identically distributed) and the generating matrices $\mathbf{H}_i = \mathbf{H} = E\mathbf{D}_i$ are identical and nonrandom. But in some applications, the addition rule $\mathbf{D}_i$ depends on the total history of previous trials [see Andersen, Faries and Tamura (1994) and Bai, Hu and Shen (2002)]; then the general generating matrix $\mathbf{H}_i$ is the conditional expectation of $\mathbf{D}_i$ given $\mathcal{F}_{i-1}$. Therefore, the general generating matrices $\{\mathbf{H}_i\}$ are usually random. In this paper, we consider this general case. Examples are considered in Section 5.

A GFU model is said to be *homogeneous* if $\mathbf{H}_i = \mathbf{H}$ for all $i = 1, 2, 3, \dots$. In the literature, research is focused on asymptotic properties of $\mathbf{Y}_n$ for homogeneous GFU. First-order asymptotics for homogeneous GFU models are determined by the generating matrices $\mathbf{H}$. In most cases, $\mathbf{H}$ is an irreducible nonnegative matrix, for which the maximum eigenvalue is unique and positive (called the maximal eigenvalue in the literature) and its corresponding left eigenvector has positive components. In some cases, the entries of $\mathbf{H}$ may not be all nonnegative (e.g., when there is no replacement after the draw), and we may assume that the matrix $\mathbf{H}$ has a unique maximal eigenvalue λ with associated left eigenvector $\mathbf{v} = (v_1, \dots, v_K)$ with $\sum v_i = 1$. Under the following assumptions:

(i) $\Pr\{D_{qk} = 0, k = 1, \dots, K\} = 0$ for every $q = 1, \dots, K$,
(ii) $D_{qk} \geq 0$ for all $q, k = 1, \dots, K$,
(iii) $\mathbf{H}$ is irreducible,

Athreya and Karlin (1967, 1968) prove that

$$(1.1)\qquad \frac{N_{nk}}{n} \to v_k \quad \text{and} \quad \frac{Y_{nk}}{\sum_{q=1}^{K} Y_{nq}} \to v_k$$

almost surely as $n \to \infty$.

Let λ_1 be the eigenvalue with a second largest real part, associated with a right eigenvector ξ. If $\lambda > 2\operatorname{Re}(\lambda_1)$, Athreya and Karlin (1968) show that

$$(1.2)\qquad n^{-1/2}\mathbf{Y}_n\xi' \to N(0, c)$$

in distribution, where c is a constant. When $\lambda = 2\operatorname{Re}(\lambda_1)$ and λ_1 is simple, then (1.2) holds when $n^{-1/2}$ is replaced by $1/\sqrt{n\ln(n)}$. Asymptotic results under various addition schemes are considered in Freedman (1965), Mahmoud and Smythe (1991), Holst (1979) and Gouet (1993).

Homogeneity of the generating matrix is often not the case in clinical trials, where patients may exhibit a drift in characteristics over time. Examples are given in Altman and Royston (1988), Coad (1991) and Hu and Rosenberger (2000). Bai and Hu (1999) establish the weak consistency and the asymptotic normality of $\mathbf{Y}_n$ under GFU models with nonhomogeneous generating matrices $\mathbf{H}_i$. [In that paper, it is assumed that $\mathbf{H}_i = E\mathbf{D}_i$, so $\mathbf{H}_i$ are fixed (not random) matrices.] They consider the following GFU model (GFU1): $\sum_{k=1}^{K} D_{qk}(i) = c_1 > 0$, for all $q = 1, \ldots, K$ and $i = 1, \ldots, n$, the total number of balls added at each stage is a positive constant. They assume there is a nonnegative matrix $\mathbf{H}$ such that

$$(1.3)\qquad \sum_{i=1}^{\infty} \frac{\alpha_i}{i} < \infty,$$

where $\alpha_i = \|\mathbf{H}_i - \mathbf{H}\|_\infty$.

In clinical trials, N_{nk} represents the number of patients assigned to the treatment k in the first n trials. Doubtless, the asymptotic distribution and asymptotic variance of $\mathbf{N}_n = (N_{n1}, \ldots, N_{nK})$ is of more practical interest than the urn compositions to sequential design researchers. As Athreya and Karlin [(1967), page 275] said, "It is suggestive to conjecture that $(N_{n1}, \ldots, N_{nK})$ properly normalized is asymptotically normal. This problem is open." The problem has stayed open for decades due to mathematical complexity. One of our main goals of this paper is to present a solution to this problem.

Smythe (1996) defined the extended Pólya urn (EPU) (homogeneous) models, satisfying $\sum_{k=1}^{K} E(D_{qk}) = c_1 > 0, q = 1, \ldots, K$; that is, the expected total number of balls added to the urn at each stage is a positive constant. For EPU models, Smythe (1996) established the weak consistency and the asymptotic normality of $\mathbf{Y}_n$ and $\mathbf{N}_n$ under the assumptions that the eigenvalues of the generating matrix $\mathbf{H}$ are simple. The asymptotic variance of $\mathbf{N}_n$ is a more important and difficult proposition [Rosenberger (2002)]. Recently, Hu and Rosenberger (2003) obtained an explicit relationship between the power and the variance of $\mathbf{N}_n$ in an adaptive design. To compare the randomized urn models with other adaptive designs, one just has to calculate and compare their variances. Matthews and Rosenberger (1997) obtained the formula for asymptotic variance for the randomized play-the-winner rule ($K = 2$) which was initially proposed by Wei and Durham (1978). A general formula for asymptotic variance of $\mathbf{N}_n$ was still an open problem [Rosenberger (2002)].

In this paper, we

(i) show the asymptotic normality of $\mathbf{N}_n$ for general $\mathbf{H}$;
(ii) obtain a general and explicit formula for the asymptotic variance of $\mathbf{N}_n$;
(iii) show the strong consistency of both $\mathbf{Y}_n$ and $\mathbf{N}_n$; and
(iv) extend these results to nonhomogeneous urn model with random generating matrices $\mathbf{H}_i$.

The paper is organized as follows. The strong consistency of $\mathbf{Y}_n$ and $\mathbf{N}_n$ is proved in Section 2 for both homogeneous and nonhomogeneous EPU models. Note that the GFU1 is a special case of EPU. The asymptotic normality of $\mathbf{Y}_n$ for homogeneous and nonhomogeneous EPU models is shown in Section 3 under the assumption (1.3). We consider cases where the generating matrix $\mathbf{H}$ has a general Jordan form. In Section 4, we consider the asymptotic normality of $\mathbf{N}_n = (N_{n1}, \dots, N_{nK})$ for both homogeneous and nonhomogeneous EPU models. Further, we obtain a general and explicit formula for the asymptotic variance of $\mathbf{N}_n$.

The condition (1.3) in a nonhomogeneous urn model is widely satisfied in applications. In some applications [e.g., Bai, Hu and Shen (2002)], the generating matrix $\mathbf{H}_i$ may be estimates of some unknown parameters updated at each stage, for example, $\hat{\mathbf{H}}_i$ at ith stage. In these cases, we usually have $\alpha_i = O(i^{-1/2})$ in probability or $O(i^{-1/4})$ almost surely, so the condition (1.3) is satisfied. Also (1.3) is satisfied for the case of Hu and Rosenberger (2000). Some other applications are considered in Section 5.

2. Strong consistency of $\mathbf{Y}_n$ and $\mathbf{N}_n$. Using the notation defined in the Introduction, $\mathbf{Y}_n$ is a sequence of random K-vectors of nonnegative elements which are adaptive with respect to $\{\mathcal{F}_n\}$, satisfying

$$E(\mathbf{Y}_i|\mathcal{F}_{i-1}) = \mathbf{Y}_{i-1}\mathbf{M}_i, \tag{2.1}$$

where $\mathbf{M}_i = \mathbf{I} + a_{i-1}^{-1}\mathbf{H}_i$, $\mathbf{H}_i = E(\mathbf{D}_i|\mathcal{F}_{i-1})$ and $a_i = \sum_{j=1}^K Y_{ij}$. Without loss of generality, we assume $a_0 = 1$ in the following study.

In the sequel, we need the following assumptions.

ASSUMPTION 2.1. The generating matrix $\mathbf{H}_i$ satisfies

$$H_{qk}(i) \ge 0 \qquad \text{for all } k, q \quad \text{and}$$

$$\sum_{k=1}^K H_{qk}(i) = c_1 \qquad \text{for all } q = 1, \dots, K, \tag{2.2}$$

almost surely, where $H_{qk}(i)$ is the (q,k)-entry of the matrix $\mathbf{H}_i$ and c_1 is a positive constant. Without loss of generality, we assume $c_1 = 1$ throughout this work.

ASSUMPTION 2.2. The addition rule $\mathbf{D}_i$ is conditionally independent of the drawing procedure $\mathbf{X}_i$ given $\mathcal{F}_{i-1}$ and satisfies

$$E\big(D_{qk}^{2+\delta}(i)|\mathcal{F}_{i-1}\big) \le C < \infty \qquad \text{for all } q,k = 1,\dots,K \text{ and some } \delta > 0. \tag{2.3}$$

Also we assume that

$$\operatorname{cov}\big[(D_{qk}(i), D_{ql}(i))|\mathcal{F}_{i-1}\big] \to d_{qkl} \qquad \text{for all } q,k,l = 1,\dots,K, \tag{2.4}$$

where $\mathbf{d}_q = (d_{qkl})_{k,l=1}^K$, $q = 1,\dots,K$, are some $K \times K$ positive definite matrices.

REMARK 2.1. Assumption 2.1 defines the EPU model [Smythe (1996)]; it ensures that the number of expected balls added at each stage is a positive constant. So after n stages, the total number of balls, a_n, in the urn should be very close to n (a_n/n converges to 1).

The elements of the addition rule are allowed to take negative values in the literature, which corresponds to the situation of withdrawing balls from the urn. But, to avoid the dilemma that there are no balls to withdraw, only diagonal elements of $\mathbf{D}_i$ are allowed to take negative values, which corresponds to the case of drawing without replacement.

To investigate the limiting properties of $\mathbf{Y}_n$, we first derive a decomposition. From (2.1), it is easy to see that

$$\begin{aligned}\mathbf{Y}_n &= \big(\mathbf{Y}_n - E(\mathbf{Y}_n|\mathcal{F}_{n-1})\big) + \mathbf{Y}_{n-1}\mathbf{M}_n \\ &= \mathbf{Q}_n + \mathbf{Y}_{n-1}\mathbf{G}_n + \mathbf{Y}_{n-1}(\mathbf{M}_n - \mathbf{G}_n) \\ &= \mathbf{Y}_0\mathbf{G}_1\mathbf{G}_2\cdots\mathbf{G}_n + \sum_{i=1}^n \mathbf{Q}_i\mathbf{B}_{n,i} + \sum_{i=1}^n \mathbf{Y}_{i-1}(\mathbf{M}_i - \mathbf{G}_i)\mathbf{B}_{n,i} \\ &= \mathbf{S}_1 + \mathbf{S}_2 + \mathbf{S}_3,\end{aligned} \tag{2.5}$$

where $\mathbf{Q}_i = \mathbf{Y}_i - E(\mathbf{Y}_i|\mathcal{F}_{i-1})$, $\mathbf{G}_i = \mathbf{I} + i^{-1}\mathbf{H}$ and $\mathbf{B}_{n,i} = \mathbf{G}_{i+1}\cdots\mathbf{G}_n$ with the convention that $\mathbf{B}_{n,n} = \mathbf{I}$ and $\mathcal{F}_0$ denotes the trivial σ-field.

We further decompose $\mathbf{S}_3$ as follows:

$$\begin{aligned}\mathbf{S}_3 &= \sum_{i=1}^n \mathbf{Y}_{i-1}(a_{i-1}^{-1}\mathbf{H}_i - i^{-1}\mathbf{H})\mathbf{B}_{n,i} \\ &= \sum_{i=1}^n a_{i-1}^{-1}\mathbf{Y}_{i-1}(\mathbf{H}_i - \mathbf{H})\mathbf{B}_{n,i} + \sum_{i=1}^n \frac{\mathbf{Y}_{i-1}}{a_{i-1}}\frac{i - a_{i-1}}{i}\mathbf{H}\mathbf{B}_{n,i} \\ &= \mathbf{S}_{31} + \mathbf{S}_{32}.\end{aligned} \tag{2.6}$$

To estimate the above terms in the expansion, we need some preliminary results. First, we evaluate the convergence rate of a_n. To this end, we have the following theorem.

THEOREM 2.1. *Under Assumptions* 2.1 *and* 2.2, (a) $a_n/n \to 1$ *a.s. as* $n \to \infty$, *and* (b) $n^{-\kappa}(a_n - n) \to 0$ *a.s. for any* $\kappa > 1/2$.

PROOF. Let $e_i = a_i - a_{i-1}$ for $i \geq 1$. By definition, we have $e_i = \mathbf{X}_i\mathbf{D}_i\mathbf{1}$, where $\mathbf{X}_i$ is the result of the ith draw, multinomially distributed according to the urn composition at the previous stages; that is, the conditional probability that the ith draw is a ball of type k (the kth component of $\mathbf{X}_i$ is 1 and other components are 0) given previous status is $Y_{i-1,k}/a_{i-1}$.

From Assumptions 2.1 and 2.2, we have

$$E(e_i|\mathcal{F}_{i-1}) = 1 \tag{2.7}$$

and

$$\begin{aligned} E(e_i^2) &= E[E(e_i^2|\mathcal{F}_{i-1})] = E[E(\mathbf{1}'\mathbf{D}_i'\mathbf{X}_i'\mathbf{X}_i\mathbf{D}_i\mathbf{1}|\mathcal{F}_{i-1})] \\ &= \mathbf{1}'E[E(\mathbf{D}_i'\mathbf{X}_i'\mathbf{X}_i\mathbf{D}_i|\mathcal{F}_{i-1})]\mathbf{1} \\ &= \mathbf{1}'E\big[E\big(\mathbf{D}_i' \operatorname{diag}(a_{i-1}^{-1}\mathbf{Y}_{i-1})\mathbf{D}_i|\mathcal{F}_{i-1}\big)\big]\mathbf{1} \\ &= \sum_{q=1}^{K}\sum_{k=1}^{K}\sum_{l=1}^{K} E\big[(a_{i-1}^{-1}Y_{i-1,q})E\big(D_{qk}(i)D_{ql}(i)|\mathcal{F}_{i-1}\big)\big] \\ &\leq CK^2, \end{aligned} \tag{2.8}$$

so that

$$a_n - n = a_0 + \sum_{i=1}^{n}(e_i - 1) = 1 + \sum_{i=1}^{n}(e_i - E(e_i|\mathcal{F}_{i-1})) \tag{2.9}$$

forms a martingale sequence.

From Assumption 2.2 and $\kappa > 1/2$, we have

$$\sum_{i=1}^{\infty} E\left(\left(\frac{e_i - 1}{i^{\kappa}}\right)^2 \middle| \mathcal{F}_{i-1}\right) < \infty.$$

By three series theorem for martingales, this implies that the series

$$\sum_{i=1}^{\infty} \frac{e_i - 1}{i^{\kappa}}$$

converges almost surely. Then, by Kronecker's lemma,

$$\frac{1}{n^{\kappa}} \sum_{i=1}^{n}(e_i - 1) \to 0$$

almost surely. This completes the proof for conclusion (b) of the theorem.

The conclusion (a) is a consequence of conclusion (b). The proof of Theorem 2.1 is then complete. □

ASSUMPTION 2.3. Assume that (1.3) holds almost surely. Suppose that the limit generating matrix $\mathbf{H}$, $K \times K$, is irreducible.

This assumption guarantees that $\mathbf{H}$ has the Jordan form decomposition

$$\mathbf{T}^{-1}\mathbf{H}\mathbf{T} = \mathbf{J} = \begin{pmatrix} 1 & 0 & \cdots & 0 \\ 0 & \mathbf{J}_1 & \cdots & 0 \\ \cdots & \cdots & \cdots & \cdots \\ 0 & 0 & \cdots & \mathbf{J}_s \end{pmatrix} \qquad \text{with } \mathbf{J}_t = \begin{pmatrix} \lambda_t & 1 & 0 & \cdots & 0 \\ 0 & \lambda_t & 1 & \cdots & 0 \\ \vdots & \cdots & \ddots & \ddots & \vdots \\ 0 & 0 & \cdots & \lambda_t & 1 \\ 0 & 0 & 0 & \cdots & \lambda_t \end{pmatrix},$$

where 1 is the unique maximal eigenvalue of the matrix $\mathbf{H}$. Denote the order of $\mathbf{J}_t$ by ν_t and $\tau = \max\{\mathrm{Re}(\lambda_1), \dots, \mathrm{Re}(\lambda_s)\}$. We define $\nu = \max\{\nu_t : \mathrm{Re}(\lambda_t) = \tau\}$.

Moreover, the irreducibility of $\mathbf{H}$ also guarantees that the elements of the left eigenvector $\mathbf{v} = (v_1, \dots, v_p)$ associated with the positive maximal eigenvalue 1 are positive. Thus, we may normalize this vector to satisfy $\sum_{i=1}^{p} v_i = 1$.

REMARK 2.2. Condition (1.3) in Assumption 2.3 is very mild, just slightly stronger than $\alpha_i \to 0$, for example, if the nonhomogeneous generating matrix $\mathbf{H}_i$ converges to a generating matrix $\mathbf{H}$ with a rate of $\log^{-1-c} i$ for some $c > 0$.

What we consider here is the general case where the Jordan form of the generating matrix $\mathbf{H}$ is arbitrary, relaxing the constraint of a diagonal Jordan form as usually assumed in the literature [see Smythe (1996)].

In some conclusions, we need the convergence rate of $\mathbf{H}_i$ as described in the following assumption.

ASSUMPTION 2.4.

$$\|\mathbf{H}_i - E\mathbf{H}_i\| = \begin{cases} O(i^{-1/2}), & \text{if } \tau \neq \frac{1}{2}, \\ O(i^{-1/2}\log^{-1/2}(i+1)), & \text{if } \tau = \frac{1}{2}, \end{cases} \tag{2.10}$$

where $\|(a_{ij})\| = \sqrt{\sum_{ij} E a_{ij}^2}$, for any random matrix (a_{ij}).

A slightly stronger condition is

$$\|\mathbf{H}_i - E\mathbf{H}_i\| = o(i^{-1/2}). \tag{2.11}$$

REMARK 2.3. This assumption is trivially true if $\mathbf{H}_i$ is nonrandom. It is also true when $\mathbf{H}_i$ is a continuously differentiable matrix function of status at stage i, such as $\mathbf{Y}_i$, $\mathbf{N}_i$ or the relative frequencies of the success, and so on. These are true in almost all practical situations.

For further studies, we define

$$V_n = \begin{cases} \sqrt{n}, & \text{if } \tau < 1/2, \\ \sqrt{n}\log^{\nu-1/2} n, & \text{if } \tau = 1/2, \\ n^{\tau}\log^{\nu-1} n, & \text{if } \tau > 1/2. \end{cases}$$

THEOREM 2.2. *Under Assumptions* 2.1–2.3, *for some constant* M,

$$E\|\mathbf{Y}_n - E\mathbf{Y}_n\|^2 \le MV_n^2. \tag{2.12}$$

From this, for any $\kappa > \tau \vee \frac{1}{2}$, *we immediately obtain* $n^{-\kappa}(\mathbf{Y}_n - E\mathbf{Y}_n) \to 0$, *a.s., where* $a \vee b = \max(a, b)$. *Also, if* $\kappa = 1$ *or the condition* (1.3) *is strengthened to*

$$\sum_{i=1}^{\infty} \frac{\alpha_i}{\sqrt{i}} < \infty, \tag{2.13}$$

then $E\mathbf{Y}_n$ *in the above conclusions can be replaced by* $n\mathbf{v}$. *This implies that* $n^{-1}\mathbf{Y}_n$ *almost surely converges to* $\mathbf{v}$, *the same limit of* $n^{-1}E\mathbf{Y}_n$, *as* $n \to \infty$.

PROOF. Without loss of generality, we assume $a_0 = 1$ in the following study. For any random vector, write $\|\mathbf{Y}\| := \sqrt{E\mathbf{Y}\mathbf{Y}'}$. Define $\mathbf{y}_n = (y_{n,1}, \dots, y_{n,K}) = \mathbf{Y}_n\mathbf{T}$. Then, (2.12) reduces to

$$\|\mathbf{y}_n - E\mathbf{y}_n\| \le MV_n. \tag{2.14}$$

In Theorem 2.1, we have proved that $\|a_n - n\|^2 \le CK^2n$ [see (2.9) and (2.8)]. Noticing that $Ea_n = n + 1$, the proof of (2.12) further reduces to showing that, for any $j = 2, \dots, K$,

$$\|y_{n,j} - Ey_{n,j}\| \le MV_n. \tag{2.15}$$

We shall prove (2.15) by induction.

Suppose n_0 is an integer and M a constant such that

$$\begin{gathered} \sum_{i=n_0}^{\infty} \frac{\alpha_i}{i} < \varepsilon, \qquad \sum_{i=n_0}^{\infty} \frac{\log^{\nu} i}{i^{5/4}} < \varepsilon, \\ M = \frac{C_1 + C_2 + C_3 + C_4 + C_5 + (C_3 + 2C_5)M_0}{1 - 3\varepsilon}, \end{gathered} \tag{2.16}$$

where $\varepsilon < 1/4$ is a prechosen small positive number, $M_0 = \max_{n \le n_0}\{\|y_{n,j} - Ey_{n,j}\|/V_n\}$ and the constants C's are absolute constants specified later.

We shall complete the proof by induction. Consider $m > n_0$ and assume that $\|\tilde{\mathbf{y}} - E\tilde{\mathbf{y}}_n\| \le MV_n$ for all $n_0 \le n < m$.

By (2.5) and (2.6), we have

$$
(2.17)\quad
\begin{aligned}
y_{m,j} &= \mathbf{y}_0\widetilde{\mathbf{B}}_{m,0,j} + \sum_{i=1}^{m}\widetilde{\mathbf{Q}}_i\widetilde{\mathbf{B}}_{m,i,j} + \sum_{i=1}^{m}\frac{\mathbf{y}_{i-1}}{i}\mathbf{W}_i\widetilde{\mathbf{B}}_{m,i,j}\\
&\quad + \sum_{i=1}^{m}\left(\frac{i-a_{i-1}}{i}\right)\frac{\mathbf{y}_{i-1}}{a_{i-1}}\mathbf{J}\widetilde{\mathbf{B}}_{m,i,j} + \sum_{i=1}^{m}\left(\frac{i-a_{i-1}}{i}\right)\frac{\mathbf{y}_{i-1}}{a_{i-1}}\mathbf{W}_i\widetilde{\mathbf{B}}_{m,i,j},
\end{aligned}
$$

where $\widetilde{\mathbf{Q}}_i = \mathbf{Q}_i\mathbf{T}$, $\mathbf{W}_i = \mathbf{T}^{-1}(\mathbf{H}_i - \mathbf{H})\mathbf{T}$ and

$$
(2.18)\quad
\begin{aligned}
\widetilde{\mathbf{B}}_{n,i} &= \mathbf{T}^{-1}\mathbf{B}_{n,i}\mathbf{T} = (\mathbf{I} + (i+1)^{-1}\mathbf{J})\cdots(\mathbf{I} + n^{-1}\mathbf{J})\\
&= \begin{pmatrix}
\prod_{j=i+1}^{n}\left(1+\frac{1}{j}\right) & 0 & \cdots & 0\\
0 & \prod_{j=i+1}^{n}(\mathbf{I}+j^{-1}\mathbf{J}_1) & \cdots & 0\\
\cdots & \cdots & \cdots & \cdots\\
0 & 0 & \cdots & \prod_{j=i+1}^{n}(\mathbf{I}+j^{-1}\mathbf{J}_s)
\end{pmatrix},
\end{aligned}
$$

and $\widetilde{\mathbf{B}}_{m,i,j}$ is the jth column of the matrix $\widetilde{\mathbf{B}}_{m,i}$.

In the remainder of the proof of the theorem, we shall frequently use the elementary fact that

$$
(2.19)\quad \prod_{j=i+1}^{n}\left(1+\frac{\lambda}{j}\right) = \left(\frac{n}{i}\right)^{\lambda}\psi(n,i,\lambda),
$$

where $\psi(n,i,\lambda)$ is uniformly bounded (say $\le \psi$) and tends to 1 as $i\to\infty$. In the sequel, we use $\psi(n,i,\lambda)$ as a generic symbol, that is, it may take different values at different appearances and is uniformly bounded (by ψ, say) and tends to 1 as $i\to\infty$. Based on this estimation, one finds that the $(h,h+\ell)$-element of the block matrix $\prod_{i=j+2}^{n}(\mathbf{I}+i^{-1}\mathbf{J}_t)$ is asymptotically equivalent to

$$
(2.20)\quad \frac{1}{\ell!}\left(\frac{j}{n}\right)^{-\lambda_t}\log^{\ell}\left(\frac{n}{j}\right)\psi(n,j,\lambda_t),
$$

where λ_t is the eigenvalue of $\mathbf{J}_t$.

By (2.17) and triangular inequality, we have

$$
\begin{aligned}
&\|y_{m,j} - Ey_{m,j}\| \\
&\quad \le \|\mathbf{y}_0\widetilde{\mathbf{B}}_{m,0,j}\| + \left\|\sum_{i=1}^{m} \widetilde{\mathbf{Q}}_i \widetilde{\mathbf{B}}_{m,i,j}\right\| \\
&\qquad + \left\|\sum_{i=1}^{m}\left(\frac{i-a_{i-1}}{i}\right)\frac{\mathbf{y}_{i-1}}{a_{i-1}}\mathbf{J}\widetilde{\mathbf{B}}_{m,i,j}\right\| + \sum_{i=1}^{m}\left\|\left(\frac{i-a_{i-1}}{i}\right)\frac{\mathbf{y}_{i-1}}{a_{i-1}}\mathbf{W}_i\widetilde{\mathbf{B}}_{m,i,j}\right\| \\
&\qquad + \sum_{i=1}^{m}\left\|\frac{\mathbf{y}_{i-1}\mathbf{W}_i - E\mathbf{y}_{i-1}\mathbf{W}_i}{i}\widetilde{\mathbf{B}}_{m,i,j}\right\|.
\end{aligned} \tag{2.21}
$$

Consider the case where $1 + \nu_1 + \cdots + \nu_{t-1} < j \le 1 + \nu_1 + \cdots + \nu_t$. Then, by (2.20) we have

$$
\|\mathbf{y}_0\widetilde{\mathbf{B}}_{m,0,j}\| \le C_1 |m^{\lambda_t}| \log^{\nu_t - 1} m \le C_1 V_m. \tag{2.22}
$$

Since the elements of $E(\widetilde{\mathbf{Q}}_i^*\widetilde{\mathbf{Q}}_i)$ are bounded, we have

$$
\begin{aligned}
\left\|\sum_{i=1}^{m}\widetilde{\mathbf{Q}}_i\widetilde{\mathbf{B}}_{m,i,j}\right\| &= \left\{\sum_{i=1}^{m}\widetilde{\mathbf{B}}_{m,i,j}^* E(\widetilde{\mathbf{Q}}_i^*\widetilde{\mathbf{Q}}_i)\widetilde{\mathbf{B}}_{m,i,j}\right\}^{1/2} \\
&\le C_2\left\{\sum_{i=1}^{m}(m/i)^{2\operatorname{Re}(\lambda_t)}\log^{2\nu_t-2}(m/i)\right\}^{1/2} \\
&\le C_2 V_m,
\end{aligned} \tag{2.23}
$$

for all m and some constant C_2.

Noticing that $a_{i-1}^{-1}\|\mathbf{y}_{i-1}\|$ is bounded, for $\tau \ne \frac{1}{2}$, we have

$$
\begin{aligned}
&\left\|\sum_{i=1}^{m}\left(\frac{i-a_{i-1}}{i}\right)\frac{\mathbf{y}_{i-1}}{a_{i-1}}\mathbf{J}\widetilde{\mathbf{B}}_{m,i,j}\right\| \\
&\quad \le \sum_{i=1}^{m}\left\|\left(\frac{i-a_{i-1}}{i}\right)\frac{\mathbf{y}_{i-1}}{a_{i-1}}\mathbf{J}\widetilde{\mathbf{B}}_{m,i,j}\right\| \\
&\quad \le \sum_{i=1}^{m} C_3 i^{-1/2}(|\lambda_t|+1)\left(\frac{m}{i}\right)^{\operatorname{Re}(\lambda_t)}\log^{\nu_t-1}\left(\frac{m}{i}\right) \\
&\quad \le \frac{C_3}{\sqrt{m}}\sum_{i=1}^{m}\left(\frac{m}{i}\right)^{\tau+1/2}\log^{\nu_t-1}\left(\frac{m}{i}\right) \le C_3 V_m,
\end{aligned}
$$

for all m and some constant C_3.

Now we estimate this term for the case $\tau = \frac{1}{2}$. We have

$$\left\| \sum_{i=1}^{m} \left(\frac{i - a_{i-1}}{i} \right) \frac{\mathbf{y}_{i-1}}{a_{i-1}} \mathbf{J}\widetilde{\mathbf{B}}_{m,i,j} \right\|$$

$$\leq \left\| \sum_{i=1}^{m} \left(\frac{i - a_{i-1}}{i} \right) E \frac{\mathbf{y}_{i-1}}{a_{i-1}} \mathbf{J}\widetilde{\mathbf{B}}_{m,i,j} \right\| + \sum_{i=1}^{m} \left\| \left(\frac{i - a_{i-1}}{i} \right) \left[\frac{\mathbf{y}_{i-1}}{a_{i-1}} - E \frac{\mathbf{y}_{i-1}}{a_{i-1}} \right] \mathbf{J}\widetilde{\mathbf{B}}_{m,i,j} \right\|.$$

First, we have

$$\left\| \sum_{i=1}^{m} \left(\frac{i - a_{i-1}}{i} \right) E \frac{\mathbf{y}_{i-1}}{a_{i-1}} \mathbf{J}\widetilde{\mathbf{B}}_{m,i,j} \right\|^2$$

$$= \sum_{ik} \left(\frac{E(i - a_{i-1})(k - a_{k-1})}{ik} \right) E \frac{\mathbf{y}_{i-1}}{a_{i-1}} \mathbf{J}\widetilde{\mathbf{B}}_{m,i,j} E \frac{\mathbf{y}_{k-1}}{a_{k-1}} \mathbf{J}\widetilde{\mathbf{B}}_{m,k,j}$$

$$\leq C_3^2 \sum_{i \leq k \leq m} \frac{\|i - a_{i-1}\|^2}{ik} \left(\frac{m}{i} \right)^{1/2} \log^{\nu_t - 1} \left(\frac{m}{i} \right) \left(\frac{m}{k} \right)^{1/2} \log^{\nu_t - 1} \left(\frac{m}{k} \right)$$

$$\leq C_3^2 m \log^{2\nu_t - 2} m \sum_{i \leq k \leq m} \frac{1}{k\sqrt{ik}}$$

$$\leq C_3^2 m \log^{2\nu_t - 1} m.$$

Here we point out a fact that for any $\mu > 1$, there is a constant $C_\mu > 0$ such that

$$E|a_n - n|^\mu \leq C_\mu n^{\mu/2}.$$

This inequality is an easy consequence of the Burkholder inequality. [The Burkholder inequality states that if $X_1, \ldots, X_n$ is a sequence of martingale differences, then for any $p > 1$, there is a constant $C = C(p)$ such that $E|\sum_{i=1}^{m} X_i|^p \leq C_p E(\sum_{i=1}^{n} E(|X_i^2||\mathcal{F}_{i-1})^{p/2}$.]

By using $\frac{1}{a_{i-1}} = \frac{1}{i} + \frac{i - a_{i-1}}{i a_{i-1}}$ and the above inequality, we have

$$\sum_{i=1}^{m} \left\| \left(\frac{i - a_{i-1}}{i} \right) \left[\frac{\mathbf{y}_{i-1}}{a_{i-1}} - E \frac{\mathbf{y}_{i-1}}{a_{i-1}} \right] \mathbf{J}\widetilde{\mathbf{B}}_{m,i,j} \right\|$$

$$\leq \sum_{i=1}^{m} \left\| \left(\frac{i - a_{i-1}}{i} \right) \left[\frac{\mathbf{y}_{i-1} - E\mathbf{y}_{i-1}}{i} \right] \mathbf{J}\widetilde{\mathbf{B}}_{m,i,j} \right\| + \sum_{i=1}^{m} \frac{\|(i - a_{i-1})^2\|}{i^2} \|\mathbf{J}\widetilde{\mathbf{B}}_{m,i,j}\|$$

$$\leq C_3 \sum_{i=1}^{m} \left[\frac{\|\mathbf{y}_{i-1} - E\mathbf{y}_{i-1}\|}{i^{5/4}} + P(|a_{i-1} - i| \geq i^{3/4}) + \frac{1}{i} \right] \|\mathbf{J}\widetilde{\mathbf{B}}_{m,i,j}\|$$

$$\begin{aligned}
&\le C_3 M_0 \sum_{i=1}^{n_0} \frac{V_i}{i^{5/4}} \left(\frac{m}{i}\right)^{1/2} \log^{\nu_t - 1}\left(\frac{m}{i}\right) \\
&\quad + C_3 M \sum_{i=n_0+1}^{m} \frac{V_i}{i^{5/4}} \left(\frac{m}{i}\right)^{1/2} \log^{\nu_t - 1}\left(\frac{m}{i}\right) \\
&\quad + C_3 \sum_{i=1}^{m} i^{-1} \left(\frac{m}{i}\right)^{1/2} \log^{\nu_t - 1}\left(\frac{m}{i}\right) \\
&\le (C_3 M_0 + C_3 + \varepsilon M)\sqrt{m} \log^{\nu - 1} m.
\end{aligned}$$

Combining the above four inequalities, we have proved

$$\left\| \sum_{i=1}^{m} \left(\frac{i - a_{i-1}}{i}\right) \frac{\mathbf{y}_{i-1}}{a_{i-1}} \mathbf{J}\widetilde{\mathbf{B}}_{m,i,j} \right\| \le (C_3 M_0 + C_3 + \varepsilon M) V_m. \tag{2.24}$$

By (1.3) and the fact that $a_{i-1}^{-1}\|\mathbf{y}_{i-1}\|$ is bounded, we have

$$\begin{aligned}
&\sum_{i=1}^{m} \left\| \left(\frac{i - a_{i-1}}{i}\right) \frac{\mathbf{y}_{i-1}}{a_{i-1}} \mathbf{W}_i \widetilde{\mathbf{B}}_{m,i,j} \right\| \\
&\qquad \le \sum_{i=1}^{m} C_4 i^{-1/2} \alpha_i \left(\frac{m}{i}\right)^{\mathrm{Re}(\lambda_t)} \log^{\nu_t - 1}\left(\frac{m}{i}\right) \\
&\qquad \le C_4 \sqrt{m} \sum_{i=1}^{m} \frac{\alpha_i}{i} \left(\frac{m}{i}\right)^{\tau - 1/2} \log^{\nu_t - 1}\left(\frac{m}{i}\right) \\
&\qquad \le C_4 V_m.
\end{aligned} \tag{2.25}$$

Next, we show that

$$\begin{aligned}
&\sum_{i=1}^{m} \left\| \frac{\mathbf{y}_{i-1}\mathbf{W}_i - E\mathbf{y}_{i-1}\mathbf{W}_i}{i} \widetilde{\mathbf{B}}_{m,i,j} \right\| \\
&\qquad \le \sum_{i=1}^{m} \left\| \frac{\mathbf{y}_{i-1} - E\mathbf{y}_{i-1}}{i} \mathbf{W}_i \widetilde{\mathbf{B}}_{m,i,j} \right\| \\
&\qquad\quad + \sum_{i=1}^{m} \left\| \frac{E(\mathbf{y}_{i-1} - E\mathbf{y}_{i-1})\mathbf{W}_i}{i} \widetilde{\mathbf{B}}_{m,i,j} \right\| \\
&\qquad\quad + \sum_{i=1}^{m} \left\| \frac{E\mathbf{y}_{i-1}}{i} (\mathbf{W}_i - E\mathbf{W}_i) \widetilde{\mathbf{B}}_{m,i,j} \right\| \\
&\qquad \le \big(2\varepsilon M + C_5(2M_0 + 1)\big) V_m.
\end{aligned} \tag{2.26}$$

By (1.3) and the induction assumption that $\|\mathbf{y}_{i-1} - E\mathbf{y}_{i-1}\| \le M\sqrt{i}$,

$$\begin{aligned}
&\sum_{i=1}^{m} \left\| \frac{\mathbf{y}_{i-1} - E\mathbf{y}_{i-1}}{i} \mathbf{W}_i \widetilde{\mathbf{B}}_{m,i,j} \right\| \\
&\qquad \le \sum_{i=1}^{n_0} M_0 V_i i^{-1} \alpha_i \left(\frac{m}{i}\right)^{\mathrm{Re}(\lambda_t)} \log^{\nu_t - 1}\left(\frac{m}{i}\right) \\
&\qquad\quad + \sum_{i=n_0+1}^{m} M V_i i^{-1} \alpha_i \left(\frac{m}{i}\right)^{\mathrm{Re}(\lambda_t)} \log^{\nu_t - 1}\left(\frac{m}{i}\right) \\
&\qquad \le (C_5 M_0 + \varepsilon M) V_m.
\end{aligned}$$

By Jensen's inequality, we have

$$\begin{aligned}
\sum_{i=1}^{m} \left\| E \frac{\mathbf{y}_{i-1} - E\mathbf{y}_{i-1}}{i} \mathbf{W}_i \widetilde{\mathbf{B}}_{m,i,j} \right\| &\le \sum_{i=1}^{m} \left\| \frac{\mathbf{y}_{i-1} - E\mathbf{y}_{i-1}}{i} \mathbf{W}_i \widetilde{\mathbf{B}}_{m,i,j} \right\| \\
&\le (C_5 M_0 + \varepsilon M) V_m.
\end{aligned}$$

The estimate of the third term is given by

$$\begin{aligned}
&\sum_{i=1}^{m} \left\| \frac{E\mathbf{y}_{i-1}}{i} (\mathbf{W}_i - E\mathbf{W}_i) \widetilde{\mathbf{B}}_{m,i,j} \right\| \\
&\quad \le C_5 \sum_{i=1}^{m} \|\mathbf{W}_i - E\mathbf{W}_i\| \left(\frac{m}{i}\right)^{\mathrm{Re}(\lambda_t)} \log^{\nu_t - 1}\left(\frac{m}{i}\right) \\
&\quad \le \begin{cases} C_5 \sum_{i=1}^{m} i^{-1/2} \left(\frac{m}{i}\right)^{\mathrm{Re}(\lambda_t)} \log^{\nu_t - 1}\left(\frac{m}{i}\right), & \text{if } \tau \ne \frac{1}{2}, \\ C_5 \sum_{i=1}^{m} i^{-1/2} \log^{-1/2}(i+1) \left(\frac{m}{i}\right)^{\mathrm{Re}(\lambda_t)} \log^{\nu_t - 1}\left(\frac{m}{i}\right), & \text{if } \tau = \frac{1}{2}, \end{cases} \qquad (2.27) \\
&\quad \le C_5 V_m.
\end{aligned}$$

The above three estimates prove the assertion (2.26).

Substituting (2.22)–(2.26) into (2.21), we obtain

$$\|y_{n,j} - Ey_{n,j}\| \le \big(3\varepsilon M + C_1 + C_2 + C_3 + C_4 + C_5 + (C_3 + 2C_5) M_0\big) V_m \le M V_m.$$

We complete the proof of (2.15) and thus of (2.12).

Since $\kappa > \tau \vee 1/2$, we may choose κ_1 such that $\kappa > \kappa_1 > \tau \vee 1/2$. By (2.12), we have

$$\|\mathbf{Y}_n - E\mathbf{Y}_n\|^2 \le M n^{2\kappa_1}.$$

From this and the standard procedure of subsequence method, one can show that

$$n^{-\kappa}(\mathbf{Y}_n - E\mathbf{Y}_n) \to 0 \qquad \text{a.s.}$$

To complete the proof of the theorem, it remains to show the replacement of $E\mathbf{Y}_n$ by $n\mathbf{v}$, that is, to show that $\|y_{n,j}\| \le MV_n$ if (2.13) holds and that $\|y_{n,j}\| = o(n)$ under (1.3). Here the latter is for the convergence with $\kappa = 1$.

Following the lines of the proof for the first conclusion, we need only to change $Ey_{m,j}$ on the left-hand side of (2.21) and replace $E\mathbf{y}_{i-1}\mathbf{W}_i$ on the right-hand side of (2.21) by 0. Checking the proofs of (2.22)–(2.26), we find that the proofs of (2.22)–(2.26) remain true. Therefore, we need only show that

$$\begin{aligned}
&\sum_{i=1}^{m}\left\|\frac{E\mathbf{y}_{i-1}\mathbf{W}_i}{i}\widetilde{\mathbf{B}}_{m,i,j}\right\| \\
&\quad\le \sum_{i=1}^{m}\alpha_i\left(\frac{m}{i}\right)^{\mathrm{Re}(\lambda_t)}\log^{\nu_t-1}\left(\frac{m}{i}\right) \\
&\quad= \begin{cases}\sqrt{m}\displaystyle\sum_{i=1}^{m}\frac{\alpha_i}{\sqrt{i}}\left(\frac{m}{i}\right)^{\tau-1/2}\log^{\nu_t-1}\left(\frac{m}{i}\right)\le O(V_m), & \text{if (2.13) holds,}\\ m\displaystyle\sum_{i=1}^{m}\frac{\alpha_i}{i}\left(\frac{m}{i}\right)^{\tau-1}\log^{\nu_t-1}\left(\frac{m}{i}\right)\le \varepsilon m, & \text{if (1.3) holds.}\end{cases}
\end{aligned}\tag{2.28}$$

This completes the proof of this theorem. □

Recall the proof of Theorem 2.2 and note that ε can be arbitrarily small; with a slight modification to the proof of Theorem 2.2, we have in fact the following corollary.

COROLLARY 2.1. *In addition to the conditions of Theorem* 2.2, *assume* (2.11) *is true. Then, we have*

$$\mathbf{y}_{n,-} - E\mathbf{y}_{n,-} = \sum_{i=1}^{n}\widetilde{\mathbf{Q}}_i\widetilde{\mathbf{B}}_{n,i,-} + o_p(V_n), \tag{2.29}$$

where $\mathbf{y}_{n,-} = (y_{n,2},\dots,y_{n,K})$ *and* $\widetilde{\mathbf{B}}_{n,i,-} = (\widetilde{\mathbf{B}}_{n,i,2},\dots,\widetilde{\mathbf{B}}_{n,i,K})$.

Furthermore, if (2.13) *is true,* $E\mathbf{y}_{n,-}$ *in* (2.29) *can be replaced by* 0.

PROOF. Checking the proof of Theorem 2.2, one finds that the term estimated in (2.22) is not necessary to appear on the right-hand side of (2.21). Thus, to prove (2.29), it suffices to improve the right-hand sides of (2.24)–(2.26) as to εV_m. The modification for (2.24) and (2.25) can be done without any further conditions, provided one notices that the vector $\mathbf{y}_{i-1}$ in these inequalities can be replaced by $(0, \mathbf{y}_{i-1,-})$. The details are omitted. To modify (2.26), we first note that (2.27) can be trivially modified to εV_m if the condition (2.10) is strengthened to (2.11). The other two estimates for proving (2.26) can be modified easily without any further assumptions. □

Note that

$$\mathbf{N}_n = \sum_{i=1}^{n} \mathbf{X}_i = \sum_{i=1}^{n} (\mathbf{X}_i - E(\mathbf{X}_i | \mathcal{F}_{i-1})) + \sum_{i=1}^{n} \frac{\mathbf{Y}_{i-1}}{a_{i-1}}.$$

Since $(\mathbf{X}_i - E(\mathbf{X}_i | \mathcal{F}_{i-1}))$ is a bounded martingale difference sequence, we have

$$n^{-\kappa} \sum_{i=1}^{n} (\mathbf{X}_i - E(\mathbf{X}_i | \mathcal{F}_{i-1})) \to 0 \qquad \text{a.s.}$$

for any $\kappa > 1/2$. Also,

$$n^{-\kappa} \left\| \sum_{i=1}^{n} \mathbf{Y}_{i-1} |i^{-1} - a_{i-1}^{-1}| \right\| \le n^{-\kappa} \sum_{i=1}^{n} i^{-1+\kappa_1} \frac{|a_{i-1} - i|}{i^{\kappa_1}} \to 0 \qquad \text{a.s.}$$

In view of these relations and Theorem 2.2, we have established the following theorem for the strong consistency of $\mathbf{N}_n$.

THEOREM 2.3. *Under the assumptions of Theorem* 2.2, $n^{-\kappa}(\mathbf{N}_n - E\mathbf{N}_n) \to 0$, *a.s. for any* $\kappa > \tau \vee 1/2$. *Also, in the above limit,* $E\mathbf{N}_n$ *can be replaced by* $n\mathbf{v}$ *if* $\kappa = 1$ *or* (2.13) *is true. This implies that* $n^{-1}\mathbf{N}_n$ *almost surely converges to* $\mathbf{v}$, *the same limit of* $n^{-1} E\mathbf{N}_n$, *as* $n \to \infty$.

3. Asymptotic normality of $\mathbf{Y}_n$. In the investigation of the asymptotic normality of the urn composition, we first consider that of a_n, the total number of balls in the urn after n stages.

THEOREM 3.1. *Under Assumptions* 2.1–2.3, $n^{-1/2}(a_n - n)$ *is asymptotically normal with mean* 0 *and variance* σ_{11}, *where* $\sigma_{11} = \sum_{q=1}^{K} \sum_{k=1}^{K} \sum_{l=1}^{K} v_q d_{qkl}$.

PROOF. From Theorems 2.1 and 2.2, we have that $\mathbf{Y}_n / a_n \to \mathbf{v}$ a.s. Similar to (2.8), we have

$$n^{-1} \sum_{i=1}^{n} \operatorname{var}(e_i | \mathcal{F}_{i-1}) \to \sum_{q=1}^{K} \sum_{k=1}^{K} \sum_{l=1}^{K} v_q d_{qkl} \qquad \text{a.s.}$$

Assumption 2.2 implies that $\{e_i - E(e_i | \mathcal{F}_{i-1})\}$ satisfies the Lyapunov condition. From the martingale CLT [see Hall and Heyde (1980)], Assumptions 2.1–2.3 and the fact that

$$a_n - n = 1 + \sum_{i=1}^{n} (e_i - E(e_i | \mathcal{F}_{i-1})),$$

the theorem follows. □

THEOREM 3.2. *Under the assumptions of Theorem* 2.2, $V_n^{-1}(\mathbf{Y}_n - E\mathbf{Y}_n)$ *is asymptotically normal with mean vector* 0 *and variance–covariance matrix* $\mathbf{T}^{-1*}\Sigma\mathbf{T}^{-1}$, *where* Σ *is specified later,* $V_n^2 = n$ *if* $\tau < 1/2$ *and* $V_n^2 = n\log^{2\nu-1} n$ *if* $\tau = 1/2$. *Here* τ *is defined in Assumption* 2.3.

Also, if (2.13) *holds, then* $E\mathbf{Y}_n$ *can be replaced by* $n\mathbf{v}$.

PROOF. To show the asymptotic normality of $\mathbf{Y}_n - E\mathbf{Y}_n$, we only need to show that of $(\mathbf{Y}_n - E\mathbf{Y}_n)\mathbf{T} = \mathbf{y}_n - E\mathbf{y}_n$.

From the proof of Theorem 3.1, we have

$$y_{n,1} - Ey_{n,1} = a_n - n - 1 = \sum_{i=1}^{n}(e_i - E(e_i|\mathcal{F}_{i-1})).$$

From Corollary 2.1, we have

$$\mathbf{y}_{n,-} - E\mathbf{y}_{n,-} = \sum_{i=1}^{n} \widetilde{Q}_i\widetilde{\mathbf{B}}_{n,i,-} + o_p(V_n).$$

Combining the above estimates, we get

$$\mathbf{y}_n - E\mathbf{y}_n = \begin{pmatrix} \sum_{i=1}^{n}(e_i - E(e_2|\mathcal{F}_{i-1})) \\ \sum_{i=1}^{n-1}\widetilde{\mathbf{Q}}_i\widetilde{\mathbf{B}}_{n,i,2} \\ \cdots\cdots\cdots \\ \sum_{i=1}^{n-1}\widetilde{\mathbf{Q}}_i\widetilde{\mathbf{B}}_{n,i,K} \end{pmatrix} + o(V_n). \tag{3.1}$$

Again, Assumption 2.2 implies the Lyapunov condition. Using the CLT for martingale sequence, as was done in the proof of Theorem 2.3 of Bai and Hu (1999), from (3.1), one can easily show that $V_n^{-1}(\mathbf{y}_n - E\mathbf{y}_n)$ tends to a K-variate normal distribution with mean 0 and variance–covariance matrix $\begin{pmatrix}\sigma_{11} & \Sigma_{12}\\ \Sigma_{12}' & \Sigma_{22}\end{pmatrix}$. The variance–covariance matrix Σ_{22} of the second to the Kth elements of $V_n^{-1}(\mathbf{y}_n - E\mathbf{y}_n)$ can be found in (2.17) of Bai and Hu (1999).

By Theorem 3.1, for the case $\tau = 1/2$, $V_n = \sqrt{n}\log^{\nu-1/2} n$, $\sigma_{11} = 0$ and $\Sigma_{12} = 0$. When $\tau < 1/2$, $V_n = \sqrt{n}$, $\sigma_{11} = \sum_{q=1}^{K}\sum_{k=1}^{K}\sum_{l=1}^{K} v_q d_{qkl}$. Now, let us find Σ_{12}. Write $\mathbf{T} = (\mathbf{1}', \mathbf{T}_1, \ldots, \mathbf{T}_s) = (\mathbf{1}', \mathbf{T}_-)$, $\mathbf{T}_j = (\mathbf{t}_{j1}', \ldots, \mathbf{t}_{j\nu_j}')$ and $\widetilde{\mathbf{B}}_{n,i,-} = \mathbf{T}^{-1}\mathbf{B}_{n,i}\mathbf{T}_- = (\widetilde{\mathbf{B}}_{n,i,2}, \ldots, \widetilde{\mathbf{B}}_{n,i,K})$, where $\mathbf{1} = (1, \ldots, 1)$ throughout this paper. Then the vector Σ_{12} is the limit of

$$
\begin{aligned}
& n^{-1}\sum_{i=1}^{n}\operatorname{cov}[(e_i,\mathbf{Q}_i)|\mathcal{F}_{i-1}]\mathbf{T}\widetilde{\mathbf{B}}_{n,i,-} \\
&\quad = n^{-1}\sum_{i=1}^{n}\mathbf{1}\operatorname{cov}[(\mathbf{D}_i'\mathbf{X}_i',\mathbf{X}_i\mathbf{D}_i)|\mathcal{F}_{i-1}]\mathbf{T}\widetilde{\mathbf{B}}_{n,i,-} \\
&\quad = n^{-1}\sum_{i=1}^{n}\mathbf{1}\left(\sum_{q=1}^{K}v_q\mathbf{d}_q+\mathbf{H}^*(\operatorname{diag}(\mathbf{v})-\mathbf{v}^*\mathbf{v})\mathbf{H}\right)\mathbf{T}\widetilde{\mathbf{B}}_{n,i,-}+o_p(1) \qquad (3.2)\\
&\quad = \mathbf{1}\left(\sum_{q=1}^{K}v_q\mathbf{d}_q+\mathbf{H}^*(\operatorname{diag}(\mathbf{v})-\mathbf{v}^*\mathbf{v})\mathbf{H}\right)\mathbf{T}n^{-1}\sum_{i=1}^{n}\widetilde{\mathbf{B}}_{n,i,-}+o_p(1) \\
&\quad = \mathbf{1}\left(\sum_{q=1}^{K}v_q\mathbf{d}_q\right)\mathbf{T}n^{-1}\sum_{i=1}^{n}\widetilde{\mathbf{B}}_{n,i,-}+o_p(1),
\end{aligned}
$$

where the matrices $\mathbf{d}_q$ are defined in (2.4). Here we have used the fact that $\mathbf{1}\mathbf{H}^*(\operatorname{diag}(\mathbf{v})-\mathbf{v}^*\mathbf{v})=\mathbf{1}(\operatorname{diag}(\mathbf{v})-\mathbf{v}^*\mathbf{v})=0$.

By elementary calculation and the definition of $\widetilde{\mathbf{B}}_{n,i,-}$, we get

$$
\begin{aligned}
& n^{-1}\sum_{i=1}^{n}\widetilde{\mathbf{B}}^*_{n,i,-} \\
&= \begin{pmatrix}
0 & \cdots & 0 \\
n^{-1}\sum_{i=1}^{n}\prod_{j=i+1}^{n}(\mathbf{I}+j^{-1}\mathbf{J}_1) & \cdots & 0 \\
\vdots & \cdots & \vdots \\
0 & \cdots & n^{-1}\sum_{i=1}^{n}\prod_{j=i+1}^{n}(\mathbf{I}+j^{-1}\mathbf{J}_s)
\end{pmatrix}. \qquad (3.3)
\end{aligned}
$$

In the hth block of the quasi-diagonal matrix

$$
n^{-1}\sum_{i=1}^{n}\prod_{j=i+1}^{n}(\mathbf{I}+j^{-1}\mathbf{J}_h),
$$

the $(g,g+\ell)$-element $(0\le\ell\le\nu_h-1)$ has the approximation

$$
n^{-1}\sum_{i=1}^{n}\frac{1}{\ell!}\left(\frac{n}{i}\right)^{\lambda_h}\log^{\ell}\left(\frac{n}{i}\right)(1+o(1))\to\left(\frac{1}{1-\lambda_h}\right)^{\ell+1}. \qquad (3.4)
$$

Combining (3.2)–(3.4), we get an expression of Σ_{12}.

Therefore, $n^{-1/2}(\mathbf{y}_n - E\mathbf{y}_n)$ has an asymptotically joint normal distribution with mean 0 and variance–covariance matrix Σ. Thus, we have shown that

$$n^{-1/2}(\mathbf{Y}_n - E\mathbf{Y}_n) \to N(0, (\mathbf{T}^{-1})^* \Sigma \mathbf{T}^{-1})$$

in distribution.

When (2.13) holds, $\mathbf{y}_n - n\mathbf{e}_1$ has the same approximation of the right-hand side of (3.1). Therefore, in the CLT, $E\mathbf{Y}_n$ can be replaced by $n\mathbf{v}$. Then, we complete the proof of the theorem. □

EXAMPLE 3.1. Consider the most important case in application, where $\mathbf{H}$ has a diagonal Jordan form and $\tau < 1/2$. We have

$$\mathbf{T}^{-1}\mathbf{H}\mathbf{T} = \mathbf{J} = \begin{pmatrix} 1 & 0 & \cdots & 0 \\ 0 & \lambda_1 & \cdots & 0 \\ \cdots & \cdots & \cdots & \cdots \\ 0 & 0 & \cdots & \lambda_{K-1} \end{pmatrix},$$

where $\mathbf{T} = (\mathbf{1}', \mathbf{t}_1', \dots, \mathbf{t}_{K-1}')$. Now let

$$\mathbf{R} = \sum_{j=1}^{K} v_j \mathbf{d}_j + \mathbf{H}^*(\operatorname{diag}(\mathbf{v}) - \mathbf{v}^*\mathbf{v})\mathbf{H}.$$

The variance–covariance matrix $\Sigma = (\sigma_{ij})_{i,j=1}^K$ has the following simple form: $\sigma_{11} = \mathbf{1R1}' = \sum_{q=1}^K \sum_{k=1}^K \sum_{l=1}^K v_q d_{qkl}$, $\sigma_{1j} = (1 - \lambda_{j-1})^{-1} \mathbf{1R}\mathbf{t}_{j-1}' = (1 - \lambda_{j-1})^{-1} \sum_{k=1}^K v_k \mathbf{1d}_k \mathbf{t}_{j-1}'$, $j = 2, \dots, K$, and

$$\sigma_{ij} = (1 - \lambda_{i-1} - \bar{\lambda}_{j-1})^{-1} (\mathbf{t}_{i-1}^*)' \mathbf{R} \mathbf{t}_{j-1}'.$$

4. Asymptotic normality of $\mathbf{N}_n$. Now, $\mathbf{N}_n = (N_{n1}, \dots, N_{nK})$, where N_{nk} is the number of times a type-k ball is drawn in the first n draws:

$$\mathbf{N}_n = (N_{n1}, \dots, N_{nK}) = \mathbf{N}_{n-1} + \mathbf{X}_n = \sum_{i=1}^n \mathbf{X}_i,$$

where the vectors $\mathbf{X}_i$ are defined as follows: If a type-k ball is drawn in the ith stage, then define the draw outcome $\mathbf{X}_i$ as the vector whose kth component is 1 and all others are 0. Therefore $\mathbf{1X}_i' = 1$ and $\mathbf{1N}_n' = n$. We shall consider the limiting property of $\mathbf{N}_n$.

THEOREM 4.1 (for the EPU urn). *Under the assumptions of Corollary* 2.1, $V_n^{-1}(\mathbf{N}_n - E\mathbf{N}_n)$ *is asymptotically normal with mean vector* 0 *and variance–covariance matrix* $\mathbf{T}^{-1*}\tilde{\Sigma}\mathbf{T}^{-1}$, *where* $\tilde{\Sigma}$ *is specified later,* $V_n^2 = n$ *if* $\tau < 1/2$ *and* $V_n^2 = n \log^{2\nu-1} n$ *if* $\tau = 1/2$. *Here* τ *is defined in Assumption* 2.3.

Furthermore, if (2.13) *holds, then* $E\mathbf{N}_n$ *can be replaced by* $n\mathbf{v}$.

PROOF. At first we have

$$\mathbf{N}_n = \sum_{i=1}^{n}(\mathbf{X}_i - E(\mathbf{X}_i|\mathcal{F}_{i-1})) + \sum_{i=1}^{n} E(\mathbf{X}_i|\mathcal{F}_{i-1})$$
$$= \sum_{i=1}^{n}(\mathbf{X}_i - \mathbf{Y}_{i-1}/a_{i-1}) + \sum_{i=0}^{n-1}\mathbf{Y}_i/a_i. \tag{4.1}$$

For simplicity, we consider the asymptotic distribution of $\mathbf{N}_n\mathbf{T}$. Since the first component of $\mathbf{N}_n\mathbf{T}$ is a nonrandom constant n, we only need consider the other $K-1$ components. From (2.29) and (4.1), we get

$$\begin{aligned}
\mathbf{N}_n\mathbf{T}_- &= \sum_{i=1}^{n}(\mathbf{X}_i - \mathbf{Y}_{i-1}/a_{i-1})\mathbf{T}_- + \sum_{i=0}^{n-1}\mathbf{y}_{i,-}/a_i \\
&= \sum_{i=1}^{n}(\mathbf{X}_i - \mathbf{Y}_{i-1}/a_{i-1})\mathbf{T}_- \\
&\quad + \sum_{i=0}^{n-1}\mathbf{y}_{i,-}/(i+1) + \sum_{i=0}^{n-1}\frac{\mathbf{y}_{i,-}}{a_i}\left(\frac{i+1-a_i}{i+1}\right) \\
&= \sum_{i=1}^{n}(\mathbf{X}_i - \mathbf{Y}_{i-1}/a_{i-1})\mathbf{T}_- + \sum_{i=0}^{n-1}\frac{\mathbf{y}_{i,-}}{a_i}\left(\frac{i+1-a_i}{i+1}\right) + \mathbf{y}_{0,-} \\
&\quad + \sum_{i=1}^{n-1}\frac{1}{i+1}\left[\sum_{j=1}^{i}\widetilde{\mathbf{Q}}_j\widetilde{\mathbf{B}}_{i,j,-} + E\mathbf{y}_{i,-} + o_p(V_i)\right] \\
&= \sum_{i=1}^{n}(\mathbf{X}_i - \mathbf{Y}_{i-1}/a_{i-1})\mathbf{T}_- \\
&\quad + \sum_{j=1}^{n-1}\widetilde{\mathbf{Q}}_j\left(\sum_{i=j}^{n-1}\frac{1}{i+1}\widetilde{\mathbf{B}}_{i,j,-}\right) + \sum_{i=1}^{n-1}\frac{E\mathbf{y}_{i,-}}{i+1} + o(V_n) \\
&= \sum_{i=1}^{n}(\mathbf{X}_i - \mathbf{Y}_{i-1}/a_{i-1})\mathbf{T} + \sum_{j=1}^{n-1}\widetilde{\mathbf{Q}}_j\widehat{\mathbf{B}}_{n,j,-} + \sum_{i=1}^{n-1}\frac{E\mathbf{y}_{i,-}}{i+1} + o_p(V_n),
\end{aligned} \tag{4.2}$$

where $\widetilde{\mathbf{B}}_{i,j} = \mathbf{T}^{-1}\mathbf{G}_{j+1}\cdots\mathbf{G}_i\mathbf{T}$, $\widehat{\mathbf{B}}_{n,j} = \sum_{i=j}^{n-1}\frac{1}{i+1}\widetilde{\mathbf{B}}_{i,j}$, the matrices with a minus sign in subscript denote the submatrices of the last $K-1$ columns of their corresponding mother matrices. Here, in the fourth equality, we have used the fact that $\sum_{i=0}^{n-1}\frac{\mathbf{y}_{i,-}}{a_i}(\frac{i+1-a_i}{i+1}) = o_p(\sqrt{n})$ which can be proven by the same approach as showing (2.24) and (2.28).

In view of (4.2), we only have to consider the asymptotic distribution of the martingale

$$\mathbf{U}=\sum_{i=1}^{n}(\mathbf{X}_i-\mathbf{Y}_{i-1}/a_{i-1})\mathbf{T}_- + \sum_{j=1}^{n-1}\widetilde{\mathbf{Q}}_j\widehat{\mathbf{B}}_{n,j,-}.$$

We now estimate the asymptotic variance–covariance matrix of $V_n^{-1}\mathbf{U}$. For this end, we need only consider the limit of

$$\begin{aligned}\widetilde{\Sigma}_n = V_n^{-2}\Bigg[&\sum_{j=1}^{n}E(\mathbf{q}_j^*\mathbf{q}_j|\mathcal{F}_{j-1})+\sum_{j=1}^{n-1}E(\mathbf{q}_j^*\widetilde{\mathbf{Q}}_j\widehat{\mathbf{B}}_{n,j,-}|\mathcal{F}_{j-1})\\ &+\sum_{j=1}^{n-1}E(\widehat{\mathbf{B}}_{n,j,-}^*\widetilde{\mathbf{Q}}_j^*\mathbf{q}_j|\mathcal{F}_{j-1})+\sum_{j=1}^{n-1}E(\widehat{\mathbf{B}}_{n,j,-}^*\widetilde{\mathbf{R}}_j\widehat{\mathbf{B}}_{n,j,-}|\mathcal{F}_{j-1})\Bigg],\end{aligned}\tag{4.3}$$

where $\mathbf{q}_j=(\mathbf{X}_j-\mathbf{Y}_{j-1}/a_{j-1})\mathbf{T}_-$ and $\widetilde{\mathbf{R}}_j=E(\widetilde{\mathbf{Q}}_j^*\widetilde{\mathbf{Q}}_j|\mathcal{F}_{j-1})=\mathbf{T}^*\mathbf{R}_j\mathbf{T}$.

From Theorem 3.1, we know that

$$E(\mathbf{q}_j^*\mathbf{q}_j|\mathcal{F}_{j-1})\to\mathbf{T}_-^*(\operatorname{diag}(\mathbf{v})-\mathbf{v}^*\mathbf{v})\mathbf{T}_-=\mathbf{T}_-^*\operatorname{diag}(\mathbf{v})\mathbf{T}_- \qquad \text{as } j\to\infty,$$

since $\mathbf{v}\mathbf{T}_-=0$. This estimate implies that

$$\begin{aligned}&V_n^{-2}\sum_{j=1}^{n}E(\mathbf{q}_j^*\mathbf{q}_j|\mathcal{F}_{j-1})\\ &\quad\to\widetilde{\Sigma}_1=\begin{cases}\mathbf{T}_-^*\operatorname{diag}(\mathbf{v})\mathbf{T}_-, & \text{if } \tau<1/2,\\ 0, & \text{if } \tau=1/2,\end{cases}\quad \text{as } j\to\infty.\end{aligned}\tag{4.4}$$

Because $\widetilde{\mathbf{Q}}_j=[\mathbf{X}_j\mathbf{D}_j-(\mathbf{Y}_{j-1}/a_{j-1})\mathbf{H}_j]\mathbf{T}$,

$$\begin{aligned}&V_n^{-2}\sum_{j=1}^{n-1}E(\mathbf{q}_j^*\widetilde{\mathbf{Q}}_j\widehat{\mathbf{B}}_{n,j,-}|\mathcal{F}_{j-1})\\ &\quad=V_n^{-2}\sum_{j=1}^{n-1}\mathbf{T}_-^*E(\mathbf{X}_j-\mathbf{Y}_{j-1}/a_{j-1})^*\\ &\qquad\qquad\times[\mathbf{X}_j\mathbf{D}_j-(\mathbf{Y}_{j-1}/a_{j-1})\mathbf{H}_j|\mathcal{F}_{j-1}]\mathbf{T}\widehat{\mathbf{B}}_{n,j,-}\\ &\quad=\mathbf{T}_-^*\operatorname{diag}(\mathbf{v})\mathbf{H}\mathbf{T}\left(V_n^{-2}\sum_{j=1}^{n-1}\widehat{\mathbf{B}}_{n,j,-}\right)+o(1).\end{aligned}\tag{4.5}$$

From (2.18), we have

$$n^{-1}\sum_{j=1}^{n-1}\widehat{\mathbf{B}}_{n,j}=n^{-1}\sum_{j=1}^{n-1}\sum_{i=j}^{n-1}\frac{1}{i+1}(\mathbf{I}+(j+1)^{-1}\mathbf{J})\cdots(\mathbf{I}+i^{-1}\mathbf{J}).\tag{4.6}$$

Based on (2.18)–(2.20), we have that the $(h, h+\ell)$-element of the block matrix

$$n^{-1}\sum_{j=1}^{n}\sum_{i=j}^{n-1}\frac{1}{i+1}(\mathbf{I}+(j+1)^{-1}\mathbf{J}_t)\cdots(\mathbf{I}+i^{-1}\mathbf{J}_t)$$

has a limit obtained by

$$n^{-1}\sum_{j=1}^{n}\sum_{i=j}^{n-1}\frac{1}{i+1}\frac{1}{\ell!}\left(\frac{j}{i}\right)^{-\lambda_t}\log^{\ell}\left(\frac{i}{j}\right)(1+o(1))$$

$$\rightarrow\frac{1}{\ell!}\int_0^1\int_0^u u^{-1}\left(\frac{v}{u}\right)^{-\lambda_t}\log^{\ell}\left(\frac{u}{v}\right)du\,dv \tag{4.7}$$

$$=\left(\frac{1}{1-\lambda_t}\right)^{\ell+1}.$$

Substituting this into (4.6) and then (4.5), when $V_n^2=n$, we obtain that

$$V_n^{-2}\sum_{j=1}^{n-1}E(\mathbf{q}_j^*\widetilde{\mathbf{Q}}_j\widehat{\mathbf{B}}_{n,j,-}|\mathcal{F}_{j-1})\rightarrow\widetilde{\Sigma}_2=\mathbf{T}_-^*\operatorname{diag}(\mathbf{v})\mathbf{H}\mathbf{T}\widetilde{\mathbf{J}},$$

where $\widetilde{\mathbf{J}}$ is a $K\times(K-1)$ matrix whose first row is 0 and the rest is a block diagonal matrix, the t-block is $\nu_t\times\nu_t$ and its $(h, h+\ell)$-element is given by the right-hand side of (4.7). The matrix $\widetilde{\Sigma}_2$ is obviously 0 when $V_n^2=n\log^{2\nu-1}n$.

Note that the third term in (4.3) is the complex conjugate transpose of the second term; thus we have also got the limit of the third term, that is, $\widetilde{\Sigma}_2^*$.

Now, we compute the limit $\widetilde{\Sigma}_3$ of the fourth term in (4.3). By Assumption 2.2, the matrices $\mathbf{R}_j$ in (4.3) converge to $\mathbf{R}$. Then, the fourth term in (4.3) can be approximated by

$$\Bigg[V_n^{-2}\sum_{j=1}^{n}\sum_{i=j+1}^{n-1}\frac{1}{i}\prod_{r=j+1}^{i}(\mathbf{I}+r^{-1}\mathbf{J}_g^*)\mathbf{T}_g^*\mathbf{R}\mathbf{T}_h \times\sum_{i=j+1}^{n-1}\frac{1}{i}\prod_{r=j+1}^{n}(\mathbf{I}+r^{-1}\mathbf{J}_h)\Bigg]_{g,h=1}^{s}. \tag{4.8}$$

Similar to (4.7), we can show that the (w,t)-element of the (g,h)-block of the matrix in (4.8) is approximately

$$V_n^{-2}\sum_{w'=0}^{w-1}\sum_{t'=0}^{t-1}\sum_{j=1}^{n-1}\sum_{i=j}^{n-1}\sum_{m=j}^{n-1}\frac{(i/j)^{\bar{\lambda}_g}(m/j)^{\lambda_h}\log^{w'}(i/j)\log^{t'}(m/j)}{(i+1)(m+1)(w')!(t')!} \times[\mathbf{T}_g^*\mathbf{R}\mathbf{T}_h]_{(w-w',t-t')}, \tag{4.9}$$

where $[\mathbf{T}_g^*\mathbf{R}\mathbf{T}_h]_{(w',t')}$ is the (w',t')-element of the matrix $[\mathbf{T}_g^*\mathbf{R}\mathbf{T}_h]$. Here, strictly speaking, in the numerator of (4.9), there should be factors $\psi(i,j,w')$ and

$\psi(m, j, t')$. Since for any j_0, the total contributions of terms with $j \le j_0$ is $o(1)$ and the ψ's tend to 1 as $j \to \infty$, we may replace the ψ's by 1.

For fixed w, w', t and t', if $\lambda_g \neq \lambda_h$ or $\mathrm{Re}(\lambda_g) < 1/2$, we have

$$\frac{1}{n}\sum_{j=1}^{n-1}\sum_{i=j}^{n-1}\sum_{m=j}^{n-1}\frac{(i/j)^{\bar\lambda_g}(m/j)^{\lambda_h}\log^{w'}(i/j)\log^{t'}(m/j)}{(i+1)(m+1)(w')!(t')!}$$

$$\to \sum_{\ell=0}^{w'}\frac{(t'+\ell)!}{\ell!(t')!}(1-\bar\lambda_g)^{-(w'-\ell+1)}(1-\bar\lambda_g-\lambda_h)^{-(t'+\ell+1)} + \sum_{\ell=0}^{t'}\frac{(w'+\ell)!}{\ell!(w')!}(1-\lambda_h)^{-(t'-\ell+1)}(1-\bar\lambda_g-\lambda_h)^{-(w'+\ell+1)}. \tag{4.10}$$

Thus, when $\tau < 1/2$, if we split $\widetilde\Sigma_3$ into blocks, then the (w,t)-element of the (g,h)-block $\Sigma_{g,h}$ $(\nu_g \times \nu_h)$ of $\widetilde\Sigma_3$ is given by

$$\sum_{w'=0}^{w-1}\sum_{t'=0}^{t-1}\Bigg[\sum_{\ell=0}^{w'}\frac{(t'+\ell)!}{\ell!(t')!}(1-\bar\lambda_g)^{-(w'-\ell+1)}(1-\bar\lambda_g-\lambda_h)^{-(t'+\ell+1)} + \sum_{\ell=0}^{t'}\frac{(w'+\ell)!}{\ell!(w')!}(1-\lambda_h)^{-(t'-\ell+1)}(1-\bar\lambda_g-\lambda_h)^{-(w'+\ell+1)}\Bigg] \times [\mathbf{T}_g^*\mathbf{R}\mathbf{T}_h]_{(w-w',t-t')}. \tag{4.11}$$

When $\tau = 1/2$, $\Sigma_{g,h} = 0$ if $\lambda_g \neq \lambda_h$ or if $\mathrm{Re}(\lambda_g) < 1/2$. Now, we consider $\Sigma_{g,h}$ with $\lambda_g = \lambda_h$ and $\mathrm{Re}(\lambda_g) = 1/2$. If $w' + t' < 2\nu - 2$, then

$$\sum_{j=1}^{n-1}\sum_{i=j}^{n-1}\sum_{\ell=j}^{n-1}\frac{(i/j)^{\bar\lambda_g}(\ell/j)^{\lambda_g}\log^{w'}(i/j)\log^{t'}(\ell/j)}{(i+1)(\ell+1)(w')!(t')!}$$

$$\le \sum_{j=1}^{n-1}\sum_{i=j}^{n-1}\sum_{\ell=j}^{n-1}\frac{\log^{w'+t'}n}{j\sqrt{i\ell}(w')!(t')!} \le n\log^{w'+t'+1}n = o(V_n^2).$$

When $w' = t' = \nu - 1$ which implies $w = t = \nu = \nu_g = \nu_h$, by Abelian summation, we have

$$V_n^{-2}\sum_{j=1}^{n-1}\sum_{i=j}^{n-1}\sum_{\ell=j}^{n-1}\frac{(i/j)^{\bar\lambda_g}(\ell/j)^{\lambda_g}\log^{\nu-1}(i/j)\log^{\nu-1}(\ell/j)}{(i+1)(\ell+1)[(\nu-1)!]^2} \to |\lambda_g|^{-2}[(\nu-1)!]^{-2}(2\nu-1)^{-1}. \tag{4.12}$$

Hence, for this case, $\Sigma_{g,h}$ has only one nonzero element which is the one on the right-lower corner of $\Sigma_{g,h}$ and given by

$$|\lambda_g|^{-2}[(\nu-1)!]^{-2}(2\nu-1)^{-1}[\mathbf{T}_g^*\mathbf{R}\mathbf{T}_h]_{(1,1)}. \tag{4.13}$$

Combining (4.3), (4.4), (4.7), (4.11) and (4.12), we obtain an expression of $\widetilde{\Sigma}$. □

Now we consider one of the most important special cases, where the matrix **H** has a diagonal Jordan form and $\tau < 1/2$.

COROLLARY 4.1. *Suppose the assumptions of Corollary* 2.1 *hold with* $\tau < 1/2$ *and*

$$\mathbf{T}^{-1}\mathbf{H}\mathbf{T} = \mathbf{J} = \begin{pmatrix} 1 & 0 & \cdots & 0 \\ 0 & \lambda_1 & \cdots & 0 \\ \cdots & \cdots & \cdots & \cdots \\ 0 & 0 & \cdots & \lambda_{K-1} \end{pmatrix},$$

where $\mathbf{T} = (\mathbf{1}', \mathbf{t}_1', \dots, \mathbf{t}_{K-1}')$. *Now let*

$$a_{ij} = (\mathbf{t}_{i-1}^*)'(\mathrm{diag}(\mathbf{v}) - \mathbf{v}^*\mathbf{v})\mathbf{t}_{j-1}',$$

$$b_{ij} = \lambda_{j-1}(1-\lambda_{j-1})^{-1}(\mathbf{t}_{i-1}^*)'(\mathrm{diag}(\mathbf{v}) - \mathbf{v}^*\mathbf{v})\mathbf{t}_{j-1}'$$

and

$$c_{ij} = [(1-\bar{\lambda}_{i-1})^{-1} + (1-\lambda_{j-1})^{-1}](1-\bar{\lambda}_{i-1}-\lambda_{j-1})^{-1}(\mathbf{t}_{i-1}^*)'\mathbf{R}\mathbf{t}_{j-1}',$$

for $i, j = 2, \dots, K$. *Then* $n^{-1/2}(\mathbf{N}_n - E\mathbf{N}_n)$ *is asymptotically normal with mean vector* 0 *and variance–covariance matrix* $(\mathbf{T}^{-1})^*\widetilde{\Sigma}\mathbf{T}^{-1}$, *where* $\widetilde{\Sigma} = (\widetilde{\sigma}_{ij})_{i,j=1}^K$ *has the following simple form*:

$$\widetilde{\sigma}_{11} = \widetilde{\sigma}_{1j} = \widetilde{\sigma}_{i1} = 0 \quad \textit{and} \quad \widetilde{\sigma}_{ij} = a_{ij} + b_{ij} + \bar{b}_{ji} + c_{ij}$$

for $i, j = 2, \dots, K$.

5. Applications.

5.1. *Adaptive allocation rules associated with covariates.* In clinical trials, it is usual that the probability of success (here we assume that the subject response is dichotomous) may depend upon some observable covariates on the patients, that is, $p_{ik} = p_k(\xi_i)$, where ξ_i are covariates observed on the patient i and the result of the treatment at the ith stage. Here $p_{ik} = P(T_i = 1 | X_i = k, \xi_i)$, for $i = 1, \dots, n$ and $k = 1, \dots, K$, where $X_i = k$ indicates that a type-k ball is drawn at the ith stage and $T_i = 1$ if the response of the subject i is a success and 0 otherwise. Thus, for a given ξ_i, the addition rule could be $\mathbf{D}(\xi_i)$ and the generating matrices $\mathbf{H}_i = \mathbf{H}(\xi_i) = E\mathbf{D}(\xi_i)$.

Assume that $\xi_1, \dots, \xi_n$ are i.i.d. random vectors and let $\mathbf{H} = E\mathbf{H}(\xi_1)$. The asymptotic properties of the urn composition $\mathbf{Y}_n$ are considered by Bai and Hu (1999). Based on the results in Sections 2 and 4, we can get the corresponding asymptotic results of the allocation number of patients $\mathbf{N}_n$. Here we illustrate the results by considering the case $K = 2$.

Consider the generalized play-the-winner rule [Bai and Hu (1999)] and let $E(p_k(\xi_i)) = p_k$, $k = 1, 2$. Then the addition rule matrices are denoted by

$$\mathbf{D}(\xi_i) = \begin{pmatrix} d_1(\xi_i) & 1 - d_1(\xi_i) \\ 1 - d_2(\xi_i) & d_2(\xi_i) \end{pmatrix} \quad \text{and} \quad \mathbf{H} = \begin{pmatrix} p_1 & q_1 \\ q_2 & p_2 \end{pmatrix},$$

where $0 \le d_k(\xi_i) \le 1$ and $q_k = 1 - p_k$ for $k = 1, 2$.

It is easy to see that $\lambda = 1$, $\lambda_1 = p_1 + p_2 - 1$, $\tau = \max(0, \lambda_1)$ and $\mathbf{v} = (q_2/(q_1 + q_2), q_1/(q_1 + q_2))$. Further, we have

$$\mathbf{R} = \frac{(a_1 q_2 + a_2 q_1)(q_1 + q_2) + q_1 q_2 (p_1 - q_2)^2}{(q_1 + q_2)^2} \begin{pmatrix} 1 & -1 \\ -1 & 1 \end{pmatrix},$$

$$\mathbf{T} = \begin{pmatrix} 1 & q_1 \\ 1 & -q_2 \end{pmatrix} \quad \text{and} \quad \mathbf{T}^{-1} = \frac{1}{q_1 + q_2} \begin{pmatrix} q_2 & q_1 \\ 1 & -1 \end{pmatrix},$$

where $a_k = \operatorname{Var}(d_k(\xi_1))$. For the case $\tau < 1/2$, we have that $V_n = n$ and the values corresponding to Corollary 4.1 are

$$a_{22} = q_1 q_2, \qquad b_{22} = \frac{(1 - q_1 - q_2) q_1 q_2}{q_1 + q_2},$$

$$c_{22} = \frac{2[(a_1 q_2 + a_2 q_1)(q_1 + q_2) + q_1 q_2 (p_1 - q_2)^2]}{(q_1 + q_2)(1 - 2(p_1 + p_2 - 1))}.$$

So

$$\tilde{\sigma}_{22} = q_1 q_2 + \frac{2(1 - q_1 - q_2) q_1 q_2}{q_1 + q_2} + \frac{2[(a_1 q_2 + a_2 q_1)(q_1 + q_2) + q_1 q_2 (p_1 - q_2)^2]}{(q_1 + q_2)(1 - 2(p_1 + p_2 - 1))}.$$

From Theorem 2.3 and Corollary 4.1, we have

$$n^{\delta}\left(\frac{\mathbf{N}_n}{n} - \mathbf{v}\right) \to 0 \quad \text{a.s. for any } \delta < 1/2 \quad \text{and} \quad n^{1/2}\left(\frac{\mathbf{N}_n}{n} - \mathbf{v}\right) \to N(0, \Sigma_1)$$

in distribution, where

$$\Sigma_1 = (\mathbf{T}^{-1})^* \begin{pmatrix} 0 & 0 \\ 0 & \tilde{\sigma}_{22} \end{pmatrix} \mathbf{T}^{-1} = \frac{\tilde{\sigma}_{22}}{(q_1 + q_2)^2} \begin{pmatrix} 1 & -1 \\ -1 & 1 \end{pmatrix}.$$

For the randomized play-the-winner rule [Wei and Durham (1978)], we have $a_k = p_k q_k$, $k = 1, 2$. Then we have

$$\tilde{\sigma}_{22} = \frac{(5 - 2(q_1 + q_2)) q_1 q_2}{2(q_1 + q_2) - 1}.$$

This result agrees with that of Matthews and Rosenberger (1997).

For the case $\tau = 1/2$, $V_n = n \log n$ and the value corresponding to (4.11) is

$$\tilde{\sigma}_{22} = 4[(a_1 q_2 + a_2 q_1)(q_1 + q_2) + q_1 q_2 (p_1 - q_2)^2].$$

We have

$$(n \log n)^{-1/2}(\mathbf{N}_n - n\mathbf{v}) \to N(0, \Sigma_2)$$

in distribution, where

$$\Sigma_2 = \frac{4[(a_1q_2 + a_2q_1)(q_1 + q_2) + q_1q_2(p_1 - q_2)^2]}{(q_1 + q_2)^2} \begin{pmatrix} 1 & -1 \\ -1 & 1 \end{pmatrix}.$$

For the case of the randomized play-the-winner rule, we have

$$\Sigma_2 = \frac{4q_1q_2}{(q_1 + q_2)^2} \begin{pmatrix} 1 & -1 \\ -1 & 1 \end{pmatrix}.$$

5.2. *Clinical trials with time trend in adaptive designs.* Time trends are present in many sequential experiments. Hu and Rosenberger (2000) have studied time trend in adaptive designs and applied to a neurophysiology experiment. It is important to know the asymptotic behavior of the allocation number of patients in these cases.

In Section 5.1, $p_{ik} = P(T_i = 1 | X_i = k)$, where $X_i = k$ if the kth element of $\mathbf{X}_i$ is 1. There may be a drift in patient characteristics over time, for example, $\lim_{i \to \infty} p_{ik} = p_k$ [Hu and Rosenberger (2000)]. Then the results in Sections 2, 3 and 4 are applicable here. For the case $K = 2$, we can get similar results as in Section 5.1.

The results in this paper may also apply for GFU model with homogeneous generating matrix with a general Jordan form as well as $\tau = 1/2$. In these cases, the results of Smythe (1996) are not applicable.

5.3. *Urn models for multi-arm clinical trials.* For multi-arm clinical trials, Wei (1979) proposed the following urn model (as an extension of the randomized play-the-winner rule of two treatments): Starting from $\mathbf{Y}_0 = (Y_{01}, \dots, Y_{0K})$, when a type k splits (randomly from the urn), we assign the patient to the treatment k and observe the patient's response. A success on treatment k adds a ball of type k to the urn and a failure on treatment k adds $1/(K-1)$ ball for each of the other $K - 1$ types. Let p_k be the probability of success of treatment k, $k = 1, 2, \dots, K$, and $q_k = 1 - p_k$. The generating matrix for this urn model is

$$\mathbf{H} = \begin{pmatrix} p_1 & (K-1)^{-1}q_1 & \cdots & (K-1)^{-1}q_1 \\ (K-1)^{-1}q_2 & p_2 & \cdots & (K-1)^{-1}q_2 \\ \cdots & \cdots & \cdots & \cdots \\ (K-1)^{-1}q_K & (K-1)^{-1}q_K & \cdots & p_K \end{pmatrix}.$$

The asymptotic properties of $\mathbf{Y}_n$ can be obtained from Athreya and Karlin (1968) and Bai and Hu (1999). From Theorem 4.1 in Section 4, we obtain the asymptotic normality of $\mathbf{N}_n$ and its asymptotic variance.

Recently, Bai, Hu and Shen (2002) proposed an urn model which adds balls depending on the success probabilities of each treatment. Write $\mathbf{N}_n = (N_{n1}, \dots, N_{nK})$ and $\mathbf{S}_n = (S_{n1}, \dots, S_{nK})$, where N_{nk} denotes the number of times that the kth treatment is selected in the first n stages, and S_{nk} denotes the number of successes of the kth treatment in the N_{nk} trials, $k = 1, \dots, K$. Define $\mathbf{R}_n = (R_{n1}, \dots, R_{nK})$ and $M_n = \sum_{k=1}^{K} R_{nk}$, where $R_{n,k} = \frac{S_{nk}+1}{N_{nk}+1}$, $k = 1, \dots, K$. The generating matrices are

$$\mathbf{H}_{i+1} = \begin{pmatrix} p_1 & \dfrac{R_{i2}}{M_i - R_{i1}} q_1 & \cdots & \dfrac{R_{iK}}{M_i - R_{i1}} q_1 \\ \dfrac{R_{i1}}{M_i - R_{i2}} q_2 & p_2 & \cdots & \dfrac{R_{iK}}{M_i - R_{i2}} q_2 \\ \cdots & \cdots & \cdots & \cdots \\ \dfrac{R_{i1}}{M_i - R_{iK}} q_K & \dfrac{R_{i2}}{M_i - R_{iK}} q_K & \cdots & p_K \end{pmatrix}.$$

In this case, $\mathbf{H}_i$ are random matrices and converge to

$$\mathbf{H} = \begin{pmatrix} p_1 & \dfrac{p_2}{M - p_1} q_1 & \cdots & \dfrac{p_K}{M - p_1} q_1 \\ \dfrac{p_1}{M - p_2} q_2 & p_2 & \cdots & \dfrac{p_K}{M - p_2} q_2 \\ \cdots & \cdots & \cdots & \cdots \\ \dfrac{p_1}{M - p_K} q_K & \dfrac{p_2}{M - p_K} q_K & \cdots & p_K \end{pmatrix}$$

almost surely, where $M = p_1 + \cdots + p_K$.

Bai, Hu and Shen (2002) considered the convergences of $\mathbf{Y}_n/n$ and $\mathbf{N}_n/n$. The asymptotic distributions of $\mathbf{Y}_n$ and $\mathbf{N}_n$ can be obtained from Theorems 3.2 and 4.1 in this paper. From Lemma 3 of Bai, Hu and Shen (2002) we have $\alpha_i = o(i^{-1/4})$ almost surely, so the condition (1.3) is satisfied.

Acknowledgments. Special thanks go to anonymous referees for the constructive comments, which led to a much improved version of the paper. We would also like to thank Professor W. F. Rosenberger for his valuable discussions which led to the problem of this paper.

REFERENCES

ALTMAN, D. G. and ROYSTON, J. P. (1988). The hidden effect of time. *Statist. Med.* **7** 629–637.

ANDERSEN, J., FARIES, D. and TAMURA, R. N. (1994). Randomized play-the-winner design for multi-arm clinical trials. *Comm. Statist. Theory Methods* **23** 309–323.

ATHREYA, K. B. and KARLIN, S. (1967). Limit theorems for the split times of branching processes. *Journal of Mathematics and Mechanics* **17** 257–277.

ATHREYA, K. B. and KARLIN, S. (1968). Embedding of urn schemes into continuous time branching processes and related limit theorems. *Ann. Math. Statist.* **39** 1801–1817.

BAI, Z. D. and HU, F. (1999). Asymptotic theorem for urn models with nonhomogeneous generating matrices. *Stochastic Process. Appl.* **80** 87–101.

BAI, Z. D., HU, F. and SHEN, L. (2002). An adaptive design for multi-arm clinical trials. *J. Multivariate Anal.* **81** 1–18.

COAD, D. S. (1991). Sequential tests for an unstable response variable. *Biometrika* **78** 113–121.

FLOURNOY, N. and ROSENBERGER, W. F., eds. (1995). *Adaptive Designs*. IMS, Hayward, CA.

FREEDMAN, D. (1965). Bernard Friedman's urn. *Ann. Math. Statist.* **36** 956–970.

GOUET, R. (1993). Martingale functional central limit theorems for a generalized Pólya urn. *Ann. Probab.* **21** 1624–1639.

HALL, P. and HEYDE, C. C. (1980). *Martingale Limit Theory and Its Application.* Academic Press, London.

HOLST, L. (1979). A unified approach to limit theorems for urn models. *J. Appl. Probab.* **16** 154–162.

HU, F. and ROSENBERGER, W. F. (2000). Analysis of time trends in adaptive designs with application to a neurophysiology experiment. *Statist. Med.* **19** 2067–2075.

HU, F. and ROSENBERGER, W. F. (2003). Optimality, variability, power: Evaluating response-adaptive randomization procedures for treatment comparisons. *J. Amer. Statist. Assoc.* **98** 671–678.

JOHNSON, N. L. and KOTZ, S. (1977). *Urn Models and Their Applications.* Wiley, New York.

MAHMOUD, H. M. and SMYTHE, R. T. (1991). On the distribution of leaves in rooted subtree of recursive trees. *Ann. Appl. Probab.* **1** 406–418.

MATTHEWS, P. C. and ROSENBERGER, W. F. (1997). Variance in randomized play-the-winner clinical trials. *Statist. Probab. Lett.* **35** 193–207.

ROSENBERGER, W. F. (1996). New directions in adaptive designs. *Statist. Sci.* **11** 137–149.

ROSENBERGER, W. F. (2002). Randomized urn models and sequential design (with discussion). *Sequential Anal.* **21** 1–21.

SMYTHE, R. T. (1996). Central limit theorems for urn models. *Stochastic Process. Appl.* **65** 115–137.

WEI, L. J. (1979). The generalized Pólya's urn design for sequential medical trials. *Ann. Statist.* **7** 291–296.

WEI, L. J. and DURHAM, S. (1978). The randomized play-the-winner rule in medical trials. *J. Amer. Statist. Assoc.* **73** 840–843.

ZELEN, M. (1969). Play the winner rule and the controlled clinical trial. *J. Amer. Statist. Assoc.* **64** 131–146.

COLLEGE OF MATHEMATICS AND STATISTICS
NORTHEAST NORMAL UNIVERSITY
CHANGCHUN
CHINA
AND
DEPARTMENT OF STATISTICS
AND APPLIED PROBABILITY
NATIONAL UNIVERSITY OF SINGAPORE
SINGAPORE

DEPARTMENT OF STATISTICS
HALSEY HALL
UNIVERSITY OF VIRGINIA
CHARLOTTESVILLE, VIRGINIA 22904-4135
USA
E-MAIL: fh6e@virginia.edu

Probab. Theory Relat. Fields 131, 528–552 (2005)
Digital Object Identifier (DOI) 10.1007/s00440-004-0384-5

Zhidong Bai · Tailen Hsing

The broken sample problem

Dedicated to Professor Xiru Chen on His 70th Birthday

Received: 20 February 2002 / Revised version: 16 June 2004
Published online: 12 September 2004 –

Abstract. Suppose that $(X_i, Y_i), i = 1, 2, \dots, n$, are iid. random vectors with uniform marginals and a certain joint distribution F_ρ, where ρ is a parameter with $\rho = \rho_o$ corresponds to the independence case. However, the X's and Y's are observed separately so that the pairing information is missing. Can ρ be consistently estimated? This is an extension of a problem considered in DeGroot and Goel (1980) which focused on the bivariate normal distribution with ρ being the correlation. In this paper we show that consistent discrimination between two distinct parameter values ρ_1 and ρ_2 is impossible if the density f_ρ of F_ρ is square integrable and the second largest singular value of the linear operator $h \to \int_0^1 f_\rho(x, \cdot)h(x)dx,\ h \in L^2[0, 1]$, is strictly less than 1 for $\rho = \rho_1$ and ρ_2. We also consider this result from the perspective of a bivariate empirical process which contains information equivalent to that of the broken sample.

1. Introduction

Consider a family of bivariate distributions with a parameter ρ and let F_ρ be the joint cdf. One can think of ρ as a measure of association such as the correlation. We assume that the parameter space contains a specific value ρ_o which corresponds to the independence of the marginals. Let $(X_i, Y_i), i = 1, 2, \dots, n$, be iid. random vectors from this distribution. However, we assume an incomplete or "broken" sample in which the X's and Y's are observed separately, and the information on the pairing of the two sets of observations is lost. Our goal is to investigate the consistent discrimination of the F_ρ, where consistency in this paper refers to weak consistency. In DeGroot and Goel (1980), the problem of estimating the correlation of a bivariate normal distribution based on a broken sample was considered. They showed that the Fisher information at $\rho = 0$ is equal to 1 for all sample

Z. Bai: North East Normal University, China and Department of Statistics and Applied Probability, National University of Singapore, Singapore.
e-mail: stabaizd@leonis.nus.edu.sg. Research supported by NSFC Grant 201471000 and the NUS Grant R-155-000-040-112.

T. Hsing: Texas A&M University and Department of Statistics, Texas A&M University, College Station, Texas, USA.
e-mail: thsing@stat.tamu.edu. Research supported by the Texas Advanced Research Program.

Mathematics Subject Classification (2000): primary: 60F99, 62F12

*Key words or phrases:*Consistent estimation – Empirical process – Gaussian process – Kullback-Leibler information sizes, which leads to the conjecture that consistent estimation is not possible (if the parameter space contains a neighborhood of 0). However, they failed to give a definitive conclusion.

Since the marginal distributions can be consistently estimated with the broken sample, in order for the problem stated here to make sense we need for ρ to be either not present, or at least not identifiable, in the marginal distributions. With that consideration in mind, we assume without loss of generality that the marginal distributions are uniform [0, 1], for we may otherwise consider $(F_X(X_i), F_Y(Y_i))$ where F_X and F_Y are the marginal distributions of X and Y respectively. Thus, the distribution under ρ_o is the uniform distribution on $[0, 1] \times [0, 1]$. The main purpose of this paper is to try to understand whether it is possible to consistently discriminate two distinct parameter values ρ_1 and ρ_2 based on the broken sample, that is, whether there exists a sequence of statistics T_n of the broken sample, where n refers to the sample size, taking values in $\{\rho_1, \rho_2\}$ and such that

$$\lim_{n\to\infty} P_{\rho_i}(T_n = \rho_i) = 1, \; i = 1, 2. \tag{1}$$

Here and in the sequel, P_ρ denotes probability computation when the true parameter is ρ. The condition under which consistent discrimination rules do not exist turns out to be remarkably simple. Let f_ρ be the density of F_ρ. We will show in Theorem 1 that ρ_1 and ρ_2 cannot be consistently discriminated if for $\rho = \rho_1$ and ρ_2, $\int_0^1 \int_0^1 f_\rho^2(x, y)dxdy < \infty$ and the second largest singular value of the linear operator $h \to \int_0^1 f_\rho(x, \cdot)h(x)dx, \; h \in L^2[0, 1]$, is strictly less than 1.

To give some insight into this result, we consider the two-dimensional empirical process

$$F_n(x, y) = n^{-1}\left(\sum_{i=1}^n I_{\{X_i \le x\}}, \sum_{i=1}^n I_{\{Y_i \le y\}}\right), \quad x, y \in [0, 1],$$

which contains all the existing information in the broken sample. It is straightforward to verify that the standardized empirical process $\mathbf{Z}_n(x, y) = n^{1/2}(F_n - EF_n)$ converges weakly to a Gaussian process $\mathbf{Z} = (Z_1, Z_2)$ in the space $D[0, 1] \times D[0, 1]$ where the Z_i are marginally Brownian bridge with

$$\begin{aligned} \operatorname{cov}(Z_1(x_1), Z_1(x_2)) &= x_1 \wedge x_2 - x_1 x_2 \\ \operatorname{cov}(Z_2(y_1), Z_2(y_2)) &= y_1 \wedge y_2 - y_1 y_2 \\ \operatorname{cov}(Z_1(x), Z_2(y)) &= F_\rho(x, y) - xy. \end{aligned}$$

Let $\mathcal{P}_\rho$ henceforth denote the probability distribution of the limiting Gaussian process $\mathbf{Z}$ described above under parameter value ρ. Note that the standardization does not involve ρ, so it is reasonable to argue that most of the information about ρ in F_n carries over to $\mathbf{Z}$. We also remark in passing that the weak convergence implies that ρ is identifiable in the broken sample setting so long as it is identifiable in the bivariate distribution F_ρ. Suppose that for two given parameter values ρ_1 and ρ_2, $\mathcal{P}_{\rho_1}$ and $\mathcal{P}_{\rho_2}$ are equivalent, also called mutually absolutely continuous and denoted by $\mathcal{P}_{\rho_1} \equiv \mathcal{P}_{\rho_2}$ here. Then it is clearly not possible to discriminate between the two models with probability one based on **Z**. Theorem 3 shows that the same conditions of Theorem 1 plus some additional minor regularity condition ensure that $\mathcal{P}_{\rho_i} \equiv \mathcal{P}_{\rho_o}, i = 1, 2$ and hence $\mathcal{P}_{\rho_1} \equiv \mathcal{P}_{\rho_2}$.

To demonstrate the results, we will revisit the bivariate normal problem in DeGroot and Goel (1980) and show that consistent discrimination of any two bivariate normal distributions with different correlations in $(-1, 1)$ is impossible. We will also consider other examples for which ρ can be consistently discriminated or even estimated.

2. Main results and examples

We assume that F_ρ has a density f_ρ, and write

$$A(\rho) = \int_0^1 \int_0^1 f_\rho^2(x, y)dxdy.$$

Define the linear operator

$$T_\rho : h \to \int_0^1 f_\rho(x, \cdot)h(x)dx, \; h \in L^2[0, 1].$$

Suppose $A(\rho) < \infty$. Then T_ρ is a Hilbert-Schmidt operator and admits the singular-value decomposition (cf. Riesz and Sz.-Nagy, 1955). Since 1 is necessarily a singular value of T_ρ with singular value functions equal to the constant function 1, the singular-value decomposition can be written as

$$T_\rho = 1 + \sum_{i=1}^{\infty} \lambda_{i,\rho} \psi_{i,\rho} \otimes \phi_{i,\rho},$$

where, with

$$(\psi_{i,\rho} \otimes \phi_{i,\rho})h = \left(\int_0^1 \psi_{i,\rho}(x)h(x)dx \right) \phi_{i,\rho},$$

we have $1 \geq \lambda_{1,\rho} \geq \lambda_{2,\rho} \geq \cdots \geq 0$,

$$\int_0^1 \psi_{i,\rho}(x)dx = \int_0^1 \phi_{i,\rho}(y)dy = 0,$$

and

$$\int_0^1 \psi_{i,\rho}(x)\psi_{j,\rho}(x)dx = \int_0^1 \phi_{i,\rho}(y)\phi_{j,\rho}(y)dy = \delta_{i,j}.$$

Equivalently, we can write

$$f_\rho(x, y) = 1 + \sum_{i=1}^{\infty} \lambda_{i,\rho} \psi_{i,\rho}(x)\phi_{i,\rho}(y). \tag{2}$$

Thus,

$$A(\rho) = 1 + \sum_{i=1}^{\infty} \lambda_{i,\rho}^2 < \infty. \tag{3}$$

Define the following condition:

(HS) $A(\rho) < \infty$ where $\lambda_{1,\rho}$ is strictly less than 1.

Theorem 1. *Assume that the condition (HS) holds for $\rho = \rho_1, \rho_2$. Then there does not exist a consistent discrimination rule for ρ_1 versus ρ_2 based on the broken sample.*

Remark 1. The condition (HS) is a not a stringent one, and is satisfied by the majority of the commonly used bivariate statistical models. However, it will be demonstrated in a number of examples below that the condition (HS) can be violated, and for each of those examples consistent discrimination rules do exist. Hence a natural question is whether the violation of the condition (HS) necessarily implies the existence of consistent discrimination rules. We conjecture that the answer is affirmative, but we have not been able to show that.

At the heart of the proof of Theorem 1 is the following result, which deserves prominent attention in its own right. Denote by $g_{n,\rho}(\boldsymbol{x}, \boldsymbol{y})$ the density of the broken sample, i.e.

$$g_{n,\rho}(\boldsymbol{x}, \boldsymbol{y}) = \frac{1}{n!} \sum_{\pi} \prod_{i=1}^{n} f_\rho(x_i, y_{\pi_i}),$$

where the summation is taken over all permutations π of $1, \dots, n$. By assumption, $g_{n,\rho_o}(\boldsymbol{x}, \boldsymbol{y}) \equiv 1$. As a result, $g_{n,\rho}(\boldsymbol{x}, \boldsymbol{y})$ can also be viewed as a likelihood ratio.

Theorem 2. *Let the condition (HS) hold for some ρ. Then*

$$\lim_{n \to \infty} P_{\rho_o}(g_{n,\rho}(\boldsymbol{X}, \boldsymbol{Y}) \le x) = P(\xi \le x) \textit{ for all } x,$$

where

$$\xi = \prod_{k=1}^{\infty} (1 - \lambda_{k,\rho}^2)^{-1/2} \exp\left(-\frac{1}{2} \sum_{k=1}^{\infty} \frac{\lambda_{k,\rho}}{1 - \lambda_{k,\rho}} U_k^2 + \frac{1}{2} \sum_{k=1}^{\infty} \frac{\lambda_{k,\rho}}{1 + \lambda_{k,\rho}} V_k^2 \right),$$

with the U_i, V_i denoting iid. standard normal random variables.

Observe that $\log \xi$ is a constant plus a weighted average of independent χ^2 random variables.

To give some insight into the conclusion of Theorem 1, we present the following perspective. Define

$$f_{\rho,\delta} = \delta \wedge f_\rho, \; \delta > 0,$$

Theorem 3. *Suppose that the condition (HS) holds for some ρ, and that each $\delta > 0$, $f_{\rho,\delta}$ is square integrable in the Riemann sense on $[0,1] \times [0,1]$. Then $\mathcal{P}_\rho \equiv \mathcal{P}_{\rho_o}$ (see section 1 for notation). Thus, the class of probability distributions $\mathcal{P}_\rho$ for which F_ρ satisfy these conditions are mutually equivalent.*

It is well-known that the probability distributions of any two Gaussian processes with the same sample space are equivalent if and only of the Kulback-Leibler information between the two is finite (cf. Hajék (1958)). Our proof therefore is based on the derivation of the Kulback-Leibler information between $\mathcal{P}_\rho$ and $\mathcal{P}_{\rho_o}$ in terms of f_ρ, where we show under the conditions stated in Theorem 3 that the Kulback-Leibler information between $\mathcal{P}_\rho$ and $\mathcal{P}_{\rho_o}$ is equal to

$$\sum_{i=1}^{\infty} \frac{\lambda_{i,\rho}^2}{1-\lambda_{i,\rho}^2} < \infty.$$

The proofs of Theorems 1-3 are collected in section 3. We now present a few examples for both cases for which consistent estimation is possible and is not possible.

Example A. First we revisit the setting of DeGroot and Goel (1980). Let (U, V) have the bivariate normal distribution with standard marginals and correlation ρ and denote by ϕ_ρ the joint pdf. It is well known (see Cramér, 1946) that

$$\phi_\rho(u,v) = \sum_{k=0}^{\infty} \frac{\rho^k}{k!} H_k(u)\phi(u)H_k(v)\phi(v),$$

where ϕ is the standard normal pdf and $H_k(u) = (-1)^k e^{u^2/2}\frac{d^k}{du^k}e^{-u^2/2}$ is the k-th Hermite polynomial. Let f_ρ be the pdf of $(\Phi(U), \Phi(V))$. Then

$$f_\rho(x,y) = \sum_{k=0}^{\infty} \frac{\rho^k}{k!} H_k(Q(x))H_k(Q(y))$$

where Φ and Q are the standard normal cdf and quantile function, respectively. It is easy to check that (HS) holds for each ρ where $\lambda_{i,\rho} = |\rho|^i$. Thus, the question posed by DeGroot and Goel (1980) is completely answered.

Example B. Suppose that $y(x)$ is a monotone function such that $P_\rho(Y_1 = y(X_1)) = c(\rho)$. Then

$$\hat{c}(\rho) := n^{-1} \sum_{i=1}^{n}\sum_{j=1}^{n} I(Y_j = y(X_i))$$

is obviously $\sqrt{n}$-consistent for $c(\rho)$. In this case, of course, (X_1, Y_1) does not have a joint density. One such example (cf. Chan and Loh, 2001) is to let J_i be iid. with $P(J_1 = 1) = 1 - P(J_1 = 0) = \rho$ where $\rho \in [0, 1]$ and

$$X_i = J_iU_i + (1-J_i)V_i, \quad Y_i = J_iU_i + (1-J_i)W_i$$

where $U_i, V_i, W_i, 1 \le i \le n$ are iid; in this case, $P_\rho(Y = X) = \rho$.

Example C. Let

$$X_i = \rho U_i + (1-\rho)V_i, \; Y_i = \rho U_i + (1-\rho)W_i \tag{4}$$

where $\rho \in (0, 1)$ and the U_i, V_i, W_i are iid. standard Cauchy. In this case $A(\rho) = \infty$ and ρ can be consistently estimated. The intuition here is that when a large value of X is observed, the probability that it is due to a large U is ρ and the probability that it is due to a large V is $1-\rho$. Thus, the probability of finding a matching Y for a large X is roughly ρ. Indeed the following can be proved.

Theorem 4. *Let (X_i, Y_i) be defined by (4) and k_n and ε_n be positive constants such that $k_n \to \infty$, $k_n\varepsilon_n \to 0$ and $n\varepsilon_n/k_n \to \infty$. Then $A(\rho) = \infty$ for all $\rho \in (0, 1)$ and*

$$\hat{\rho}_n := \frac{1}{k_n}\sum_{i=1}^{k_n}\sum_{j=1}^{n} I(Y_j/X_{(i)} \in (1-\varepsilon_n, 1+\varepsilon_n)) \xrightarrow{p} \rho$$

where $X_{(i)}$ is the i-th largest value of $X_1, \ldots, X_n$.

The proof of Theorem 4 is given in section 3.4. This example can be easily extended to other heavy-tailed scenarios (cf. Resnick, 1987).

Example D. For $\rho \in [0, 1]$, define the density

$$f_\rho(x, y) = \rho^{-1}I(0 < x, y \le \rho) + (1-\rho)^{-1}I(\rho < x, y < 1), \; x, y \in (0, 1).$$

In this case, $\rho = 0, 1$ correspond to independence and $\rho = 1/2$ to maximal dependence. Let

$$g(x) = \sqrt{\frac{1-\rho}{\rho}}\, I(0 < x \le \rho) - \sqrt{\frac{\rho}{1-\rho}}\, I(\rho < x < 1).$$

Then $\int_0^1 g(x)dx = 0$, $\int_0^1 g^2(x)dx = 1$ and $\int_0^1\int_0^1 g(x)f_\rho(x, y)g(y)dxdy = 1$ so that $\lambda_{1,\rho} = 1$. Consistent discrimination between any two distinct values ρ_1, ρ_2 is trivial; an obvious such rule is $T_n = \rho_1$ if $\sum_{i=1}^n I_{\{X_i \le \rho_1\}} = \sum_{i=1}^n I_{\{Y_i \le \rho_1\}}$ and $T_n = \rho_2$ otherwise. However, it is not clear whether a consistent estimator exists.

3. Proofs

We will prove Theorems 2, 1, 3, and 4 in the subsections 3.1, 3.2, 3.3, and 3.4, respectively. For simplicity of notation, where no confusion is likely, we will sometimes suppress the reference to ρ in $\lambda_{i,\rho}$, $\psi_{i,\rho}$ and $\phi_{i,\rho}$.

3.1. Proof of Theorem 2

We need the following lemma.

Lemma 5. *Assume that the condition (HS) holds for some* ρ. *Then*

$$\lim_{n\to\infty} E_{\rho_o} g_{n,\rho}^2(\boldsymbol{X},\boldsymbol{Y}) = \prod_{k=1}^{\infty}(1-\lambda_{k,\rho}^2)^{-1} < \infty.$$

Proof. Clearly,

$$E_{\rho_o} g_{n,\rho}(\boldsymbol{X},\boldsymbol{Y}) = \frac{1}{n!}\sum_{\pi}\int\int\prod_{i=1}^{n} f_\rho(x_i, y_{\pi_i})dx_i dy_i = 1. \tag{5}$$

Also,

$$\begin{aligned}
E_{\rho_o} g_{n,\rho}^2(\boldsymbol{X},\boldsymbol{Y}) &= \frac{1}{(n!)^2}\sum_{\pi,\pi'}\int\cdots\int\prod_{i=1}^{n} f_\rho(x_i, y_{\pi_i}) f_\rho(x_i, y_{\pi_i'})dx_i dy_i \\
&= \frac{1}{(n!)^2}\sum_{\pi,\pi'}\int\cdots\int\prod_{i=1}^{n}\sum_{k=0}^{\infty}\lambda_k^2\phi_k(y_{\pi_i})\phi_k(y_{\pi_i'})dy_i \\
&= \frac{1}{n!}\sum_{\pi}\int\cdots\int\prod_{i=1}^{n}\sum_{k=0}^{\infty}\lambda_k^2\phi_k(y_i)\phi_k(y_{\pi_i})dy_i.
\end{aligned}$$

It is easy to verify that for a given permutation π

$$\int\cdots\int\prod_{i=1}^{n}\sum_{k=0}^{\infty}\lambda_k^2\phi_k(y_i)\phi_k(y_{\pi_i})dy_i = \prod_{t=1}^{\ell}\sum_{k=0}^{\infty}\lambda_k^{2i_t},$$

if the permutation π consists of ℓ cycles of sizes $i_1, \cdots, i_\ell$ ($\sum i_t = n$). Therefore, if j_t denotes the number of t among $\{i_1, \cdots, i_\ell\}$, then

$$E_{\rho_o} g_{n,\rho}^2(\boldsymbol{X},\boldsymbol{Y}) = \sum_{\ell=1}^{n}\sum_{\substack{j_1+j_2\cdots+j_n=\ell \\ j_1+2j_2\cdots+nj_n=n}}\prod_{t=1}^{n}\frac{1}{j_t!}\Big(\frac{1}{t}\sum_{k=0}^{\infty}\lambda_k^{2t}\Big)^{j_t} \tag{6}$$

In fact, it is easy to see from this that that $E_{\rho_o} g_{n,\rho}^2(\boldsymbol{X},\boldsymbol{Y})$ is the coefficient of z^n in the Taylor expansion of the function $\prod_{k=0}^{\infty}(1-z\lambda_k^2)^{-1}$. Choose $r \in (1, \lambda_1^{-2})$ and consider the Cauchy integral

$$\frac{1}{2\pi i}\oint_{|z|=r}\frac{1}{z^{n+1}}\prod_{k=0}^{\infty}(1-z\lambda_k^2)^{-1}dz$$

whose absolute value is less than

$$r^{-n}(r-1)^{-1}\prod_{k=1}^{\infty}(1-r\lambda_k^2)^{-1} \to 0.$$

By the Cauchy integral theorem, we conclude that the integral equals

$$E_{\rho_o} g^2_{n,\rho}(\boldsymbol{X}, \boldsymbol{Y}) - \prod_{k=1}^{\infty}(1-\lambda_k^2)^{-1}.$$

Hence, $E_{\rho_o} g^2_{n,\rho}(\boldsymbol{X}, \boldsymbol{Y}) \to \prod_{k=1}^{\infty}(1-\lambda_k^2)^{-1}$. □

Proof of Theorem 2. Suppose that $X_i, Y_i, i = 1, \cdots, n$ are $2n$ iid. random variables uniformly distributed over (0, 1). Write $W(x, y) = f_\rho(x, y) - 1 = \sum_{k=1}^{\infty} \lambda_k \psi_k(x)\phi_k(y)$. Then we have

$$\frac{1}{n!}\sum_{\pi}\prod_{i=1}^{n} f_\rho(X_i, Y_{\pi_i}) = \frac{1}{n!}\sum_{\pi}\prod_{i=1}^{n}(1 + W(X_i, Y_{\pi_i})) = 1 + \sum_{t=1}^{n} Q_t,$$

where

$$Q_t = \frac{1}{n!}\sum_{\pi}\sum_{1\le i_1<\cdots<i_t\le n}\prod_{j=1}^{t} W(X_{i_j}, Y_{\pi_{i_j}}).$$

We first simplify Q_t. Suppose $(s_1, \cdots, s_t)$ is a given ordered subset of $\{1, \cdots, n\}$. Since there are $(n-t)!$ different permutations such that $(\pi_{i_1}, \cdots, \pi_{i_t}) = (s_1, \cdots, s_t)$,

$$Q_t = \frac{(n-t)!}{n!}\sum_{1\le i_1<\cdots<i_t\le n}\sum_{s_1,\cdots,s_t}\prod_{j=1}^{t} W(X_{i_j}, Y_{s_j}).$$

It is easy to verify that all Q_t are of mean 0 and uncorrelated with each other, and

$$EQ_t^2 = \frac{1}{t!}\sum_{s_1,\cdots,s_t} E\prod_{j=1}^{t}\sum_{k=1}^{\infty}\lambda_k^2\phi_k(Y_j)\phi_k(Y_{s_j}).$$

Observe that the expectation in the above summation is 0 if $\{s_1, \cdots, s_t\} \neq \{1, \cdots, t\}$ due to the orthogonality of the functions ϕ_k's. Thus, for each non zero term, $\{s_1, \cdots, s_t\}$ can be considered as a permutation of $\{1, \cdots, t\}$. Classify the permutations by the numbers of cycles of sizes $j = 1, \cdots, t$. Suppose the number of cycles of size j is ν_j, then we have $\nu_1 + 2\nu_2 + \cdots + t\nu_t = t$ and

$$EQ_t^2 = \frac{1}{t!}\sum_{\nu_1+2\nu_2+\cdots+t\nu_t=t}\prod_{j=1}^{t}\frac{1}{\nu_j!}\Big(\frac{1}{j}\sum_{k=1}^{\infty}\lambda_k^{2j}\Big)^{\nu_j}.$$

Similar to the derivation of the limit of the left hand side of (6), we have

$$EQ_t^2 = \frac{1}{2\pi i t!}\oint_{|z|=r}\frac{1}{z^{t+1}}\prod_{k=1}^{\infty}(1-z\lambda_k^2)^{-1}dz.$$

By choosing $r \in (1, \lambda_1^{-2})$, one sees that

$$EQ_t^2 \le \frac{1}{t!}r^{-t}\prod_{k=1}^{\infty}(1-r\lambda_k^2)^{-1}.$$

Thus, it follows that

$$\sum_{t=T}^{n} Q_t \to 0 \text{ in } L_2 \tag{7}$$

for any (slow) $T = T_n \to \infty$. Thus, to prove the claim of the theorem, it suffices to deal with the joint distributional convergence of $(Q_1, \cdots, Q_t)$ for each fixed t.

By central limit theorem and the orthonormality of the ψ_k and ϕ_k, it follows that for any fixed positive integer m, we have

$$\left(n^{-1/2}\sum_{i=1}^{n}\psi_k(X_i), n^{-1/2}\sum_{i=1}^{n}\phi_k(Y_i), k=1,\cdots,m\right) \xrightarrow{d} N(0, I_{2m}) \tag{8}$$

In addition, by the Marcinkiewitz law of large numbers, with probability 1 we have

$$n^{-1}\sum_{i=1}^{n}\psi_{k_1}(X_i)\psi_{k_2}(X_i) \to \delta_{k_1,k_2},\ n^{-1}\sum_{i=1}^{n}\phi_{k_1}(Y_i)\phi_{k_2}(Y_i) \to \delta_{k_1,k_2}, \tag{9}$$

and for $s \geq 3$,

$$n^{-s/2}\sum_{i=1}^{n}\psi_{k_1}(X_i)\cdots\psi_{k_s}(X_i) \to 0,\ n^{-s/2}\sum_{i=1}^{n}\phi_{k_1}(Y_i)\cdots\phi_{k_s}(Y_i) \to 0. \tag{10}$$

Write $W_m(x, y) = \sum_{k=1}^{m} \lambda_k \psi_k(x)\phi_k(y)$ and define

$$\begin{aligned} Q_{mt} &= \frac{(n-t)!}{n!} \sum_{1\le i_1<\cdots<i_t\le n} \sum_{s_1,\cdots,s_t} \prod_{j=1}^{t} W_m(X_{i_j}, Y_{s_j}) \\ &= \sum_{k_1,\cdots,k_t=1}^{m} \prod_{j=1}^{t} \lambda_{k_j} \frac{(n-t)!}{n!} \sum_{1\le i_1<\cdots<i_t\le n} \sum_{s_1,\cdots,s_t} \prod_{j=1}^{t} \psi_{k_j}(X_{i_j})\phi_{k_j}(Y_{s_j}) \\ &= \sum_{k_1,\cdots,k_t=1}^{m} \prod_{j=1}^{t} \lambda_{k_j} \frac{(n-t)!}{t!n!} \left(\sum_{i_1,\cdots,i_t} \prod_{j=1}^{t} \psi_{k_j}(X_{i_j})\right)\left(\sum_{s_1,\cdots,s_t} \prod_{j=1}^{t} \phi_{k_j}(Y_{s_j})\right). \end{aligned} \tag{11}$$

Note that for fixed t, $(n-t)!/n! \sim n^{-t}$ in the above. To find the limit distribution of Q_{mt}, let us consider some special cases first. By (8), we have

$$Q_{m1} = \sum_{k=1}^{m} \lambda_k n^{-1} \sum_{i=1}^{n} \psi_k(X_i) \sum_{i=1}^{n} \phi_k(Y_i) \xrightarrow{d} \sum_{k=1}^{m} \lambda_k U_k V_k$$

and by (8) and (9)

$$Q_{m2} = \sum_{k_1,k_2=1}^{m} \lambda_{k_1}\lambda_{k_2}\frac{1}{2n(n-1)}\Big[\Big(\sum_{i=1}^{n}\psi_{k_1}(X_i)\Big)\Big(\sum_{i=1}^{n}\psi_{k_2}(X_i)\Big)-\sum_{i=1}^{n}\psi_{k_1}(X_i))\psi_{k_2}(X_i)\Big]$$
$$\Big[\Big(\sum_{i=1}^{n}\phi_{k_1}(Y_i)\Big)\Big(\sum_{i=1}^{n}\phi_{k_2}(Y_i)\Big)-\sum_{i=1}^{n}\phi_{k_1}(Y_i))\phi_{k_2}(Y_i)\Big]$$
$$\xrightarrow{d} \frac{1}{2}\sum_{k_1,k_2=1}^{m}\lambda_{k_1}\lambda_{k_2}(U_{k_1}U_{k_2}-\delta_{k_1,k_2})(V_{k_1}V_{k_2}-\delta_{k_1,k_2})$$
$$= \frac{1}{2}\left((\sum_{k=1}^{m}\lambda_k U_k V_k)^2-\sum_{k=1}^{m}\lambda_k^2(U_k^2+V_k^2-1)\right).$$

For $t=3$, we have

$$n^{-3/2}\sum_{i_1,i_2,i_3}\psi_{k_1}(X_{i_1})\psi_{k_2}(X_{i_2})\psi_{k_3}(X_{i_3})$$
$$= n^{-3/2}\prod_{j=1}^{3}\sum_{i=1}^{n}\psi_{k_j}(X_i) - n^{-3/2}\sum_{i=1}^{n}\psi_{k_1}(X_i)\psi_{k_2}(X_i)\sum_{i=1}^{n}\psi_{k_3}(X_i)$$
$$-n^{-3/2}\sum_{i=1}^{n}\psi_{k_1}(X_i)\psi_{k_3}(X_i)\sum_{i=1}^{n}\psi_{k_2}(X_i) - n^{-3/2}\sum_{i=1}^{n}\psi_{k_2}(X_i)\psi_{k_3}(X_i)\sum_{i=1}^{n}\psi_{k_1}(X_i)$$
$$+n^{-3/2}\sum_{i=1}^{n}\psi_{k_1}(X_i)\psi_{k_2}(X_i)\psi_{k_3}(X_i)$$
$$\xrightarrow{d} U_{k_1}U_{k_2}U_{k_3}-\delta_{k_1,k_2}U_{k_3}-\delta_{k_1,k_3}U_{k_2}-\delta_{k_2,k_3}U_{k_1}.$$

Similarly,

$$n^{-3/2}\sum_{s_1,s_2,s_3}\phi_{k_1}(Y_{s_1})\phi_{k_2}(Y_{s_2})\phi_{k_3}(Y_{s_3})$$
$$\xrightarrow{d} V_{k_1}V_{k_2}V_{k_3}-\delta_{k_1,k_2}V_{k_3}-\delta_{k_1,k_3}V_{k_2}-\delta_{k_2,k_3}V_{k_1}.$$

Therefore,

$$Q_{m3} \xrightarrow{d} \frac{1}{6}\left[\left(\sum_{k=1}^{m}\lambda_k U_k V_k\right)^3 - 3\left(\sum_{k=1}^{m}\lambda_k U_k V_k\right)\left(\sum_{k=1}^{m}\lambda_k^2(U_k^2+V_k^2-1)\right)\right.$$
$$\left.+6\sum_{k=1}^{m}\lambda_k^3 U_k V_k\right].$$

In general, by the inclusion-exclusion identity and noticing (10), we have for a given $\mathbf{k} = (k_1, \cdots, k_t)$,

$$
\begin{aligned}
n^{-t/2} & \sum_{i_1,\cdots,i_t} \prod_{j=1}^{t} \psi_{k_j}(X_{i_j}) \\
& \xrightarrow{d} \prod_{j=1}^{t} U_{k_j} - \sum_{\{j_1,j_2\}\subset\{1,\cdots,t\}} \delta_{k_{j_1},k_{j_2}} \prod_{j\notin\{j_1,j_2\}} U_{k_j} \\
& + \sum_{\{j_1,j_2\}\cup\{j_3,j_4\}\subset\{1,\cdots,t\}} \delta_{k_{j_1},k_{j_2}}\delta_{k_{j_3},k_{j_4}} \prod_{j\notin\{j_1,\cdots,j_4\}} U_{k_j} \\
& - \sum_{\{j_1,j_2\}\cup\{j_3,j_4\}\cup\{j_5,j_6\}\subset\{1,\cdots,t\}} \delta_{k_{j_1},k_{j_2}}\delta_{k_{j_3},k_{j_4}}\delta_{k_{j_5},k_{j_6}} \prod_{j\notin\{j_1,\cdots,j_6\}} U_{k_j} + \cdots \\
& =: S_{t,\mathbf{k},1} - S_{t,\mathbf{k},2} + \cdots + (-1)^{[t/2]} S_{t,\mathbf{k},[t/2]+1},
\end{aligned}
$$

where the sum in $S_{t,\mathbf{k},1+\ell}$, $\ell \geq 1$, runs over all possible ℓ pairs of indices $\{\{j_1, j_2\}, \cdots, \{j_{2\ell-1}, j_{2\ell}\}\}$, in which the 2ℓ indices are distinct, with the understanding that, for example, $\{\{j_1, j_2\}, \{j_3, j_4\}\}$ and $\{\{j_1, j_3\}, \{j_2, j_4\}\}$ are two different partitions. By symmetry, we have

$$
\begin{aligned}
n^{-t/2} & \sum_{s_1,\cdots,s_t} \prod_{j=1}^{t} \phi_{k_j}(Y_{s_j}) \\
& \xrightarrow{d} \prod_{j=1}^{t} V_{k_j} - \sum_{\{j_1,j_2\}\subset\{1,\cdots,t\}} \delta_{k_{j_1},k_{j_2}} \prod_{j\notin\{j_1,j_2\}} V_{k_j} \\
& + \sum_{\{j_1,j_2\}\cup\{j_3,j_4\}\subset\{1,\cdots,t\}} \delta_{k_{j_1},k_{j_2}}\delta_{k_{j_3},k_{j_4}} \prod_{j\notin\{j_1,\cdots,j_4\}} V_{k_j} \\
& - \sum_{\{j_1,j_2\}\cup\{j_3,j_4\}\cup\{j_5,j_6\}\subset\{1,\cdots,t\}} \delta_{k_{j_1},k_{j_2}}\delta_{k_{j_3},k_{j_4}}\delta_{k_{j_5},k_{j_6}} \prod_{j\notin\{j_1,\cdots,j_6\}} V_{k_j} + \cdots \\
& =: T_{t,\mathbf{k},1} - T_{t,\mathbf{k},2} + \cdots + (-1)^{[t/2]} T_{t,\mathbf{k},[t/2]+1}.
\end{aligned}
$$

Substituting these two limits into (11), we get

$$
Q_{mt} \xrightarrow{d} \frac{1}{t!} \sum_{\ell=0}^{[t/2]} \sum_{\ell'=0}^{[t/2]} (-1)^{\ell+\ell'} \sum_{k_1,\cdots,k_t=1}^{m} \prod_{j=1}^{t} \lambda_{k_j} S_{t,\mathbf{k},\ell+1} T_{t,\mathbf{k},\ell'+1}. \tag{12}
$$

Then, letting $m \to \infty$, we get the limit distribution of Q_t.

We now proceed to simplify the limiting distribution of Q_t. First note that

$$\begin{aligned}
&\sum_{k_1,\cdots,k_t=1}^{m} \prod_{j=1}^{t} \lambda_{k_j} S_{t,\mathbf{k},1} T_{t,\mathbf{k},1} \\
&\quad = \sum_{k_1,\cdots,k_t=1}^{m} \prod_{j=1}^{t} \lambda_{k_j} \prod_{j=1}^{t} U_{k_j} \prod_{j=1}^{t} V_{k_j} \\
&\quad = \left(\sum_{k=1}^{m} \lambda_k U_k V_k \right)^t \to \left(\sum_{k=1}^{\infty} \lambda_k U_k V_k \right)^t \quad \text{as } m \to \infty. \qquad (13)
\end{aligned}$$

Next,

$$\begin{aligned}
&\sum_{k_1,\cdots,k_t=1}^{m} \prod_{j=1}^{t} \lambda_{k_j} S_{t,\mathbf{k},1} T_{t,\mathbf{k},\ell+1} \\
&\quad = \sum_{\cup_{s=1}^{\ell}\{j_{2s-1},j_{2s}\}\subset\{1,\cdots,t\}} \sum_{k_1,\cdots,k_t=1}^{m} \prod_{j=1}^{t} \lambda_{k_j} U_{k_j} \prod_{s=1}^{\ell} \delta_{k_{2s-1},k_{2s}} \prod_{j\notin\{j_1,\cdots,j_{2s}\}} V_{k_j} \\
&\quad = \frac{t!}{2^{\ell}\ell!(t-2\ell)!} \left(\sum_{k=1}^{m} \lambda_k^2 U_k \right)^{\ell} \left(\sum_{k=1}^{m} \lambda_k^2 U_k V_k \right)^{t-2\ell} \\
&\quad \to \frac{t!}{2^{\ell}\ell!(t-2\ell)!} \left(\sum_{k=1}^{\infty} \lambda_k^2 U_k^2 \right)^{\ell} \left(\sum_{k=1}^{\infty} \lambda_k U_k V_k \right)^{t-2\ell}. \qquad (14)
\end{aligned}$$

More generally, for $\ell, \ell' \geq 1$,

$$\begin{aligned}
&\sum_{k_1,\cdots,k_t=1}^{m} \prod_{j=1}^{t} \lambda_{k_j} S_{t,\mathbf{k},\ell+1} T_{t,\mathbf{k},\ell'+1} \\
&\quad = \sum_{\cup_{s=1}^{\ell}\{j_{2s-1},j_{2s}\}} \sum_{\cup_{s'=1}^{\ell'}\{i_{2s'-1},i_{2s'}\}} \sum_{k_1,\ldots,k_t=1}^{m} \prod_{j=1}^{t} \lambda_{k_j} \prod_{s=1}^{\ell} \delta_{k_{j_{2s-1}},k_{j_{2s}}} \prod_{s'=1}^{\ell'} \delta_{k_{i_{2s'-1}},k_{i_{2s'}}} \\
&\qquad \prod_{j\notin\{j_1,\cdots,j_{2\ell}\}} U_{k_j} \prod_{i\notin\{i_1,\cdots,i_{2\ell'}\}} V_{k_i}. \qquad (15)
\end{aligned}$$

To classify various products in its expression, we first introduce some notation. Let $(u_1, u_2, \ldots, u_h)$ be a sequence of distinct integers from $\{1, 2, \cdots, t\}$. It is said to be a

- a cycle of length h (even) if $\{\{v_1, v_2\}, \cdots, \{v_{h-1}, v_h\}\} \subset \{\{j_1, j_2\}, \cdots, \{j_{2\ell-1}, j_{2\ell}\}\}$ and $\{\{v_2, v_3\}, \cdots, \{v_{h-2}, v_{h-1}\}, \{v_h, v_1\}\} \subset \{\{i_1, i_2\}, \cdots, \{i_{2\ell-1}, i_{2\ell}\}\}$, where $(v_1, v_2, \ldots, v_h)$ is any permutation of $(u_1, u_2, \ldots, u_h)$.
- a vv-chain of length h (even) if $\{\{u_1, u_2\}, \cdots, \{u_{h-1}, u_h\}\} \subset \{\{j_1, j_2\}, \cdots, \{j_{2\ell-1}, j_{2\ell}\}\}$ and $\{\{u_2, u_3\}, \cdots, \{u_{h-2}, u_{h-1}\}\} \subset \{\{i_1, i_2\}, \cdots, \{i_{2\ell-1}, i_{2\ell}\}\}$ and $u_1, u_h \notin \{i_1, \cdots, i_{2\ell'}\}$.

- a uu-chain of length h (even) if $\{\{u_2, u_3\}, \cdots, \{u_{h-2}, u_{h-1}\}\} \subset \{\{j_1, j_2\}, \cdots, \{j_{2\ell-1}, j_{2\ell}\}\}$ and $\{\{u_1, u_2\}, \cdots, \{u_{h-1}, u_h\}\} \subset \{\{i_1, i_2\}, \cdots, \{i_{2\ell-1}, i_{2\ell}\}\}$ and $u_1, u_h \notin \{j_1, \cdots, j_{2\ell}\}$.
- a uv-chain of length h (odd) if $\{\{u_1, u_2\}, \cdots, \{u_{h-2}, u_{h-1}\}\} \subset \{\{j_1, j_2\}, \cdots, \{j_{2\ell-1}, j_{2\ell}\}\}$ and $\{\{u_2, u_3\}, \cdots, \{u_{h-1}, u_h\}\} \subset \{\{i_1, i_2\}, \cdots, \{i_{2\ell-1}, i_{2\ell}\}\}$ and $u_1 \notin \{i_1, \cdots, i_{2\ell'}\}$, $u_h \notin \{j_1, \cdots, j_{2\ell}\}$, or alternatively.

An integer $u \le t$ is called a singleton if $u \notin \{j_1, \cdots, j_{2\ell}\} \cup \{i_1, \cdots, i_{2\ell'}\}$. A singleton corresponds to a factor $\sum_{k=1}^m \lambda_k U_k V_k$. A singleton can be considered as a uv-chain of length 1. In the sequel, we shall not specify singletons.

Observe that for each given partitions $\{\{j_1, j_2\}, \cdots, \{j_{2\ell-1}, j_{2\ell}\}\}$ and $\{\{i_1, i_2\}, \cdots, \{i_{2\ell'-1}, i_{2\ell'}\}\}$, the set of numbers $\{1, 2, , \ldots, t\}$ can be uniquely partitioned into disjoint sets each of which is a cycle or contains a uu, vv or a uv-chain. As a simple illustration, let $t = 12$ and consider the partitions

$$\{\{j_1, j_2\}, \cdots, \{j_{2\ell-1}, j_{2\ell}\}\} = \{\{1, 2\}, \{3, 4\}, \{5, 6\}, \{7, 8\}, \{10, 11\}\},$$

and

$$\{\{i_1, i_2\}, \cdots, \{i_{2\ell'-1}, i_{2\ell'}\}\} = \{\{2, 3\}, \{5, 7\}, \{6, 8\}, \{11, 12\}\}.$$

Then $(5, 6, 7, 8)$ is a cycle, $(1, 2, 3, 4)$ is a vv-chain, 9 is a uv-chain (which is also a singleton), and $(10, 11, 12)$ is a uv-chain. Then it is easy to see that for this example,

$$\sum_{k_1,\ldots,k_t=1}^m \prod_{j=1}^t \lambda_{k_j} \prod_{s=1}^{\ell} \delta_{k_{j_{2s-1}}, k_{j_{2s}}} \prod_{s'=1}^{\ell'} \delta_{k_{i_{2s'-1}}, k_{i_{2s'}}} \prod_{j \notin \{j_1,\cdots,j_{2\ell}\}} U_{k_j} \prod_{i \notin \{i_1,\cdots,i_{2\ell'}\}} V_{k_i}$$
$$= \left(\sum_{k=1}^m \lambda_k^4\right)\left(\sum_{k=1}^m \lambda_k^4 V_k^2\right)\left(\sum_{k=1}^m \lambda_k U_k V_k\right)\left(\sum_{k=1}^m \lambda_k^3 U_k V_k\right).$$

Observe that each cycle, vv, uu and uv-chain of length h produce factors $\sum_{k=1}^m \lambda_k^h$, $\sum_{k=1}^m \lambda_k^h V_k^2$, $\sum_{k=1}^m \lambda_k^h U_k^2$ and $\sum_{k=1}^m \lambda_k^h U_k V_k$, respectively, in such a computation.

In general, for any given partitions $\{\{j_1, j_2\}, \cdots, \{j_{2\ell-1}, j_{2\ell}\}\}$ and $\{\{i_1, i_2\}, \cdots, \{i_{2\ell'-1}, i_{2\ell'}\}\}$, denote by ν_{h1}, the numbers of cycles of length h, and ν_{h2}, nu_{h3} and ν_{h4} the number of vv, uu and uv-chains of length h, respectively. Note that $\sum_{h,i} h\nu_{hi} = t$. Then

$$\sum_{k_1,\ldots,k_t=1}^m \prod_{j=1}^t \lambda_{k_j} \prod_{s=1}^{\ell} \delta_{k_{j_{2s-1}}, k_{j_{2s}}} \prod_{s'=1}^{\ell'} \delta_{k_{i_{2s'-1}}, k_{i_{2s'}}} \prod_{j \notin \{j_1,\cdots,j_{2\ell}\}} U_{k_j} \prod_{i \notin \{i_1,\cdots,i_{2\ell'}\}} V_{k_i}$$
$$= \prod_{h \text{ even}} \left(\sum_{k=1}^m \lambda_k^h\right)^{\nu_{h1}} \left(\sum_{k=1}^m \lambda_k^h V_k^2\right)^{\nu_{h2}} \left(\sum_{k=1}^m \lambda_k^h U_k^2\right)^{\nu_{h3}} \prod_{h \text{ odd}} \left(\sum_{k=1}^m \lambda_k^h U_k V_k\right)^{\nu_{h4}}.$$

In addition to $\sum_{h,i} h\nu_{hi} = t$, we should also have the constraints $\nu_{h1} = \nu_{h2} = \nu_{h3} = 0$ for odd h, $\nu_{h4} = 0$ for even h, $\sum_h (2\nu_{h3} + \nu_{h4}) = t - 2\ell$ and $\sum_h (2\nu_{h2} + \nu_{h4}) = t - 2\ell'$. These constraints come directly from the definition; for example, each uv-chain and vv-chain correspond to one and two elements in $1, \ldots, t$, respectively, that are not in $\{\{i_1, i_2\}, \cdots, \{i_{2\ell'-1}, i_{2\ell'}\}\}$.

While each j-i combination, namely, $\{\{j_{2s-1}, j_{2s}\}, s = 1, \cdots, \ell\}$ and $\{\{i_{2s-1}, i_{2s}\}, s = 1, \cdots, \ell'\}$, generates a unique partition of $1, \ldots, t$, the converse is not true. For our proof it is easier to start from the partitions of $1, \ldots, t$ and compute the number of j-i combinations that correspond to the partition. Given a set of ν_{hi}, $h = 1, \cdots, t$, $i = 1, 2, 3, 4$, partition $\{1, 2, \cdots, t\}$ into groups $G_{hi}^{(s)}$ of sizes h, $s = 1, 2, \cdots, \nu_{hi}$, $h = 1, \cdots, t$, $i = 1, 2, 3, 4$. The number of such partitions is

$$\frac{t!}{\prod_{h=1}^{t}\prod_{i=1}^{4} \nu_{hi}!(h!)^{\nu_{hi}}}.$$

For a given partition of $G_{hi}^{(s)}$, we compute the number of possible j-i pairs, namely, sets $\{\{j_{2s-1}, j_{2s}\}, s = 1, \cdots, w\}$ and $\{\{i_{2s-1}, i_{2s}\}, s = 1, \cdots, w'\}$, using elements of $G_{hi}^{(s)}$.

First consider any $G_{h1}^{(s)}$ with $h \geq 2$ even. Observe that there is a one-one correspondence between the circular permutations of the elements of $G_{h1}^{(s)}$ and the j-i pairs. Hence there correspond $(h-1)!$ possible j-i pairs.

Next consider any $G_{h2}^{(s)}$ with $h \geq 2$ even. For each permutation $(u_1, \cdots, u_h)$ of elements of $G_{h1}^{(s)}$, we construct the i-j pairs by putting $\{u_1, u_2\}, \cdots, \{u_{h-1}, u_h\}$ into the j-pairs, $\{u_2, u_3\}, \cdots, \{u_{h-2}, u_{h-1}\}$ into the i-pairs, and leaving u_1, u_h as the unpaired i's. Since the reverse permutation gives the same j-i partition, we have $h!/2$ ways to construct all possible j-i pairs.

Symmetrically, for each $G_{h3}^{(s)}$ with $h \geq 2$ even, we have $h!/2$ ways to construct all possible j-i pairs.

Finally consider any $G_{h4}^{(s)}$ with $h \geq 2$ odd. For each permutation $(u_1, \cdots, u_h)$ of elements of $G_{h1}^{(s)}$, we construct the i-j pairs by putting $\{u_1, u_2\}, \cdots, \{u_{h-2}, u_{h-1}\}$ into the j-pairs and $\{u_2, u_3\}, \cdots, \{u_{h-1}, u_h\}$ into the i-pairs and leaving u_1 as an unpaired i and u_h as an unpaired j. Thus, there are $h!$ ways to construct the possible $j - i$ pairs.

Put all integers in $G_{13}^{(s)}$ as both unpaired j's as well as the unpaired i's. Combining these, by (15), we have

$$\begin{aligned}
&\sum_{k_1, \cdots, k_t=1}^{m} \prod_{j=1}^{t} \lambda_{k_j} S_{t,\mathbf{k},\ell+1} T_{t,\mathbf{k},\ell'+1} \\
&= \sum{}^{*} t! \prod_{h \text{ even}} \frac{1}{\nu_{h1}!\nu_{h2}!\nu_{h3}!} \left(\frac{1}{h}\sum_{k=1}^{m} \lambda_k^h\right)^{\nu_{h1}} \left(\frac{1}{2}\sum_{k=1}^{m} \lambda_k^h V_k^2\right)^{\nu_{h2}} \\
&\quad \left(\frac{1}{2}\sum_{k=1}^{m} \lambda_k^h U_k^2\right)^{\nu_{h3}} \prod_{h \text{ odd}} \frac{1}{\nu_{h4}!} \left(\sum_{k=1}^{m} \lambda_k^h U_k V_k\right)^{\nu_{h4}} \\
&\to \sum{}^{*} t! \prod_{h \text{ even}} \frac{1}{\nu_{h1}!\nu_{h2}!\nu_{h3}!} \left(\frac{1}{h}\sum_{k=1}^{\infty} \lambda_k^h\right)^{\nu_{h1}} \left(\frac{1}{2}\sum_{k=1}^{\infty} \lambda_k^h V_k^2\right)^{\nu_{h2}}
\end{aligned}$$

$$\left(\frac{1}{2}\sum_{k=1}^{\infty}\lambda_k^h U_k^2\right)^{\nu_{h3}} \prod_{h \text{ odd}} \frac{1}{\nu_{h4}!}\left(\sum_{k=1}^{\infty}\lambda_k^h U_k V_k\right)^{\nu_{h4}},$$

where $\sum^*$ runs over all possible integers ν_{hi} subject to $\sum_{h,i} h\nu_{hi} = t$, $\nu_{h1} = \nu_{h2} = \nu_{h3} = 0$ for odd h, $\nu_{h4} = 0$ for even h, $\sum_h(2\nu_{h3} + \nu_{h4}) = t - 2\ell$ and $\sum_h(2\nu_{h2} + \nu_{h4}) = t - 2\ell'$.

Note that the results in (13) – (15) can also be written in the above form by letting $\ell = 0$ or $\ell' = 0$ or $\ell = \ell' = 0$. In addition, we have

$$(-1)^{\ell+\ell'} = (-1)^{t+\sum_{h=1}^t(\nu_{h2}+\nu_{h3}+\nu_{h4})}.$$

Thus,

$$Q_t \xrightarrow{d} (-1)^t \sum{}^{**} \prod_{h \text{ even}} \frac{1}{\nu_{h1}!\nu_{h2}!\nu_{h3}!}\left(\frac{1}{h}\sum_{k=1}^{\infty}\lambda_k^h\right)^{\nu_{h1}}\left(-\frac{1}{2}\sum_{k=1}^{\infty}\lambda_k^h V_k^2\right)^{\nu_{h2}}$$
$$\left(-\frac{1}{2}\sum_{k=1}^{\infty}\lambda_k^h U_k^2\right)^{\nu_{h3}} \prod_{h \text{ odd}} \frac{1}{\nu_{h4}!}\left(-\sum_{k=1}^{\infty}\lambda_k^h U_k V_k\right)^{\nu_{h4}},$$

where $\sum^{**}$ runs over all possible integers ν_{hi} subject to $\sum_{h,i} h\nu_{hi} = t$, $\nu_{h1} = \nu_{h2} = \nu_{h3} = 0$ for odd h, $\nu_{h4} = 0$ for even h. We further simplify it as

$$Q_t \xrightarrow{d} (-1)^t \sum{}^{***} \prod_{h \text{ even}} \frac{1}{\nu_{h0}!}\left(\sum_{k=1}^{\infty}\lambda_k^h\left(\frac{1}{h} - \frac{1}{2}U_k^2 - \frac{1}{2}V_k^2\right)\right)^{\nu_{h0}}$$
$$\prod_{h \text{ odd}} \frac{1}{\nu_{h4}!}\left(-\sum_{k=1}^{\infty}\lambda_k^h U_k V_k\right)^{\nu_{h4}}$$
$$=: \xi_t,$$

where the $\sum^{***}$ is taken for $\sum_h h(\nu_{h0} + \nu_{h4}) = t$.

Note that ξ_t is the coefficient of z^t in the Taylor expansion of

$$H(z) = \exp\left(\sum_{\substack{h\ge 2\\ \text{even}}}(-z)^h\sum_{k=1}^{\infty}\lambda_k^h\left(\frac{1}{h} - \frac{1}{2}U_k^2 - \frac{1}{2}V_k^2\right) - \sum_{\substack{h\ge 1\\ \text{odd}}}(-z)^h\sum_{k=1}^{\infty}\lambda_k^h U_k V_k\right).$$

Straightforward algebra shows that

$$\sum \xi_t = H(1) = \prod_{k=1}^{\infty}(1 - \lambda_{k,\rho}^2)^{-1/2}$$
$$\exp\left(-\frac{1}{2}\sum_{k=1}^{\infty}\frac{\lambda_{k,\rho}^2}{1 - \lambda_{k,\rho}^2}(U_k^2 + V_k^2) + \sum_{k=1}^{\infty}\frac{\lambda_{k,\rho}}{1 - \lambda_{k,\rho}^2}U_k V_k\right).$$

Making an orthogonal transformation with $U_k' = (U_k + V_k)/\sqrt{2}$, $V_k' = (U_k - V_k)/\sqrt{2}$, the right hand side is easily seen to have the same distribution as ξ described in the theorem. This concludes the proof. □

3.2. Proof of Theorem 1

Let $\mathcal{T}_n$ be a consistent discrimination rule for ρ_1 versus ρ_2. Write, for any fixed $M \in (0, \infty)$,

$$P_{\rho_1}(\mathcal{T}_n = \rho_1) = P_{\rho_1}(\mathcal{T}_n = \rho_1, g_{n,\rho_1}(\boldsymbol{X}, \boldsymbol{Y}) > Mg_{n,\rho_2}(\boldsymbol{X}, \boldsymbol{Y})) \\ + P_{\rho_1}(\mathcal{T}_n = \rho_1, g_{n,\rho_1}(\boldsymbol{X}, \boldsymbol{Y}) \leq Mg_{n,\rho_2}(\boldsymbol{X}, \boldsymbol{Y})).$$

Suppose that $P_{\rho_1}(g_{n,\rho_1}(\boldsymbol{X}, \boldsymbol{Y}) > Mg_{n,\rho_2}(\boldsymbol{X}, \boldsymbol{Y})) \not\to 1$. Since $P_{\rho_1}(\mathcal{T}_n = \rho_1) \to 1$, we have $P_{\rho_1}(\mathcal{T}_n = \rho_1, g_{n,\rho_1}(\boldsymbol{X}, \boldsymbol{Y}) \leq Mg_{n,\rho_2}(\boldsymbol{X}, \boldsymbol{Y})) \not\to 0$. Thus

$$\limsup_{n\to\infty} P_{\rho_2}(\mathcal{T}_n = \rho_1) \geq \limsup_{n\to\infty} M^{-1} P_{\rho_1}(\mathcal{T}_n = \rho_1, g_{n,\rho_1}(\boldsymbol{X}, \boldsymbol{Y}) \\ \leq Mg_{n,\rho_2}(\boldsymbol{X}, \boldsymbol{Y})) > 0.$$

This contradicts the assumption that $\mathcal{T}_n$ is consistent. Hence, we must have

$$P_{\rho_1}(g_{n,\rho_1}(\boldsymbol{X}, \boldsymbol{Y}) > Mg_{n,\rho_2}(\boldsymbol{X}, \boldsymbol{Y})) \to 1 \text{ for all } M \in (0, \infty). \tag{16}$$

Now,

$$\begin{aligned} E_{\rho_o} g^2_{n,\rho_1}(\boldsymbol{X}, \boldsymbol{Y}) &\geq \int g^2_{n,\rho_1}(\boldsymbol{x}, \boldsymbol{y}) I\big(g_{n,\rho_1}(\boldsymbol{x}, \boldsymbol{y}) > Mg_{n,\rho_2}(\boldsymbol{x}, \boldsymbol{y})\big) d\boldsymbol{x} d\boldsymbol{y} \qquad (17) \\ &\geq M \int g_{n,\rho_1}(\boldsymbol{x}, \boldsymbol{y}) g_{n,\rho_2}(\boldsymbol{x}, \boldsymbol{y}) I\big(g_{n,\rho_1}(\boldsymbol{x}, \boldsymbol{y}) > Mg_{n,\rho_2}(\boldsymbol{x}, \boldsymbol{y})\big) d\boldsymbol{x} d\boldsymbol{y} \\ &\geq Md \int g_{n,\rho_1}(\boldsymbol{x}, \boldsymbol{y}) I\big(g_{n,\rho_1}(\boldsymbol{x}, \boldsymbol{y}) > Mg_{n,\rho_2}(\boldsymbol{x}, \boldsymbol{y}),\ g_{n,\rho_2}(\boldsymbol{x}, \boldsymbol{y}) > d\big) d\boldsymbol{x} d\boldsymbol{y} \\ &\geq Md \int g_{n,\rho_1}(\boldsymbol{x}, \boldsymbol{y}) I\big(g_{n,\rho_1}(\boldsymbol{x}, \boldsymbol{y}) > Mg_{n,\rho_2}(\boldsymbol{x}, \boldsymbol{y})\big) d\boldsymbol{x} d\boldsymbol{y} \\ &\quad - Md \int g_{n,\rho_1}(\boldsymbol{x}, \boldsymbol{y}) I\big(g_{n,\rho_2}(\boldsymbol{x}, \boldsymbol{y}) \leq d\big) d\boldsymbol{x} d\boldsymbol{y}. \end{aligned}$$

Choose $M = M_n \to \infty$ so slowly that (16) still holds when M is replaced by M_n. Then choose $d = d_n \to 0$ so slowly such that $M_n d_n \to \infty$. By Theorem 2 and the fact that ξ has a continuous distribution,

$$P_{\rho_o}(g_{n,\rho_2}(\boldsymbol{X}, \boldsymbol{Y}) \leq d) \to 0.$$

Hence, by Lemma 5 and the Cauchy-Schwarz inequality,

$$\int g_{n,\rho_1}(\boldsymbol{x}, \boldsymbol{y}) I\big(g_{n,\rho_2}(\boldsymbol{x}, \boldsymbol{y}) \leq d\big) d\boldsymbol{x} d\boldsymbol{y} \to 0. \tag{18}$$

It follows from (16), (18), and the choices of M, d that the last expression in (17) tends to ∞, which implies that

$$E_{\rho_o} g^2_{n,\rho_1}(\boldsymbol{X}, \boldsymbol{Y}) \to \infty.$$

This contradicts Lemma 5 and concludes the proof. □

3.3. Proof of Theorem 3

We begin by mentioning the following result due to Hajék (1958) specialized and simplified to our setting. See also Grenander (1981) and Rozanov (1971). Let $\mathcal{F}$ be the σ-field of the product space $D[0, 1] \times D[0, 1]$. Let Q_1 and Q_2 be two probability measures on $(D[0, 1] \times D[0, 1], \mathcal{F})$ which each correspond to the distribution of a Gaussian process. Let $\mathcal{F}_n$ be a sequence of sub σ-fields of $\mathcal{F}$ with $\mathcal{F} = \sigma(\cup_n \mathcal{F}_n)$. The Kulback-Leibler information number between Q_1, Q_2 with respect to $\mathcal{F}_n$ is

$$J_n = E_{Q_1}(-\log q_n) + E_{Q_2}(\log q_n)$$

where q_n is the Radon-Nikodym derivative of the absolutely continuous part of Q_2 with respect to Q_1 on $(\Omega, \mathcal{F}_n)$.

Theorem 6. *Q_1 and Q_2 are equivalent on $(D[0, 1] \times D[0, 1], \mathcal{F})$ iff $\sup_n J_n < \infty$.*

Proof of Theorem 3. First we show that

$$\lim_{m\to\infty} \min_{1\le i\le m} \max_{1\le j\le m} m \int_{x=\frac{i-1}{m}}^{\frac{i}{m}} \int_{y=\frac{j-1}{m}}^{\frac{j}{m}} f_\rho(x, y)dxdy = 0. \tag{19}$$

To do that, choose δ_m and split the integral in (19) into two parts according to $f_\rho < \delta_m$ or not. Then

$$\begin{aligned}
&\min_{1\le i\le m} \max_{1\le j\le m} m \int_{x=\frac{i-1}{m}}^{\frac{i}{m}} \int_{y=\frac{j-1}{m}}^{\frac{j}{m}} f_\rho(x, y)dxdy \\
&\le \frac{\delta_m}{m} + \frac{m}{\delta_m} \min_i \int_{(i-1)/m}^{i/m} \int_0^1 f_\rho^2(x, y)dxdy. \qquad (20)
\end{aligned}$$

By the condition (HS),

$$\min_i \int_{(i-1)/m}^{i/m} \int_0^1 f_\rho^2(x, y)dxdy \to 0.$$

Thus, we may select a sequence δ_m such that $\delta_m/m \to 0$ so slowly that the second term of (20) tends to 0, which proves (19). It follows from (19) that there exists a sequence $i_m \in \{1, \ldots, m\}$ such that

$$\lim_{m\to\infty} \max_{1\le j\le m} mP\left(\frac{i_m - 1}{m} < X \le \frac{i_m}{m}, \frac{j-1}{m} < Y \le \frac{j}{m}\right) = 0. \tag{21}$$

We will apply Theorem 6 by letting $\mathcal{F}_m = \sigma\{Z_1(\frac{i-1}{m}) - Z_1(\frac{i}{m}), Z_2(\frac{i-1}{m}) - Z_2(\frac{i}{m}) : i \in \{1, \ldots, m\} - \{i_m\}\}$ and $\mathcal{F} = \sigma\{Z_1(t), Z_2(t), t \in (0, 1)\}$. Since almost all paths of the Gaussian process are continuous, we have $\mathcal{F} = \sigma(\cup \mathcal{F}_m)$. The joint pdf of

$$\begin{aligned}
\boldsymbol{U} = \Bigg(&Z_1\left(\frac{i-1}{m}\right) - Z_1\left(\frac{i}{m}\right), Z_2\left(\frac{i-1}{m}\right) \\
&- Z_2\left(\frac{i}{m}\right) : i \in \{1, \ldots, m\} - \{i_m\}\Bigg)
\end{aligned}$$

is $(2\pi)^{-(m-1)/2}|\Sigma_{m,\rho}|^{-1/2}\exp\left(-\frac{1}{2}\boldsymbol{u}'\Sigma_{m,\rho}^{-1}\boldsymbol{u}\right)$ where $\Sigma_{m,\rho}$ is the covariance matrix of $\boldsymbol{U}$, namely

$$\Sigma_{m,\rho} = \begin{bmatrix} A_m & B_{m,\rho} \\ B'_{m,\rho} & A_m \end{bmatrix}$$

where A_m $((m-1)\times(m-1))$ is the auto-covariance of $(Z_k(\frac{i-1}{m}) - Z_k(\frac{i}{m}), \in \{1,\dots,m\} - \{i_m\}, k = 1, 2)$, and $B_{m,\rho}$ $((m-1)\times(m-1))$ is the cross-covariance of $(Z_1(\frac{i-1}{m}) - Z_1(\frac{i}{m}), i \in \{1,\dots,m\} - \{i_m\})$ with $(Z_2(\frac{i-1}{m}) - Z_2(\frac{i}{m}) : i \in \{1,\dots,m\} - \{i_m\})$. Clearly, $A_m = \frac{1}{m}I_{m-1} - \frac{1}{m^2}\mathbf{1}\mathbf{1}'$ and

$$B_{m,\rho} = \left[\mathcal{P}_\rho\left(\frac{i-1}{m} < X \le \frac{i}{m}, \frac{j-1}{m} < Y \le \frac{j}{m}\right)\right.$$
$$\left. -\frac{1}{m^2},\ i, j \in \{1,\dots,m\} - \{i_m\}\right], \tag{22}$$

where $\mathbf{1}$ the $(m-1)$-vector of all elements 1. It is well-known (cf. Parzen, 1963) that the Kulback-Leibler information between $\mathcal{P}_\rho$ and $\mathcal{P}_{\rho_o}$ with respect to $\mathcal{F}_m$ is

$$J_m = \frac{1}{2}\{\mathrm{tr}(\Sigma_{m,\rho_o}^{-1}\Sigma_{m,\rho}) + \mathrm{tr}(\Sigma_{m,\rho}^{-1}\Sigma_{m,\rho_o}) - 4(m-1)\}.$$

By assumption $B_{m,\rho_o} = 0$ and hence

$$\mathrm{tr}(\Sigma_{m,\rho_o}^{-1}\Sigma_{m,\rho}) = 2(m-1).$$

As a result,

$$J_m = \frac{1}{2}\mathrm{tr}(\Sigma_{m,\rho}^{-1}\Sigma_{m,\rho_o}) - (m-1)$$

which, by Lemma 8 below, is equal to

$$\begin{aligned}\mathrm{tr}\{[I_{m-1} - A_m^{-1}B_{m,\rho}A_m^{-1}B'_{m,\rho}]^{-1}\} - (m-1) &= \sum_{i=1}^{m-1}(1-\lambda_{m,i}^2)^{-1} - (m-1)\\ &= \sum_{i=1}^{m-1}\frac{\lambda_{m,i}^2}{1-\lambda_{m,i}^2},\end{aligned}$$

where $\lambda_{m,1} \ge \dots \ge \lambda_{m,m-1} \ge 0$ are the singular values of $A_m^{-1/2}B_{m,\rho}A_m^{-1/2}$. Note that these are the canonical correlations between $(Z_1(\frac{i-1}{m}) - Z_1(\frac{i}{m}), i \in \{1,\dots,m\} - \{i_m\})$ and $(Z_2(\frac{i-1}{m}) - Z_2(\frac{i}{m}) : i \in \{1,\dots,m\} - \{i_m\})$ and so $\lambda_{m,1} \le 1$. It follows from (24) of Lemma 7 that

$$J_m \to \sum_{i=1}^{\infty}\frac{\lambda_i^2}{1-\lambda_i^2} < \infty$$

where $\lambda_i = \lambda_{i,\rho}$. By Theorem 6, we have $\mathcal{P}_\rho \equiv \mathcal{P}_{\rho_o}$. This concludes the proof. □

Lemma 7. *Assume that the conditions of Theorem 3 hold. Let $\lambda_{m,i}$ be as defined in the proof of Theorem 3 and $\lambda_i = \lambda_{i,\rho}$. Then*

$$\lim_{m\to\infty}\sum_{i=1}^{\infty}(\lambda_{m,i}-\lambda_i)^2 = 0, \tag{23}$$

where $\lambda_{m,i} = 0$ if $i \geq m$, and

$$\lim_{m\to\infty}\sum_{i=1}^{m-1}\frac{\lambda_{m,i}^2}{1-\lambda_{m,i}^2} = \sum_{i=1}^{\infty}\frac{\lambda_i^2}{1-\lambda_i^2} < \infty. \tag{24}$$

Proof. We first prove (23). Let A_m and $B_{m,\rho}$ be as given in the proof of Theorem 3. Note that

$$A_m^{-1} = (m^{-1}(I_{m-1} - m^{-1}\mathbf{l}\mathbf{l}'))^{-1} = m(I_{m-1} + \mathbf{l}\mathbf{l}') = mM\begin{bmatrix} m & 0 \\ 0 & I_{m-2}\end{bmatrix}M',$$

where M is an $(m-1)\times(m-1)$ orthogonal matrix with the first column $\mathbf{l}/\sqrt{m-1}$. This implies that

$$A_m^{-1/2} = \sqrt{m}M\begin{bmatrix}\sqrt{m} & 0 \\ 0 & I_{m-2}\end{bmatrix}M',$$

Clearly, $\{\lambda_{m,i}\}$ are the singular values of

$$\begin{aligned} A_m^{-1/2}B_{m,\rho}A_m^{-1/2} &= mM\begin{bmatrix}\sqrt{m} & 0 \\ 0 & I_{m-2}\end{bmatrix}M'B_{m,\rho}M\begin{bmatrix}\sqrt{m} & 0 \\ 0 & I_{m-2}\end{bmatrix}M' \\ &= mM\begin{bmatrix}\frac{m}{m-1}\mathbf{l}'B_{m,\rho}\mathbf{l} & \sqrt{\frac{m}{m-1}}\mathbf{l}'B_{m,\rho}M_1 \\ \sqrt{\frac{m}{m-1}}M_1'B_{m,\rho}\mathbf{l} & M_1'B_{m,\rho}M_1\end{bmatrix}M', \end{aligned}$$

where M_1 consists of the last $(m-2)$ columns of M. Thus, $\lambda_{m,1},\cdots,\lambda_{m,m-1}$ are also the singular values of

$$m\begin{bmatrix}\frac{m}{m-1}\mathbf{l}'B_{m,\rho}\mathbf{l} & \sqrt{\frac{m}{m-1}}\mathbf{l}'B_{m,\rho}M_1 \\ \sqrt{\frac{m}{m-1}}M_1'B_{m,\rho}\mathbf{l} & M_1'B_{m,\rho}M_1\end{bmatrix}.$$

Let $\eta_{m,1} \geq \cdots \geq \eta_{m,m-1}$ be the singular values of $mM'B_{m,\rho}M$. By Lemma A1, we have

$$\begin{aligned}\sum_{i=1}^{m-1}(\lambda_{m,i}-\eta_{m,i})^2 &\leq m^2\Big[(\mathbf{l}'B_{m,\rho}\mathbf{l})^2 + \frac{(\sqrt{m}-1)^2}{m-1}\mathbf{l}'B_{m,\rho}M_1M_1'B_{m,\rho}'\mathbf{l}\Big] \\ &\leq m^2\Big[(\mathbf{l}'B_{m,\rho}\mathbf{l})^2 + \mathbf{l}'B_{m,\rho}B_{m,\rho}'\mathbf{l}\Big].\end{aligned}$$

We shall show the right hand side of the above inequality tends to 0. By (21) and (22),

$$\begin{aligned}\mathbf{1}'B_{m,\rho}\mathbf{1} &= \operatorname{cov}\left(I\left(\frac{i_m-1}{m}<X\le\frac{i_m}{m}\right), I\left(\frac{i_m-1}{m}<Y\le\frac{i_m}{m}\right)\right)\\ &= P\left(\frac{i_m-1}{m}<X\le\frac{i_m}{m}, \frac{i_m-1}{m}<Y\le\frac{i_m}{m}\right)-m^{-2}=o(m^{-1}).\end{aligned}$$

Similarly,

$$\begin{aligned}\mathbf{1}'B_{m,\rho}B'_{m,\rho}\mathbf{1} &= \sum_{j\ne i_m}\operatorname{cov}^2\left(I\left(\frac{i_m-1}{m}<X\le\frac{i_m}{m}\right), I\left(\frac{j-1}{m}<Y\le\frac{j}{m}\right)\right)\\ &\le 2\sum_{j\ne i_m}\left[P^2\left(\frac{i_m-1}{m}<X\le\frac{i_m}{m}, \frac{j-1}{m}<Y\le\frac{j}{m}\right)+m^{-4}\right]\\ &\le 2o(m^{-1})\sum_{j\ne i_m}P\left(\frac{i_m-1}{m}<X\le\frac{i_m}{m}, \frac{j-1}{m}<Y\le\frac{j}{m}\right)+o(m^{-3})\\ &= o(m^{-2})\end{aligned}$$

where the second inequality follows again from (21). Thus, to complete the proof of the lemma, we need only show that

$$\sum_{i=1}^{\infty}(\eta_{m,i}-\lambda_i)^2\to 0,$$

where $\eta_{m,i}=0$ if $i\ge m$. Let the singular decomposition of $mB_{m,\rho}$ be given by

$$mB_{m,\rho}=\sum_{\substack{1\le i\le m\\ i\ne i_m}}\eta_{m,i}\boldsymbol{\psi}_{m,i}\boldsymbol{\phi}'_{m,i}, \tag{25}$$

where $\boldsymbol{\psi}_{m,i}=(\psi_{m,i,1}, i\in\{1,\dots,m\}-\{i_m\})'$ and $\boldsymbol{\phi}_{m,i}=(\phi_{m,i,1}, i\in\{1,\dots,m\}-\{i_m\})'$. Now, define

$$p_m(x,y)=\sum_{\substack{1\le i,j\le m\\ i,j\ne i_m}}(\xi_{i,j}-1)I_{[\frac{i-1}{m},\frac{i}{m})\times[\frac{j-1}{m},\frac{j}{m})}(x,y), \tag{26}$$

where $\xi_{i,j}=m^2P(\frac{i-1}{m}<X\le\frac{i}{m}, \frac{j-1}{m}<Y\le\frac{j}{m})$, and also define

$$\begin{aligned}\psi_{m,i}(x) &= \sqrt{m}\sum_{\substack{1\le k\le m\\ k\ne i_m}}\psi_{m,i,k}I_{[\frac{k-1}{m},\frac{k}{m})}(x),\\ \phi_{m,i}(y) &= \sqrt{m}\sum_{\substack{1\le k\le m\\ k\ne i_m}}\phi_{m,i,k}I_{[\frac{k-1}{m},\frac{k}{m})}(y).\end{aligned}$$

Note that

$$\int_0^1 \psi_{m,i}(x)dx = \int_0^1 \phi_{m,i}(y)dy = 0,$$

$$\int_0^1 \psi_{m,i}(x)\psi_{m,j}(x)dx = \int_0^1 \phi_{m,i}(y)\phi_{m,j}(y)dy = \delta_{i,j}.$$

With these, it is easy to verify that the singular-value decomposition (25) can be rewritten as the singular-value decomposition of the linear transformation T_{p_m} : $g \to \int p_m(\cdot, y)g(y)dy$ as

$$p_m(x, y) = \sum_{\substack{1 \le i \le m \\ i \ne i_m}} \eta_{m,i}\psi_{m,i}(x)\phi_{m,i}(y)$$

For any $\delta > 0$, define

$$\xi_{i,j,\delta} = m^2 \int_{\frac{i-1}{m}}^{\frac{i}{m}} \int_{\frac{j-1}{m}}^{\frac{j}{m}} f_{\rho,\delta}(x, y)dxdy.$$

Thus, by Lemma A1 and (26), we have

$$\begin{aligned}
&\sum_{i=1}^{\infty}(\eta_{m,i} - \lambda_i)^2 \\
&\le \int_0^1 \int_0^1 (p_m(x, y) - f_\rho(x, y) + 1)^2 dxdy \\
&= \int_0^1 \int_0^1 I\left(x \text{ or } y \in \left[\frac{i_m - 1}{m}, \frac{i_m}{m}\right)\right)(f_\rho(x, y) - 1)^2 dxdy \\
&+ \sum_{\substack{1 \le i,j \le m \\ i,j \ne i_m}} \int_{\frac{i-1}{m}}^{\frac{i}{m}} \int_{\frac{j-1}{m}}^{\frac{j}{m}} (\xi_{i,j} - f_\rho(x, y))^2 dxdy.
\end{aligned}$$

The first term on the right-hand side tends to 0 as m tends to ∞ since f_ρ is square integrable. By the triangle inequality, the second term is bounded by $3(B_{m,\delta,1} + B_{m,\delta,2} + B_{m,\delta,3})$ where

$$B_{m,\delta,1} = \sum_{\substack{1 \le i,j \le m \\ i,j \ne i_m}} \int_{\frac{i-1}{m}}^{\frac{i}{m}} \int_{\frac{j-1}{m}}^{\frac{j}{m}} (\xi_{i,j,\delta} - f_{\rho,\delta}(x, y))^2 dxdy,$$

$$B_{m,\delta,2} = \sum_{\substack{1 \le i,j \le m \\ i,j \ne i_m}} \int_{\frac{i-1}{m}}^{\frac{i}{m}} \int_{\frac{j-1}{m}}^{\frac{j}{m}} (\xi_{i,j} - \xi_{i,j,\delta})^2 dxdy$$

$$B_{m,\delta,3} = \sum_{\substack{1 \le i,j \le m \\ i,j \ne i_m}} \int_{\frac{i-1}{m}}^{\frac{i}{m}} \int_{\frac{j-1}{m}}^{\frac{j}{m}} (f_\rho(x, y) - f_{\rho,\delta}(x, y))^2 dxdy.$$

For each fixed δ, $B_{m,\delta,1} \to 0$ as m tends to ∞ by Riemann integrability of $f^2_{\rho,\delta}$. Applying the Cauchy-Schwarz inequality,

$$B_{m,\delta,2} = m^2 \sum_{\substack{1\le i,j\le m\\ i,j\ne i_m}} \left(\int_{\frac{i-1}{m}}^{\frac{i}{m}} \int_{\frac{j-1}{m}}^{\frac{j}{m}} (f_\rho(x,y) - f_{\rho,\delta}(x,y))dxdy\right)^2$$

$$\le \sum_{\substack{1\le i,j\le m\\ i,j\ne i_m}} \int_{\frac{i-1}{m}}^{\frac{i}{m}} \int_{\frac{j-1}{m}}^{\frac{j}{m}} (f_\rho(x,y) - f_{\rho,\delta}(x,y))^2 dxdy = B_{m,\delta,3}.$$

Now,

$$B_{m,\delta,3} = \sum_{\substack{1\le i,j\le m\\ i,j\ne i_m}} \int_{\frac{i-1}{m}}^{\frac{i}{m}} \int_{\frac{j-1}{m}}^{\frac{j}{m}} (f_\rho(x,y) - f_{\rho,\delta}(x,y))^2 dxdy$$

$$\le \int_0^1 \int_0^1 (f_\rho(x,y) - f_{\rho,\delta}(x,y))^2 dxdy.$$

Since $0 \le f_\rho(x,y) - f_{\rho,\delta}(x,y) \le f_\rho(x,y)$, by dominated convergence theorem we conclude that

$$\lim_{\delta\to\infty} \limsup_{m\to\infty} B_{m,\delta,2} = \lim_{\delta\to\infty} \limsup_{m\to\infty} B_{m,\delta,3} = 0.$$

This concluded the proof of (23).

We next prove (24). Since $\lim_{m\to\infty} \lambda_{m,i} \to \lambda_i$ for all i by (23) and $\lambda_1 < 1$, it follows that for any $\epsilon \in (0, 1-\lambda_1)$, we have $\lambda_{m,i} \le \lambda_1 + \epsilon < 1$ for all i and all large m. Thus it is straightforward to conclude from (23) and (3) that that

$$\sum_{i=1}^{m-1} \frac{\lambda^2_{m,i}}{1-\lambda^2_{m,i}} \to \sum_{i=1}^{\infty} \frac{\lambda_i^2}{1-\lambda_i^2} \le (1-\lambda_1)^{-1}(A(\rho) - 1) < \infty.$$

This concludes the proof. □

Lemma 8. *Assume that the conditions of Theorem 3 hold. Then for all large m,* $A_m - B_{m,\rho}A_m^{-1}B'_{m,\rho}$ *is invertible and hence*

$$\text{tr}(\Sigma^{-1}_{m,\rho}\Sigma_{m,\rho_o}) = 2\text{tr}\{[I_{n(m)} - A_m^{-1}B_{m,\rho}A_m^{-1}B'_{m,\rho}]^{-1}\}.$$

Proof. The first assertion follows simply from the fact that $\lambda_{m,1} \to \lambda_1 < 1$ by Lemma 7 and the condition (HS). To show the second assertion, for simplicity, we'll drop the indices m and ρ in A_m and $B_{m,\rho}$. By Theorem 8.5.11 of Harvill (1997),

$$\Sigma^{-1}_{m,\rho} = \begin{bmatrix} A & B \\ B' & A \end{bmatrix}^{-1}$$

$$= \begin{bmatrix} A^{-1} + A^{-1}B(A - B'A^{-1}B)^{-1}B'A^{-1} & -A^{-1}B(A - B'A^{-1}B)^{-1} \\ -(A - B'A^{-1}B)^{-1}B'A^{-1} & (A - B'A^{-1}B)^{-1} \end{bmatrix}.$$

Hence

$$\Sigma_{m,\rho}^{-1}\Sigma_{m,\rho_o} = \begin{bmatrix} A & B \\ B' & A \end{bmatrix}^{-1} \begin{bmatrix} A & 0 \\ 0 & A \end{bmatrix}$$
$$= \begin{bmatrix} I_n + A^{-1}B(A - B'A^{-1}B)^{-1}B' & -A^{-1}B(A - B'A^{-1}B)^{-1}A \\ -(A - B'A^{-1}B)^{-1}B' & (A - B'A^{-1}B)^{-1}A \end{bmatrix}.$$

Thus,

$$\begin{aligned}
&\mathrm{tr}(\Sigma_{m,\rho}^{-1}\Sigma_{m,\rho_o}) \\
&= \mathrm{tr}(I_n + A^{-1}B(A - B'A^{-1}B)^{-1}B' + (A - B'A^{-1}B)^{-1}A) \\
&= \mathrm{tr}(I_n) + \mathrm{tr}[(A - B'A^{-1}B)^{-1}(B'A^{-1}B - A)] + 2\mathrm{tr}[(A - B'A^{-1}B)^{-1}A] \\
&= 2\mathrm{tr}[(A - B'A^{-1}B)^{-1}A] \\
&= 2\mathrm{tr}[(I_n - A^{-1}BA^{-1}B')^{-1}].
\end{aligned}$$

□

3.4. *Proof of Theorem 4*

First,

$$A(\rho) = \int_{-\infty}^{\infty}\int_{-\infty}^{\infty} w_\rho^2(x, y)/(h(x)h(y))dxdy$$

where h is standard Cauchy pdf and

$$w_\rho(x, y) = \int_{-\infty}^{\infty} h(u)h\left(\frac{x - \rho u}{1 - \rho}\right)h\left(\frac{y - \rho u}{1 - \rho}\right)du$$

For $(x, y) \in [j, j + 1]^2$, $j \neq 0$,

$$w_\rho(x, y) \geq c\int_{|\rho u - j| < 1} h(u)du \geq c/j^2.$$

Hence,

$$A(\rho) \geq c\sum_{j\neq 0}\int_j^{j+1}\int_j^{j+1}\frac{1}{j^4 h(x)h(y)}dxdy = \infty.$$

Next we prove the consistency of the estimator $\hat{\rho}_n$. For convenience, drop n in k_n and ε_n. Let $i*$ be the index of $X_{(i)}$. Write

$$\hat{\rho}_n := S_n + R_n$$

where

$$S_n = \frac{1}{k}\sum_{i=1}^{k} I(Y_{i*}/X_{(i)} \in (1 - \varepsilon, 1 + \varepsilon))$$

and

$$R_n = \frac{1}{k}\sum_{i=1}^{k}\sum_{1\le j\le n, j\ne i*} I(Y_j/X_{(i)} \in (1-\varepsilon, 1+\varepsilon)).$$

Using the fact that, given $X_{(k+1)} = z$, $X_{(1)}, \ldots, X_{(k)}$ are distributed as the order statistics of an iid. sample with pdf $h(x)I(x > z)/\int_z^\infty h(u)du$, we have

$$E(R_n|X_{(k+1)}) = n\int_{X_{(k+1)}}^{\infty} h(x)dx \int_{x(1-\varepsilon)}^{x(1+\varepsilon)} h(y)dy/\int_{X_{(k+1)}}^{\infty} h(u)du$$

It is therefore easy to see that

$$\begin{aligned} E[E(R_n|X_{(k+1)})I(X_{(k+1)} > 1)] &\le cn\varepsilon E[I(X_{(k+1)} > 1)/X_{(k+1)}] \\ &\le cn\varepsilon k/n \to 0. \end{aligned}$$

Since $I(X_{(k+1)} \le 1) \xrightarrow{p} 0$, we conclude that $R_n \xrightarrow{p} 0$. Next we will show that $S_n \xrightarrow{p} \rho$. Using the fact that, given $X_{(k+1)} = z$, $(X_{(1)}, Y_{(1)}), \ldots, (X_{(k)}, Y_{(k)})$ are distributed as iid. with pdf $w_\rho(x,y)I(x > z)/\int_z^\infty h(u)du$, we obtain

$$\begin{aligned} E(S_n|X_{(k+1)}) &= \int_{X_{(k+1)}}^{\infty}\int_{x(1-\varepsilon)}^{x(1+\varepsilon)} w_\rho(x,y)dxdy/\int_{X_{(k+1)}}^{\infty} h(x)dx \\ &= \int_{X_{(k+1)}}^{\infty} h(x)P(Y \in [x(1-\varepsilon), x(1+\varepsilon)]|X \\ &= x)dx/\int_{X_{(k+1)}}^{\infty} h(x)dx. \end{aligned}$$

Let u_n be constants such that $\varepsilon u_n \to \infty$ and $u_n = o(n/k)$ so that $P(X_{(k+1)} < u_n) \to 0$. It is easy to show (cf. Resnick, 1987) that

$$\lim_{x\to\infty} P(Y_1/X_1 \in (1-\delta(x), 1+\delta(x))|X_1 = x) = \rho.$$

for any $\delta(x)$ with satisfying $\delta(x) \to 0$ and $x\delta(x) \to \infty$. Hence, for $X_{(k+1)} > u_n$ we have

$$E(S_n|X_{(k+1)}) \sim \rho$$

and, similarly,

$$\mathrm{var}(S_n|X_{(k)}) \le (1/k)E(S_n|X_{(k+1)}) \sim \rho/k.$$

It is then straightforward to conclude from these that $S_n \xrightarrow{p} \rho$. □

Appendix

The following technical result was applied in the proof of Lemma 7. The proof can be found in Lemma 2.7 of Bai (1999).

Lemma A1. (i) *Let A and B be $m \times n$ matrices with singular values λ_i and η_i (both in descending order) respectively. Then,*

$$\sum_{i=1}^{m\wedge n}(\lambda_i - \eta_i)^2 \le tr[(A-B)(A-B)'].$$

(ii) *Let $\phi(s,t)$ and $\psi(s,t)$ be square integrable functions on $[0,1]\times[0,1]$ and let T_ϕ and T_ψ be two linear operators from $L^2[0,1]$ into itself defined by*

$$T_\phi(g) = \int_0^1 \phi(\cdot, y)g(y)dy \text{ and } T_\psi(g) = \int_0^1 \psi(\cdot, y)g(y)dy.$$

Let the λ_i and η_i be the singular values (both in descending order) of T_ϕ and T_ψ, respectively. Then,

$$\sum_{i=1}^{\infty}(\lambda_i - \eta_i)^2 \le \int_0^1\int_0^1 [\phi(s,t) - \psi(s,t)]^2 dsdt.$$

□

References

Bai, Z.D.: Methodologies in spectral analysis of large dimensional random matrices. A rev. Statistica Sinica **9**, 611–677 (1999)

Chan, H.-P., Loh, W.-L.: A file linkage problem of DeGroot and Goel revisited. Statistica Sinica **11**, 1031–1045 (2001)

Cramér, H.: Mathematical methods of statistics. Princeton University Press, 1946

DeGroot, M.H., Goel, P.K.: Estimation of the correlation coefficient from a broken sample. Ann. Statist. **8**, 264–278 (1980)

Feller, W.: An Introduction to Probability and Its Applications. vol 2. Wiley, 1971

Grenander, U.: Abstract Inference. Wiley, 1981

Hajék, J.: On a property of normal distribution of any stochastic process (in Russian). Czechoslovak Math. J. **8**(83), 610–618 (1958), (An English translation appeared in American Mathematical Society Translations in Probability and Statistics, 1961)

Harvill, D.A.: Matrix Algebra from a Statistician's Perspective. Springer, 1997

Parzen, E.: Probability density functionals and reproducing kernel Hilbert spaces. M. Rosenblatt (ed.), Proceedings of the Symposium on Time Series Analysis, Wiley, 1963, pp. 155–169

Resnick, S.: Extreme Values, Regular Variation, and Point Processes. Springer, 1987

Riesz, F., Sz.-Nagy, B.: Functional Analysis. Translated from the 2d French ed. by Leo F. Boron. Ungar, 1955

Rozanov, J.A.: Infinite Dimensional Gaussian Distributions. English translation published by American Mathematical Society, 1971